G. Lubec (ed.)

Protein Expression in Down Syndrome Brain

SpringerWienNewYork

Prof. Dr. Gert Lubec
Universität-Kinderklinik
Währinger Gürtel 18-20
A-1090 Wien, Austria

Printed in Austria

Typesetting: Best-Set Typesetter Ltd., Hong Kong
Printing: A. Holzhausen's Nfg., A-1140 Wien

Printed on acid-free and chlorine-free bleached paper

SPIN: 10850847

CIP data applied for

With 57 Figures

ISBN 3-211-83732-9 (hard cover) Springer-Verlag Wien New York
ISBN 3-211-83704-3 (soft cover) Springer-Verlag Wien New York

Preface

When we worked on Down Syndrome brain in the past we have been focusing on adult brain. This was a major step forwards as most work on Down Syndrome was carried out on fibroblasts or other tissues and, moreover, we introduced proteomics to identify and quantify brain protein expression. We considered evaluation of brain protein expression in Down Syndrome brain by and by more important than gene hunting at the nucleic acid level realizing the long unpredictable way from RNA to protein.

The availability of fetal samples along with the proteomic appproach stimulated and reinforced studies on Down Syndrome brain. And indeed, it was found out that some observations on aberrant protein expression in adult Down Syndrome brain could not be verified in the fetal samples indicating that neurodegeneration in adult Down Syndrome brain may have been responsible rather than trisomy 21. Using brains from the early second trimester of gestation led to the generation of a series of clues for the understanding of aberrant wiring of the brain in Down Syndrome and enabled the determination of altered key functions in early life; e.g. undetectably low drebrin was observed in Down Syndrome cortex, an integral constituent and marker for dendritic spines, main effectors of cross-talk between neurons. In addition, evaluation of the nature of the neuronal deficits in terms of neurotransmission markers could be established as well as neuronal density in fetal Down Syndrome cortex. Last not least the gene dosage hypothesis has been challenged, as none of several genes encoded on chromosome 21 was overexpressed at the protein level.

All this work answers basic questions but again, rises many additional questions including those on new experimental treatments.

I do appreciate the input of contributors to this book, who addressed major topics and also the many co-operators worldwide, who kindly assist ed us with know how or antibodies etc., with Michael Fountoulakis from F. Hoffmann-La Roche, Basle, in particular. I am grateful for the dedicated work of my research unit of young scientists and I am highly indebted to the sponsor of the Down Syndrome project, the Red Bull Company, Salzburg.

This book not only represents the most recent and unique outcome of protein expression in fetal brain but also stands for a common effort of universities and privat bag to fight a successful battle against a devastating and humiliating disorder, reflecting a dream to defeat Down Syndrome.

Vienna, August 2001 **G. Lubec**

This book is dedicated to Dietrich Mateschitz, Salzburg, Austria

Contents

Decreased alpha-endosulfine, an endogenous regulator of ATP-sensitive potassium channels, in brains from adult Down syndrome patients

S. H. Kim and **G. Lubec**

Department of Pediatrics, University of Vienna, Vienna, Austria

Summary. Alpha-endosulfine has the ability to block ATP-sensitive potassium (K_{ATP}) channels and stimulate insulin release in β cells like sulfonylurea. Alpha-endosulfine is expressed in a wide range of tissue, including brain and endocrine tissues. Although K_{ATP} channels are also present in brain and its regulators have been reported to be involved in the release of neurotransmitters such as acetylcholine that plays an important role in cognitive function, the neurobiological role of α-endosulfine has not been studied yet. We examined the expression levels of α-endosulfine protein in frontal cortex and cerebellum from patients with Down syndrome (DS) showing Alzheimer's disease (AD) pathology using Western blotting. In frontal cortex, α-endosulfine was detected in all of 10 controls, but only 1 (from female) out of 8 DS with weak density. In cerebellum, α-endosulfine was also detected in all of 9 controls, but only 1 (from male) out of 6 DS with weak density. The considerably decreased α-endosulfine could result in the continuous opening of K_{ATP} channels and the subsequent decrease of neurotransmitters release associated with cognition. This study is of significance providing evidence for a biological role of α-endosulfine in brain and α-endosulfine protein could be a pharmacological target for therapeutic intervention.

Introduction

Alpha-endosulfine, an endogenous regulator for K_{ATP} channels, has the sulfonylurea-like ability to inhibit (to close) K_{ATP} channels and thereby stimulate insulin secretion in β cells. Alpha-endosulfine was purified with β-endosulfine from ovine brain (Virsolvy-Vergine et al., 1992) and also isolated from porcine brain (Virsolvy-Vergine et al., 1996). A partial cDNA sequence of α-endosulfine was determined and showed strong homology with a bovine brain cyclic AMP-regulated phosphoprotein 19-kDa protein called ARPP-19, but the two proteins appeared to be encoded by different genes (Peyrollier et al., 1996). The human ENSA gene encoding alpha-endosulfine co-localizes on human chromosome 14 (14q24.3–q31) with a locus for susceptibility to type 1 diabetes (Heron et al., 1999). Alpha-endosulfine is suggested to be a protein

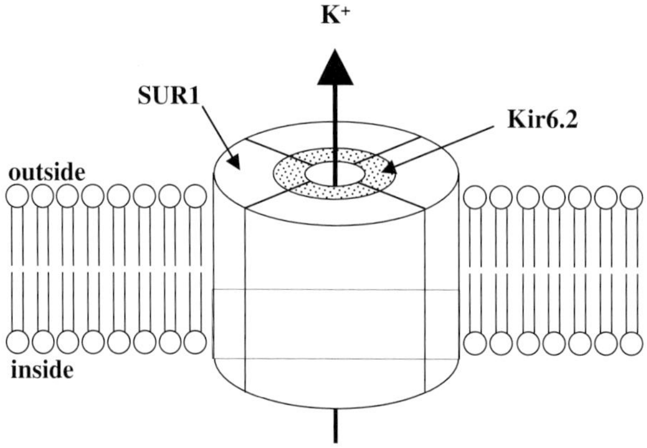

Fig. 1. The structure of K_{ATP} channels. K_{ATP} channels controlling K^+ efflux are large octameric complexes of two subunits, Kir6.2 forming the channel pore and SUR1 possessing sensitivity, which co-assemble with a 4:4 stoichiometry

kinase A-dependent phosphorylated intracellular protein since there is no obvious signal peptide sequence at the N terminus and it contains a cAMP-dependent phosphorylation site in the C terminus, as observed for ARPP-19 (Heron et al., 1998). However it blocked K_{ATP} currents when applied both intracellularly and extracellularly. This suggests it is able to act at more than one site by its uptake into cell facilitated by a transport protein or by endocytosis. The tissue distribution of α-endosulfine that is expressed in a wide range of tissues including muscle and brain, with low levels in the pancreas resembles that of the K_{ATP} channel and is consistent with the ability of α-endosulfine to regulate K_{ATP} channel activity (Inagaki et al., 1995; Sakura et al., 1995).

K_{ATP} channels are large octameric complexes of two structurally unrelated types of subunits, one of the inwardly rectifying K+ channel family (Kir6.2) and a regulatory sulfonylurea receptor subunit from the ATP-binding cassette transporter family (SUR1) (Clement et al., 1997; Schwappach et al., 2000; Shyng and Nichols, 1997). Kir6.2 forms the channel pore and possesses intrinsic ATP sensitivity (Inagaki et al., 1995; Tucker et al., 1997) while SUR1 endows sensitivity to K_{ATP} channel openers (Aguilar-Bryan et al., 1995) as shown in Fig. 1. K_{ATP} channels are present in the brain with high density in substantia nigra, globus pallidus, cerebellum, and cortex (Miller, 1990; Mourre et al., 1990). They are also present in hippocampus both presynaptically and postsynaptically (Mourre et al., 1990) and are particularly situated in mossy fibers that are associated with glutamate release (Mourre et al., 1991). They are sensitive to external concentrations of glucose (Amoroso et al., 1990) since it could change the intracellular level of the ATP/ADP ratio, close K_{ATP} channels leading to cellular depolarization, action potential generation, Ca^{2+} entry and insulin secretion in β cells (Lazdunski, 1996). In brains, the regulation of K_{ATP} channels by modulators (openers and blockers of K_{ATP} channels such as antidiabetic sulfonylureas) as well as extracellular glucose levels could control the release of neurotransmitters (Lopez and Sannita,

1997; Sakuta, 1995). In DS brain, impaired glucose metabolism and neurotransmitters deficits have been well-known and increase of circulating glucose has been reported to enhance learning and memory processes in the broad range of subjects including healthy aged humans and individuals with severe cognitive pathologies (Gold, 1995).

In spite of the important relationship between α-endosulfine and K_{ATP} channels in brain, the involvement and biological function of α-endosulfine have not been studied yet in human brain with cognitive defects. Here we report expression levels of α-endosulfine protein in brain regions of patients with Down syndrome (DS) showing the neurotransmitter deficits and impaired glucose metabolism in order to provide evidence for the involvement of α-endosulfine in cognitive defect.

Methods

Brain samples

Postmortem human brain samples were obtained from Dr. N. J. Cairns (MRC London Brain Bank for Neurodegenerative Diseases, Institute of Psychiatry, King's College, UK). All the DS patients were karyotyped and possessed trisomy 21. A formal cognitive assessment of dementia in DS was not performed. In all DS brains there were abundant and extensive beta-amyloid deposits, neurofibrillary tangles and neuritic plaques and were shown to have AD-like neuropathology. Normal brains obtained from individuals with no history of neurological or psychiatric illness were used as controls. Two brain regions (frontal cortex and cerebellum which are regularly affected in DS) of patients with DS and controls were used for the studies at the protein level (Table 1). The major cause of death was bronchopneumonia in DS, and heart disease in controls. Coronal slices were immediately frozen and stored at −70°C until use.

Western blotting

Brain tissues were suspended in homogenate buffer consisting of 10 mM Tris-HCl (pH 7.5), 150 mM NaCl, 0.05% (v/v) Tween 20, 4% CHAPS, 1 mM EDTA and a mixture of protease inhibitors, 1 mM PMSF and 1 μg/ml of each pepstatin A, chymostatin, leupeptin and antipain. The suspension was homogenized in a Potter-Elvehjem homogenizer at 4°C and centrifuged at 10,000 ×g for 10 minutes. The BCA protein assay kit (Pierce, USA)

Table 1. Autopsy data used in this study

	n (male:female)	Age (years)	Postmortem interval (hours)
Frontal cortex			
control	10 (7:3)	62.60 ± 8.02	37.30 ± 18.39
DS	8 (6:2)	55.88 ± 7.97	32.38 ± 20.70
Cerebellum			
control	9 (6:3)	63.11 ± 8.33	38.89 ± 18.77
DS	6 (4:2)	55.83 ± 9.37	28.00 ± 23.21

determined the concentration of protein in the supernatant. Brain samples (10 μg) were mixed with the same volume of sample buffer (125.5 mM Tris, 70 mM SDS, 0.001% bromophenol blue, 20% glycerol, 2% 2-mercaptoethanol, pH 6.8), incubated at 95°C for 15 minutes and loaded onto a 12.5% ExcelGel SDS homogeneous gel (Amersham Pharmacia Biotech, Sweden). The electrophoresis was performed with Multiphor II Electrophoresis System (Amersham Pharmacia Biotech, Sweden). Proteins separated on the gel were transferred onto PVDF membrane (Millipore, Bedford, USA) and the membrane was blocked in blocking buffer (10 mM Tris-HCl, pH 7.5, 150 mM NaCl, 0.1% Tween 20, 1% non-fat dry milk). The membrane was incubated for 4 hours at room temperature with 1:2,000 diluted purified rabbit polyclonal IgGs for α-endosulfine (kindly supplied by Prof. A. C. Nairn in Laboratory of Molecular and Cellular Neuroscience, Rockefeller University, NY, USA). After 3 times washing for 15 minutes with blocking buffer, membranes were probed with a 1:1,000 dilution of secondary antibody, anti-rabbit IgG(H+L), coupled to horseradish peroxidase (Southern Biotechnology Associates, Inc., Alabama, USA) for 1 hour. The membrane was washed 3 times for 15 minutes and developed with the Western blot chemiluminescence reagents (NEN™ Life Science Products, Inc., USA).

Results

Expression levels of α-endosulfine protein in brains from patients with Down syndrome

Western blotting analysis showed that 19-kDa α-endosulfine protein was extremely decreased or not present in both frontal cortex and cerebellum of patients with DS compared to controls (Fig. 2). In frontal cortex, α-endosulfine was detected in all of 10 controls with strong density and in 3 (2

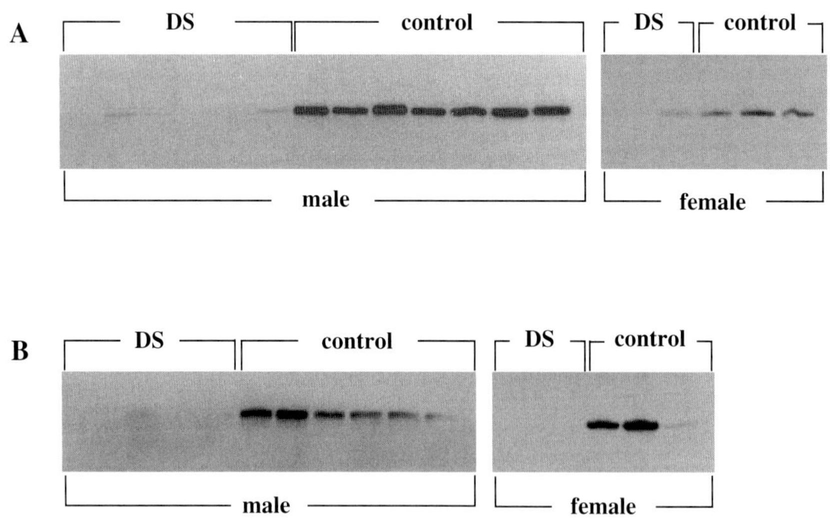

Fig. 2. Immunoblot for α-endosulfine in frontal cortex (**A**) and cerebellum (**B**) of patients with DS and controls. Denatured proteins (10 μg) were loaded, separated on a 12.5% homogeneous gel and transferred onto PVDF membrane. The membrane was incubated with rabbit antisera against α-endosulfine. Immunoreactive α-endosulfine was detected in 19-kDa using chemiluminescence reagents

from male and 1 from female) out of 8 DS with very weak density compared to controls. In cerebellum, α-endosulfine was also detected in all of 9 controls and in 3 (from male) out of 6 DS with very weak density compared to controls.

Discussion

In previous work, we observed the decreased mRNA sequence showing the high homology with ARPP-19 by gene hunting technique. We also observed the significantly deceased ARPP-19 protein and protein kinase A activities in DS and AD. Since α-endosulfine showed strong homology with ARPP-19 and it is also a phosphoprotein regulated by protein kinase A, we examined α-endosulfine at the protein level.

Alpha-endosulfine possessing the sulfonylureas-like ability to block K_{ATP} channels is expressed in brain and the tissue distribution of α-endosulfine resembles that of K_{ATP} channels. K_{ATP} channels are activated by potassium channel openers such as diazoxide, cromakalim, pinacidil, nicorandil or aprikalim (Edwards and Weston, 1993; Kozlowski et al., 1989; Quast, 1992) and inhibited by the antidiabetic sulfonylureas (or blockers) such as tolbutamide, glibenclamide and glipizide (Sturgess et al., 1985). Although openers and blockers clearly exert their respective actions by association with different types of receptor sites, there is an excellent correlation for a series of sulfonylureas between their efficacy for K_{ATP} channels blockade and binding to their drug receptor (Schmid-Antomarchi et al., 1987).

Recently these modulators of K_{ATP} channel activity have been reported to be involved in the neurotransmitter release. Tolbutamide and glibenclamide were shown to stimulate acetylcholine release from striatal cholinergic interneurons by blockade of K_{ATP} channel activity (Lee et al., 1997). The blockade of K_{ATP} channels by glibenclamide also suggested to trigger the release of γ-aminobutyric acid (GABA) on GABAergic interneurons, which in turn, inhibits serotoninergic neurons (Haj-Dahmans et al., 1993). Sulfonylureas were also reported to enhance $GABA_A$ synaptic potentials in rat midbrain dopamine neurons (Grenhoff and Johnson, 1996). On the contrary, diazoxide enhanced the excitatory glutamatergic currents and reduced GABAergic currents in hippocampal neurons (Crepel et al., 1993). Cromakalim showed to prevent glutamate-induced neuronal death in hippocampal neurons, but this effect was abolished by glibenclamide (Lauritzen et al., 1997).

K_{ATP} channels could also be regulated by the concentration of intracellular ATP that is essentially controlled through glucose metabolism (Lopez and Sannita, 1997). Circulating glucose concentrations regulate many brain functions including learning and memory (Korol and Gold, 1998). Extensive evidence indicated that relatively modest increases in circulating glucose concentrations enhance learning and memory processes in healthy aged humans and individuals with severe cognitive pathologies such as DS and AD (Gold, 1995). Several studies have reported the impaired metabolic functions in DS brain. Older demented adults with DS showed lower than normal rates of absolute glucose metabolism in temporal and parietal association neocortical areas using positron emission tomography with [^{18}F] fluorodeoxyglucose, a well-established method for the measurement of regional cerebral glucose metabolism rates (Schapiro et al., 1992). Although older adults with DS without cognitive impairment maintain regional cerebral glucose metabolism within the normal range for many years, repeated measurements showed a progressive decline of glucose metabolism over time and after the onset of dementia.

On the basis of these studies, the mechanism underlying glucose effects on memory could be related with the release of neurotransmitters via K_{ATP} channels. Both local and systemic glucose availability could influence nigral GABA release via an effect on K_{ATP} channels and inhibition of GABA release may in part mediate the hyperexcitability associated with hypoglycemia (During et al., 1995).

Several neurotransmitter deficits have been studied in DS. The cholinergic deficits were observed in DS to show significant reduction of the specific cholinergic marker enzyme, choline acetyltransferase, which catalyzes the synthesis of acetylcholine (Godridge et al., 1987; Kish et al., 1990; Yates et al., 1983). In murine trisomy 16 that is considered to serve as a model of the human trisomy 21 brain, brain choline acetyltransferase activity was reduced to 67% of control and steady-state choline uptake was reduced to 35% (Fiedler et al., 1994). GABA was reduced in the hippocampus and temporal cortex in DS patients with neocortical neurofibrillary tangles and these results are consistent with losses of GABAergic neurons (Reynolds and Warner, 1988). Substantial deficits of noradrenaline, 5-hydroxytryptamine and 5-hydroxyindoleacetic acid were observed in the hippocampus, temporal and frontal cortex of DS patients (Reynold and Godridge, 1985). Although these neurotransmitter deficits could result from the damage to cholinergic, noradrenergic and serotonin neurotransmitter systems observed in DS (Mann et al., 1985), the neurobiological function of endogenous regulators could not be excluded in the control of the neurotransmitter release via ionic channels.

We observed that α-endosulfine protein was extremely decreased or not present in frontal cortex and cerebellum of patients with DS and in this part we observed the impaired glucose metabolism by showing the significantly deceased activity of phosphoglucose isomerase (Labudova et al., 1999). We propose that decreased α-endosulfine could inhibit the neurotransmitters release resulting from the continuous opening of K_{ATP} channels in DS with AD-like pathology together with impaired glucose metabolism. This study is meaningful for suggesting a biological role of α-endosulfine in brain and α-endosulfine could serve as a diagnostic marker for DS and AD. It furthermore provides a tentative target for pharmacological treatment particularly due to the plausible role as potassium channel regulator.

Acknowledgements

We thank Dr. A. C. Nairn (Laboratory of Molecular and Cellular Neuroscience, Rockefeller University, NY, USA) for kindly providing the antibody for α-endosulfine and Dr. N. J. Cairns (Brain Bank, Institute of Psychiatry, King's College, UK) for providing the postmortem brain samples. We are highly indebted to the Red Bull Company, Salzburg, for financial support of the study.

References

Aguilar-Bryan L, Nichols CG, Wechsler SW, Clement JP4th, Boyd AE3rd, Gonzalez G, Herrera-Sosa H, Nguy K, Bryan J, Nelson DA (1995) Cloning of the beta cell high-affinity sulfonylurea receptor: a regulator of insulin secretion. Science 268: 423–426

Amoroso S, Schmid-Antomarchi H, Fosset M, Lazdunski M (1990) Glucose, antidiabetic sulfonylureas and neurotransmitter release. Role of ATP-sensitive K^+ channels. Science 247: 852–854

Clement JP4th, Kunjilwar K, Gonzalez G, Schwantstecher M, Panten U, Aguilar-Bryan L, Bryan J (1997) Association and stoichiometry of K(ATP) channel subunits. Neuron 18: 827–838

Crepel V, Rovira C, Ben-Ari Y (1993) The K^+ channel opener diazoxide enhances glutamatergic currents and reduces GABAergic currents in hippocampal neurons. J Neurophysiol 69: 494–503

During MJ, Leone P, Davis KE, Kerr D, Sherwin RS (1995) Glucose modulates rat substantia nigra GABA release in vivo via ATP-ensitive potassium channels. J Clin Invest 95: 2403–2408

Edwards G, Weston AH (1993) The pharmacology of ATP-sensitive potassium channels. Ann Rev Pharmacol Toxicol 33: 597–637

Fiedler JL, Epstein CJ, Rapoport SI, Caviedes R, Caviedes P (1994) Regional alteration of cholingeric function in central neurons of trisomy 16 mouse fetuses, an animal model of human trisomy 21 (Down syndrome). Brain Res 658: 27–32

Godridge H, Reynolds GP, Czudek C, Calcutt NA, Benton M (1987) Alzheimer-like neurotransmitter deficits in adult Down's syndrome brain tissue. J Neurol Neurosurg Psychiatry 50: 775–778

Gold PE (1995) Role of glucose in regulating the brain and cognition. Am J Clin Nutr [Suppl] 61: 987S–995S

Grenhoff J, Johnson SW (1996) Sulfonylureas enhance $GABA_A$ synaptic potentials in rat midbrain dopamine neurons. Acta Physiol Scand 156: 147–148

Haj-Dahmans S, Laporte AM, Vantalon V, Fattaccini CM, Hamon M, Lanfumey L (1993) Possible involvement of K(ATP) channels in the control of 5-HT neurons. Brain Res 614: 270–278

Heron L, Virsolvy A, Peyrollier K, Gribble F, Le Cam A, Ashcroft FM, Bataille D (1998) Human α-endosulfine, a possible regulator of sulfonylurea-sensitive K_{ATP} channel: molecular cloning, expression and biological properties. Proc Natl Acad Sci USA 95: 8387–8391

Heron L, Virsolvy A, Apiou F, Le Cam A, Bataille D (1999) Isolation, characterization, and chromosomal localization of the human ENSA gene that encodes alpha-endosulfine, a regulator of beta-cell K(ATP) channels. Diabetes 48: 1873–1876

Inagaki N, Gonoi T, Clement JP4th, Namba N, Inazawa J, Gonzalez G, Aguilar-Bryan L, Seino S, Bryan J (1995) Reconstitution of IKATP: an inward rectifier subunit plus sulfonylurea receptor. Science 270: 1166–1170

Kish SJ, Distefano LM, Dozic S, Robitaille Y, Rajput A, Deck JH, Hornykiewicz O (1990) [^{3}H] Vesamicol binding in human brain cholinergic deficiency disorders. Neurosci Lett 117: 347–352

Korol DL, Gold PE (1998) Glucose, memory, and aging. Am J Clin Nutr [Suppl] 67: 764S–771S

Kozlowski RZ, Hales CN, Ashford MLJ (1989) Dual effects of diazoxide on ATP-K^+ currents recorded from an insulin-secreting cell line. Br J Pharmacol 97: 1039–1050

Labudova O, Cairns N, Kitzmuller E, Lubec G (1999) Impaired brain glucose metabolism in patients with Down syndrome. J Neural Transm [Suppl] 57: 247–256
Lazdunski M (1996) Ion channel effects of antidiabetic sulfonylureas. Horm Metab Res 28: 488–495
Lauritzen I, De Weille JR, Lazdunski M (1997) The potassium channel opener (−)-cromakalim prevents glutamate-induced cell death in hippocampal neurons. J Neurochem 69: 1570–1579
Lee K, Brownhill V, Richardson PJ (1997) Antidiabetic sulphonylureas stimulate acetylcholine release from striatal cholinergic interneruons through inhibition of K(ATP) channel activity. J Neurochem 69: 1774–1776
Lopez L, Sannita WG (1997) Glucose availability and the electrophysiology of the human visual system. Clin Neurosci 4: 336–340
Mann DM, Yates PO, Marcyniuk B, Ravindra CR (1985) Pathological evidence for neurotransmitter deficits in Down's syndrome of middle age. J Ment Defic Res 29: 125–135
Miller RJ (1990) Glucose-regulated potassium channels are sweet news for neurobiologists. Trends Neurosci 13: 197–199
Mourre C, Widmann C, Lazdunski M (1990) Sulfonylurea binding sites associated with ATP-regulated K^+ channels in the central nervous system: autoradiographic analysis of their distribution and ontogenesis, and of their localization in mutant mice cerebellum. Brain Res 519: 29–43
Mourre C, Widmann C, Lazdunski M (1991) Specific hippocampal lesions indicate the presence of sulfonylurea binding sites associated to ATP-sensitive K^+ channels both post-synpatically and on mossy fibers. Brain Res 540: 340–344
Peyrollier K, Heron L, Virsolvy-Vergine A, Le Cam A, Bataille D (1996) Alpha endosulfine is a novel molecule, structurally related to a family of phosphoproteins. Biochem Biophys Res Commun 223: 583–586
Quast U (1992) Potassium channel openers: pharmacological and clinical aspects. Fundam Clin Pharmacol 6: 279–293
Reynolds GP, Godridge H (1985) Alzheimer-like brain monoamie deficits in adults with Down's syndrome. Lancet ii: 1368–1369
Reynolds GP, Warner CE (1988) Amino acid neurotransmitter deficits in adult Down's syndrome brain tissue. Neurosci Lett 94: 224–227
Sakuta H (1995) ATP-sensitive potassium channel and hormone/neuropeptide. Nippon Naibunpi Gakkai Zasshi 71: 579–586
Sakura H, Ammala C, Smith PA, Gribble FM, Ashcroft FM (1995) Cloning and functional expression of the cDNA encoding a novel ATP-sensitive potassium channels subunit expressed in pancreatic beta-cells, brain, heart and skeletal muscle. FEBS Lett 377: 338–344
Schapiro MB, Haxby JV, Grady CL (1992) Nature of mental retardation and dementia in Down syndrome: study with PET, CT, and neuropsychology. Neurobiol Aging 13: 723–734
Schmid-Antomarchi H, De Weille JR, Fosset M, Lazdunski M (1987) The receptor for antidiabetic sulfonylureas controls the activity of the ATP-modulated K^+ channel in insulin-secreting cells. J Biol Chem 262: 15840–15844
Schwappach B, Zerangue N, Jan YN, Jan LY (2000) Molecular basis for K(ATP) assembly: transmembrane interactions mediate association of a K^+ channel with an ABC transporter. Neuron 26: 155–167
Shyng S, Nichols CG (1997) Octameric stoichiometry of the K_{ATP} channel complex. J Gen Physiol 110: 655–664
Sturgess NC, Hales CN, Cook DL, Ashford MLJ (1985) The sulphonylurea receptor may be an ATP-sensitive K^+ channel. Lancet ii: 474–475
Tucker SJ, Gribble FM, Zhao C, Trapp S, Ashcroft FM (1997) Truncation of Kir6.2 produces ATP-sensitive K^+ channels in the absence of the sulfonylurea receptor. Nature 387: 179–183

Virsolvy-Vergine A, Leray H, Kuroki S, Lupo B, Dufour M, Bataille D (1992) Endosulfine, an endogenous peptidic ligand for the sulfonylurea receptor: purification and partial characterization from ovine brain. Proc Natl Acad Sci USA 89: 6629–6633

Virsolvy-Vergine A, Salazar G, Sillard R, Denoroy L, Mutt V, Bataille D (1996) Endosulfine, endogenous ligand for the sulfonylurea receptor: isolation from porcine brain and partial structural determination of the alpha form. Diabetologia 39: 135–141

Yates CM, Simpson J, Gordon A, Maloney AF, Allison Y, Ritchie IM, Urquhart A (1983) Catecholamines and cholinergic enzymes in pre-senile and senile Alzheimer-type dementia and Down's syndrome. Brain Res 280: 119–126

Authors' address: Prof. Dr. G. Lubec, Department of Pediatrics, University of Vienna, Währinger Gürtel 18, A-1090 Vienna, Austria, e-mail: gert.lubec@akh-wien.ac.at

Developmental instability of the cerebellum and its relevance to Down syndrome

B. L. Shapiro

Departments of Oral Science and Laboratory Medicine and Pathology and Institute of Human Genetics, University of Minnesota, Minneapolis, MN, U.S.A.

Summary. It has been recognized for many years that cerebellar abnormalities are frequently observed in association with Down syndrome (DS). An important question to be asked about these and other findings in DS is whether their occurrence (i) is attributable to specific loci on the triplicated chromosome or chromosomal segment or (ii) derives from exaggerated responses secondary to the genetic imbalance resulting from trisomy (Ts). Recently, similar cerebellar alterations were observed in subjects with DS and in Ts65Dn mice (Baxter et al., 2000), mice segmentally trisomic for a portion of chromosome 16, which is homologous for loci on the long arm of human chromosome 21. It was concluded by these authors that the occurrence of similar cerebellar changes in DS and in the DS mouse model resulted from triplication of these homologous loci in the two trisomic organisms, i.e. cerebellar development is affected similarly by homologous loci in each species. They wrote that their study of Ts65Dn mice "correctly predicts an analagous pathology in humans" . . . and that . . . "The candidate region of genes on chromosome 21 affecting cerebellar development in DS is therefore delimited to the subset of genes whose orthologs are at dosage imbalance in Ts65Dn mice, providing the first localization of genes affecting a neuroanatomical phenotype in DS." Findings described in this review suggest otherwise — that cerebellar findings in DS and in the Ts65Dn mouse are a result of exaggerated vulnerability in general of the cerebellum to disturbing events and that liability to expression of response(s) is exacerbated by trisomy. This conclusion is based on the following: (i) the cerebellum has an extended postnatal development; (ii) numerous genetic, environmental, epigenetic and metabolic conditions express cerebellar changes similar to those observed in Down syndrome; (iii) most if not all chromosomal imbalance syndromes express similar cerebellar abnormalities; (iv) the cerebellum is particularly sensitive to diverse toxic agents which may act prenatally, postnatally and/or in the mature organism; and (v) cerebellar abnormalities similar to those found in Ts65Dn mice have been described in Ts19 mice which have no segments homologous to any segment of human chromosome 21. An unavoidable conclusion from the review is that triplication of specific loci on 21q is an unlikely explanation for the cerebellar findings in DS. A simple positive control, in which the effect of triplication of loci other than those in question on a specific phenotype, should be used in experiments comparing human and experimental trisomies. As pointed out many years ago by Lorke and his coworkers (Lorke et al., 1989; Lorke, 1994; Lorke and Albrecht, 1994) similar phenotypic findings in the presence of different trisomies in the same species would suggest that the trisomic state itself rather than the gene content of a particular trisomy is responsible for the genesis of traits at issue.

Introduction

That the brain in Down syndrome (DS), particularly the cerebellum, expresses abnormalities has been recognized for many years. The question of whether cerebellar changes from normal are specific or characteristic findings in DS due to specific loci on chromosome 21, or are the result of diminished capacity of the aneuploid system to ward off traumatizing events during development is the subject of this review.

Examination of the brain in DS began soon after its initial description by Down in 1866. In all reports the brain in DS was described as being smaller than normal, a finding supported by Crome (1965) who noted also that "the neuropathological changes in Down's syndrome are usually diffuse and non-specific." Many of the reported abnormalities have been uncertain. This may be one of the reasons for numerous unconfirmed and contradictory statements in the published neuropathological descriptions of the brain in DS. Maldevelopment of the brainstem and cerebellum has been described by various authors and Crome et al. (1966) showed that the reduction of weight of the cerebellum and brainstem was disproportionately greater than reduction of total brain weight. Numerous studies including some very recent ones have confirmed that weight and volume of the cerebellum are reduced in DS and these will be discussed in some detail. Baxter et al. (2000) reported that cerebellar volume and cerebellar granular cells are significantly reduced in Ts65Dn mice. These are mice trisomic for the distal segment of mouse chromosome 16, which is homologous for much of the long arm of human chromosome 21. They reported that cerebellar granule cell density is also reduced in DS and concluded from these findings that the loci (in triplicate) responsible for cerebellar maldevelopment is DS are those loci that exist (also in triplicate) on the homologous segment of mouse Ts65Dn.

The purpose of the discussion to follow is a review of (i) cerebellar findings in DS and (ii) the generalized susceptibility of the cerebellum to a wide variety of genetic and environmental traumas. These facts about the cerebellum and its pathobiolgy indicate that any suggestion of specificity of cerebellar findings in DS is unlikely to be true.

The human cerebellum

The cerebellum is located just below and behind the cerebral hemispheres above the medulla oblongata and occupies between a quarter and a third of cranial capacity in humans. It receives sensory, motor, perceptual and cognitive information from all parts of the nervous system and sends information to all parts of the motor system, except the basal ganglia. The cerebellum functions as a coordination center for posture and fine control of voluntary movement, speech and equilibrium. Evidence suggests that the cerebellum is critical also for thought, behavior and emotion (Ciesielski et al., 1994). It is present in all vertebrates and most primitive prevertebrates (Bastian et al., 1999).

The first cerebellar structures in the human begin developing about 32 days post fertilization and its development is not complete until after birth. At about twelve weeks post fertilization, in the cerebellar cortex, an external granular layer and deep to this a layer of Purkinje cells can be seen. During later stages, cells of the external granular layer migrate inward to mix with the Purkinje cells and postnatally while Purkinje cells mature, the granular cells migrate further to form the internal granular layer. During the first and second postnatal years the external granular layer disappears. Ultimately the cerebellar cortex is comprised of three layers: the outermost molecular layer, the Purkinje cell layer and the internal granular cell layer. Full myelinization of cerebellar connections does not occur until the second year of life. Thus, development of the cerebellum is conspicuously late, cell migrations and fiber connections continue to occur after birth and after development of the remainder of the motor nervous system. The cerebellum grows later and faster than the cerebrum and achieves adult cell number by about 18 months of age (Dobbing and Sands, 1973). (Timing of cerebellar development in the rat and mouse, organisms used for many studies on the cerebellum, is discussed under *Cerebellar sensitivity*.)

Cerebellum and Down syndrome

Observations of the brain in DS were made soon after its account by Down more than 125 years ago. Several excellent summaries of neuropathology in DS have been written (Crome, 1965; Zellweger, 1977; Gullotta et al., 1981; Coyle et al., 1986). Evaluation of the brain in DS has been based on gross and microscopic neuropathology and in more recent years on computer tomography (CT), ultrasonography and magnetic resonance imaging (MRI). All studies agree that after the second in utero trimester the brain in DS is, on average, somewhat smaller than normal and has been variously described in terms of size, volume, weight or hypoplasia, depending on the measurement examined (Table 1). Davidoff (1928) did postmortem examinations on 10 individuals with the clinical diagnosis of DS. He reported that the entire brain weighed less than normal in most cases, but that decreased weight of the brain stem and cerebellum in DS was disproportionately greater. Crome et al. (1966) examined autopsy material from 19 subjects with DS and found the average brain weight of subjects with DS to be about 76% of normal while brainstem and cerebellum taken together were on the average 66% of normal. Schmidt-Sidor et al. (1990) examined brain growth in subjects with DS and

Table 1. Cerebellar size/weight/volume in Down syndrome

Author(s) (year)	Age	Method	Conclusion
Rosati and Guariglia (1999)	11–16 wks i.u.	ultrasound	no difference
Hill et al. (1991)	15–20 wks i.u.	ultrasound	no difference
Schmidt-Sidor et al. (1990)	15–22 wks i.u.	weight	no difference
Ieshima et al. (1984)	birth–37 yrs	CT	DS ↓
Rotmensch et al. (1997)	2nd trimester	ultrasound	DS ↓ (real but slight)
Winter et al. (2000)	2nd trimester	ultrasound	DS ↓
Schmidt-Sidor et al. (1990)	birth–60 mos	weight	DS ↓ (after 2 mos)
Davidoff LM (1928)	7 mos–42 yrs	weight	DS ↓
Crome (1996)	1–17 yrs	weight	DS ↓
Tamaraz et al. (1987)	12–35 yrs	MRI	4/7 w. DS ↓
Jernigan and Bellugi (1990)	14–17 yrs	MRI	DS ↓
Jernigan et al. (1993)	10–20 yrs	MRI	DS ↓
Weis et al. (1991)	30–45 yrs	MRI	DS ↓
Raz et al. (1995)	adults	MRI	DS ↓
Aylward et al. (1997)	adults	MRI	DS ↓

DS Down syndrome, *CT* computer tomography, *MRI* magnetic resonance imaging; ↓ DS less than "control" or "normal"

controls aged 15–22 weeks in utero and from birth to 60 months. In the prenatal material skull shape, brain growth, brain maturation and size of the cerebellum and brainstem in DS were similar to controls both by gross and light microscopic examination. The majority of fetuses in both groups were brachycephalic. Any dysmorphic features associated with DS were difficult to discern in the fetuses. Gross brain shape in newborns was similar in DS and controls. Differences between the two groups became apparent after 3–5 months postnatally. At this time, in the DS sample, frontal lobes were reduced in size and the brainstem and cerebellum were smaller. Brain weight in newborns with DS was within the lower normal range for brain weight at different ages between 2 and 60 months. About 70% of individuals with DS were below the normal range (Schmidt-Sidor et al., 1990).

Among a number of differences from controls, subjects with DS from 0 months to 37 years showed small cerebella when examined by morphometric CT (Ieshima et al., 1984). Several studies evaluated the brain in DS by MRI: Four of 7 subjects with DS (ages 12–35 years) were found to have a small cerebellum (Tamaraz et al., 1987). Jernigan and Bellugi (1990) and Jernigan et al. (1993) found cerebral size reduced in young adults with DS but cerebellar hypoplasia was even more dramatic. They noted no distinctive dysmorpholgy in the subjects with DS but found reduced cerebral and cerebellar volumes. The cerebellar and brainstem volumes were dramatically more reduced than was the cerebral volume, consistent with the findings of Crome et al. (1966). Weis et al. (1993) found smaller volumes for the whole brain, cerebral cortex, white matter and cerebellum in subjects with DS. Raz et al. (1995) also used MRI morphometry to compare subjects with DS and controls. They found that subjects with DS had substantially smaller cerebral and cerebellar hemispheres, ventral pons, mammillary bodies and hippocampal formations. In the cerebellar vermis of subjects with DS they observed smaller lobules VI to VIII without appreciable differences in other regions. Aylward et al. (1997) studied 40 adults with DS using MRI scans. Thirty of them were matched on age, sex and race with cognitively normal subjects. They determined volumes of cerebellum, total brain and intracranial region. Subjects with DS had significantly smaller cerebellar volumes than matched controls, even after adjusting for total brain volume or total intracranial volume. They concluded that although cerebellar volumes are disproportionately small in individuals with DS, they do not diminish significantly with age in either group and do not undergo age-related atrophy that is different from controls. Volume reduction in the cerebellum did not appear to be specifically responsible for the age-related decline in fine-motor control observed in adults with DS. It is of some interest that even though mean brain and cerebellar volumes were reduced in subjects with DS, variances were greater in the sample with DS ($p < 0.05$). In the MRI study of Weis et al. (1991) of 7 adult subjects with DS and 7 controls, while mean cerebellar volume in the DS sample was 81.2% of the controls ($p = 0.05$) a greater variance in the DS sample was observed ($p < 0.02$). (See Shapiro, 1989, 1992 concerning the significance of increased variance in DS.)

Several in utero studies of the brain have made use of ultrasound. Hill et al. (1991) examined the transverse cerebellar diameter in fetuses with trisomy 21 at in utero ages of 15–20 weeks and found no differences from controls. They concluded that cerebellar hypoplasia in DS probably occurs after the second trimester. This was borne out by Rosati and Guariglia (1999) who found no difference from controls at 11–16 in utero weeks and Rotmensch et al. (1997) and Winter et al. (2000) who observed slight but statistically significant cerebellar reduction in fetuses with DS during the second trimester. Transverse cerebellar diameter was less than the normal range in only 2 of 34 fetuses with trisomy 21 from 11 to 16 weeks of gestation in the study of Rosati and Guariglia (1999). In the study of Rotmensch et al. (1997), 42 second trimester fetuses with trisomy 21 were compared with 1,161 karyotypically normal fetuses. The average transcerebellar diameter was 0.67 to 0.87 mm less in the sample with DS ($p < 0.005$) but the difference was considered too small to be clinically useful in prenatal screening (Rotmensch et al., 1997).

Neuropathologic (Crome et al., 1966), tomographic (Ieshima et al., 1984) and magnetic resonance imaging (Weis et al., 1991) and ultrasound (Winter et al., 2000) studies of cerebellar size in infants, children and adults have demonstrated that DS is associated with small cerebellar size. In addition to cerebellar size differences, various references to morphological differences from normal have been made.

Crome (1965) noted that the neuropathological changes in DS are usually diffuse and nonspecific. Many abnormalities have been reported but according to Crome (1965) they have been inadequately described or have not been confirmed. Accordingly, Crome (1965) suggested that reports of specific abnormalities of the brain in DS should be treated with reserve. In his thorough review of neuropathology in DS Zellweger (1977) cited Adams and Sidman

(1968) who wrote that "No two neuropathologists have been able to agree on the morbid anatomy of the brain in DS" and added "some find it of normal size and shape, most do not. Some find histological abnormalities of the cortex and white matter, others do not. The structural alterations encountered ... vary widely" (Zellweger, 1977). Gullotta et al. (1981) contrasted brain findings among different chromosomal abnormalities (qv). They described the frequency of cerebellar dysplasias (or nerve cell heterotopias) in which cell clusters of different shape and size occur, particularly in the vermis or in the white substance. These observations were found in 16% of their sample of subjects with DS but also in a large per cent of controls (40%) and in subjects with trisomy 18 and trisomy 13 (qv). Seventeen per cent of subjects with trisomy 18 and 22% of subjects with trisomy 13 subjects as well. They suggested that cerebellar heterotopias are frequently found in normal newborns and disappear progressively with increasing age. Gullotta et al. (1981) concluded that: "It thus seems reasonable to assume that structural alterations of the central nervous system ... are neither constant nor obligatorily accompanying a specific chromosomal syndrome ..." and that "the different types of microscopical anomalies described in the CNS of patients with Down syndrome ... cannot be regarded as specific or constant." Coyle et al. (1986) described also neurohistologic changes in DS but emphasized that none were pathognomonic. Among the changes reported in DS were reductions in neuronal number/density in cochlear nuclei, and cerebellar and cerebral cortex (granule cells). Baxter et al. (2000) examined histological sections of the cerebellum in autopsy material from 8 subjects with DS and 8 age-matched controls, aged 2 to 59 years. They found that cell density in the internal granular layer was about 70% of that of controls and attributed this difference from normal to gene(s) on a particular segment of chromosome 21 (Baxter et al., 2000).

There probably exists an overall reduction in brain size in DS with a disproportionately greater reduction in the cerebellum. There appears to be general agreement that the cerebellum in DS is reduced relative to the karyotypically normal population. Other reported cerebellar abnormalities have not been confirmed. The question to be asked at this point is whether cerebellar hypoplasia (and other reported cerebellar findings in DS) is particular to DS. I turn now to cerebellar findings in other conditions.

Cerebellar abnormalities in other conditions

Diffuse hypoplasia and other microscopic abnormalities of the cerebellum occur in a wide variety of neurological and systemic disorders in which cerebellar pathology is only one of many findings.

Conditions with apparently normal karyotypes

In the OMIM (2000) 246 distinct conditions are listed in which an associated cerebellar finding occurs and 82 in which cerebellar hypoplasia is the primary abnormality. These do not include conditions with an apparent chromosomal abnormality (qv). Loci associated with abnormal cerebellar findings occur on autosomes, the X chromosome and the mitochondrial chromosome. Both recessive and dominant traits are described. Some of the better known conditions in which cerebellar hypoplasia occurs are: Joubert and related syndromes (Chance et al., 1999), carbohydrate-deficient glycoprotein syndrome (de Lonlay et al., 2000; Patterson, 1999), congenital muscular dystrophies (Barkovich, 1998), and metabolic disorders (Steinlin et al., 1998). Several excellent reviews of these and other conditions associated with cerebellar hypoplasia have been published (Arts et al., 1995; Ramaekers et al., 1997; Steinlin, 1998; Steinlin et al., 1998; de Lonlay et al., 2000). [Cerebellar hypoplasia and other abnormalities associated with exposure to toxic agents are discussed in a subsequent section of this review (qv).]

Conditions other than Down syndrome with abnormal karyotypes

Often, findings such as transverse palmar crease or epicanthal fold and other abnormalities which do occur in association with various chromosomal anomalies, are thought to be characteristic of DS in which they are also frequently present (Shapiro, 1994). This misimpression of specificity of particular findings to DS results from the relative commonness of DS and the rarity of most other chromosomal abnormality syndromes. As will become apparent in the following discussion this misimpression is true also of reported cerebellar abnormalities in DS, i.e. cerebellar abnormalities observed in subjects with DS are found in conditions associated with diverse chromosomal aberrations.

As can be seen in Table 2, cerebellar neuropathology (size, weight, hypoplasia and dysplasia) occurs in association with trisomy of chromosomes 13 and 18 (Gullotta et al., 1981) as well as in partial trisomy of all (or part) of 1q, 3p or q, 9p, 10p, 13q and with deletion of all (or part) of 4p, 5p, 6p, and 9p. It has been described in a case of translocation of Xp; 11p and in a complex rearrangement involving chromosomes 2, 5, 16 and 17 (Maserati et al., 1999). Arts et al. (1995) described a case of an infant with a translocation leading to a 46,XY,−5,+der(5),t(5;10)(p15.3;p12.2) karyotype with a resultant partial monosomy for 5p and trisomy for 10p. They observed extreme hypoplasia of the cerebellum and brainstem by MRI. At autopsy, the 3 day old patient of Alashari et al. (1995) with a terminal deletion of 6p had a markedly hypoplastic cerebellum and absent cerebellar vermix. In addition to their subjects with DS, Tamaraz et al. (1987) using MRI noted cerebellar hypoplasia in patients with trisomy 13, trisomy 18, and in a patient with a partial deletion of 5p. Thurmond et al. (1986), using ultrasonography, found transverse cerebellar diameter to be greater than 2 standard deviations below the normal mean in their 5 cases with trisomy 18. In addition to their study of DS (qv), Gullatto et al. (1981) observed cerebellar dysplasia (nerve cell heterotopias) in which cell clusters of abnormal shape and size were found in the vermis or white substance in 5 of 23 subjects with trisomy 13 and 8 of 47 with trisomy 18. They were even more common in premature births. These lesions were thought by

Table 2. Cerebellar pathology in chromosomal abnormalities other than Down syndrome

Author(s) (year)	Abnormality	Cerebellar pathology
Rorke et al. (1968)	47, +13	heterotopia
	47, +18	heterotopia
Gullatto et al. (1981)	47, +13	cerebellar dysplasia (5 of 23)
	47, +18	cerebellar dysplasia (8 of 47)
Schnizel (1984)	dup(1) (q25 → qter)	cerebellar hypodysplasia &/or dysplasia in all cases
	dup(3) (pter → p21)	
	dup(3) (q21 → qter)	
	del(4) (pter → p16)	
	del(9) (pter → p21)	
	dup(9p)	
	dup(13) (q14-qter)	
Nakamura et al. (1986)	47, +18	hypoplasia, dec. weight, heterotopia, all cases all cases reduced internal granular cell layer thickness
Ferrer et al. (1986)	47, +13	cortical cerebellar dysplasia, granule cell heterotopia & immaturity
Tamraz et al. (1987)	47, +13	cerebellar hypoplasia
	47, +18	
	46, del5p	
Thurmond et al. (1989)	47, +18	5 cases: diameter > 2 SD below mean
Arts et al. (1995)	46, −5p, +10p	extremely hypoplastic
Alashari et al. (1995)	46, del(6)p23	markedly hypoplastic
Dollfuss et al. (1998)	t(X;11) (p22.32;p12)	hypoplasia
Rosati and Guaraglia (1999)	47, +13	hypoplasia (11–16 wks i.u.) 2 of 4 < 10th percentile
Wax et al. (2000)	46, +1q21 . . . > 1qter	hypoplasia

them to be transitory and rarely seen beyond the fifth month of postnatal life. Nakamura et al. (1986) examined 9 infants from 0–177 days of age with trisomy 18 by CT, and by gross and microscopic pathology. In addition to various cerebral anomalies, 6 of the 9 showed cerebellar hypoplasia by CT and weight of the cerebellum and brainstem relative to the whole brain was reduced from normal in all subjects. In all but one case, thickness of the cerebellar inner granule layer was reduced (an observation attributed by Baxter and Reeves and coworkers to loci on chromosome 21 because the same occurred in DS and in Ts65Dn mice). Cerebellar heterotopia was common also in these autosomal trisomies.

Based on their findings with trisomies 13, 18 and 21 and the literature Gullatto et al. (1981) concluded (as stated earlier in this review) that, "It thus seems reasonable to assume that structural alterations of the central nervous system . . . are neither constant nor obligatorily accompanying a specific chromosomal syndrome" and that "(t)he different types of microscopical anomalies described in the CNS of patients with Down's syndrome . . . cannot be regarded as specific or constant." (Gullotta et al., 1981).

Cerebellar hypersensitivity

In the preceding sections I have emphasized that abnormalities of the cerebellum are associated with numerous and distinct human diseases and conditions. In this section I illustrate that the cerebellum is particularly sensitive to diverse toxic agents. Ordinarily, organs and/or other structures are most vulnerable to disturbance during specific time periods, usually coinciding with the differentiation, proliferation, migration or organization of their cell components. The cerebellum seems to be particularly sensitive to disturbing agents not only during these most sensitive developmental periods but afterwards as well. This sensitivity spans species, genotypes, agents, mode of administration and most importantly, transcends the period of its developmental growth spurt. Because of their particular time of appearance and/or migration and maturation, cerebellar granule cells and Purkinje cells are particularly sensitive to toxic agents (Fonnum and Lock, 2000).

The particular vulnerability of the cerebellum rests, to some extent, on its lengthy postnatal duration of development in comparison with other parts of the brain and nervous system. The sequence of events is comparable among species although, as expected, time scales are generally different (Rice and Barrone, 2000). In the rat, an organism in which much research into cerebellar hypersensitivity has been conducted, prenatal days 1–10 and 11–20 are approximately parallel to the first 2 trimesters in the human, respectively. Postnatal days 1–10 in the rat are parallel to the third human trimester and the first year or two of postnatal age in the human. Insults in the rat beyond 10 postnatal days are about equivalent to beyond 2 years in the human. In the cerebellum the Purkinje and granule cells are the most important targets for toxic substances. Purkinje cells appear in the rat brain on prenatal day 14–16 and complete final mitoses before birth. Granule cells appear postnatally and therefore express extensive development after birth (Freeman, 1995; Fonnum and Lock, 2000). In the rat, cerebellar volume increases more than 20-fold during the first 3 postnatal weeks (Freeman, 1995).

In Table 3 I cite numerous examples of cerebellar hypersensitivity to toxic insults. As can be seen, this sensitivity occurs in response to hypoxia, ethanol and numerous other chemicals, and viruses in species as divergent as the human, rodents, fish, cattle, sheep, chicks and cats as well as in vitro in cultured cerebellar slices and cells. Response(s) in the cerebellum to these insults include: hypoplasia, volume and/or weight decrease and dysplasia of granule cell, Purkinje cell and molecular layers of the cortex. The list of citations is not meant to be comprehensive but is fairly complete between 1995 and the present and does illustrate the particular sensitivity of the cerebellum. Fonnum and Lock (2000) have recently written a review of the cerebellum as a target for toxic substances. Little overlap exists in cited references in the two reviews; much of the literature on the subject is covered in the two.

Cerebellar vulnerability to ethanol is particularly informative. Human and experimental animal studies over more than 20 years have shown clearly that exposure to alcohol in developing organisms can lead to a distinct pattern of

Table 3. Cerebellar sensitivity to toxic insults

Agent/time of exposure	Species	Cerebellar effect	Author(s) year
cytochrome oxidase deficiency, fetal	human	hypoplasia (*only area mentioned*)	Lincke et al. (1996)
hypoxia, mid gestation	sheep	reductions: vol. density Purkinje cells; molecular layer width; area inner granule cell layer, white matter & total cross sectional area (*hippocampus & cerebrum much less*)	Rees et al. (1999)
reduced i.u. placental blood flow, 2nd 1/2 preg.	guinea pig	reduction Purkinje neurons, molecular layer, internal granular layer, white matter (*hippocampus & body weight*)	Mallard et al. (2000)
methyliodide (antioxidant toxicity), in vitro	rat	target in the brain: granule cells (*other organs*)	Chamberlain et al. (1999)
mercury vapor, adult	rat	conspicuous loss of Purkinje and granule cells reduction granule cell layer; may be only or primary target in CNS for Hg vapor (*specific to cerebellum*)	Sørensen et al. (2000)
ethanol, adult	human	significant reduction in volume (*only region cited*)	Nicolás (2000)
ethanol, fetal	human	"especially susceptible" (*also basal ganglia, corpus callosum, espcially susceptible*)	Roebuck (1998)
ethanol/thiamine defy, adult	human	reduced volume, Purkinje cells selectively vulnerable to thiamine defy. molecular layer (*not with alcohol alone*)	Baker (1999)
ethanol chronic, adult	human	volumes loss (*and mammilary body*)	Shear et al. (1996)

ethanol, drinking, 4–22 mos.	rat	reduction vol. molecular & granular layer (*specific to cerebellum*)	Rintala et al. (1997)
ethanol, nursing	rat	degen. Purkinje cells, delayed myelinization (*no mention of other changes*)	Hekmatpanah et al. (1994)
ethanol vapor, neonatal	rat	Purkinje & granule cells affected (*postnatal 4–5; not after*)	Heaton et al. (1999)
ethanol, neonatal	rat	"significantly" reduced Purkinje & granule cells (*1st postnatal week but not 2nd*)	Moore et al. (1999)
ethanol vapor, 4–9 days	rat	Purkinje cell reduction (*postnatal day 5; not after 10*)	Miki et al. (1999)
ethanol binge-like, 1st 10 days, 2nd 10 days, pn 4–9 days	rat	reduced Purkinje & granule cells (*cortex & hippocampus also but primariily cerebellum*)	Maier et al. (1999)
ethanol, pre- & post natal	rat	reduced weight, DNA & protein (*gest 6–21, pn 4–12 only cortex & cerebellar weight reduced*)	Miller (1996)
ethanol chronic, adult	rat	decrease iron concentration (*no change in liver or cerebrum*)	Xia et al. (1999)
ethanol binge-like, pregnant 1st 2nd 3rd trimester equiv.	rat	weight to body weight reduced; 3rd trimester Most deficient (*forebrain whole brain*)	Maier et al. (1997)
gadodiamide (MRI contrast medium), lateral ventricle injection, adult	rat	vermis hypertrophy, hyperplasia (*cerebellum . . . brain area most sensitive*)	Ray et al. (1998)
essential fatty acid defy larvae	flounder	significant dec. cerebellar vol; other brain parts no difference (*cerebellum only*)	Furuita et al. (1998)
hyperbilirubinemia postnatal d. 3, 7, 17, 30	rat	prominent hypoplasia (*specific to cerebellum*)	Conlee and Shapiro (1997)

(continued)

Table 3. *Continued*

Agent/time of exposure	Species	Cerebellar effect	Author(s) year
bromodeoxyuridine (mitotic marker) embryonic & postnat.	rat	reduced size relative to cerebrum (*also body, brain, behavior depending on age*) (*@E 11–13 & P 10; effects vary w. developmental age*)	Kolb et al. (1999)
DNA alkylation, pregnant	guinae pig	hypoplasia (*42–47 days gran. cell devel*) (*42–47 days gestation w. devel of granule cells*)	Mehl et al. (2000)
ALL radio-, chemo-Rx 2–5 yrs	human	hypoplasia in all (*sole abnormaity*)	Ciesielski et al. (1994)
irradiation postnatal 5–8 days	rat	hypoplasia, depletion of granule cells, Purkinje cell dysplasia, near birth Rx much more severe than sev'l. days after birth (*no effects on other brain areas*)	Ferguson et al. (1996)
L-2-chloropropionic a. (pesticide), slices, adult	rat	selectively toxic to granule cells (see paper) (*selectively toxic*)	Williams et al. (1999)
phenytoin newborn, p.n. day 2–4	mice	neurotoxic: Purkinje & externalgranule cells migration inhibited (*primary site*)	Ohmori et al. (1999)
dexamethasone postnatal day 3 or 7	rats	reduced weight (*cortex & brainstembut primarily cerebellum*)	Ferguson and Holson (1999)
aminopropionitrile (toxic amino acid derivative), perinatal	human	hypoplasia (*other organs; brain only cerebellum*)	Dembinski et al. (1997)
methylazomenthanol (antimitotic), pn 4–7	rat	hypoplasia (*specific; no overt dysmorphology in other brain regions*)	Freeman et al. (1995)
methylazoxymethanol postnatal 5–8 days	rat	hypoplasia, depletion of granule cells, Purkinje cell dysplasia, near birth Rx much more severe than sev'l. days after birth (*no effects on other brain areas*)	Ferguson et al. (1996)

hyperglycemia, neonatal	human	hypoplasia	Shield (2000)
valproic a. prenatal	rat	reduced volume, cell numbers: Purkinje cells	Ingram (2000)
propylthiouracil (induced hypothyroidism), pre- & postnatal	rat	reduced volume, reduced cellular nuclei volume (*no other site mentioned*)	Mwangi (1998)
trichlorfon (organophosphrous agent), pregnant	guinae pig	loss of weight (*other brain areas also; cerebellum most vulnerable*)	Hjelde et al. (1998)
congenital CMV	human	hypoplasia (*other lesions too*)	Berkovich and Linden (1994)
congenital CMV	human	hypoplasia (*3rd trimester exposure — other lesions also*)	Steinlen et al. (1996)
bovine virus diarrhea calves & subadult	cattle	"markedly" hypoplasia, heterotopia (*most frequent brain lesion*)	Hewickertraut (1995)
parovirus, in utero	cat	agenesis/hypoplasia (*hydranencephaly*)	Sharp et al. (1999)
Borna disease virus, neonat.	rat	Purkinje cell loss (75%); reduced size (*target*)	Eisenman et al. (1999)
Borna disease virus, perinatal	rat	reduction granule & Purkinje cells	Weissenbock (1999)
Borna disease virus, postnatal days 1 & 15	rat	hypoplasia day 1, not day 15	Rubin et al. (1999)
Borna disease virus, neonatal	rat	reduced no granule & Purkinje cells, disrupted architecture (*hippocampus too*)	Hornig et al. (1999)
Kasba (Chuzan) virus, suckling	mcust	hypoplasia (*other sites also*)	Yamaguchi et al. (1999)
Aino virus, embryos	chick	hypoplasia (*other abnormalities but primarily cerebellum*)	Kitano et al. (1997)

congenital defects termed the fetal alcohol syndrome. Early descriptions did not reveal specific patterns of brain malformation, but more recent quantitative MRI studies have indicated particular vulnerability of basal ganglia, the corpus callosum and the cerebellum (Roebuck et al., 1998). A number of studies with the rat suggest that cerebellar Purkinje cells are the most vulnerable cell type when exposed to appropriate doses of ethanol sometime in or around the first postnatal week (human third trimester equivalent), at the time of major cerebellar growth spurt (Pierce et al., 1999; Miki et al., 1999; Hekmatpanah et al., 1994; Heaton et al., 1999). However, Moore et al. (1999) and Maier et al. (1999) with rats, demonstrated that granule cell layers are also markedly reduced in response to ethanol at this time. Maier (1997) showed that both cell types are hypersensitive to binge-like exposures of ethanol during the first, second and third trimester equivalents in rats. The outermost, molecular layer, was shown to be affected by alcohol in adult humans (Baker et al., 1999) and rats (Rintala, 1997). Cerebellar sensitivity to ethanol has been shown with adult exposure also in the rat (Rintala, 1997; Xia, 1999) and human (Nicolas et al., 2000; Baker et al., 1999; Shear, 1996). It would seem that the cerebellum is, in general, particularly susceptible to alcohol exposure. In addition to agents which affect Purkinje cells or Purkinje cells and granule layer cells, several agents [such as the antioxidant, methyl iodide (Chamberlain et al., 1999) and the pesiticide, L-2-chloroproprionic acid (Simpson et al., 1996; Williams et al., 1999)] are specific to granule cells with little affect on Purkinje cells. Interestingly, L-2-chloroproprionic is only toxic to adult animals and not to animals during development.

A most unusual demonstration of cerebellar hypersensitivity was described by Ciesielski et al. (1994). Because of its protracted course of development they hypothesized that cerebellar consequences would result from radiation and chemotherapy used on children treated for acute lymphoblastic leukemia (ALL). Thirteen survivors of ALL, all with age of onset from 2 to 5 years exposed to a uniform protocol of radiation and intrathecal methotrexate were compared with 10 age matched controls using neuropsychological scores and MRI. Cerebellar hypoplasia was consistently observed in the survivors of ALL, but not in control subjects.

Reduced cerebellar size resulted from exposure to all the agents listed in Table 3. In addition to the more recent citations cited, Lorke et al. (1989) in their review of trisomy and murine cerebellar development stated that "the developing cerebellum reacts similarly to other insults, e.g. hypothyroidism . . . , undernutrition . . . , corticosteroid-treatment . . . , alcohol-intoxication . . . , zinc deficiency . . . , or surgical deafferentiation . . ." Two additional examples of cerebellar hypersensitivity involve enhanced stimulatory responses. Ye et al. (1996) studied the in vivo action of insulin-like growth factor on cerebellar development in transgenic mice in whom an insulin-like growth factor was highly expressed in the cerebellum. Among brain regions examined the cerebellum exhibited the greatest increase in postnatal size, beginning during the second postnatal week, when compared with littermate controls. By 50 days of life cerebellar weight and DNA content were increased by 90% and 143%, respectively, compared with littermate controls. Total number of granule and Purkinje cells was increased by 82% and 20%, respectively. The greatest increase was in granule cell number. A second example of excessive response of the cerebellum was reaction to the MRI imaging agent, gadodiamide. When injected into the lateral ventricle of rats several cerebellar responses were observed including hyperplasia. The cerebellum was the brain area most sensitive to gadodiamide (Ray et al., 1998).

This lengthy review should leave no doubt that the cerebellum is a particularly sensitive area of the body. Structural effects in response to a wide range of substances including numerous and disparate conditions, among them, chromosomal aberrations also is not surprising. I now turn to cerebellar findings in murine trisomy.

The cerebellum and murine trisomy

Innovative mouse models for studying human aneuploid conditions have been developed over the last two and a half decades. In progeny of mice carrying Robertsonian fusion chromosomes, trisomies for each of the 19 mouse acrocentric autosomes can be generated (Gropp, 1974). Only trisomy (Ts) of chromosomes 16 and 19 survive to birth. Newborns with Ts16 die in the first day of life and Ts19 mice live a few postnatal weeks (Gearhart et al., 1986). An important development for DS research was the generation of a viable mouse with partial Ts16 from an induced translocation (t16;17) known as Ts65Dn, which can survive until adulthood (Davisson et al., 1993). Ts65Dn mice have an extra copy of the distal portion of mouse chromosome 16, a segment homologous to distal human chromosome 21q (except for the most distal portion of 21q). The resultant mouse is partially (or segmentally) trisomic for a portion of mouse chromosome 16 containing genes thought to be important in the production of the DS phenotype in the human [but, see Shapiro (1999) concerning the so-called Down syndrome critical region]. In addition to distal mouse 16, human chromosome 21 genes have been conserved also in mouse chromosomes 10 and 17. Mouse chromosome 16 has conserved segments from at least 3 other human chromosomes in addition to 21: 3, 16 and 22 (Wiltshire et al., 1999).

Gearhart et al. (1986) described the general features of mouse trisomics. These included: few survive to term (animals with trisomy of chromosome 19, the smallest mouse autosome, survive a few weeks postnatally); they express growth retardaton and hypoplasia; those that survive past midgestation express transient subcutaneous edema; and specific malformations occur with particular trisomies. These authors suggested that specific anomalies may result from gene dosage effects of specific loci. I would add that different anomalies may also result from the amount of superfluous DNA in the different trisomies. In any case, the central nervous system, the cardiovascular system and the craniofacial complex are commonly affected in all trisomic conditions.

In order to study the effect(s) of extra chromosomal material on neural development, Lorke and coworkers, in a series of carefully conducted and analyzed papers, described cerebellar abnormalities in the Ts19 mouse (Lorke et al., 1989; Lorke and Klimascheski, 1992; Lorke and Albrecht, 1994). The cerebellum was the most markedly hypoplastic area of the brain in these mice (Lorke and Albrecht, 1994). Among the cerebellar cortical layers significantly reduced in the Ts19 mice were the internal granular layer and the molecular layer (Lorke et al., 1989). These authors noted that comparable cerebellar alterations were also observed in human trisomy 21 (Lorke and Neidermauntel, 1993) and mouse trisomy 16 (Lorke and Albrecht, 1994). Lorke and his coworkers concluded that cerebellar changes "seem to be a general characteristic of autosomal trisomies" (Lorke and Niedermauntel, 1993). Although loci involved in human trisomy 21 are not at issue in the studies of Lorke et al. their findings do shed some light on the effect of trisomy in general. Lorke commented that, "Ts19 mice help to identify the effects of gene imbalance, as abnormal features seen both in Ts16 and Ts19 mice are probably due to general effects of trisomy" (Lorke, 1992) and comparable cerebellar findings in human Ts21, Ts19 and Ts65Dn "most likely represent the nonspecific results of gene imbalance." Despite these thorough analyses by Lorke and his coworkers, Baxter et al. (2000) in their study of the cerebellum in Ts65Dn mice concluded otherwise.

Baxter et al. (2000) stated that, "Cerebellar volume is significantly reduced in Ts65Dn mice due to reduction of both the internal granular layer and the molecular layer of the cerebellum. Granule cell number is further reduced by a decrease in cell density in the internal granule layer. . . . Reduction in granule cell density in Ts65Dn mice correctly predicts an analogous pathology in humans; a significant reduction in granule cell density in the DS cerebellum is reported here for the first time. The candidate region of genes on chromosome 21 affecting cerebellar development in DS is therefore delimited to the subset of genes whose orthologs are at dosage imbalance in Ts65Dn mice, providing the first localization of genes affecting a neuroanatomical phenotype in DS."

As has been discussed above and elsewhere, cerebellar abnormalities in Ts65Dn mice and human trisomy 21 (DS) are seen also in human trisomy 13 and 18 and in the Ts19 mouse. For some reason that is not clear the authors (Baxter et al., 2000) attribute these very nonpecific findings to genetic loci on human 21q and murine Ts65Dn.

Discussion

Similar cerebellar findings in DS and in the Ts65Dn mouse prompted Baxter et al. (2000) to attribute these observations to responsible genes in common on human chromosome 21 and mouse chromosome Ts65Dn. The potential importance of their paper is captured in its title: "Discovery and genetic localization of Down syndrome cerebellar phenotypes using the Ts65Dn mouse." (Baxter et al., 2000). These findings and the conclusion, warranted the thorough review presented in these pages. This examination has indicated that the cerebellum is a particularly sensitive structure. Cerebellar abnormalities occur frequently in association with most, if not all, chromosomal imbalance syndromes and in a wide variety of metabolic and developmental conditions as well as in response to diverse toxic challenges (Tables 2 and 3). A partial explanation for this vulnerability of the cerebellum is its protracted postnatal development. Neonatal and early postnatal susceptibility is not the whole story since the cerebellum shows also amplified vulnerability throughout life as shown by abnormal cerebellar responses to various agents in the adult human and rat (Table 3). An explanation for a general cerebellar sensitivity is not readily apparent (Fonnum and Lock, 1990), but that it does exist may contribute to an understanding of the pathogenesis of many findings associated with DS.

It is not unreasonable that a quest for a simple explanation for clinical findings in DS persists. This is the most common postnatally viable human aneuploid condition. It is probably the single most common cause of major mental retardation. It predisposes to conditions, such as a range of congenital malformations, acute leukemias, Alzeimers disease among others which, when they are expressed in the general population, impose an important medical burden. The attribution of most findings in DS to the proximate presence of a hundred or so genes on chromosome 21 which occur in triplicate has been naive. As I have discussed many times, the clinical picture that permits the diagnosis of DS is not comprised of signs, such as the cerebellar abnormalities, which lend themselves to explanation by triplication of single gene products. Without exception these signs, none of which is unique to DS and all of which are found in the general population and in other conditions, are the result of complex interactions of numerous gene products under diverse environmental influences. The frequently cited statement of H. L. Mencken that "For every complex problem, there is a solution that is simple, neat, and wrong" is appropriate in this context. The attribution of clinical findings in DS to specific loci is easy and is wrong. Similar findings in different species with homologous (to a segment of human 21q) segmental trisomies, especially when these particular findings occur also in organisms that possess trisomy of nonhomologous segments, cannot be evidence for gene action of loci on the trisomic segment(s). Lorke and his coworkers recognized this many years ago. In 1992 they wrote that "Ts19 mice help to identify the effects of gene imbalance, as abnormal features seen both in Ts16 and Ts19 mice are probably due to general effects of trisomy" (Lorke and Klimaschewski, 1992). They added that, "Abnormal features shared by all these trisomics (mouse Ts16, mouse Ts19 and human Ts21) are most probably the result of general effects, whereas alterations that are only found in human Ts21 and mouse Ts16, but not in mouse Ts19, are most likely attributable to specific gene dosage effects" (Lorke and Klimaschewski, 1992). Similar cerebellar anomalies reported in murine Ts16 and Ts19 and human Ts21, Ts18 and Ts13 cannot be the result of human Ts21/mouse Ts16 specific loci as proposed by Reeves and Baxter and their coworkers.

The work of Vekemans and Trasler (1986) and Zlotogora (1989, 1992) provides insight into the issue of pathogenesis of many abnormalities found associated with aneuploidy. Vekemans and Trasler (1986) examined palatal development in mouse strains with differing liability to cleft palate. They compared the incidence of clefts in 4 strains of mice with trisomy 19 and their euploid littermates. Their study indicated that palatal development in trisomy 19 fetuses is significantly delayed compared with euploid littermates and that the delay is in direct relationship with the observed pattern of palatal development in normal littermates. This suggested to them that "the increased expression of major congenital malformations associated with aneuploidy is due to a change in their underlying liability" because of aneuploidy and not specific loci on chromosome 19.

Zlotogora et al. (1989) reported that the incidence of imperforate anus was 5½ times greater in newborns with DS than in all newborns. In a follow up Zlotogora (1992) reported on the parallel between prevalence in the general population and that in subjects with DS. He found a parallel between the prevalence of anorectal malformations in 3 general populations [Jerusalem (0.065%), California (0.035%) and Spain (0.020%)] and in the subpopulation of each with DS [Jerusalem (2.2%) and Spain (0.36%)], California (1.1%). He found that these data supported "the theory . . . (of) . . . decreased developmental and physiological buffering against genetic and environmental forces" (Zlotogora, 1992) responsible for increased prevalence of congenital malformations in DS. The studies of Vekemans and Trasler (1986), Zlotogora (1992) and Langenbeck et al. (1984) among others provided data supporting the notion of decreased developmental homeostasis as the basis for clinical findings in chromosome abnormality syndromes. A contrary view was expressed by Baxter et al. (2000) who considered a complex trait such as cerebellar hypoplasia in DS due to specific loci on chromosome 21.

Baxter et al. (2000) described "precise neuroanatomical parallels between the DS and the Ts65Dn cerebellum." and concluded from their findings that the "candidate region of genes on chromosome 21 affecting cerebellar development in DS is . . . delimited by the subset of genes whose orthologs are at dosage imbalance in Ts65Dn mice, providing the first localization of genes affecting a neuroanatomical phenotype in DS." This conclusion is surprising in light of the earlier and extensive work with Ts19 mice by Lorke and his coworkers. As long ago as 1989 Lorke suggested that comparable neurological findings in human Ts21, mouse Ts16 and Ts19 "most likely represent the nonspecific results of gene imbalance and is similar to the comparable alterations observed after postnatal undernourishment." He and Niedermauntel (1993) concluded another paper by stating that: "Trisomics . . . seem to be associated with more cell heterotopias in species that also show this abnormality in euploid organisms, although relatively rarely. This observation lends support to the observation that trisomy leads to an amplified prevalence of abnormalities also occurring in the general population" and speaks against monogenic origins of most of the clinical findings found in Down syndrome.

References

Alashari M, Chen M, Poskanzer L (1995) Partial deletion of chromosome 6p — autopsy findings in a premature infant and review of the literature. Pediatr Pathol & Lab Med 15: 941–947

Arts WFM, Hofstee Y, Drejer GF, Beverstodk GC, Oosterwijk JC (1995) Cerebellar and brainstem hypoplasia in a child with partial trisomy for the short arm of chromosome 5 and partial trisomy for the short arm of chromosome 10. Neuropediatrics 26: 41–44

Aylward EH, Habbak R, Warren AC, Pulsifer MB, Barta PE, Jerram M, Pearlson GD (1997) Cerebellar volume in adults with Down syndrome. Arch Neurol 54: 209–212

Baker KG, Harding AJ, Halliday GM, Kril jjarper CG (1999) Neuronal loss in functional zones of the cerebellum of chronic alcoholics with and without Wernicke's encephalopathy. Neurosci 91: 429–438

Barkovich AJ (1998) Neuroimaging manifestations and classification of congenital muscular dystrophies. Am J Neuroradiol 19: 1389–1396

Bastion AJ, Mugnaini E, Thach WY (1999) "Cerebellum" in fundamental neuroscience. Academic Press, San Diego, pp 973–992

Baxter LL, Moran TH, Richtsmeier JT, Troncoso J, Reeves RH (2000) Discovery and genetic localization of Down syndrome cerebellar phenotypes. Hum Mol Genet 9: 195–202

Chamberlain MP, Sturgess NC, Lock EA, Reed CJ (1999) Methyl iodide toxicity in rat cerebellar cells in vitro: the role of glutathione. Toxicology 139: 27–37

Chance PF, Cavalier L, Satran D, Pellegrino JE, Koenig M, Dobyns WB (1999) Clinical nosologic and genetic aspects of Joubert and related syndromes. J Child Neurol 14: 660–666

Ciesielski KT, Yanofsky R, Ludwig RN, Hill DE, Hart BL, Astur RS, Snyder T (1994) Hypoplasia of the cerebellar vermis and cognitive defects in survivors of childhood leukemia. Arch Neurol 51: 985

Conlee JW, Shapiro SM (1997) Development of cerebellar hypoplasia in jaundiced Gunn rats — a quantitative light microscoopic analysis. Acta Neuropathol 93: 450–460

Coyle JT, Oster-Granite ML, Gearhart JD (1986) The neurobiologic consequences of Down syndrome. Brain Res Bull 16: 773–787

Crome L (1965) The pathology of Down's disease. In: Hilliard LT, Kirman BH (eds) Mental deficiency, 2nd ed. J & A Churchill Ltd, London

Crome L, Cowie V, Slater E (1966) A statistical note on cerebellar and brain-stem weight in mmongolism. J Ment Def Res 10: 69–72

Davidoff LM (1928) The brain in mongolian idiocy: a report of ten cases. Arch Neurol Psychiatr 20: 1229–1257

Davisson MT, Schmidt C, Reeves RH, Irving NG, Akeson EC, Harris BS, Bronson RT (1993) Segmental trisomy as a mouse model for Down syndrome. Prog Clin Biol Res 384: 117–133

de Lonlay P, Cormier-Daire V, Vuillaumier-Barrot S, Cuer M, Durand G, Munnich A, Saudubray JM, Seta N (2000) "Carbohydrate-deficient glycoprotein" syndrome. Arch Pediatrie 7: 173–184

Dembinski J, Heyl W, Steidel K, Hermanns B, Hörnchen H, Schröder W (1997) The Cantrell-sequence: a result of maternal exposure to aminopropionitriles? Am J Perinatal 14: 567–571

Desouza N, Chaudhuri R, Bingham J, Cox T (1994) MRI in cerebellar hypoplasia. Neuroradiology 36: 148–151

Desouza N, Chaudhuri R, Bingham J, Cox T (1994) MRI in cerebellar hypoplasia. Neuroradiol 36: 148–151

Dobbing J, Sands J (1973) Quantitative growth and development of human brain. Arch Dis Childhood 48: 757–767

Dollfus H, Joannyflinois O, Docofenzy M, Veyre L, Joannyflinois L, Khoury M, Jonveaux P, Abitbol M, Dufier JL (1998) Gillespie syndrome phenotype with a t(X-11) (p22.32-p-12) de novo translocation. Am J Ophthalmol 125: 397–399
Down JLH (1866) Observation of an ethnic classification of idiots. London Hosp Clin Lect Rep 3: 259
Eisenman LM, Brothers R, Tran MH, Kean RB, Dickson GM, Dietzschold B, Hooper DC (1999) Neonatal Borna disease virus infection in the rat causes a loss of Purkinje cells in the cerebellum. J Neurovirol 5: 181–189
Ferguson SA, Holson RR (1999) Neonatal dexamethasone on day 7 causes mild hyperactivity and cerebellar stunting. Neurotoxicol Teratol 21: 71–76
Ferguson SA, Paule MG, Holson RR (1996) Functional effects of methylozomethanol — induced cerebellar hypoplasia in rats. Neurotoxicol Teratol 18: 529–537
Ferrer I, Ribalta T, Fabregues I, Pineda M, Cusi V (1986) A Golgi study of cerebellar malformation in 13 trisomy. Clin Neuropathol 5: 53–59
Fonnum F, Lock EA (2000) Cerebellum as a target for toxic substances. Toxicol Lett 112–113: 9–16
Freeman JH, Barone Jr H, Stanton ME (1995) Disruption of cerebellar maturation by an antimitotic agent impairs the ontogeny of eyeblink conditioning in rats. J Neurosci 15: 7301–7314
Furuita H, Takeuchi T, Uematsu K (1998) Effects of eicosapentaenic and docoshexaenoic acids on growth, survival and brain development of larval Japanese flounder (paralichthysolivaceus). Aquaculture 161: 269–279
Gearhart JD, Davisson MT, Oster-Granite (1986) Autosomal aneuploidy in mice: generation and developmental consequences. Brain Res Bull 16: 789–801
Gropp A, Giers D, Kolbus U (1974) Trisomy in the fetal backcross progeny of male and female metacentric heterozygotes in the mouse. I. Cytogenet Cell Genet 13: 511–535
Gullotta F, Rehder H, Gropp A (1981) Descriptive neuropathology of chromosomal disorders in man. Hum Genet 57: 337–344
Heaton MB, Mitchell JJ, Paiva M (1999) Ethanol-induced alterations in neurotrophin expression in developing cerebellum: relationship to periods of temporal susceptibility. Alcoholism 23: 1637–1642
Hekmatpanah J, Haghighat N, Adams CR (1994) Alcohol consumption by nursing rats and its effect on the cerebellum of offspring. Alcohol Alcoholism 29: 535 ff
Hewickertrautwein M, Liess B, Trautwein G (1995) Brain lesions in calves following transplacental infection with bovine-virus diarrhoea virus. Zentralbl Veterinarmed — Reihe B 42: 65–77
Hill LH, Rivello D, Peterson C, Marchese S (1991) The transverse cerebellar diameter in the second trimester is unaffected in Down syndrome. Am J Obstet Gynecol 164: 101–103
Hjelde T, Mehl A, Schanke TM, Fonnum F (1998) Teratogenic effects of trichlorfon (metrifonate) on the guinae-pig-determination of the effective dose and the sensitive period. Neurochem Int 32: 469–477
Hornig M, Weisenböck — Horsscroft N, Lipkin I (1999) An infection-based model of neurodevelopmental damage. Proc Natl Acad Sci USA 96: 12102–12107
Ieshima A, Kisa T, Yoshino K, Takashima S, Takeshita K (1984) A morphometric CT study of Down's syndrome showing small posterior fossa and calcification of basal ganglia. Neuroradiology 26: 493–498
Ingram JL, Peckham SM, Tisdale B, Rodier PM (2000) Prenatal exposure of rats to valproic acic reproduces the cerebellar anomalies associated with autism. Neurotoxicol Teratol 22: 319–324
Jernigan TL, Bellugi U (1990) Anolalous brain morphology on magnetic resonance images in Williams syndrome and Down syndrome. Arch Neurol 47: 529–533
Jernigan TL, Bellugi U, Sowell E, Doherty S, Hesselink JR (1993) Cerreral morphologic distinctions between Williams and Down syndromes. Arch Neurol 50: 186–191

Kitano Y, Yasuda N, Shimizu T, Ohzono H, Iwamoto T (1997) Teratogenicity of Aino virus in the chick embryo. Res Vet Sci 62: 195–198
Kitano Y, Ohzono H, Yasuda N, Shimizu T (1996) Hydranencephaly, cerebellar hypoplasia, and myopathy in chick embryos infected with aino virus. Veterinary Pathol 33: 672–681
Kolb B, Pedersen B, Ballermann M, Gibb R, Whishaw IQ (1999) Embryonic and postnatal injections of bromodeoxyuridine produce age-dependent morphological and behavioral abnormalities. J Neurosci 19: 2337–2346
Lincke CR, van den Bogert C, Nijtmans LGJ, Wanders RJA, Tamminga P, Barth PG (1996) Cerebellar hypoplasia in respiratory chain dysfunction. Neuropediat 27: 216–218
Lorke DE (1994) Developmental characteristics of trisomy 19 mice. Acta Anat 150: 159–169
Lorke DE, Klimaschewski (1992) Postnatal development of the locus coeruleus in trisomy 19 mice: morphological and morphometric study. Brain Res Bull 28: 923–930
Lorke DE, Neidermauntel (1993) Foliation of the cerebellar vermis in trisomy 19 mice. Brain Res Bull 32: 215–222
Lorke DE, Albrecht H (1994) Altered postnatal development of the visual cortex in trisomy 19 mice. Brain Res Bull 34: 563–570
Lorke DE, Stan A, Lierse W (1989) Morphogenesis of the cerbellum of trisomy 19 mice. Biomed Res [Suppl] 3: 385–396
Maier SE, Chen W-J A, Miller JA, West JR (1997) Fetal alcohol exposure and temporal vulnerability — regional differences in alcohol-inducedmicroencephaly as a function of the timing of binge-like alcohol exposure during rat brain development. Alcoholism 21: 1418–1428
Maier SE, Miller JA, Blackwell JM, West JR (1999) Fetal alcohol exposure and temporal vulnerability: regional differences in cell loss as a function of the timing of binge-like alcohol exposure during brain development. Alcohol Clin Exp Res 23: 726–734
Mallard C, Loeliger M, Copolov D, Rees S (2000) Reduced number of neurons in the hippocampus and the cerebellum in the postnatal guinae-pig following intrauterine growth-restriction. Neurosci 100: 327–333
Maserati E, Verri A, Seghezzi L, Tupler R, Federico A, Tiepolo L, Maraschio P (1999) Cerebellar dysgenesis and mental retardation associated with a complex chromosome rearrangement. Ann Genet 420: 210–214
Mehl A, Rolseth V, Gordon S, Bjoraas M, Seeberg E, Fonnum F (2000) Brain hypoplasia caused by exposure to trichlorfon and dichlorvos during development can be ascribed to DNA alkylation damage and inhibition of DNA alkyltransferase repair. Neurotoxicol 21: 165–173
Miki T, Harris S, Wilce P, Takeuchi Y, Bedi KS (1999) The effect of timing of ethanol exposure during early postnatal life on total number of Purkinje cells in rat cerebellum. J Anat 194: 423–431
Miller MW (1996) Effect of early exposure to ethanol on the protein and DNA contents of specific brain regions in the rat. Brain Res 734: 286–294
Moore DB, Walker DW, Heaton MB (1999) Neonatal ethanol exposure alters bcl-2 family in mRNA levels in the rat cerebellar vermis. Alcoholism 23: 1251–1261
Mwangi DK (1998) Effect of propylthiouracil induced hypothyroidism in developing rat cerebellum. Comparison of cerebellar parameters in five day old normal and treated rat pups. East African Med J 75: 602–608
Nakamura Y, Hashimoto T, Sasaguri Y, Yamana K, Tanaka S, Morodomi T, Murakami T, Machara F, Nakshima T, Fukuda S, et al (1986) Brain anomalies found in 18 trisomy: CT scanning, morphologic and morphometric study. Clin Neuropathol 5: 47–52

Nicolás JM, Fernández-Sola J, Robért J, Antúnez E, Cofán M, Cardenal C, Sacanella E, Estruch R, Urbano-Márquez R (2000) High ethanol intake and malnutriton in alcoholic cerebellar shrinkage. Qjm-Monthly J Assoc Physicians 93: 449–456

Ohmori H, Ogura Y, Yasuda M, Nakamura S, Hatta T, Kawano K, Michikawa T, Yamashita K, Mikoshiba K (1999) Developmental neurotoxicity of phenytoin on granule cells and Purkinje cells in mouse cerebellum. J Neurochem 72: 1497–1506

Online Mendelian Inheritance in Man, OMIM (TM) (2000) McKusick-Nathans Institute for Genetic Medicine, Johns Hopkins University (Baltimore, MD) and National Center for Biotechnology Information, National Library of Medicine (Bethesda, MD) World Wide Web URL: http://www.ncbi.nlm.nih.gov/omim/

Patterson MC (1999) Screening for "prelysosomal disorders": carbohydrate-deficient glycoprotein syndromes. Child Neurol [14 Suppl]: S16–S22

Pierce DR, Williams DK, Light KE (1999) Purkinje cell vulnerability to developmental exposure in the rat cerebellum. Alcohol 23: 1650–1659

Ramaekers VT, Heimann G, Reul J, Thron A, Jacken J (1997) Genetic disorders and cerebellar structural abnormalities in childhood. Brain 120: 1739–1751

Ray DE, Holton JL, Nolan CC, Cavanagh JB, Harpur ES (1998) Neurotoxic potential of gadodiamide after injection into the lateral cerebral ventricle of rats. Am J Neuroradiol 19: 1455–1462

Raz N, Torres IJ, Briggs SD, Spencer WD, Thorton AE, Loken WJ, Gunning FM, Mcquain JD, Driesen NR, Acker JD (1995) Selective neuroanatomic abnormalities in Down's syndrome and their cognitive correlates: evidence from MRI morphometry. Neurology 45: 356–366

Rees S, Breen S, Loeliger M, Mccrabb G, Harding R (1999) Hypoxemia near midgestation has long-term effects on fetal brain development. J Neuropathol Exp Neurol 58: 932–945

Rice D, Barone S (2000) Critical periods of vulnerability for the developing nervous system: evidence from humans and animal models. Environ Health Perspect 108 [Suppl 3]: 511–533

Rintala J, Jaatinen P, Lu W, Sarviharjum M, Eriksson CJP, Laippala P, Kiianmaak K, Hervonen A (1997) Effects of lifelong ethanol consumption on cerebellar layer volumes in AA and ANA rats. Alcohol Clin Exp Res 21: 311–317

Roebuck TM, Mattson SN, Riley EP (1998) A review of the neuroanatomical findings in children with fetal alcohol syndrome or prenatal exposure to alcohol. Alcohol Clin Exp Res 22: 339–344

Rorke LB, Fogelson MH, Riggs HE (1968) Cerebellar heterotopia in infancy. Dev Med Child Neurol 10: 644–650

Rosati P, Guariglia L (1999) Cerebellar hypoplasia: could it be a sonogrphic finding of abnormal fetal karyotype in early pregnancy? Fetal Diagnosis Ther 14: 365–367

Rotmensch S, Goldstein I, Liberati M, Shalev J, Ben-Rafael Z, Copel JA (1997) Fetal transcerebellar diameter in Down syndrome. Obstet Gynecol 89: 534–537

Rubin SA, Bautista JR, Moran TH, Schwart GJ, Carbone KM (1999) Viral teratogenesis: brain developmental damage associated with maturation state at time of infection. Brain Res Dev Brain Res 112: 237–244

Schmidt-Sidor B, Wisniewski KE, Shepard TH, Sersen EA (1990) Brain growth in Down syndrome subjects 15 to 22 weeks of gestational age and birth to 60 months. Clin Neuropathol 9: 181–190

Schnizel A (ed) (1984) Catalogue of unbalanced chromosome aberrations in man. Walter de Gruyter, Berlin New York

Shapiro BL (1983) Down syndrome — a disruption of homeostasis. J Med Genet 14: 241–269

Shapiro BL (1989) The pathogenesis of aneuploid phenotypes: the fallacy of explanatory reductionism. Am J Med Genet 33: 146–150

Shapiro BL (1992) Development of human autosomal aneuploid phenotypes (with an emphasis on Down syndrome). Acta Zool Fenn 191: 97–105

Shapiro BL (1994) The environmental basis of he Down syndrome. Dev Med Child Neurol 36: 84–90

Shapiro BL (1999) The Down syndrome critical region. J Neural Transm [Suppl] 57: 41–60

Sharp NJ, Davis BJ, Guy JS, Cullen JM, Steingold SF, Kornegay JN (1999) Hydranencephaly and cerebellar hypoplasia in two kittens attributed to intrauterine parvovirus infection. J Comp Pathol 121(1): 39–53

Shear PK, Sullivan EV, Lane B, Pfefferbaum A (1996) Mamillary body and cerebellar shrinkage in chronic alcoholics with and without amnesia. Alcohol Clin Exp Res 20: 1489–1495

Simpson MG, Wyatt I, Jones HB, Gyte HJ, Widdowsson ES, Lock EA (1996) Neuropathological changes in rat brains following oral administration of 2-chloroproprionic acid. Neurotoxicol 17: 471–480

Sørensen FW, Larsen JO, Eide R, Schionning JD (2000) Neuron loss in cerebellar cortex of rats exposed to mercry vapor: a stereological study. Acta Neuropathol 100: 95–100

Steinlin M (1998) Non-progressive congenital ataxias. Brain Dev 20: 199–208

Steinlin MI, Nadal D, Eich GF, Martin E, Boltshauser EJ (1996) Late intrauterine cytomegalovirus infection — clinical and neuroimaging findings. Pediat Neurol 15: 249–253

Steinlin M, Blaser S, Boltshauser E (1998) Cerebellar involvement in metabolic disorders — a pattern-recognition approach. Neuroradiology 40: 347–354

Steinlin M, Blaser S, Boltshauser E (1998) Cerebellar involvementin metaboloic disorders — a pattern-recognition approach [Article]. Neuroradiology 40: 347–354

Tamaraz JC, Rethoré M-O, Iba-Zizen M-T, Lejeune J, Cabanis EA (1987) Contribution of magnetic resonance imaging to the knowledge of CNS malformations related to chromosomal aberrations. Hum Genet 76: 265–273

Thurmond AS, Nelson DW, Lowensohn RI, Young WP, Davis L (1989) Enlarged cisterna magna in trisomy 18: prenatal ultrasonic diagnosis. Am J Obst Gynecol 161: 83–85

Vekamans M, Trasler T (1986) Liability to cleft palate in trisomy 19 mouse embryos. J Craniofac Genet Dev [Suppl 2]: 235–240

Wax JR, Benn P, Steinfeld JD, Ingardia CJ (2000) Prenatally diagnosed sacrococcygeal teratoma: a unique expression of trisomy 1q. Cancer Genet Cytogenet 117: 84–86

Weis S (1991) Morphometry and magnetic resonance imaging of the human brain in normal controls and Down's syndrome. Anat Rec 231: 593–598

Weis S, Weber G, Neuhold A, Rett A (1991) Down syndrome: MR quantification of brain structures and comparison with normal control subjects. AJNR 12: 1207–1211

Williams RE, Jones P, Lock EA, Bachelard HS (1999) Biochemical and neurotoxilogical effects of L-2-chloropropionic acid on rodent brain. J Neurochem 73: 362–371

Wiltshire T, Pletcher M, Cole SE, Villanueva M, Birren B, Lehozcky J, Dewar K, Reeves RH (1999) Perfect conserved linkage across the entire mouse chromosome 10 region homologous to human chromosome 21. Genome Res 9: 1214–1222

Winter TC, Ostrovsky AA, Komaniski CA, Uhrich SB (2000) Cerebellar and frontal lobe hypoplasia in fetuses with trisomy 21: usefulness as combined US markers. Radiology 214: 533–538

Xia JM, Simonyi A, Sun GY (1999) Chronic ethanol and iron administration on iron content, neuronal nitric oxide synthase, and superoxide dismutase in rat cerebellum. Alcohol Clin Exp Res 23: 702–707

Yamaguchi R, Naitoh Y, Uchida K, Hirano N, Wada Tateyama S (1996) Encephalopathy in suckling mice infected with Kasba (Chuzan) virus. J Comp Pathol 120: 347–256

Ye P, Xing YZ, Dai ZH, Dercole AJ (1996) In vivo actions of insulin-like growth factor-1 (IGF-1) on cerebellum development in transgenic mice — evidence that IGF-1 increases proliferation of granule cell progenitors. Brain Res Dev Brain Res 95: 44–54

Zellweger H (1977) Down syndrome. In: Vinken PJ, Bruyn GW, Myrianthopoulos NC (eds) Handbook of clinical neurology: congenital malformations of the brain and skull, v 31, ch 10. Elsevier, Amsterdam, pp 367–469

Zlotogora J (1992) Relation between the prevalence of anorectal malformations in children with Down syndrome and their prevalence in the general population. Am J Med Genet 44: 850

Zlotogora J, Abu-Dalu K, Lernan O, Sagi M, Voss R, Cohen T (1989) Anorectal malformations and Down syndrome. Am J Med Genet 34: 330–331

Authors' address: Dr. B. L. Shapiro, Departments of Oral Science and Laboratory Medicine and Pathology and Institute of Human Genetics, University of Minnesota, 17-220 Moos Tower, 515 Delaware Street SE, Minneapolis, MN 55455, U.S.A., e-mail: Burt@mailbox.mail.umn.edu

Expression of the multidrug resistance P glycoprotein (Pgp) and multidrug resistance associated protein (MRP1) in Down syndrome brains

E. Engidawork[1], **J. C. Roberts**[2], **R. Hardmeier**[1], **R. J. Scheper**[3], and **G. Lubec**[1]

[1] Department of Pediatrics, University of Vienna, Vienna, Austria
[2] Department of Medical Chemistry, University of Utah, Salt Lake City, USA
[3] Department of Pathology, Free University Hospital, Amsterdam, The Netherlands

Summary. Transport by ATP-dependent efflux pumps such as P glycoprotein (Pgp) and multidrug resistance associated protein (MRP), encoded by multidrug resistant (MDR) associated genes, is an increasingly recognized mechanism by which cells maintain substrate homeostasis and evade drug therapy. Pgp and MRP are members of the so-called ATP binding cassette (ABC) transporters superfamily, which are associated with many biological processes in both prokaryotes and eukaryotes, as well as clinical problems. The observation of upregulated sequences that are homologous to the Mycobacterium smegmatis phage resistance (mpr) gene and putative ABC transporters subunits in fetal Down syndrome (DS) using the gene hunting technique, subtractive hybridization formed the Rationale for this study. The expression of Pgp and MRP1 is therefore investigated in different brain regions of controls and adult DS patients with western blot technique. No apparent changes were observed between controls and DS in levels of Pgp in all brain regions examined. By contrast, MRP1 detection using the rat monoclonal antibody (MRPr1) produced a significant elevation in DS temporal cortex ($P < 0.01$) and parietal cortex ($P < 0.05$). Although MRP1 detected with the mouse monoclonal antibody (MRPm6) tended to increase in most of the regions of DS brain, it failed to reach significance level. Age or postmortem interval did not correlate with protein levels in both controls as well as DS. Taken together, the current data provide evidence for the presence of MDR related pumps in different regions of the human brain. In addition, overexpression of MRP1 in DS brain may have some relevance to the disorder either by deranging substrate homeostasis or limiting drug access.

Introduction

The adenosine triphospate binding cassette (ABC) transporters superfamily encompasses a broad range of functionally diverse transport proteins from bacteria to humans that utilize the energy of ATP hydrolysis to extrude substrates out of the cell (Higgins, 1992). Members of this superfamily have a general structure consisting of two sets of transmembrane domains and two ATP binding sites (Higgins, 1992). Multidrug resistance (MDR)-1 P glycoprotein (Pgp) and multidrug resistance associated protein (MRP) belong to the ABC transporters superfamily and are implicated in chemotherapeutic failure when expressed in tumor cells (Cole et al., 1992; Gottesman and Pastan, 1993; Loe et al., 1996). Recent studies show that Pgp and MRP confer a drug-permeability barrier function to brain cells constituting one mechanism by which the brain is protected from potentially toxic circulating exogenous as well as endogenous agents (Rao et al., 1999). In addition, they are found to be associated with channel activity that appears to be distinct from the multidrug resistance activity (Gottesman and Pastan, 1993; Higgins, 1992; Loe et al., 1996).

Human Pgp is encoded by a gene family comprising two MDR genes (MDR1 and MDR2/3) localized to the long arm of chromosome 7 (Gottesman and Pastan, 1993). Whilst MDR-1 is responsible for multidrug resistance, MDR-2/3 is involved in phosphatidylcholine secretion (Ling, 1997). Pgp was the first protein implicated in the multidrug resistance phenomenon often associated with tumor cells that overexpress the glycoprotein (Regina et al., 1998). However, expression is not restricted to malignant cells, rather extends to normal tissues with secretory and excretory functions, including adrenal cortex, kidney, liver, and gut as well as capillary endothelial cells of the brain that form the blood brain barrier and testis (Fromm, 2000). This expression profile of Pgp in normal cells clearly indicates the importance of the glycoprotein in biodisposition of a variety of lipophilic and cationic drugs (Fromm, 2000; Rao et al., 1999).

MRP, another member of the MDR transporters subfamily, was discovered in multidrug resistant tumor cells that did not express any of the structurally related Pgps (Cole et al., 1992). It has approximately 15% amino acid sequence homology to Pgp and functions as an ATP-dependent export pump to transport predominantly organic anions, such as cysteinyl leukotrienes, glutathione disulfide and various other glutathione conjugates (Cole et al., 1992; Ischikawa, 1992). The human MRP gene is designated as MRP1 and encoded on 16p13. So far, five different homologues of the MRP1 gene have been identified and denoted as MRP2-MRP6 (Kool et al., 1997, 1999). Distribution of MRPs somewhat overlaps with that of MDR1 and found to be expressed in a variety of tissues, including the brain (Cole et al., 1992; Zaman et al., 1993), with expression of MRP2 and MRP3 being scanty in whole brain (Kool et al., 1997; Zhang et al., 2000). MRP1 was shown to be expressed in the blood brain barrier endothelial cells (Huai-Yun et al., 1998; Regina et al., 1998) and choroid plexus epithelium (Rao et al., 1999), which are major detoxification sites within the brain. This suggests that MRPs together with Pgp, may regulate secretion and reabsorption of substrates and thereby maintain substrate homeostasis of the central nervous system (Rao et al., 1999).

Down syndrome (DS) is a chromosomal disorder considered as the most frequent cause of mental retardation. Performing gene hunting in fetal DS brain using the method of subtractive hybridization, we found a sequence upregulated, and homologous to Myobacterium smegmatis phage resistance (mpr) gene (Accession No. U50335) and putative ABC transporter subunits (Accession No. U75364) (Labudova and Lubec, unpublished observation). This prompted us to investigate the expression of the multidrug resistance proteins, Pgp and MRP, in different brain regions of adult controls and DS patients by western blotting. Here we demonstrate region specific aberration of MRP1 in DS, which may have relevance to disruption of intracellular substrate homeostasis and pharmacotherapy in DS brain.

Material and methods

Materials

For western blot, rat monoclonal antibody to the multidrug resistance related protein, MRP (MRPr1), mouse monoclonal antibody to MRP (MRPm6) and mouse monoclonal antibody to human P-glycoprotein (JSB-1) were obtained from Department of Pathology, Free University Hospital, Amsterdam, The Netherlands. Both MRPr1 and MRPm6 recognize intracellular epitopes of MRP1 (Flens et al., 1994). MRPr1 reacts with a segment of 168 amino acids in proximal half of the amino-termial, while MRPm6 binds to 170 amino acids in the carboxy-terminal end and part of the carboxy-proximal nucleotide-binding domain of MRP. JSB-1 reacts with a conserved cytoplasmic epitope of the plasma membrane associated Pgp (Scheper et al., 1988). Horseradish peroxidase-conjugated anti-mouse and anti-rat IgG (H+L) were purchased from ICN Biomedicals Inc. and Dako, respectively. Enhanced chemiluminescence reagent was from NEN™ Life Science Products, Inc (NEN™ Life Science Products, Inc., Boston, USA) and protein size markers (Rainbow) were from Amersham (Uppsala, Sweden). All other chemicals were obtained from commercial sources of the highest quality available.

Postmortem tissue

Five different brain regions; temporal, frontal, occipital and parietal cortex, and cerebellum of controls with no neurological or psychiatric history, and patients with DS that exhibited abundant beta-amyloid plaques and neurofibrillary tangles, were used for the present study. The samples were obtained from Dr. N. Cairns, Medical Research Council's London Brain Bank for Neurodegenerative Diseases, Department of Neuropathology, Institute of Psychiatry, London UK. The major cause of death in DS was bronchopnemonia (n = 9; three females and six males with mean age of 55.7 ± 7.5 years and postmortem interval (PMI) of 31.4 ± 19.6 hours). Heart disease was rather the cause of death in controls (n = 9; three females and six males with mean age of 54.2 ± 10.2 years and PMI of 40.4 ± 22.8 hours). All samples were stored at −70°C and the freezing chain was never interrupted.

Sample preparation

Frozen brain samples were thawed on ice and homogenized in 4 volumes of TBST buffer (100 mM Tris, 150 mM NaCl, pH 7.5) containing protease inhibitors, (1 µg/ml each of antipain, pepstatin A, and leupeptin and 10 mM of phenylmethylsulfonyl fluoride) and

0.05% Tween 20. Samples were homogenized on ice for 30 seconds (6–8 strokes) with a Potter-Elvehjem homogenizer. The brain homogenate then was centrifuged for 10 min at 10,000 × g and 4°C. To the supernatant, equal volume of sample buffer (125.5 mM Tris-HCl base, 70 mM sodium dodecyl sulfate, 0.001% Bromophenol blue, 20% glycerol, and 2% 2-mercaptoethanol, pH 6.8) was added and the mixture was heated at 95°C for 15 min. The concentration of protein was determined by the Bicinchoninic acid assay using bovine serum albumin as a standard (Pierce, USA).

Immunoblotting

Protein extracts containing 15 μg of Pgp and MRP were subjected to sodium dodecyl sulfate-polyacrylamide gel electrophoresis (50 mA for 1.15 hrs) using Multiphor II Electrophoresis system (Amersham Pharmacia Biotech AB, Sweden). Following electrophoresis, the proteins were electroblotted onto polyvinylidene fluoride membranes (Immobilon™-P, MilliPore, Bedford, USA) with a Multiphor II NovaBlot Unit (Amersham Pharmacia Biotech AB, Sweden) (192 V for 1 hr). The membranes were then washed two times for 5 min in Tris buffer (100 mM Tris-HCl, 150 mM NaCl, pH 7.5) and later blocked with blotto (1.5% non-fat dry milk in TBST (100 mM Tris-HCl, 150 mM NaCl, pH 7.5, 1 mM $MgCl_2$, 0.1% Tween-20)) overnight at 4°C followed by incubation with monoclonal mouse anti-Pgp antibody (2.5 μg/ml), rat anti-MRP antibody (1 μg/ml) and mouse anti-MRP antibody (5 μg/ml) at room temperature for 2 hours on a rocking shaker. The membranes were washed three times for 15 minutes in blocking solution. Washed membranes were incubated with the respective horseradish peroxidase conjugated secondary antibody in blotto (1:2000) for one hour at room temperature on a rocking shaker. The membranes were washed again three times for 15 minutes followed by incubation with NEN chemiluminescence reagent for 1 minute and subsequent exposure to films (Kodak Blue XB-1, Rochester, USA). Developed films were scanned and densities of the immunoreactive bands were calculated by RFLP Scan 2.1 software program (Scanalytics, Billerica, USA).

Statistical analysis

Values are expressed as mean ± standard deviation of optical density. Inter-group differences were analyzed using non-parametric Mann-Whitney U-test and within-group correlation was done using linear regression analysis. The level of significance was set at $P < 0.05$.

Results

Western blot analysis with anti-Pgp and anti-MRP monoclonal antibodies demonstrated the presence of Pgp and MRP in different regions of the human brain. Anti-Pgp antibody detected several Pgp immunoreactive (IR) bands that span in molecular size from 7 kDa to 170 kDa, with Pgp migrating at ~170 kDa (Fig. 1). As JSB-1 happens to be specific for Pgp, low molecular weight bands detected with the antibody might represent truncated forms, degradation products, or isoforms of Pgp. Both anti-MRP antibodies produced blots with the same band pattern, since they react with the same antigen at different epitopes. The antibodies reacted with several bands that range from 15 kDa to 200 kDa, among which a signal corresponding to MRP1 was

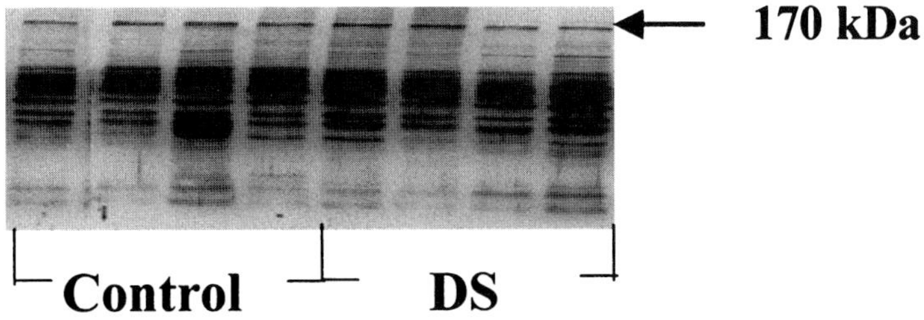

Fig. 1. Immunodetection of P glcoprotein (Pgp) in control and DS brain. Samples obtained from five different brain regions of 9 controls and 9 DS cases were homogenized and solubilzed with an appropriate buffer (see Material and methods). The supernatant (15 μg protein/lane) was subjected to SDS-PAGE and Pgp was then detected by monoclonal mouse antibody (JSB-1). The immunoblot of a representative region (frontal cortex) of 4 controls (lane 1–4) and 4 DS (lane 5–8) was shown in the upper panel

obtained at ~180 kDa (Fig. 2). The remaining bands could represent nonspecific bands as the antibodies are known to react with other antigens (Flens et al., 1994), or degradation products.

Analysis of Pgp antigen density did not reveal significant difference in expression of Pgp in any of the regions investigated between controls and DS. Levels of Pgp in temporal cortex and frontal cortex as well as in cerebellum were almost similar in controls and DS. Although mean levels of Pgp in DS occipital cortex appeared to be higher by ~117% compared to controls, the great variation among the individual values barred statistical significance. Likewise, a 23.3% increase was noted in parietal cortex that failed to reach statistical significance (Fig. 3A). MRP1 detected with MRPr1 showed aberrant expression in certain brain regions of DS. However, MRP1 detected with MRPm6 did not exhibit any detectable changes, although it tended to increase in some of the DS regions investigated. With MRPr1, levels of MRP1 were significantly elevated in DS temporal cortex and parietal cortex by ~58.5% ($P < 0.01$) and 56% ($P < 0.05$), respectively. An insignificant increase was also observed in DS frontal cortex by ~16.7% and cerebellum by ~30.8%, but a reversed pattern was noted in occipital cortex and decreased insignificantly by ~15.4% compared to controls (Fig. 3B). MRPm6 detected MRP1 exhibited a notable increase in temporal cortex by ~57.9% and cerebellum by ~54.5%, although failed to reach significance level due to a great variation among the DS individuals. In the remaining regions, levels were comparable in controls and DS brains (Fig. 3C). Neither age nor PMI was found to have any influence on levels of both MRP1 and Pgp in control as well as DS subjects.

Discussion

In the present study, we demonstrate upregulation of MRP1, at protein level, in adult DS brain confirming the initial observation in fetal DS brain made using subtractive hybridization. The detection of a 170 kDa Pgp on immunoblot is consistent with the predicted molecular weight of Pgp. Moreover, the presence of the 170 kDa band in every region examined indicates

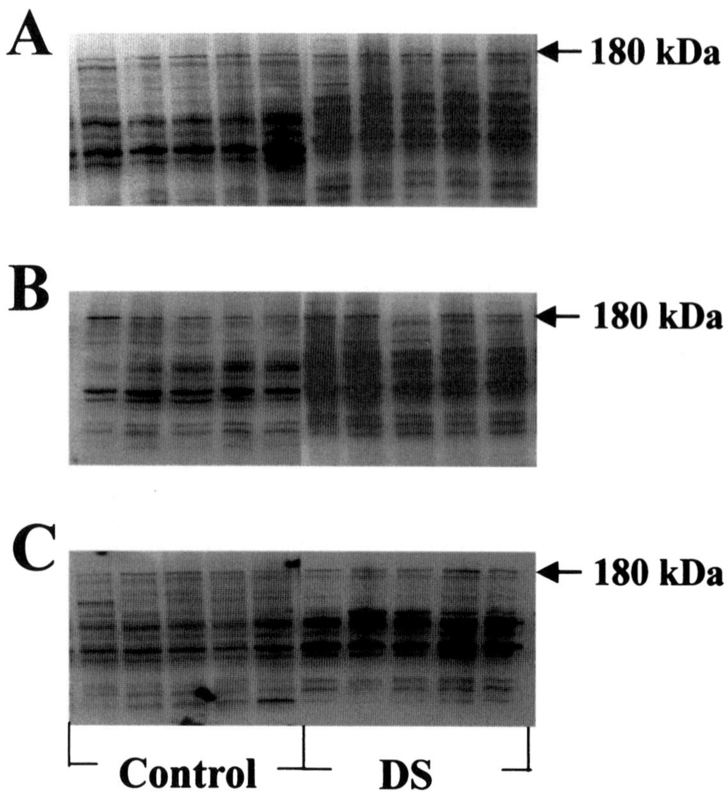

Fig. 2. Immunodetection of multidrug resistance associated protein (MRP1) in control and DS brain. Samples obtained from five different brain regions of 9 controls and 9 DS cases were homogenized and solubilzed with an appropriate buffer (see Material and methods) and the supernatant (15 μg protein/lane) was subjected to SDS-PAGE. MRP1 was then detected by monoclonal mouse antibody (MRPr1) and monoclonal rat antibody (MRPm6). *Panels A* and *B* show immunoblots obtained using MRPr1 from the temporal cortex and parietal cortex, respectively. *Panel C* shows immunoblot of the temporal cortex using MRPm6 antibody. Lanes 1–5 represent controls and lanes 6–10 DS

that cells in those region can express Pgp and may utilize it to impact the cellular bioavailability of drugs as well as metabolite levels. Although this study showed expression of Pgp in different regions of the human brain, it did not reveal any appreciable change between control and DS brains. The unchanged levels may imply that no disease-related changes are apparent in Pgp expression. Alternatively, it could have resulted from difference in relative abundance of Pgp among brain cells. In this regard, Pgp appears to be predominantly expressed in the capillary endothelial cells forming the blood brain barrier rather than in neurons and astrocytes (Regina et al., 1998; Seetharaman et al., 1998; Decleves et al., 2000). Considering the cell type dependent expression of Pgp, the comparable expression between controls and DS in the cerebral and cerebellar cells would not seem unreasonable.

MRP1 is an integral membrane N-glycosylated phosphoprotein with a calculated molecular weight of 171 kDa. Immunologically detected MRP1 has

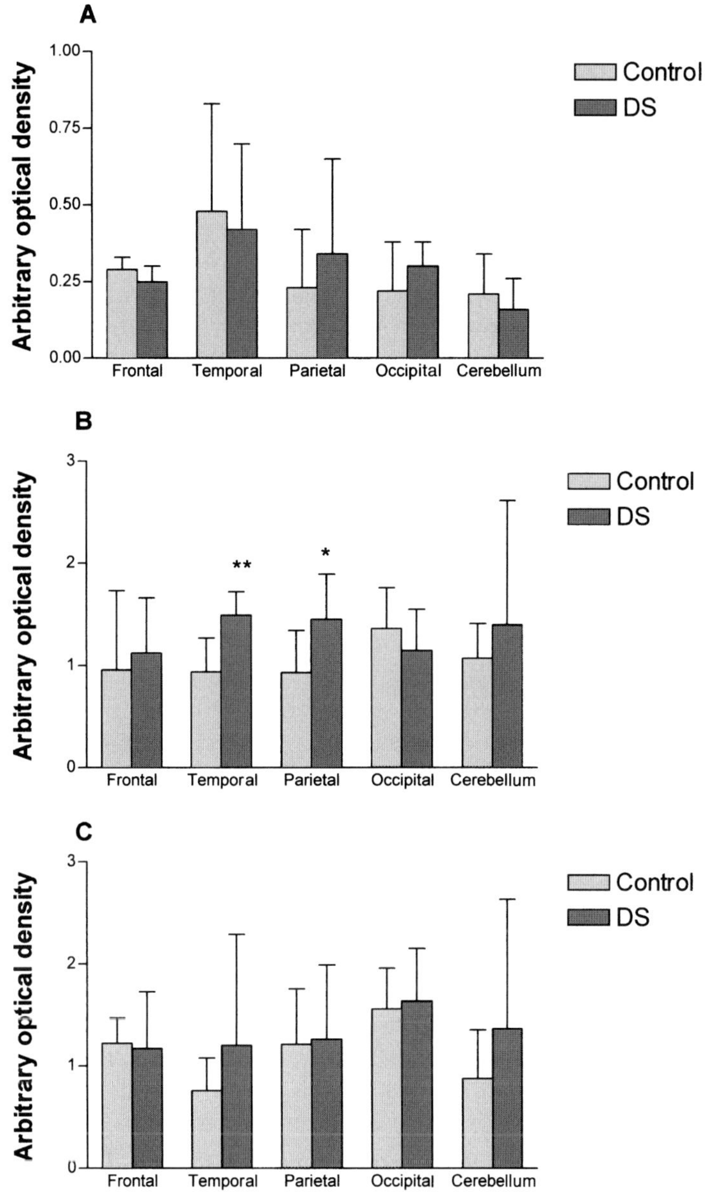

Fig. 3. Expression of the multidrug resistance related proteins, Pgp and MRP1 in different brain regions of controls and DS. Following western blot, the immunorectivity of Pgp and MRP1 from five brain regions of 6–9 controls and 8–9 DS was quantified using RFLP Scan 2.1 software program. *Panels A, B* and *C* represent Pgp, MRP1 detected with MRPr1, and with MRPm6, respectively. Values represent mean ± standard deviation of arbitrary optical density. Inter-group differences were analyzed using non-parametric Mann-Whitney U-test and the significance was set at $P < 0.05$ for a two-tailed test. $^{**}P < 0.01$, $^{*}P < 0.05$

a molecular size of 190 kDa (Loe et al., 1996), which is in agreement with our finding. Among the two antibodies used to detect MRP1, it was only MRPr1 that produced significant difference between the test groups. MRPm6, however, failed to produce significant difference owing to the greater variation among individual values in DS subjects. Otherwise the data produced is consistent with the one obtained with MRPr1 (Fig. 3B and C). One possible explanation for the different outcome may emanate from the fact that the two antibodies are directed to different epitopes. To this effect, a difference in the inherent affinity of the antibodies for the antigen was reported. The antigen displays higher affinity for the rat (MRPr1) than the mouse (MRPm6) antibody and the former detects low basal level MRP and produces a stronger staining than the latter (Flens et al., 1994), which appears to be the case in this study.

Increased expression of MRP has been found in leukemia (Wang et al., 1998), which happens to have a higher incidence in children with DS, human gliomas (Mohri et al., 2001) and DS brain (present study). MRP expression appears to be higher in astrocytes than endothelial cells and the converse is true for Pgp (Decleves et al., 2000). In our previous study, we had shown that glial acidic fibrillary protein, a marker of gliosis, was abundantly expressed in several brain regions of patients with DS (Greber-Platzer et al., 1999). Increased MRP1 expression in DS seems therefore likely to be attributed to glial proliferation. Failure to observe any apparent change in Pgp expression argues for this assumption. Another possible explanation for the observed increase could be oxidative stress. Oxidative stress, a major factor proposed for the development of brain damage in DS (Busciglio and Yankner, 1995; Iannello et al., 1999), has been demonstrated to regulate expression of MRP1 (Yamane et al., 1998) and MRP2, whose functional features are similar to MRP1 (Payen et al., 2001). Upregulation of MRP may occur by mechanisms other than gene amplifications, including positive transcriptional regulation and altered mRNA stability (Loe et al., 1996). Recently, Grandjean et al. (2001) did not find any significant difference in MRP mRNA half-life between sensitive and MDR cancer cell lines, which could exclude altered stability as a factor for upregulation. Reactive oxygen species therefore may cause increased expression through redox sensitive transcription factors that interact with cis-acting elements identified as oxidant response element like sequences in MRP1 gene (Yamane et al., 1998).

What is the significance of increased expression of MRP1 if any, in DS? Although the physiological role of MRP1 is not completely understood, available data clearly indicate the importance of MRP1 in drug penetration into the central nervous system (Ghersi-Egea and Strazielle, 2001; Rao et al., 1999) and regulation of intracellular substrate homeostasis, including glutathione (Loe et al., 1998; Rappa et al., 1997). MRP1 is responsible for the release of oxidized glutathione from cells that is considered to be an important mechanism to maintain the reduced status of intracellular thiols under oxidative stress (Yamane et al., 1998). Furthermore, MRP1 is proposed to have a protective role by extruding glutathione conjugated 4-hydroxynonenal, which is the most prevalent toxic lipid peroxidation product formed during oxidative stress (Renes et al., 2000). The aforementioned roles of MRP1 makes one to safely assume that the increased expression in DS is a cellular response initiated to counteract the deleterious effects of oxidative stress. However, the possibility that increased expression may accentuate oxidative stress induced cell damage can not be totally excluded. MRP not only transports altered drugs but also transports unaltered drugs along with glutathione (Loe et al., 1998; Rappa et al., 1997). This suggests that glutathione is normally cotransported with a yet-to-be identified endogenous substrate. Under pathological conditions like DS, oxidative stress mediated upregulation of MRP1 can lead to depletion of intracellular glutathione that weakens the cellular defense mechanism. Moreover, some members of the ABC transporters superfamily are associated with channel functions (Higgins, 1992; Loe et al., 1996) and MRP is thought to have channel function or regulate such channels (Jirsch et al., 1993). In this context, increased expression of MRP1 confers a different solute export profile that may dysregulate substrate homeostasis of the brain. The presence of hyperuricemia in individuals with DS (Epstein, 2001) clearly indicates that dysregulated substrate homeostasis is not an uncommon phenomenon in DS and lends support for this notion.

The expression of MRP in brain other than the endothelial cells of the blood brain barrier and the choroid plexus epithelium raised the possibility that other barriers might exist in the brain (Regina et al., 1998). The present study, uphold this view. Although overexpression of Pgp appears to be an important determinant in the acquisition of drug-resistance property in tumor cells, lack of change in the brain regions examined indicates that this is not the case in all brain cells. The increased expression of MRP1 in DS temporal cortex and parietal cortex points to the possibility that these regions may deny access to xenobiotics by MRP-dependent mechanism. In contrast to Pgp, which seems to be localized in the plasma membrane, MRP is distributed among various membrane compartments (Krishnamachary and Center, 1993). This distribution pattern allows MRP to cause marked changes in intracellular drug distribution thereby preventing the drug reaching its cellular target, in addition to ATP-dependent enhanced drug efflux (Loe et al., 1996). Theoretically, it is therefore imperative to rule out a drug as a substrate for MRP1 in drug development strategies or before commencing therapy.

In conclusion, the current data provide evidence for the expression of MDR related proteins in different regions of the human brain. The relative overexpression of MRP1 may result in deranged substrate fluxes in DS. In addition, it may also pose difficulty in pharmacotherapy of the DS brain.

Acknowledgements

We are highly indebted to Red Bull Company, Salzburg, Austria for the generous financial support. E.E is a scholarship holder of the Austrian Academic Exchange Program, ÖAD.

References

Busciglio J, Yankner BA (1995) Apoptosis and increased generation of active oxygen species in Down syndrome neurons in vitro. Nature 378: 776–779

Cole SPC, Bhardwaj G, Gerlach JH, Mackie JE, Grant CE, Almquist KC, Stewart AJ, Kurz EV, Duncan AMV, Deeley RG (1992) Overexpression of a transporter gene in a multidrug resistance human lung cancer cell line. Science 258: 1650–1654

Decleves X, Regina A, Laplanche JL, Roux F, Boval B, Launay JM, Scherrmann JM (2000) Functional expression of P-glycoprotein and multidrug resistance associated protein (MRP1) in primary cultures of astrocytes. J Neurosci Res 60: 594–601

Epstein CJ (2001) Down syndrome (Trisomy 21). In: Scriver CR, Beaudet AL, Sly WS, Valle D (eds) The metabolic and molecular basis of inherited disease. McGraw Hill, New York, pp 1223–1256

Flens MJ, Izquierdo MA, Scheffer GL, Fritz JM, Meijer CJLM, Scheper RJ, Zaman GJR (1994) Immunohistochemical detection of the multidrug resistance associated protein MRP in human multidrug resistant tumor cell by monoclonal antibodies. Cancer Res 54: 4557–4563

Fromm MF (2000) P-glycoprotein: a defense mechanism limiting oral bioavailability and CNS accumulation of drugs. Int J Clin Pharmacol Ther 38: 69–74

Ghersi-Egea JF, Strazielle N (2001) Brain drug delivery, drug metabolism, and multidrug resistance at the choroid plexus. Microsc Res Tech 52: 83–88

Gottesman MM, Pastan I (1993) Biochemistry of multidrug resistance mediated by the multidrug resistance transporter. Annu Rev Biochem 62: 385–427

Grandjean F, Bremaud L, Verdier M, Robert J, Ratinaud M-H (2001) Sequential gene expression of P-glycoprotein (P-gp), multidrug resistance protein (MRP) and lung resistance protein: functional activity of P-gp and MRP present in the doxorubicin-resistant human K562 cell lines. Anti-Cancer Drugs 12: 247–258

Greber-Platzer S, Lubec G, Langen H, Fountoulakis M (1999) Decreased levels of synaptosomal associated protein 25 in the brain of patients with Down syndrome and Alzheimer's disease. Electrophoresis 20: 928–934

Higgins CF (1992) ABC transporters: From microorganisms to man. Annu Rev Cell Biol 8: 67–113

Huai-Yun H, Secrest DT, Mark KS, Carney D, Brandquist C, Elmquist WF, Miller DW (1998) Expression of Multidrug-Resistance associated protein (MRP) in brain microvessel endothelial cells. Biochem Biophys Res Commun 243: 816–820

Iannello RC, Crack PJ, de Haan JW, Kola I (1999) Oxidative stress and neural dysfunction in Down syndrome. J Neural Transm 57: 257–267

Ischikawa I (1992) The ATP dependent glutathione-S export pump. Trends Biochem Sci 17: 463–468

Jirsch J, Deeley RG, Cole SPC, Stewart AJ, Fedida D (1993) Inwardly rectifying potassium channels and volume regulated anion channels in multidrug resistant small cell lung cancer. Cancer Res 53: 4156–4160

Kool M, de Haas M, Scheffer GL, Scheper RJ, van Eijk MJT, Juijin JA, Baas F, Borst P (1997) Analysis of expression CMOAT (MRP2), MRP3, MRP4, and MRP5, homologues of MRP1 in human cancer cell lines. Cancer Res 57: 3537–3547

Kool M, van der Linden M, de Haas F, Borst P (1999) Expression of human MRP6, a homologue of the multidrug resistance protein gene MRP1, in tissues and cancer cells. Cancer Res 59: 175–182

Krishnamachary N, Center MS (1993) The MRP gene associated with a non-P glycoprotein multidrug resistance encodes a 190 kDa membrane bound glycoprotein. Cancer Res 53: 3658–3661

Ling V (1997) Multidrug resistance: molecular mechanisms and clinical relevance. Cancer Chemother Pharmacol 40: S3–8

Loe DW, Deeley RG, Cole SPC (1996) Biology of the multidrug resistance associated protein, MRP. Eur J Cancer 32A: 945–957

Loe DW, Deeley RG, Cole SPC (1998) Characterization of vincristine transport by the Mr 190000 multidrug resistant protein (MRP): evidence for co-transport with reduced glutathione. Cancer Res 58: 5130–5136

Mohri M, Nitta H, Yamashita J (2001) Expression of multidrug resistance associated protein in human gliomas. J Neurooncol 49: 105–115

Payen L, Courtois A, Loewert M, Guillouzo A, Fardel O (2001) Reactive oxygen species related induction of multidrug resistance associated protein2 expression in primary hepatocytes exposed to sulforaphane. Biochem Biophys Res Commun 282: 257–263

Rao V, Dahlheimer JL, Bardgett ME, Snyder AZ, Finch RA, Sartorelli AC, Piwnica-Worms D (1999) Choroid plexus epithelial expression of MDR1 P glycoprotein and multidrug resistance associated protein contribute to the blood cerebrospinal fluid drug permeability barrier. Proc Natl Acad Sci USA 96: 3900–3905

Rappa G, Lorico A, Flavell RA, Sartorelli AC (1997) Evidence that the multidrug resistance protein (MRP) functions as a co-transporter of glutathione and natural product toxins. Cancer Res 57: 5232–5237

Regina A, Koman A, Piciotti M, El Hafny B, Center MS, Bergmann R, Couraud PO, Roux F (1998) Mrp1 multidrug resistance associated protein and P-glycoprotein expression in rat brain microvessel endothelial cells. J Neurochem 71: 705–715

Renes J, de Vries EEG, Hooiveld GJEJ, Krikken I, Jansen PLM, Müller M (2000) The multidrug resistance protein MRP1 protects against the toxicity of the major lipid peroxidation product 4-hydroxynonenal. Biochem J 350: 555–561

Scheper RJ, Bulte JW, Brakkee JGP, Quak JJ, van Der Schoot E, Balm AJM, Meijer CJLM, Broxterman HJ, Kuiper CM, Lankelma J, Pinedo HM (1988) Monoclonal antibody JSB-1 detects a highly conserved epitope on the P-glycosylation associated with multidrug resistance. Int J Cancer 42: 389–394

Seetharaman S, Barrand MA, Maskell L, Scheper RJ (1998) Multidrug resistance related transport proteins in isolated human brain microvessels and in cells cultured from these isolates. Neurochem 70: 1151–1119

Wang F, Dong Z, Luo J (1998) The clinical significance of multidrug resistance associated (MRP) gene exression in acute leukemia. Zhonghua Xve Ye Xue Za Zhi 19: 63–66

Yamane Y, Furuichi, Song R, Van NT, Mulcahy RT, Ishikawa T, Kuo MT (1998) Expression of multidrug resistance protein/GS-X pump and γ-glutamylcysteine synthetase genes is regulated by oxidative stress. J Biol Chem 273: 31075–31085

Zaman GJR, Versantvoort CHM, Smith JJM, Eijdems EWHM, de Haas M, Smith AJ, Broxterman HJ, Mulder NH, de Vries EGE, Baas F, Borst P (1993) Analysis of the expression of MRP, the gene for a new putative transmembrane drug transporter, in human multi drug resistant lung cancer cell lines. Cancer Res 53: 1747–1750

Zhang Y, Han H, Elmquist WF, Miller DW (2000) Expression of various multidrug resistance associated protein (MRP) homologues in brain microvessel endothelial cells. Brain Res 876: 148–153

Authors' address: Prof. Dr. G. Lubec, Department of Pediatrics, University of Vienna, Währinger Gürtel 18-20, A-1090 Vienna, Austria, e-mail: Gert.Lubec@akh-wien.ac.at

Deterioration of the transcriptional, splicing and elongation machinery in brain of fetal Down Syndrome

M. Freidl[1], **T. Gulesserian**[1], **G. Lubec**[1], **M. Fountoulakis**[2], and **B. Lubec**[3]

[1] Department of Pediatrics, University of Vienna, Vienna, Austria
[2] Gene Technologies, CNS Research, F. Hoffmann La Roche, Basle, Switzerland
[3] Department of Neonatology, University of Vienna, Vienna, Austria

Summary. Perturbation of brain development i.e. regulation of gene expression, differentiation, growth and migration in Down Syndrome (DS) has been reported to occur early in life pointing to impairment of the complex system of transcription and or translation and indeed, altered expression of transcription factors has been reported in adult DS brain. We therefore decided to compare the transcriptional and translational machinery in cortex of brains of controls and fetuses with Down syndrome in the second trimenon of gestation. We determined a series of transcription/translation factors by 2 D-electrophoresis followed by MALDI — identification and quantification with specific software.

The protooncogene C-CRK, CRK-like protein, elongation factor 1-alpha 1, elongation factor 2, elongation factor tu and two out of four spots representing PTB-associated splicing factor PSF were significantly downregulated in brain of fetal DS fetuses as compared to controls.

The finding of reduced transcription and translation factors may indicate deranged protein synthesis. The underlying cause for individual reduced transcription, splicing and translation factors may be explained by chromosomal imbalance or by posttranslational modifications as e.g. phosphorylation, known to be aberrant in DS. Reduced expression of transcription factors in fetal DS during early life may be responsible or reflecting impaired brain development and deficient wiring of the brain in DS.

Introduction

Brain development and differentiated neuronal functions depend on coordinated patterns of gene activation and inactivation. 30%–50% of genes in mammals are specifically expressed in the nervous system. The differential expression of these genes in distinct spatial and temporal patterns is necessary for generating the vast variety of neuronal phenotypes. The interaction and regulation of the repertoire of transcription factors — in human nervous system 1,500–2,500 transcription factors exhibit specific expression — may permit the precision and nuance of control of gene expression required for complex functional systems such as the nervous system. Transcriptional regulation may also be an important part of the machinery for higher functions of the nervous system such as learning or memory. A single short stimulus causes longlasting elevations in the synthesis of TH and other proteins in postsynaptic neurons, having been attributed to increased mRNA synthesis. From a variety of studies it is clear that de novo synthesis of mRNA and protein is required for learning and memory, therefore, an altered transcriptional state is associated with memory aquisition, just as it is associated with cellular differentiation. Thus subtle differences in the combinations of transcription factors are likely to have profound consequences for cell phenotype and may form the basis for a variety of disorders of brain (He and Rosenfeld, 1999; Arenander, 1994).

There is accumulating evidence for abnormalities of neuronal differentiation and abnormal migration in fetal and infant DS brains. It has been shown that patients with DS not only present with anatomical abnormalities (Schmidt-Sidor et al., 1990; Sylvester, 1986) but also may show nerve cell heterotopias in the white layers of the cerebellum, abnormalities in morphology and number of dendritic spines or diminution in the number of hypothalamic neurons etc. (Rehder, 1981; Becker et al., 1991, 1986; Takashima et al., 1989), findings which may well be explained by early impairment of the transcriptional machinery.

In a previous publication we have shown downregulation of the transcription factor junD in adult DS brain (Labudova et al., 1998) and we also could demonstrate a decrease of mRNA of the basic helix-loop-helix transcription factor scleraxis in adult DS brains, regulating growth and differentiation of numerous cell types (Yeghiazarian et al., 1999). Decreased ETS2 transcripts found in temporal and frontal lobe of patients with DS by our group may be involved in the pathogenesis of DS (Greber-Platzer et al., 1999) as ETS2 was proposed to play an important role in CNS development and in the maintainance of certain neuronal functions (Maraoulakou et al., 1994).

In addition we could show that c-fos, a protooncogene serving in multiple physiological processes in brain from signalling to proliferation and synaptic plasticity presented with increased expression in frontal, parietal, occipital and temporal cortex as well as in cerebellum of adult patients with DS, suggesting that the impairment of c-fos expression may be linked to increased apoptosis and neurodegeneration, overexcitation by excitatory amino acids or reactive oxygen species (Greber-Platzer et al., 1999). Studying repair mechanisms in patients with DS we showed that DNA repair genes ERCC2 and ERCC3 for nucleotide excision repair and XRCC1 for X-ray repair were increased at the transcriptional level in some brain regions paralleled by an increase of two basic markers for transcription and translation, TAF-DBP and EF1A, reflecting coordinated activation of repair, transcription and translation in DS brains (Fang-Kircher et al., 1999).

These observations of derangement of various transcription factors in DS at the mRNA level formed the basis for the present study. We decided to investigate not only the transcriptional but also the translational machinery in fetal DS brains. As mRNA levels might not necessarily reflect gene function we determined the proteome by 2 D-electrophoresis followed by MALDI — identification and quantification with a specific software to detect strongly expressed proteins both in DS and controls, identifying a battery of transcription, splicing, elongation and translation factors.

Materials and methods

Fetal brain tissue

Fetal brain samples from abortuses (cerebral cortex) of controls (n = 7; 1 female and 6 males with the mean age of 18.8 ± 2.2 weeks of gestation and post-mortem interval (PMI) of 5.5 ± 1.0 hours) and DS (n = 8; 2 females and 6 males with mean age of 19.8 ± 2.0 weeks of gestation and postmortem interval of 9.1 ± 4.4 hours) were obtained from Dr. M. Dierssen, Medical and Molecular Genetics center-IRO, Hospital Duran i Reynals, Barcelona, Spain. All samples were stored at −70°C and the freezing chain was never interrupted.

Two dimensional gel electrophoresis

Brain tissue was suspended in 0.5 ml sample buffer consisting of 40 mM tris, 5 M urea (Merck, Darmstadt, Germany), 2 M thiourea (Sigma, St. Louis, MO, USA), 4% CHAPS (Sigma), 10 mM 1,4-dithioerythrol (Merck), 1 mM EDTA (Merck) and a mixture of protease inhibitors, 1 mM PMSF and 1 ug of each pepstatin A, chymostatin, leupeptin and antipain. The suspension was sonicated for approximately 30 sec and centrifuged at 150,000 g for 45 min. The protein content in the supernatant was determined by the Coomassie blue method (Bradford, 1976).

Two-dimensional gel electrophoresis was performed essentially as reported (Langen et al., 1997). Samples of 2 mg protein were applied on immobilized pH-3-10 nonlinear gradient (IPG) strips (Amersham Pharmacia Biotechnology, Uppsala, Sweden) in sample cups at their basic and acidic ends. Focusing started at 200 V and the voltage was gradually increased to 5,000 V at 3 V/min and kept constant for a further 24 h (approximately 180,000 kVh totally). The second — dimensional separation was performed on 9–16% SDS gradient polyacrylamide gels. The gels (180 × 200 × 1.5 mm) were run at 40 mA per gel. After protein fixation for 12 hrs in 40% methanol, containing 5% phosphoric acid, the gels were stained with colloidal Coomassie blue (Novex, San Diego, CA, USA) for 48 hrs. Molecular masses were determined by running standard protein markers (Gibco, Basel, Switzerland), covering the range 10–200 kDa. pI values were used as given by the supplier of the IPG strips. Excess of dye was washed out from the gels with H20 and the gels were scanned in an Agfa DUOSCAN densitometer (resolution 200). Electronic images of the gels were recorded using Photoshop (Adobe) and Power Point (Microsoft) software. Protein spots were outlined (first automatically and then manually) and quantified using the ImageMaster 2D Elite software (Amersham Pharmacia Biotechnology). The percentage of the volume of the spots representing a certain protein was determined in comparison with the total proteins present in the 2D gel.

Matrix-associated laser desorption ionization mass spectroscopy (MALDI-MS)

MALDI-MS analysis was performed as described elsewhere (Fountoulakis and Langen, 1997) with some modifications. The spots were excised with a spot picker and placed into 96-well microtiter plates. Each spot was destained with 100 microliters of 30% acetonitrile in 50 mM ammonium bicarbonate and dried in a speed vac evaporator. Each dried gel piece was rehydrated with 4 microliters of 3 mM Tris-HCl, pH 9.0 containing 50 ng trypsin (Promega, Madison, WI, USA). After 16 hours of room temperature, seven microliters of H20 were added to each gel piece and the samples were shaken for 10 min. Four microliters of 50% acetonitrile, containing 0.3% trifluoroacetic acid and the standard peptidesdes-Arg-Bradykinin (Sigma, 904.4681 Da) and adrenocorticotrophic hormone pregnant 18–39 (Sigma, 2,465.1989 Da), in water were added to each gel piece and shaken for 10 min. The application of the samples was performed with a SymBiot I sample processor (PE Biosystems, Framingham MA, USA). 1.5 microliter of the peptide mixture was simultaneously applied with 1 microliter of matrix, consisiting of a saturated solution of alpha-cyano-4-hydroxycinnamic acid (Sigma) destained with 50% acetonitrile in 0.1 M ammonium bicarbonate and dried in a speedvac evaporator. The dried gel pieces were reswollen with 3 ul of 3 mM Tris-HCl, pH 8.8, containing 50 ng trypsin (Promega, Madison, WI, USA). After 16 h, 7 ul of water were added and left at room temperature for about 15 min. One microliter of the mixture was applied onto the dried matrix spot. The matrix consisted of 15 mg nitrocellulose (Bio-Rad) and 20 mg alpha-cyano-4-hydroxycinnamic acid (Sigma) in 50% acetonitrile, containing 0.1% trifluoroacetic acid. Samples were analyzed in a time-of-flight-mass spectrometer (REFLEX 3 Bruker Analytics, Bremen, Germany). An accelerating voltage of 20 kV was used. Pewptide matching and protein searches were performed automatically. The peptide masses were compared to the theoretical peptide masses of all available proteins of all species. Monoisotopic masses were used and the mass tolerance of 0.0025% was allowed. The algorithm used for determining the probability of a false positive match with a given MS-spectrum is described elsewhere (Berndt et al., 1999).

Statistical analysis

The software calculated the percentage of the spot volume in each particular gel after substraction of the background values. Between-group differences were calculated by non-parametric Mann-Whitney U-test and within-group-correlations were done using the Spearman rank-coefficient analysis. The level of significance was set at $P < 0.05$.

Results

As shown in Table 1 we unambiguously identified a series of spots representing transcription and translation factors. The protooncogene CRK and the CRK-like protein were reduced about twofold in DS fetal brain whereas the protooncogene-related myc far upstream element binding protein was unchanged.

The eukaryotic initiation factor 4a was also comparable between study groups. Elongation factor EF1alpha was also twofold decreased in contrast to the beta — form which was comparable between controls and DS brain.

EF-2 was represented by three spots and significantly decreased up to five fold in DS fetal brain.

The mitochondrial elongation factor Tu in turn was significantly downregulated in fetal DS.

Splicing factor PSF was represented by four spots and two out of them were significantly lower expressed in fetal DS.

The eukaryotic translation factor 3 showed a trend to decrease but levels did not reach statistical significance.

Table 1. Results of protein expression

Number	Protein name	Control Mean ± Std	Down fetal Mean ± Std	p-Value	pl	MW	Matches	Probability
P46108	Proto-Oncogene C-CRK	0.13 ± 0.04	0.25 ± 0.01	0.0022	5.52	33,908	5	2.99E-06
P46109	CRK-like protein	0.29 ± 0.21	0.10 ± 0.03	0.0012	6.73	33,869	6	7.40E-10
A53184	Myc far upstream element-binding protein-human	2.48 ± 0.78	3.45 ± 0.95	0.1255	7.69	67,663	8	8.79E-06
P38919	Eukaryotic initiation factor 4a-like nuk-34 (ha0658)	0.28 ± 0.16	0.20 ± 0.14	0.4557	6.41	47,088	6	4.97E-08
P04720	Elongation factor 1-alpha 1 (ef-1-alpha-1)	0.40 + 0.09	0.19 + 0.07	0.0012	9.72	50,451	5	1.09E-06
P24534	Elongation factor 1-beta (ef-1-beta)	2.13 ± 0.64	1.93 ± 0.63	0.6282	4.34	24,788	4	1.37E-04
P13639	Elongation factor 2 (ef-2) Spot 1	2.39 + 1.00	0.78 + 0.45	0.0317	6.81	96,246	8	1.73E-09
P13639	Elongation factor 2 (ef-2) Spot 2	4.49 + 1.57	1.09 + 1.05	0.0159	6.81	96,246	8	1.73E-09
P13639	Elongation factor 2 (ef-2) Spot 3	10.96 + 4.24	1.84 + 1.38	0.0079	6.81	96,246	8	1.73E-09
P49411	Elongation factor tu, mitochondrial precursor (p43)	3.35 + 0.94	2.14 + 1.00	0.0379	7.61	49,852	12	7.97E-20
P23246	PTB-associated splicing factor (psf) Spot 1	1.37 ± 0.75	1.97 ± 1.19	0.7922	10.25	76,215	6	4.69E-06
P23246	PTB-associated splicing factor (psf) Spot 2	0.31 ± 0.08	0.39 ± 0.27	0.6905	10.25	76,215	6	4.69E-06
P23246	PTB-associated splicing factor (psf) Spot 3	1.05 + 0.39	0.45 + 0.18	0.0173	10.25	76,215	6	4.69E-06
P23246	PTB-associated splicing factor (psf) Spot 4	2.32 + 1.21	1.23 + 0.31	0.0173	10.25	76,215	6	4.69E-06
Q13347	Eukaryotic translation factor 3 delta subunit (trip-1)	0.62 ± 0.24	0.47 ± 0.22	0.234	5.45	36,877	6	6.70E-09

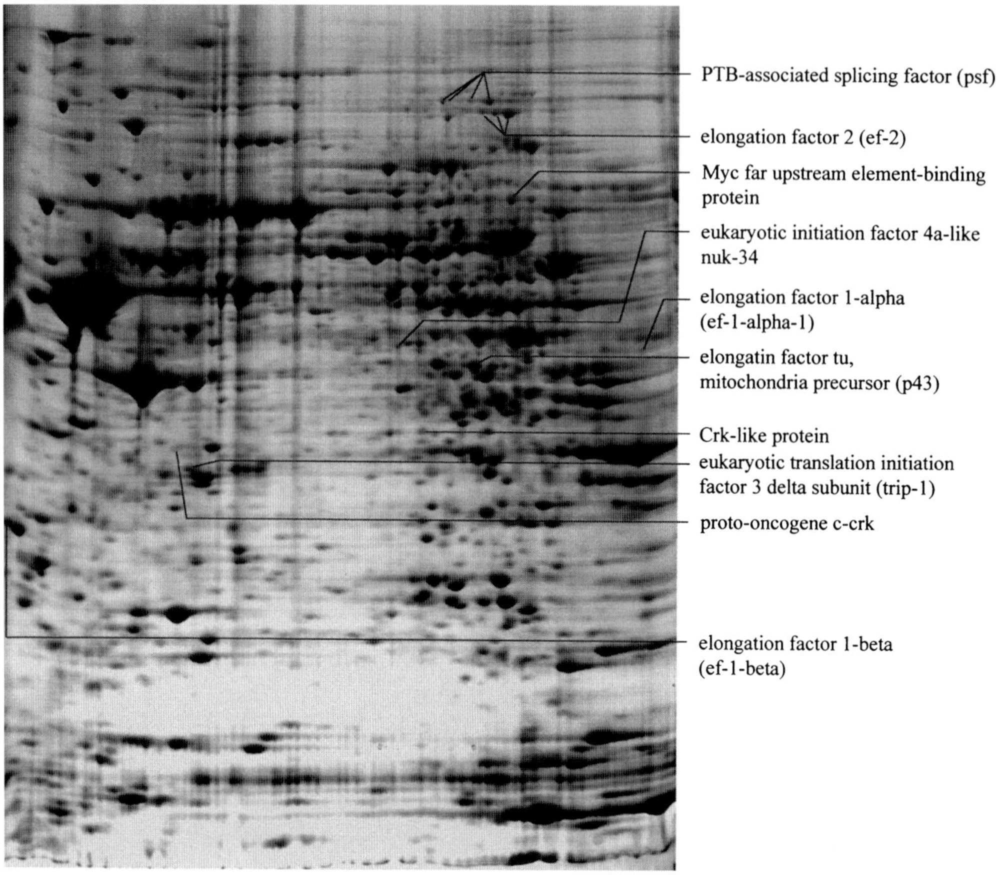

Fig. 1. Identification of TFs on two-dimensional gel electrophoresis

The identification and assignment of spots to the corresponding proteins is shown in Fig. 1.

Discussion

In our study we identified a series of transcription/elongation/translation/splicing factors expressed in brain of control and fetuses with Down Syndrome (DS).

The transcription factor and protooncogene CRK (C-CRK, syn: adaptor molecule CRK, p 38) and the CRK — like protein (CRKL) were significantly decreased in fetal cortex of patients with DS.

CRK was never described in brain before. It was originally described as the transforming component of the avian sarcoma virus CT10 and consists mainly of src-homology domains SH2 and SH3. In Man it maps to chromosome band 17p13 (Fioretos et al., 1993). It lacks an apparent catalytic domain, suggesting a function through the formation of specific complexes with other proteins mediated by the SH3 domains (Feller et al., 1995). The binding of two guanine nucleotide exchange factors Sos and C3G to CRK (and CRKL) indicates that Ras or related proteins likely play a role in signaling through crk proteins. The specific binding of crk proteins to the Abl family kinases c-Abl and Arg via the SH3 domain was reported in turn pointing to a role in signaling and phosphorylation (Feller et al., 1995). It is now widely accepted that crk-related pathways lead to the activation of the MAP kinase cascade and crk provides an alternate pathway by which growth factor receptors can signal ras (Blakesley et al., 1997). IGF-I receptor induces tyrosine phosphorylation of crk and the mitogenic effects of IGF-I are enhanced in crk-overexpressing cells. The homologue CRKL, encoded on human chromosome 22q11 containing functional domains SH2 and SH3 as well (Hoeve et al., 1993), in turn is involved in signaling and the decrease of both structures may well represent or lead to impairment of signaling and phosphorylation in fetal DS with the consequences of aberrant wiring of the developing brain.

Another transcription factor related protein, myc-far upstream element binding protein was shown for the first time in brain. The far upstream element binding protein (FBP) is one of several factors which bind to a single strand of DNA in the 5′ region of the myc gene. Duncan and coworkers (1996) showed that FBP contains a transactivation domain which directly contributes to c-myc transcription and in addition, activation is mediated by a new motif which can be negatively regulated by a repression domain of FBP. Comparable expression levels of FBP in control and DS fetuses speak against an involvement of the myc protooncogene system as FBP expression is highly correlated with c-myc expression (Bazar et al., 1995).

The eukaryotic initiation factor 4A-like nuk-34 (elF4A) with a tentative role in early transcription; it is an ATP-dependent single stranded DNA — binding protein with a sequence — independent unwinding, helicase like activity resembling other dead box family helicases. There is no references in literature but basic information is provided in Swiss Prot data (http://cxpasy.protcomc.org.au/cgi-bin/sprot-scarch-dc). Wc hcrcwith rcport the expression in human fetal brain in the early second trimester.

A major finding of our study was the observation of deranged human splicing factor PSF (syn: PTB-associated splicing factor). Patton and coworkers (1993) reported on a polypyrimidine tract-binding protein (PTB) and showcd that a complex between PTB and a 100kD protein was nccessary for pre-mRNA splicing. This highly basic protein contained two RNA binding domains and an unusual amino acid terminus rich in proline and glutamine and was subsequently termed PSF. RNA binding properties of PSF are apparently identical to those of PTB but only PSF seems to be an essential factor for early formation of the spliceosome.

Spliceosomes are the sites for the removal of sequences from pre-mRNAs and the concerted action of spliceosome factors leads to intron excision. Here, e.g. the cis-acting element PSF is necessary for splicing. PSF has not been described in human brain but in the developing mouse brain PSF was expressed in neurons and oligodendrocytes (Chanas-Sacre et al., 1999) and disappearing in adult animals. The four spots representing PSF may represent isoforms or splicing forms or posttranslational modifications and indeed two isoforms have been proposed to be produced by alternative splicing (SWISSPROT; see www address given above). Spots 3 and 4 were significantly reduced in DS fetal cortex and given the essentiality of PSF, splicing may be therefore altered which would contribute to the observation of deranged trans-acting splicing factor snRNP 70K found in brain of adult patients with DS (Seidl et al., 2001). Chromosomal imbalance may be responsible for reduced PSF protein.

The elongation machinery is run by a series of essential factors. Elongation factors 1 alpha (EF1A; EF11 human) promote the GTP-dependent binding of aminoacyl-tRNA to the A-site of ribosomes during protein biosynthesis, whereas elongation factors 1 beta (EF1B human) stimulate the GDP/GTP exchange rate and elongation factors 2 (EF2 human) promote the GTP-dependent translocation of the nascent protein chain from the A-site to the P-site of the ribosome (at the start of the microcycle the A-site is empty while the P-site is occupied by the peptidyl-tRNA (here the reader is referred to textbooks of biochemistry). Our data show that this integral process of protein synthesis is specifically deranged in fetal DS; EF1-alpha-1 (encoded on chromosome 6q14) was significantly decreased by factor 2 whereas EF1-beta (encoded on chromosome 15) did not show significant changes. EF1 alpha is a highly conserved, abundantly and ubiquitously expressed protein in the mammalian system including Man (Lee et al., 1995; Brands et al., 1986; Uetsuki et al., 1989). It is expressed in a differentiation and developmental dependent manner (Lee et al., 1995). Its decrease in fetal DS brain may be due to impaired posttranslational modification as particularly this protein is highly susceptible to ethanolamine-phosphoglycerol added to glutaminic acid residues (Whiteheart et al., 1989) and methylation (dimethyllysine, trimethyllysine formation; Dever et al., 1989). Again, reduction may be due to direct effects or indirect effects (impaired phosphorylation was described in Down Syndrome fetal brain elsewhere in the book) by chromosomal imbalance.

The cytoplasmic elongation factor 2, EF2 human (Rapp et al., 1988) (encoded on chromosome 19pter-q12), is a phosphoprotein controlling the translocation step of protein synthesis. It is the only known substrate for calmodulin dependent kinase II (Cheng et al., 1995) and EF-2 kinase completely inactivates EF2 (see above www address for EF2_human).

In our studies EF2 was presented by three spots which were remarkably and significantly reduced in fetal DS cortex. Again, deranged phosphorylation can be incriminated for the apparent reduction of unambiguously identified EF2 spots. We cannot answer the question whether these three spots were representing different isoforms or posttranslationally modified EF2 and we are still in the process to further characterize the spots in terms of posttranslational modifications, we can, however, clearly state that the EF2-system is strongly modified. A possible link between DS and Alzheimer's disease (AD) may be the observation of increased EF2-phosphorylation in the latter leading to inhibition of EF-2 with subsequently impaired protein synthesis (Johnson et al., 1992, 1993) as it is well-known that all DS patients develop AD from the fourth decade.

Unlike the elongation factors described above, the elongation factor tu (EFTU_human, syn: p 43), encoded on chromosome 16p11.2, is located in mitochondria (Woriax et al., 1995).

The function of EFTU is to promote the GTP-dependent binding of aminoacyl-tRNA to the A-site of ribosomes during protein biosynthesis (Ling et al., 1997).

EFTU directly interacts with mitochondrial tRNAs and drives the cycle of peptide elongation by forming a complex between EFTU and GDP, which in turn interacts with EFTS and exchanges GDP for GTP. At this point in the cycle ETFU:GTP complex can deliver an aminoacyl-tRNA to the A-site of the ribosome. We determined the EFTU precursor (gene name: TUFM, see above SWISSPROT address on www) and found significantly lower expression, which in turn along with other elongation factor deficits given above may be responsible for impaired protein synthesis in fetal DS cortex, although this may not account for the lower expression of a large series of reduced individual proteins in fetal DS brain.

Finally, we studied translation initiation as represented by the expression of eukaryotic translation initiation factor 3 subunit 5, one out of at least 10 different subunits of this structure (Asano et al., 1997). This factor (syn: IF35_human; EIF-3 epsilon; EIF3-P47) is an essential effector of translation by binding to the 40S ribosome and promoting the binding of methionyl-tRNA1 and mRNA and associating with complex P170-EIF3 (SWISSPROT, www address as given above). Levels were comparable between controls and DS cortex revealing that translation may not be deteriorated at this early stage of protein synthesis.

Acknowledgements

We are highly indebted to the Red Bull Company, Salzburg, Austria, for generous financial support of the study.

References

Arenander AT, De Vellis J (1994) Development of the nervous system. In: Siegel GJ, Agranoff BW, Albers RW, Molinoff PB (eds) Basic neurochemistry, 5th ed. Raven Press New York, pp 573–606

Asano K, Vornlocher HP, Richter-Cook NJ, Merrick WC, Hinnebusch AG, Hershey JW (1997) Structure of cDNAs encoding human eukaryotic initiation factor 3 subunits. Possible roles in RNA binding and macromolecular assembly. J Biol Chem 272: 27042–27052

Bazar L, Harris V, Sunitha I, Hartmann D, Avigan A (1995) A transactivator of c-myc is coordinatedly regulated with the proto-oncogene during cellular growth. Oncogene 10: 2229–2238

Becker L, Mito T, Takashima S, Onodera K (1991) Growth and development of the brain in Down Syndrome. In: Epstein CJ (ed) The morphogenesis in Down syndrome. Wiley-Liss, New York, p 133

Becker LE, Armstrong DL, Chan F (1986) Dendritic atrophy in children with Down's Syndrome. Ann Neurol 20: 520–532

Berndt P, Hobohm U, Langen H (1999) Reliable automatic protein identification from matrix assisted laser desorption/ionization mass spectrometric peptide fingerprints. Electrophoresis 20: 3521–3526

Blakesley VA, Beitner-Johnson D, Van Brocklyn JR, Rani Sh, Shen-Orr ZS, Stannard BS, Spiegel S, LeRoith D (1997) Sphingosine 1 phosphate stimulates tyrosine phosphorylation of Crk. J Biol Chem 272: 16211–16217

Bradford M (1976) A rapid and sensitive method for the quantification of microgram quantities of protein utilizing the principle of protein-dye binding. Anal Biochem 72: 248–254

Brands JHGM, Maassen JA, van Hemert FJ, Amons R, Moeller W (1986) The primary structure of the alpha subunit of human elongation factor 1. Structural aspects of guanine — nucleotide binding sites. Eur J Biochem 155: 167–171

Chanas-Sacre G, Mazy-Servais C, Wattiez R, Pirard S, Rogister B, Patton JG, Belachew S, Malgrange B, Moonen G, Leprince P (1999) Identification of PSF, the polypyrimidine tract binding protein-associated splicing factor as a developmentally regulated neuronal protein. J Neurosci Res 57: 62–73

Cheng EH, Gorelick FS, Czernik AJ, Bagaglio DM, Hait WN (1995) Calmodulin-dependent protein kinases in rat glioblastoma. Cell Growth Differ 6: 615–621

Dever TE, Costello CE, Owens CL, Rosenberry TL, Merrick WC (1989) Location of seven post-translational modifications in rabbit elongation factor 1 alpha including dimethyllysine, trimethyllysine and glycerylphosphorylethanolamine. J Biol Chem 264: 20518–20525

Duncan R, Collins I, Tomonaga T, Zhang T, Levens D (1996) A unique transactivation sequence motif is found in the carboxyl-terminal domain of the single strans binding protein FBP. Mol Cell Biol 16: 2274–2282

Fang-Kircher S, Labudova O, Kitzmueller E, Rink H, Cairns N, Lubec G (1999) Increased steady state mRNA levels of DNA-repair genes XRCC1, ERCC2 and ERCC3 in brain of patients with Down syndrome. Life Sci 64: 1689–1699

Feller SM, Knudsen B, Hanafusa H (1995) Cellular proteins binding to the first Src homology 3 (SH3) domain of the proto-oncogene c-Crk indicate Crk-specific signaling pathways. Oncogene 10: 1465–1473

Fioretos T, Heisterkamp N, Groffen J, Benjes S, Morris C (1993) CRK protooncogene maps to human chromosome band 17p13. Oncogene 8: 2853–2855

Fountoulakis M, Langen H (1997) Identification of proteins by matrix assisted laser desorption ionisation mass spectrometry following in — gel digestion in low salt, non-volatile buffer and simplified peptide recovery. Anal Biochem 250: 153–156

Gower HJ, Moore SE, Dickson G, Elsom VE, Nayak R, Walsh FS (1989) Cloning and characterisation of a myoblast cell surface antigen defined by 24,1D5 monoclonal antibody. Development 105: 723–731

Greber-Platzer S, Balcz B, Cairns N, Lubec G (1999) C-fos expression in brains of patients with Down Syndrome. In: Lubec G (ed) The molecular biology of Down syndrome. Springer, Wien New York, pp 75–86

Greber-Platzer S, Schatzmann-Turhani D, Cairns N, Balcz B, Lubec G (1999) Expression of the transcription factor ETS2 in brain of patients with Down Syndrome-evidence against the overexpression-gene dosage hypothesis. In: Lubec G (ed) The molecular biology of Down syndrome. Springer, Wien New York, pp 269–282

He X, Rosenfeld MG (1991) Mechanisms of complex transcriptional regulation: implications for brain development. Neuron 7: 183–196

Hoeve ten J, Morris C, Heisterkamp N, Groffen J (1993) Isolation and chromosomal localization of CRKL, a human crk-like gene. Oncogene 8: 2469–2474

Johnson G, Gollib J, Haroutunian V, Bierer L, Nairns AC, Merril C, Wallace W (1992) Increased phosphorylation of elongation factor 2 in Alzheimer's disease. Brain Res Mol Brain Res 15: 319–326

Johnson G, Refolo LM, Merril CR, Wallace W (1993) Altered expression and phosphorylation of amyloid precursor protein in heat shocked neuronal PC12 cells. Brain Res Mol Brain Res 19: 140–148

Labudova O, Krapfenbauer K, Moenkemann H, Rink H, Kitzmueller E, Cairns N, Lubec G (1998) Decreased transcription factor junD in brains of patients with Down syndrome. Neurosci Lett 252: 159–162

Langen H, Roeder D, Juranville J, Fountoulakis M (1997) Effect of protein application mode and acrylamide concentration on the resolution of protein spots separated by two-dimensional gel electrophoresis. Electrophoresis 18: 2085–2090

Lee S, LeBlanc A, Duttaroy A, Wang E (1995) Terminal differentiation dependent alteration in the expression of translation elongation factor 1 alpha and its sister gene S1, in neurons. Exp Cell Res 219: 589–597

Ling M, Merante F, Chen HS, Duff C, Duncan AMV, Robinson BH (1997) The human mitochondrial elongation factor Tu (EF-Tu) gene: cDNA sequence, genomic localization, genomic structure and identification of a pseudogene. Gene 197: 325–336

Maraoulakou IG, Papas TS, Green JE (1994) Diferential expression of ets-1 and ets-2 proto-oncogenes during murine embryogenesis. Oncogene 9: 1551–1565

Patton JG, Porro EB, Galceran J, Tempst P, Nadal Ginard B (1993) Cloning and characterization of PSF, a novel pre-mRNA splicing factor. Genes and Development 7: 393–406

Rapp G, Klaudiny J, Hagendorff G, Luck MR, Heinz K (1989) Complete sequence of the coding region of human elongation factor 2(EF-2) by enzymatic ampfification of cDNA from human ovarian granulosa cells. Biochem Chem Hoppe-Seyler 370: 1071–1075

Rehder H (1981) Trisomy 21: An International Symposium. In: Burgio GR, Fracaro M, Tiepolo L, Wolf U (eds) Down syndrome. Springer, Berlin Heidelberg, p 57

Schmidt-Sidor B, Wisniewski KE, Shepard TH, Sersen A (1990) Brain growth in Down syndrome subjects 15–22 weeks of gestational age and birth to 60 months. Clin Neuropathol 9: 181–195

Seidl R, Labudova O, Krapfenbauer K, Welin-Henrikssosn E, Craft J, Turhani-Schatzmann T, Achsel T, Bidmon B, Prujin GJM, Cairns N, Lubec G (2001) Deficient brain snRNP 70K in patients with Down Syndrome. Electrophoresis 22: 43–48

Sylvester PE (1986) The anterior comissure in Down's syndrome. J Mental Defic Res 30: 19–25

Takashima S, Ieshima A, Nakamura H, Becker LE (1989) Dendrites, dementia and the Down Syndrome. Brain Dev 2: 131–143

Uetsuki T, Naito A, Nagata S, Kaziro Y (1989) Isolation and characterization of the human chromosomal gene for polypeptide chain elongation factor 1 alpha. J Biol Chem 264: 5791–5798

Whiteheart SW, Shenbagarmurthy P, Chen L, Cotter RJ, Hart GW (1989) Murine elongation factor 1 alpha (EF-1alpha) is posttranslationally modified by novel amide-linked ethanolamine — phosphoglycerol moieties. J Biol Chem 264: 14334–14341

Woriax VL, Burkhart WA, Spremulli LL (1995) Cloning, sequence and expression of mammalian mitochondrial protein synthesis elongation factor Tu. Biochem Biophys Acta 1264: 347–356

Yeghiazarian K, Turhani-Schatzmann D, Labudova O, Schuller E, Olson EN, Lubec G (1999) Downregulation of the transcription factor scleraxis in brain of patients with Down syndrome. J Neural Transm 57: 305–314

Authors' address: Prof. Dr. B. Lubec, Department of Neonatology, University of Vienna, Währinger Gürtel 18, A-1090 Vienna, Austria, e-mail: gert.lubec@akh-wien.ac.at

Fetal life in Down Syndrome starts with normal neuronal density but impaired dendritic spines and synaptosomal structure

R. Weitzdoerfer[1], **M. Dierssen**[2], **M. Fountoulakis**[3], and **G. Lubec**[1]

[1] Department of Pediatrics, University of Vienna, Vienna, Austria
[2] Medical and Molecular Genetics Center-IRO, Hospital Duran i Reynals, Barcelona, Spain
[3] Gene Technologies, F. Hoffmann-La Roche, Basel, Switzerland

Summary. Information on fetal brain in Down Syndrome (DS) is limited and there are only few histological, mainly anecdotal reports and no systematic study on the wiring of the brain in early prenatal life exists. Histological methods are also hampered by inherent problems of morphometry of neuronal structures. It was therefore the aim of the study to evaluate neuronal loss, synaptic structures and dendritic spines in the fetus with Down Syndrome as compared to controls by biochemical measurements. 2 dimensional electrophoresis with subsequent mass spectroscopical identification of spots and their quantification with specific software was selected. This technique identifies proteins unambiguously and concomitantly on the same gel. Fetal cortex samples were taken at autopsy with low post-mortem time, homogenized and neuron specific enolase (NSE) determined as a marker for neuronal density, the synaptosomal associated proteins alpha SNAP [soluble N-ethylmaleimide-sensitive fusion (NSF) attachment protein], beta SNAP, SNAP 25 and the channel associated protein of synapse 110 (chapsyn 110) as markers for synaptosomal structures and drebrin (DRB) as marker for dendritic spines. NSE, chapsyn 110 and beta SNAP were comparable in the control fetus panel and in Down Syndrome fetuses.

Drebrin was significantly and remarkably reduced and not even detectable in several Down Syndrome brain samples.

Quantification of SNAP 25 revealed significantly reduced values in DS cortex and alpha SNAP was only present in half of the DS individuals.

We conclude that at the time point of about 19 weeks of gestation (early second trimester) no neuronal loss can be detected but drebrin, a marker for dendritic spines and synaptosomal associated proteins alpha SNAP and SNAP 25 were significantly reduced indicating impaired synaptogenesis. Early dendritic deterioration maybe leading to the degeneration of the dendritic tree and arborization, which is a hallmark of Down Syndrome from infancy.

Introduction

Information of fetal Down Syndrome (DS) brain is limited to a couple of morphological reports. Abnormalities in terms of neuronal differentiation, migration and synaptogenesis exist from 20th and 22nd weeks of gestation (Epstein, 1995; Wisniewski, 1994).

Becker (1991, 1986) reported that neuronal density was reduced in several brain regions including cortical areas and basal nucleus with sprouting dendritic trees and unchanged dendritic spines until the 40th week of gestation. From early infancy dendritic spines then decrease reaching statistical significance in the visual cortex at about 4 months of age and also dendritic branches decrease continuously in early infancy from about 6 months of age.

These valuable morphological studies were performed using histological techniques that did not allow morphometrical analysis but formed the basis for our studies using the proteomic approach: we selected markers for neuronal density, dendritic spines and synaptosomes and quantified these in fetal brain of controls and DS individuals at about the 19th week of gestation. The aims of the study were to show whether deficient neuronal density, reduced dendritic spines or reduced synaptosomal structures were apparent at that early fetal period. We also aimed to introduce a fair and reliable method identifying marker proteins.

Neuronal specific enolase (NSE) for neuronal density, drebrin (DRB) for dendritic spines, and SNAP 25, alpha and beta SNAP and Chapsyn 110 for synaptic density, identified unambiguously by 2D electrophoresis with mass-spectrometric (MALDI) detection and quantification by specific software. We propose neuronal density in DS to be comparable to controls but reduced synaptic and dendritic spine density as early as at the 19th weeks of gestation was observed.

Materials and methods

Fetal brain tissue and brain preparation

Abortus fetal brain sample (cerebral cortex) of controls ($n = 7$; 1 female and 6 males with mean age of 18.8 ± 2.2 weeks) and DS ($n = 8$; 2 females and 6 males with mean age of 19.8 ± 2.0 weeks) were obtained from Drs. Mara Dierssen and J. C. Ferreres the Medical and Molecular Genetics Center-IRO, Hospital Duran i Reynals, Barcelona, Spain. All samples were stored at $-70°C$ and the freezing chain was never interrupted. Brain tissue was suspended in 0.5 ml of sample buffer consisting of 40 mM Tris, 5 M urea (Merck, Darmstadt, Germany), 2 M thiourea (Sigma, St. Louis, MO, USA), 4% CHAPS (Sigma), 10 mM 1,4-dithioerythriol (Merck), 1 mM EDTA (Merck), 1 mM PMSF (Sigma) and 1 μl/mg of each pepstatin A, chymostatin, leupeptin and antipain. The suspension was sonicated for approximately 30 seconds and centrifuged further at $150{,}000 \times g$ for 45 minutes to sediment undissolved material. The protein concentration of the supernatant was determined by the Coomassie blue method (Bradford, 1976).

Two-dimensional gel electrophoresis

Two-dimensional gel electrophoresis was performed essentially as reported (Langen et al., 1997). Samples of 2 mg protein were applied on immobilized pH 3–10 nonlinear gradient strips in sample cups at their basic and acidic ends. Focusing started at 200 V and the voltage was gradually increased to 5,000 V at 3 V/min and kept constant for a further 24 h (approximately 180,000 kVh totally). The second-dimensional separation was performed on 9–16% SDS gradient polyacrylamide gels. The gels (180 × 200 × 1.5 mm) were run at 40 mA per gel. After protein fixation for 12 h in 40% methanol, containing 5% phosphoric acid, the gels were stained with colloidal Coomassie blue (Novex, San Diego, CA, USA) for 48 h. Molecular masses were determined by running standard protein markers (Gibco, Basel, Switzerland), covering the range 10–200 kDa. pI values were used as given by the supplier of the IPG strips. Excess of dye was washed out from the gels with H_2O and the gels were scanned in an Agfa DUOSCAN densitometer (resolution 200). Electronic images of the gels were recorded using Photoshop (Adobe) and PowerPoint (Microsoft) software.

Matrix-assisted laser desorption ionization mass spectroscopy (MALDI-MS)

MALDI-MS analysis was performed as described elsewhere (Fountoulakis and Langen, 1997) with some modifications. The spots were excised with a spot picker and placed into 96-well microtiter plates. Each spot was destained with 100 µl of 30% acetonitrile in 50 mM ammonium bicarbonate and dried in a speedvac evaporator. Each dried gel piece was rehydrated with 4 µl of 3 mM Tris-HCl, pH 9.0, containing 50 ng trypsin (Promega, Madison, WI, USA). After 16 h at room temperature, 7 µl of H_2O were added to each gel piece and the samples were shaken in for 10 min. Four µl of 50% acetonitrile, containing 0.3% trifluoroacetic acid and the standard peptides des-Arg-bradykinin (Sigma, 904.4681 Da) and adrenocorticotropic hormone fragment 18–39 (Sigma, 2465.1989 Da), in water were added to each gel piece and shaken for 10 min. The application of the samples was performed with a SymBiot I sample processor (PE Biosystems, Framingham, MA, USA). 1.5 µl of the peptide mixture was simultaneously applied with 1 µl of matrix, consisting of a saturated solution of α-cyano-4-hydroxycinnamic acid (Sigma) in 50% acetonitrile, containing 0.1% trifluoroacetic acid. Samples were analyzed in a time-of-flight mass spectrometer (Reflex 3, Bruker Analytics, Bremen, Germany). An accelerating voltage of 20 kV was used. Peptide matching and protein searches were performed automatically. The peptide masses were compared to the theoretical peptide masses of all available proteins from all species. Monoisotopic masses were used and a mass tolerance of 0.0025% was allowed. The algorithm used for determining the probability of a false positive match with a given MS-spectrum is described elsewhere (Berndt et al., 1999).

Quantification of proteins

Protein spots were outlined (first automatically and then manually) and quantified using the ImageMaster 2D Elite software (Amersham Pharmacia Biotechnology). The percentage of the volume of the spots representing a certain protein was determined in comparison with the total proteins present in the 2-D gel.

Statistical analysis

Software calculated the percentage of spot volume in each particular gel after subtraction of background values. Between-group differences were calculated by non-parametric

Mann-Whitney U-test. Level of significance was set at $P < 0.05$. If the spot of a protein was not detectable a value of 0.01 was inserted to make statistical analysis possible.

Results

Identification and quantification of NSE as a marker for neuronal density, drebrin for dendritic spines and alpha SNAP, beta SNAP, SNAP 25 and chapsyn 110 as synaptosomal markers

Human proteins obtained from cerebral cortex of fetal brain were applied on 2-D gel. Proteins were visualized by staining with colloidal Coomassie blue. A representative control gel with assignments of investigated proteins is shown in Fig. 1. Protein spots were analyzed by MALDI-MS, following in-gel digestion. Peptide masses were matched with the theoretical peptide masses of all proteins of the SWISS-PROT database. We used internal standards to correct

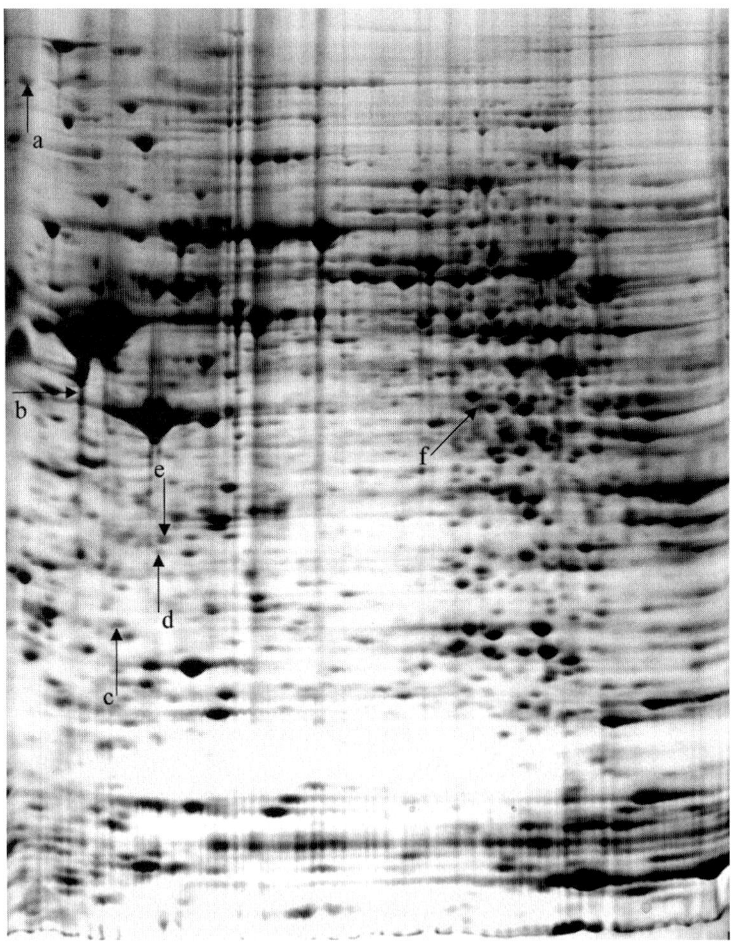

Fig. 1. Identification of proteins measured: **a** Drebrin (Q 16643) **b** NSE (P 09104) **c** SNAP 25 (P 13795) **d** alpha SNAP (P 54920) **e** beta SNAP (P 28663) **f** chapsyn 110 (Q 15700)

the measured peptide mass thus reducing the windows of mass tolerance and increasing the confidence of identification. In this study 6 proteins were quantified using the Image Master 2D Elite software. Expression levels of proteins were determined as a percentage of spot volume in the area of interest. SWISS-PROT accession numbers and names of the quantified proteins with MALDI information are given in Table 1.

Neuronal density, represented by the protein Neuron specific enolase (NSE) and represented by one spot, in DS (detected in 6 brains) were comparable to controls (detected in 4 out of 6 samples).

DRB as a marker for dendritic spines was represented by one spot and showed a statistically significant difference between control and Down individuals. While measurements in controls showed comparable results, the protein could only be detected in 1 out of 8 DS individuals. In the other 7 gels the corresponding spot was absent (see Fig. 2A and B). Statistical evaluation of these data showed a significant difference between the two groups ($P = 0.0003$).

Levels of SNAP 25 in DS compared to levels in controls showed a statistically significant reduction (P value of 0.0289) (see Fig. 2C and D).

Alpha SNAP was significantly reduced in DS cortex and even was not detected (see Fig. 2E and F) in 4 of 8 DS brains ($P = 0.0140$).

Beta SNAP represented by one spot was not significant different between control and DS individuals (detected in 7 brains).

The single spot representing chapsyn 110 showed comparable values between the two groups (detected in 6 out of 7 controls and 6 out of 8 DS samples).

Individual data of the investigated proteins are shown in Fig. 3 and 4.

Discussion

2-D-electrophoresis with subsequent mass-spectrometrical analysis of protein spots and quantification by specific software unambiguously identified NSE, synaptosomal proteins and drebrin. The major findings were that drebrin, SNAP 25 and alpha SNAP were dramatically reduced in fetal DS cortex even with absence of DRB and alpha SNAP in individual DS samples.

DRB has been studied in several species including human and is encoded on chromosome 5 (Toda et al., 1993). It is a microfilament-associated protein and the expression pattern is regulated spatially and temporally in the developing brain (Shirao, 1995) suggesting a role in morphogenesis in the developing brain. Immuneelectron microscopy revealed that DRB was highly concentrated at synapses (Shirao and Obata, 1986) and localized at the postsynaptic sites of dendrites (Shirao et al., 1989). Hayashi and Shirao clearly showed that DRB accumulated in dendritic spines and that DRB overexpression led to changes in the shape of dendritic spines rather than in dendritic shafts thus providing an excellent marker protein for dendritic spines (1999). Conversely, antisense technology knocking out drebrin in neurons, led to the attenuation of neuritic outgrowth (Toda et al., 1999).

Table 1. Identification and levels of proteins

Protein name	Accession number	pI	kDa	Matches	Probability	Control (means ± SD)	DS (means ± SD)
Neuron specific enolase	P 09104	4.79	47.467	10	1.54E-15	1.53 ± 1.28	2.14 ± 0.79
Drebrin E	Q 16643	4.22	71.837	4	3.10E-05	3.66 ± 2.25	0.06 ± 0.17***
Channel associated protein of synapse 110	Q 15700	6.38	97.895	6	2.44E-08	1.06 ± 0.69	2.09 ± 1.40
Synaptosomal associated protein 25	P 13795	4.49	23.528	5	1.10E-04	7.03 ± 1.50	5.39 ± 1.09*
Alpha soluble NSF attachment protein (SNAP alpha)	P 54920	5.14	33.836	5	3.67E-06	1.65 ± 1.08	0.46 ± 0.52*
Beta-soluble NSF attacment protein (SNAP beta)	P 28663	5.06	28.122	7	1.37E-06	2.32 ± 0.81	9 ± 2.54

* P value < 0.05, *** P value < 0.001

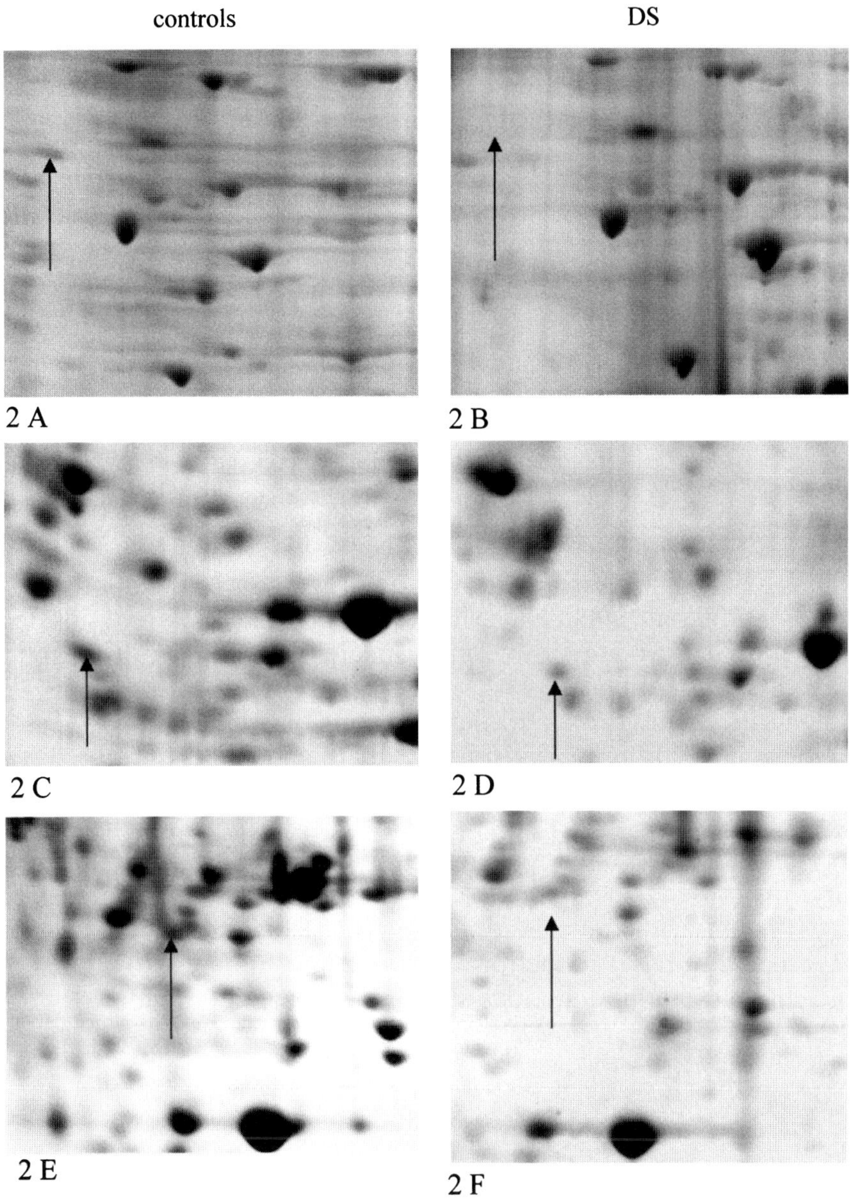

Fig. 2. A Drebrin in fetal control brain (arrow). **B** In fetal DS brain drebrin was not detected in 7 out of 8 samples. **C** SNAP 25 in control cortex (arrow). **D** SNAP 25 in fetal DS samples showed lower levels. **E** alpha SNAP was represented by one spot in control samples (arrow). **F** In DS samples alpha SNAP levels were significantly reduced and in 4 out of 8 DS patients alpha SNAP was not detected

The dramatic reduction of DRB as seen in our DS population may have led or represent deterioration of dendritic spine structure: a dendritic spine is a dynamic structure arising from the dendrite and consisting of a spine shaft and spine head. Biochemically, it consists of cytoskeleton proteins including the beta and gamma isoforms of actin, spectrin, alpha actinin, capping pro-

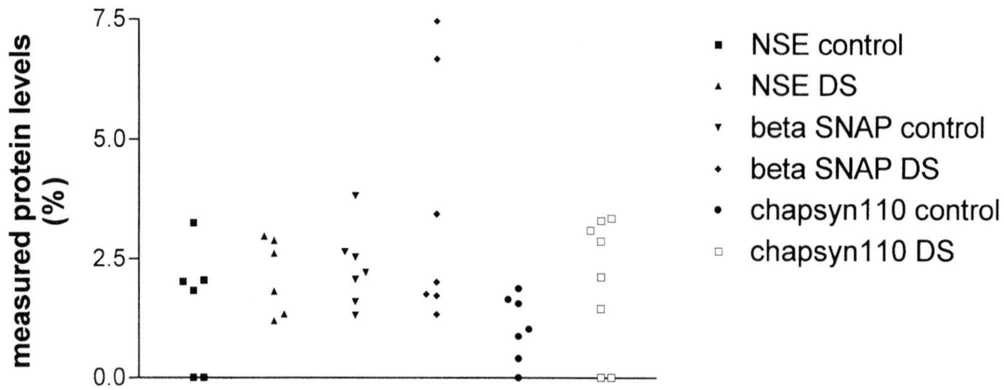

Fig. 3. Individual data of NSE, beta SNAP and chapsyn 110 showed comparable values between control and DS cortex

teins and DRB, an actin binding protein and regulator of the dendritic spine's cytoskeleton (Halpain, 2000; Hayashi et al., 1996). Reduction of this dendritic spine key protein may impair action — assembly in the dendritic spine thus leading to it's dismantling.

As dendritic spines are major determinants of brain morphogenesis, dendritic spine abnormalities (Marin-Padilla, 1976) and aberrant brain dysgenesis in DS may be at least in part, explained by deficient DRB.

As more than 90% of excitatory glutamatergic synapses in the mammalian brain occur on dendritic spines (Harris and Kater, 1994) and are clustering at actin-drebrin in dendritic spines (Allison et al., 2000), deterioration of the excitatory amino acid transmission system in DS (Risser et al., 1997) could also be assigned to DRB reduction. This mechanism may also be responsible for deficient nicotinergic functions shown in adult DS (Engidawork et al., in press).

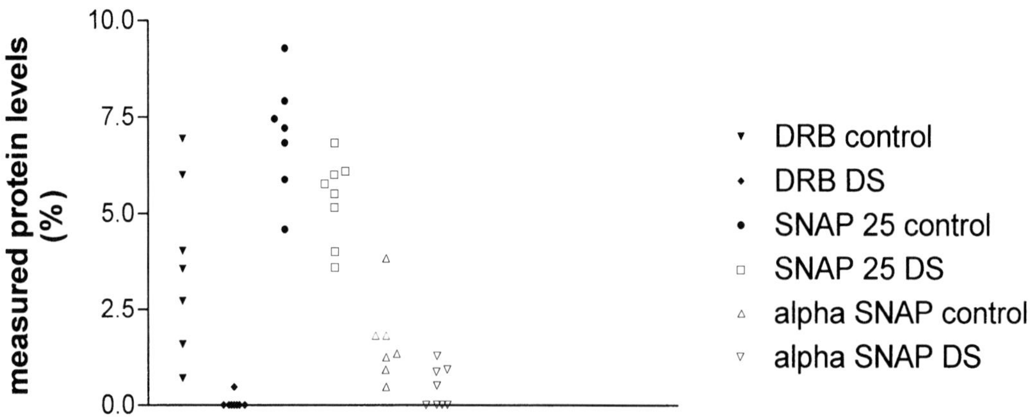

Fig. 4. Individual data of Drebrin, SNAP 25 and alpha SNAP which were significantly reduced or even absent in DS samples

A nicotinic acetylcholine receptor has been shown to bind to dendritic spines known to contain actin and drebrin, and this nicotinic receptor disperses if spines retract (Shoop et al., 2000).

As DRB is paralleling neuronal differentiation (Asada et al., 1994), decreased DRB would also be in agreement with our previous data showing dedifferentiation in adult DS brain (Kim et al., 2000).

Hatanpää et al. (1999) as well as Harigaya et al. (1996) reported reduction of DRB in brains of patients with Alzheimer's disease.

Hatanpää et al. also found comparable levels of NSE between controls and AD brains resembling our data and ruling out that drebrin is simply a consequence of general neuronal loss in both dementing diseases.

Expression of SNAP 25, a presynaptic membrane associated protein of 25 kDa mass, is known to correlate with synaptogenesis and neuronal maturation (Catsicas et al., 1991). The protein is expressed in axonal growth cone, which is a sensory structure receiving directional information from the surrounding but also a motor structure leading to axonal elongation (Kandel and Siegelbaum, 2000). Osen-Sand et al. showed that in vitro and in vivo inhibition of SNAP 25 by antisense oligonucleotides diminishes axonal extension in developing neurons (Osen-Sand et al., 1993). Based on the SNARE-hypothesis (Rothman, 1994) SNAP 25 is one of the core components of the t-SNARE (target membrane SNAP-receptor) complex consisting of syntaxin and SNAP 25 and therefore plays important role in vesicle docking and membrane fusion.

Our findings of reduced expression of SNAP 25 in fetal DS cortex indicate that early as in the 19th week of gestation synaptogenesis is impaired in DS individuals, this extends our previous results of decreased levels of SNAP 25 in the brain of adult patients with DS and Alzheimer's disease (Greber et al., 1999). It may also be responsible for shortened frontal-occipital length of DS brain (Sylvester, 1986).

Alpha and beta SNAP are soluble NSF attachment proteins necessary for the vesicle docking and fusion process. Docking and fusion of a vesicle to the target membrane needs the presence of receptors, so called vesicle-SNARES and target-SNARES. After fusion of the vesicle with the target membrane v-SNARE and t-SNARE are disassembled by NSF-protein and SNAP (Kandel and Siegelbaum, 2000).

Wang et al. reported that excess of Sec17p, the yeast homolog of alpha SNAP can reversibly arrest docking of primed vacuoles. They suggest Sec17p to stabilize the vacuolar cis-SNARE, a transient complex which is disassembled by ATP-driven displacement of Sec17p before vacuole fusion to the target membrane.

Beta SNAP is a brain specific protein that binds to synaptotagmin, a calcium binding protein and together with NSF and SNARE forms a docking and fusion complex (Schiavo et al., 1995).

Our results of reduced alpha SNAP levels in fetal DS brain but normal beta SNAP do not contradict our previous findings of decreased beta SNAP levels in brains of adult DS individuals (Yoo et al., 2001) as interpretation of beta SNAP has to take into account that no data exist on the cell type expressing the protein (neurons and/or glia) and impairment of beta SNAP levels might occur to a later time point in development.

Chapsyn 110, predominantly expressed in neuronal tissue, is a membrane-associated putative guanylate kinase, associated with the postsynaptic density (PSD) in brain. As it is a member of the PSD-95 family, Chapsyn 110 binds directly to Shaker K+ channel subunits and N-methyl-D-aspartate (NMDA) receptor and mediates their clustering in heterologous cells (Kim et al., 1995, 1996; Fukaya and Watanabe, 2000). Sans et al. conducted a study on the expression of NMDA receptor associated proteins during development in synapses of rat hippocampus. They showed increasing labeling with a Chapsyn 110 antibody of synapses from postnatal day 2 to adult correlating with increasing NR2A, a subunit of the NMDA receptor, suggesting its role in mature synapse (Sans et al., 2000).

Our results of comparable Chapsyn 110 levels in fetal DS and control brain suggest that function in the postsynaptic membrane density at this time point of development may be unaffected by DS.

We conclude that deficient drebrin, SNAP 25 and alpha SNAP in fetal DS brain of the second trimester may be responsible for and/or reflect impaired dendritic spines, axonal growth and function in the presynaptic terminal in cerebral cortex. Reduction of these proteins is most probably not simply a consequence of neuronal loss. We consider reduction of DRB, however, as a phenomenon preceding impaired arborization and atrophy of the dendritic tree, which is a morphological hallmark of DS brain.

We also suggest that deficient DRB may lead to impaired nicotinic and glutamate receptor anchoring at the head of dendritic spines in DS.

Acknowledgment

We are highly indebted to the Red Bull Company (Salzburg, Austria) for generous financial support.

References

Allison D, Chervin A, Gelfand V, Craig A (2000) Postsynaptic scaffolds of excitatory and inhibitory synapses in hippocampal neurons: maintenance of core components independent of actin filaments and microtubules. J Neurosci 20: 4545–4554

Asada H, Uyemura K, Shirao T (1994) Actin-Binding protein, drebrin, accumulates in submembraneous regions in parallel with neuronal differentiation. J Neurosci Res 38: 149–159

Becker L, Armstrong D, Chan F (1986) Dendritic atrophy in children with Down Syndrome. Ann Neurol 20: 520–526

Becker L, Mito T, Takashima S, Onodera K (1991) Growth and development of the brain in Down syndrome. Prog Clin Biol Res 373: 133–152

Berndt P, Hobolm U, Langen H (1999) Reliable automatic protein identification from matrix-assisted laser desorption/ionisation mass spectrometric peptide fingerprints. Electrophoresis 20: 3521–3526

Bradford M (1976) A rapid and sensitive method for the quantitation of microgram quantities of protein utilizing the principle of protein-dye binding. Anal Biochem 72: 248–254

Catsicas S, Larhammar D, Blomqvist A, Paolo Sanna P, Milner R, Wilson M (1991) Expression of a conserved cell-type-specific protein in nerve terminals coincides with synaptogenesis. Proc Natl Acad Sci 88: 785–789

Engidawork E, Guelesserian T, Balic N, Cairns N, Lubec G (2001) Changes in nicotinic acetylcholin receptor subunits expression in brain of patients with Down Syndrome and Alzheimer's disease. J Neural Transm (this volume)

Epstein C (1995) The metabolic and molecular bases of inherited disease. In: Scriver SR, Beaudet AL, Sly WS, Valle D (eds) Down Syndrome (Trisomy 21). McGraw Hill, New York, pp 749–794

Fountoulakis M, Langen H (1997) Identification of proteins by matrix-assisted laser desorption ionisation-mass spectrometry following in-gel digestion in low-salt, non-volatile buffer and simplified peptide recovery. Anal Biochem 250: 153–156

Fukaya M, Watanabe M (2000) Improved immunohistochemical detection of postsynaptically located PSD-95/SAP90 protein family by protease section pre-treatment: a study in the adult mouse brain. J Comp Neurol 426: 572–586

Greber S, Lubec G, Cairns N, Fountoulakis M (1999) Decreased levels of synaptosomal associated protein 25 in the brain of patients with Down Syndrome and Alzheimer's disease. Electrophoresis 20: 928–934

Halpain S (2000) Actin and the agile spine: how and why do dendritic spines dance? Trends Neurosci 23: 141–146

Hantanpaa K, Isaacs K, Shirao T, Brady D, Rapoport S (1999) Loss of proteins regulating synaptic plasticity in normal aging of the human brain and in Alzheimer's disease. J Neuropathol Exp Neurol 58: 637–643

Harris K, Kater S (1994) Dendritic spines: cellular specializations imparting both stability and flexibility to synaptic function. Ann Rev Neurosci 17: 341–371

Hayashi K, Shirao T (1999) Change in the shape of dendritic spines caused by overexpression of drebrin in cultured cortical neurons. J Neurosci 19: 3918–3925

Hayashi K, Ishikawa R, Ye LH, Ye XL, Takata K, Kohama K, Shirao T (1996) Modulatory role of drebrin on the cytoskeleton within dendritic spines in the rat cerebral cortex. J Neurosci 16: 7161–7170

Harigaya Y, Shoji M, Shirao T, Hirai S (1996) Disappearance of actin-binding protein, drebrin, from hippocampal synapses in Alzheimer's disease. J Neurosci Res 43: 87–92

Kandel E, Siegelbaum S (2000) Transmitter release. In: Kandel ER, Schwartz JH, Jessell TM (eds) Principals of neural science. McGraw-Hill, New York, pp 253–279

Kim E, Niethammer M, Rothschild A, Nung Jan Y, Sheng M (1995) Clustering of Shaker-type K+ channels by interaction with a family of membrane-associated guanylate kinases. Nature 378: 85–88

Kim E, Cho K, Rothschild A, Sheng M (1996) Heteromultimerization and NMDA receptor-clustering activity of Chapsyn 110, a member of the PSD-95 family of proteins. Neuron 17: 103–113

Kim SH, Cairns N, Yoo BC, Broers JVL, Lubec G (2000) Neuroendrocrine — specific protein C, a marker of neuronal differentiation, is reduced in brain of patients with Down Syndrome and Alzheimer's disease. Biochem Biophys Res Commun 276: 329–334

Langen H, Roder D, Juranville J, Fountoulakis M (1997) Effect of protein application mode and acrylamide concentration on the resolution of protein spots separated by two-dimensional gel electrophoresis. Electrophoresis 18: 2085–2090

Marin-Padilla M (1976) Pyramidal cell abnormalities in the motor cortex of a child with Down Syndrome: a Golgi study. J Comp Neurol 167: 63–81

Osen-Sand A, Catsicas M, Staple J, Jones K, Ayala G, Knowles J, Grenningloh G, Catsicas S (1993) Inhibition of axonal growth by SNAP-25 antisense oligonucleotides in vitro and in vivo. Nature 364: 445–448

Risser D, Lubec G, Cairns N, Herrera-Marschitz M (1997) Excitatory amino acids and monoamines in parahippocampal gyrus and frontal cortical pole in adults with Down Syndrome. Life Sci 60: 1231–1237

Rothman J (1994) Mechanisms of intracellular protein transport. Nature 372: 55–63

Sans N, Petralia S, Wang Y, Blahos J, Hell J, Wenthold R (2000) J Neurosci 20: 1260–1271

Schiavo G, Gmachl J, Stenbeck G, Söllner T, Rothman J (1995) A possible docking and fusion particle for synaptic transmission. Nature 378: 733–736

Shirao T (1995) The roles of microfilament-associated proteins, drebrins, in brain morphogenesis: a review. J Biochem 117: 231–236

Shirao T, Obata K (1986) Immunochemical homology of 3 developmentally regulated brain proteins and their developmental changes in neuronal distribution. Dev Brain Res 29: 233–244

Shirao T, Inoue H, Kano Y, Obata K (1989) Localization of a developmentally regulated neuron-specific protein S54 in dendrites as revealed by immuneelectron microscopy. Brain Res 413: 374–378

Shoop R, Yamada N, Berg D (2000) Cytoskeletal links of neuronal acetylcholine receptors containing alpha 7 subunits. J Neurosci 20: 4021–4029

Sylvester P (1986) The anterior commisure in Down syndrome. J Ment Defic Res 30: 19–25

Toda M, Shirao T, Minoshima S, Shimizu N, Toya S, Uyemura K (1993) Molecular cloning of cDNA encoding human drebrin E and chromosomal mapping of its gene. Biochem Biophys Res Commun 196: 468–472

Toda M, Shirao T, Uyemura K (1999) Suppression of an actin-binding protein, drebrin, by antisense transfection attenuates neuritic outgrowth in neuroblastoma B104 cells. Dev Brain Res 114: 193–200

Wisniewski K, Kida E (1994) Abnormal neurogenesis and synaptogenesis in Down Syndrome brain. Dev Brain Dysfunct 7: 289–301

Yoo B, Cairns N, Fountoulakis M, Lubec G (2001) Synaptosomal proteins beta-soluble N-ethylmaeimide-sensitive factor attachment protein (beta-SNAP), gamma-SNAP and synaptotagmin I in brain of patients with Down Syndrome and Alzheimer's disease. Dementia 12: 219–225

Authors' address: Prof. Dr. G. Lubec, Department of Pediatrics, University of Vienna, Währinger Gürtel 18, A-1090 Vienna, Austria, e-mail: gert.lubec@akh-wien.ac.at

Antioxidant proteins in fetal brain: superoxide dismutase-1 (SOD-1) protein is not overexpressed in fetal Down syndrome

T. Gulesserian[1], **E. Engidawork**[1], **M. Fountoulakis**[2], and **G. Lubec**[1]

[1] Department of Pediatrics, University of Vienna, Vienna, Austria
[2] F. Hoffman-La Roche, Basel, Switzerland

Summary. Exposure of living organisms to reactive oxygen species (ROS), notably oxygen free radicals and hydrogen peroxide is closely linked to the very fact of aerobic life. Oxidants, however, are not always detrimental for cell survival, indeed moderate concentrations of ROS serve as signaling molecules. To maintain this level, cells have evolved an antioxidant defense system. Disruption of this balance leads either to oxidative or reductive stress. Down syndrome (DS) is a genetic disorder associated with oxidative stress. Overexpression of superoxide dismutase-1 (SOD-1) as a result of gene loading is suggested to be responsible for this phenomenon. To examine this view, we investigated the expression of thirteen different proteins involved in the cellular antioxidant defense system in brains of control and DS fetuses by two-dimensional electrophoresis (2-DE) coupled with matrix-assisted laser desorption/ionization mass spectroscopy (MALDI-MS).

No detectable change was found in expression of SOD-1, catalase, phospholipid hydroperoxide glutathione peroxidase, glutathione reductase, antioxidant enzyme AOE372, thioredoxin-like protein and selenium binding protein between control and DS fetuses. By contrast, a significant reduction was observed in levels of glutathione synthetase ($P < 0.01$), glutathione-S-transferase mu2 ($P < 0.01$), glutathione-S-transferase p ($P < 0.05$), antioxidant protein 2 ($P < 0.05$), thioredoxin peroxidase-I ($P < 0.05$) and thioredoxin peroxidase-II ($P < 0.01$) in DS compared with controls. The data suggest that oxidative stress in fetal DS does not result from overexpression of SOD-1 protein, rather oxidative stress appears to be the consequence of low levels of reducing agents and enzymes involved in removal of hydrogen peroxide.

Introduction

Oxygen free radicals and their by-products that are capable of causing oxidative damage are collectively termed as reactive oxygen species (Sies, 1991). ROS are generated in aerobic metabolism from both endogenous and exogenous sources, mitochondrial oxidative phosphorylation being one of the major sites of endogenous production (Halliwell and Gutteridge, 1989). ROS are generally considered as dangerous for cells, particularly when produced in excess, however, moderate levels of intracellular ROS may be necessary for regulation of cellular functions including gene expression and signal transduction (Castagne et al., 1999; Sen and Packer, 1996). Mammalian cells have evolved complex mechanisms to offset oxidative challenges and to maintain a redox balance in their different subcellular compartments (Fridovich, 1978). These antioxidant defense systems include non-enzymatic antioxidants, for e.g., vitamin E, vitamin C, vitamin A and uric acid; enzymes with antioxidant properties, such as superoxide dismutase (SOD), catalase, glutathione peroxidase, glutathione reductase and thioredoxin peroxidases; and proteins and low molecular weight reducing agents including thioredoxin, glutaredoxin and glutathione. A delicate balance between the oxidant and antioxidant pathways is pivotally important for homeostasis. Any perturbation that tilts the balance towards the oxidative side leads to a phenomenon called oxidative stress (Sies, 1994).

Down syndrome (DS) is one such situation where oxidative stress is considered to be a major factor for early cortical dysgenesis and late neuronal degeneration (Busciglio and Yankner, 1995; Iannello et al., 1999). Oxidative stress in DS is proposed to result from overexpression of CuZnSOD (SOD-1), as the gene encoding this antioxidant enzyme is mapped to the region 21q22.1 of chromosome 21. Consistent with this notion, a number of studies have demonstrated elevated mRNA level or activity of SOD1 in a variety of DS cell types or organs including fetal brain (Anneren and Epstein, 1987; Brooksbank and Balazs, 1984; de Haan et al., 1997; Feaster et al., 1977). SOD catalyzes the dismutation of superoxide radicals to hydrogen peroxide, which is then metabolized to water by catalases or peroxidases. The proponents of "gene dosage effect" argue that the increased production of hydrogen peroxide, caused by overexpression of SOD-1, is not accompanied by a corresponding increase in the other hydrogen peroxide handling enzymes. Lack of such an adaptive response accelerates hydrogen peroxide formation, which can promote oxidative damage by itself or leads to production of the noxious hydroxyl radical that is capable of initiating chains of reactions, inflicting damage to cellular macromolecules. Studies performed to validate this argument have produced conflicting results. Measurements of markers for oxidation in adult DS brain did not reveal any increase in DNA adducts (Seidl et al., 1997) as well as lipid peroxidation (Hayn et al., 1996). By contrast, lipid peroxidation and protein oxidation were increased in fetal DS brain (Brooksbank and Balazs, 1984; Odetti et al., 1998). Moreover, we recently reported increased expression of SOD1 protein in brains of adults with DS, a finding that could be linked either to gene dosage effect or glial proliferation (Gulesserian et al., 2001). Thus, knowledge of the consequences of having an extra copy of this gene is far from complete. Given this, consideration of the expression pattern of SOD-1 and other antioxidant enzymes in fetal brain, which is not confounded by neurodegeneration may provide a means for better understanding the role of ROS in pathogenesis of DS. The present study is therefore undertaken to add to the current body of literature by investigating the expression of different proteins of the antioxidant defense system in cerebral cortex of normal and DS fetuses using two-dimensional electrophoresis (2-DE) followed by matrix-assisted laser desorption ionization mass spectroscopy (MALDI-MS).

Material and methods

Fetal brain tissue

Brain tissue (cerebral cortex) of aborted control fetuses with no obvious abnormalities (n = 7; 1 female and 6 males with mean gestational age of 18.8 ± 2.2 weeks) and DS (n = 9; 2 females and 7 males with mean gestational age of 19.6 ± 2.0 weeks) was used for the experiment. The brain samples were obtained from Drs. M. Dierssen and J. C. Ferreres, Medical and Molecular Genetics Center-IRO, Hospital Duran i Reynals, Barcelona, Spain. All samples were stored at −70°C and the freezing chain was never interrupted.

Two dimensional gel electrophoresis (2-DE)

Brain tissue was suspended in 0.5 ml of sample buffer consisting of 40 mM Tris, 5 M urea (Merck, Darmstadt, Germany), 2 M thiourea (Sigma, St. Louis, MO, USA), 4% CHAPS (Sigma), 10 mM 1,4-dithioerythritol (Merck), 1 mM EDTA (Merck) and a mixture of protease inhibitors, 1 mM PMSF and 1 μg of each pepstatin A, chymostatin, leupeptin and antipain. The suspension was sonicated for approximately 30 sec and centrifuged at 150,000 g for 45 min. The protein content in the supernatant was determined by the Coomassie blue method (Bradford, 1976).

2-DE was performed essentially as reported (Langen et al., 1997). Samples of 2 mg protein were applied on immobilized pH 3–10 nonlinear gradient strips in sample cups at their basic and acidic ends. Focusing started at 200 V and the voltage was gradually increased to 5,000 V at 3 V/min and kept constant for a further 24 h (approximately 180,000 kVh totally). The second-dimensional separation was performed on 9–16% SDS gradient polyacrylamide gels. The gels (180 × 200 × 1.5 mm) were run at 40 mA per gel. After protein fixation for 12 h in 40% methanol, containing 5% phosphoric acid, the gels were stained with colloidal Coomassie blue (Novex, San Diego, CA, USA) for 48 h. Molecular masses were determined by running standard protein markers (Gibco, Basel, Switzerland), covering the range 10–200 kDa. pI values were used as given by the supplier of the IPG strips. Excess of dye was washed out from the gels with H_2O and the gels were scanned in an Agfa DUOSCAN densitometer (resolution 200). Electronic images of the gels were recorded using Photoshop (Adobe) and PowerPoint (Microsoft) software.

Matrix-associated laser desorption ionization mass spectroscopy (MALDI-MS)

MALDI-MS analysis was performed as described elsewhere (Fountoulakis and Langen, 1997) with some modifications. The spots were excised with a spot picker and placed into 96-well microtiter plates. Each spot was destained with 100 μl of 30% acetonitrile in 50 mM ammonium bicarbonate and dried in a speedvac evaporator. Each dried gel piece was rehydrated with 4 μl of 3 mM Tris-HCl, pH 9.0, containing 50 ng trypsin (Promega, Madison, WI, USA). After 16 h at room temperature, 7 μl of H_2O was added to each gel piece and the samples were shaken in for 10 min. Four μl of 50% acetonitrile, containing 0.3% trifluoroacetic acid and the standard peptides des-Arg-bradykinin (Sigma, 904.4681 Da) and adrenocorticotropic hormone fragment 18–39 (Sigma, 2,465.1989 Da), in water were added to each gel piece and shaken for 10 min. The application of the samples was performed with a SymBiot I sample processor (PE Biosystems, Framingham, MA, USA). 1.5 μl of the peptide mixture was simultaneously applied with 1 μl of matrix, consisting of a saturated solution of α-cyano-4-hydroxycinnamic acid (Sigma) in 50% acetonitrile, containing 0.1% trifluoroacetic acid. Samples were analyzed in a time-of-flight mass spectrometer (Reflex 3, Bruker Analytics, Bremen, Germany). An accelerating voltage of 20 kV was used. Peptide matching and protein searches were performed automatically. The peptide masses were compared to the theoretical peptide masses of all available proteins from all species. Monoisotopic masses were used and a mass tolerance of 0.0025% was allowed. The algorithm used for determining the probability of a false positive match with a given MS-spectrum is described elsewhere (Berndt et al., 1999).

Quantification of protein spots

Protein spots were outlined (first automatically and then manually) and quantified using the ImageMaster 2D Elite software (Amersham Pharmacia Biotechnology). The percentage of the volume of the spots representing a certain protein was determined in comparison with the total proteins present in the 2-D gel.

Statistical analysis

The software calculated the percentage of the spot volume in each particular gel after subtraction of the background values. Between-group differences were calculated using non-parametric Mann-Whitney U-test. The level of significance was set at $P < 0.05$.

Results

Identification and quantification of proteins of the antioxidant defense system on 2-DE map of fetal brain

Human proteins obtained from the cerebral cortex of fetal brain were applied on 2-D gel. The protein spots were visualized by staining with colloidal Coomassie blue. The protein spots were analyzed by MALDI-MS, following in-gel digestion. The peptide masses were matched with the theoretical peptide masses of all proteins of the SWISS-PROT database. Internal standards were used to correct the measured peptide mass thus reducing the windows of mass tolerance and increasing the confidence of identification. Approximately, 389 proteins were successfully identified, among which 13 proteins involved in the antioxidant defense system were quantified using the Image Master 2D Elite software. Figure 1 shows a representative gel from control brain containing fetal proteins separated by 2-DE, with accession number given only for the investigated proteins.

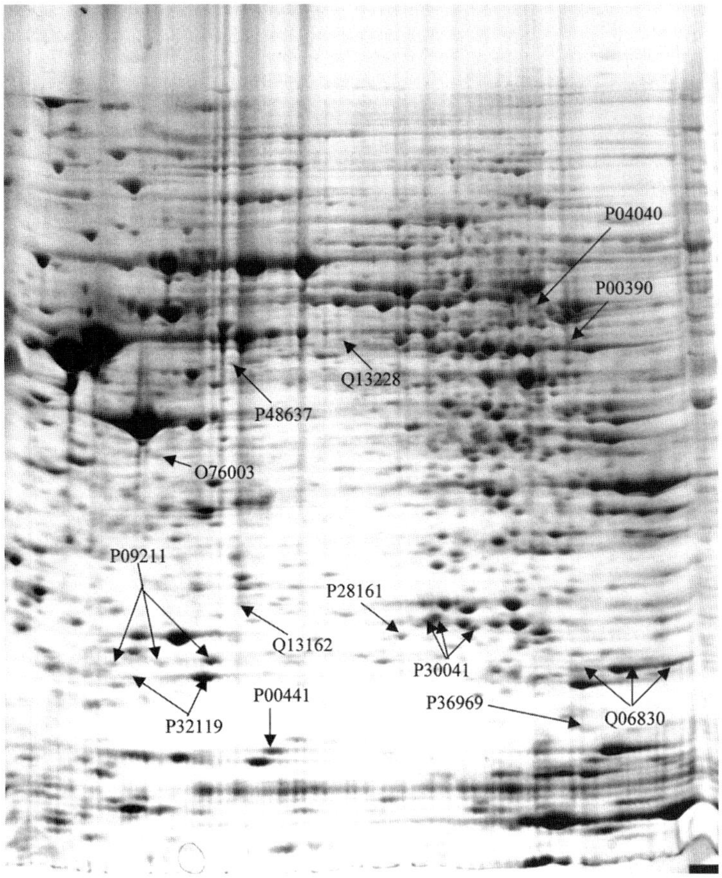

Fig. 1. Two-dimensional map of human fetal brain proteins. The proteins from cerebral cortex of fetal brain were extracted and analyzed on pH 3–10 nonlinear IPG strips, followed by a 9–16% sodiumdodecyl sulphate (SDS)-polyacrylamide gels. The gels were stained with Coomassie blue and identified by MALDI-MS. The spots representing proteins of the antioxidant defense system are designated with their SWISS-PROT number. Name of the proteins is given in Table 1

Expression pattern of the antioxidant proteins in control and DS fetal brain

The expression levels of the proteins were determined as a percentage of spot volume in the area of interest. The SWISS-PROT accession number and names of the quantified proteins along with the MALDI data and percent protein level in control and DS are given in Table 1.

SOD-1

SOD-1 is the enzyme, which dismutate superoxide anion to hydrogen peroxide and oxygen. The protein was resolved as a single spot on 2D gel (Fig. 1). Quantitative analysis of relative percent protein level with 2D Image Master

Table 1. Quantitative analysis of proteins of the antioxidant defense system in human fetal brain. The samples were analyzed and the proteins were identified and quantified as stated under Material and methods. The name and SWISS-PROT accession number, the observed Mr and pI values, and the matching peptides along with the probability as well as the relative protein levels (%) of the proteins of interest are given here. Within group analysis was performed by non-parametric Mann-Whitney U-test and the significance was set at $P < 0.05$. *$P < 0.05$, **$P < 0.01$

Protein name	Accession number	pI	MW (Da)	Matches	Probability	Control	DS
SOD-1	P00441	6.08	16,022	4	1.00E-5	2.45 ± 0.52 (n = 7)	2.59 ± 0.64 (n = 8)
Catalase	P04040	7.40	59,946	7	1.94E-08	2.62 ± 1.14 (n = 7)	2.31 ± 1.11 (n = 7)
Phospholipid hydroperoxide glutathione peroxidase	P36969	7.84	19,921	7	3.78E-10	0.09 ± 0.05 (n = 6)	0.11 ± 0.07 (n = 6)
Thioredoxin peroxidase-I	P32119	5.81	22,049	4	4.33E-05	7.08 ± 1.76 (n = 7)	4.92 ± 0.82* (n = 8)
Thioredoxin peroxidase-II	Q06830	8.19	22,324	10	9.57E-18	4.33 ± 0.76 (n = 7)	3.14 ± 0.56** (n = 6)
Antioxidant enzyme AOE372	Q13162	6.25	30,748	7	1.04E-09	0.17 ± 0.06 (n = 7)	0.10 ± 0.05 (n = 8)
Antioxidant protein 2	P30041	6.31	25,002	7	9.51E-09	1.15 ± 0.32 (n = 7)	0.66 ± 0.23* (n = 6)
Glutathione synthetase	P48637	5.80	52,523	8	6.80E-12	0.82 ± 0.42 (n = 7)	0.30 ± 0.14** (n = 8)
Glutathione reductase	P00390	7.67	52,106	6	2.00E-07	1.86 ± 0.63 (n = 7)	1.50 ± 0.81 (n = 6)
Glutathione-S-transferase mu2	P28161	6.31	25,768	8	1.56E-11	0.34 ± 0.08 (n = 7)	0.20 ± 0.04** (n = 6)
Glutathion-S-transferase p	P09211	5.38	23,438	8	5.60E-14	5.57 ± 1.31 (n = 7)	3.80 ± 0.67* (n = 7)
Thioredoxin-like protein	O76003	5.13	37,751	6	6.58E-06	0.33 ± 0.09 (n = 7)	0.42 ± 0.32 (n = 9)
Selenium binding protein	Q13228	6.57	52,906	7	8.51E-08	0.95 ± 0.47 (n = 7)	0.90 ± 0.29 (n = 8)

did not reveal any significant difference in expression of SOD-1 between controls and DS (Table 1).

Hydrogen peroxide handling enzymes

Hydrogen peroxide produced as a result of cellular metabolism, is removed by three major types of enzymes, catalases, glutathione peroxidases and peroxiredoxins (Kang et al., 1998a, b). Many proteins from a wide variety of species that show sequence similarity to yeast thioredoxin peroxidase have been identified and these homologues are referred to as peroxiredoxins (Chae et al., 1994b). Peroxiredoxins can be divided into two depending on the number of conserved cysteine residues they contain, as 1-Cys (antioxidant protein 2 (AOP2) or 2-Cys (thioredoxin peroxidases) (Kang et al., 1998a).

Whilst catalase, phospholipid hydroperoxide glutathione peroxidase and antioxidant enzyme AOE372 (a thioredoxin peroxidase) were resolved as a single spot, multiple spots had been identified for thioredoxin peroxidase-1 (2 spots), thioredoxin peroxidase-II (3 spots) and antioxidant protein 2 (3 spots) on 2-DE (Fig. 1). The spots for AOP2 were of equivalent intensity, whereas one was major and the others minor for thioredoxin peroxidase-I & II. The multiple spots may most likely represent different post-translationally modified forms of the respective protein, but the possibility that they may denote different isoforms can not be ruled out. Neither the expression of catalase, phospholipid hydroperoxide glutathione peroxidase nor that of antioxidant AOE372 showed any apparent change in DS compared with controls. However, a significant decrease was observed in expression of the remaining hydrogen peroxide handling proteins. The major spots as well as the sum of the spots consistently exhibited a significant decrease in DS in all cases, while variation was observed in expression of the minor spots (data for individual spots not shown, for sum of spots see Table 1). Accordingly, all the three spots ($P < 0.01$ & $P < 0.05$) as well as the sum of the spots ($P < 0.05$) of AOP2 (Table 1 and Fig. 2) were significantly decreased in DS. Likewise, a significant reduction was obtained in the major spots and sum of the individual spots of both thioredoxin peroxidase-1 ($P < 0.05$) and II ($P < 0.01$) (Table 1 and Fig. 2). With regard to the minor spots, among the two minor spots of thioredoxin-peroxidase-II one showed a significant decrease (second from left to right in Fig. 2) ($P < 0.05$), but the other was unchanged (first from left to right in Fig. 1). So did the minor spot of thioredoxin peroxidase-1.

Others

Glutathione reductase, glutathione-S-transferase mu 2, glutathione synthetase, thioredoxin-like protein and selenium binding protein were resolved as a single spot, whereas glutathione-S-transferase p as three spots, with one being major and the others minor. No significant difference had been noted in

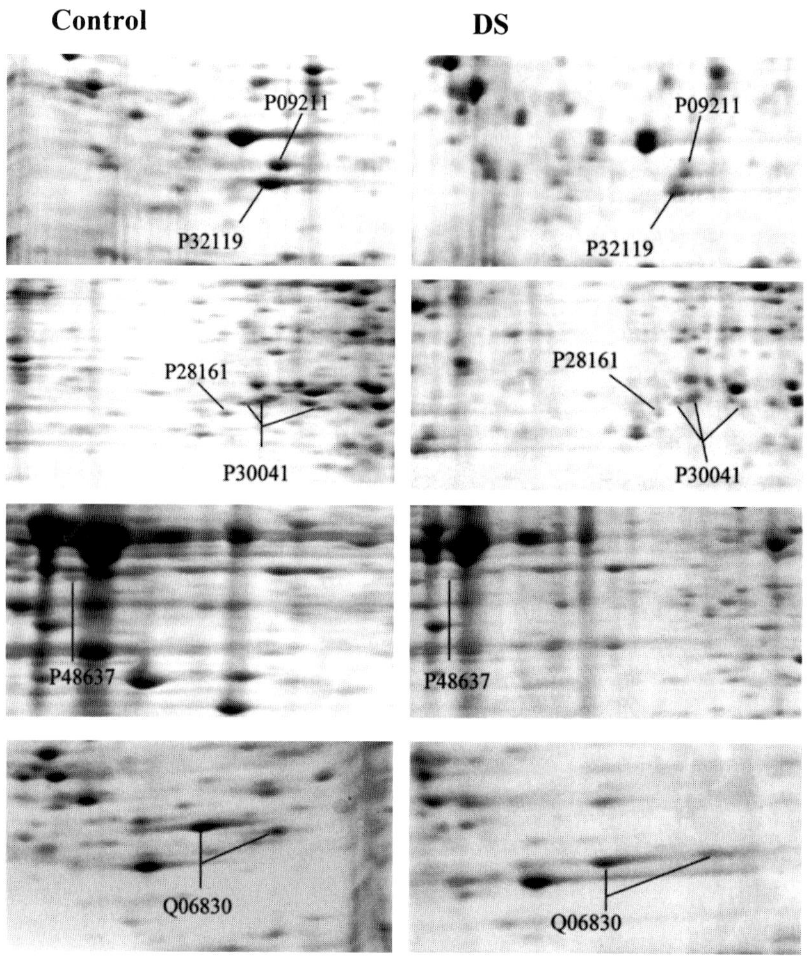

Fig. 2. Partial two-dimensional gel images of human fetal brain proteins. The gels were prepared and the proteins were analyzed as described under legend to Fig. 1. The spots corresponding to glutathione-S-transferase p (P09211), thioredoxin peroxidase-1 (P32119), glutathione-S-transferase mu2 (P28161), antioxidant protein 2 (P30041), glutathione synthetase (P48637) and thioredoxin peroxidase-II (Q06830) are indicated with their accession number in the upper panels

the spots of glutathione reductase, thioredoxin-like protein and selinium binding protein between controls and DS (Table 1). By contrast, a significant reduction was obtained in DS in expression of glutathione-S-transferase mu 2 ($P < 0.01$) and glutathione synthetase ($P < 0.01$) compared with controls (Table 1 and Fig. 2). A similar decrease was also noted in the major spot ($P < 0.01$) (data not shown) as well as sum of the spots of glutathione-S-transferase p ($P < 0.05$) (Table 1 and Fig. 2). One of the minor spots (first from left to right in Fig. 1) was not quantifiable, so quantitation was performed only for the other spot (middle one in Fig. 1). This spot did not show any detectable difference between controls and DS (data not shown).

Discussion

The imbalance in the antioxidant enzymatic pathway that protects cells from the superoxide anion toxicity has been suggested to place the developing DS fetus in an environment of increased oxidative stress (de Haan et al., 1997). This coupled with the observation of increased generation of ROS (Busciglio and Yankner, 1995) and increased lipid peroxidation (Brooksbank and Balazs, 1984; Odetti et al., 1998) in DS fetal brain, strengthened the view that oxidative cellular damage may begin in utero and continues during the life span of the individual. This paper is targeted to examine this view by investigating the expression of a wide variety of antioxidant proteins in fetal brain using 2-DE coupled with MALDI-MS, which allows unequivocal identification of proteins.

SOD-1 mRNA (de Haan et al., 1997) and activity (Brooksbank and Balazs, 1984) were shown to be increased in fetal DS brain. In addition, SOD-1 activity was elevated in various DS cell types and organs. Because of this SOD-1 is considered to be central in bringing about ROS and oxidative stress induced neuronal abnormalities associated with DS (Iannello et al., 1999). Taking this into account, the unaltered expression of SOD-1 in fetal DS was surprising, because the finding was expected to be concordant with the aforementioned studies. The disagreement between levels of mRNA and protein is not unique for SOD. Similar disagreements were reported for other proteins as well (Jin et al., 1996; Shimbara et al., 1992), suggestive of the need to take caution in drawing conclusions based solely on RNA data. Indeed, these days it is because of this very fact that proteomics is emerging as a popular technique to address biomedical questions (Banks et al., 2000; Chambers et al., 2000). With regard to protein levels and activity, increased activity not necessarily reflects increased expression. Enzyme activity can either increase or decrease without a concomitant change in expression. Recently, our group reported increased SOD-1 protein expression in adult DS brain and suggested that the increase could either be related to gene dosage effect or gliosis (Gulesserian et al., 2001). In fetal DS brain, however, SOD-1 expression was unchanged suggesting gene dosage effect is unlikely to explain the observation in adult brain. We were not able to quantify glial fibrillar acidic protein (GFAP), a marker of gliosis, in fetal brain, as the corresponding spot is too weak to be quantified. However, published findings indicate no alteration of GFAP in early infancy (Becker et al., 1991), which implies that any major change would not be apparent in fetal DS GFAP expression. If this holds true, it can offer a plausible explanation why we failed to observe detectable changes in SOD-1 protein expression in fetal DS brain. Otherwise, this finding is consistent with our previous report that gene dosage effect does not have any influence on expression of proteins encoded on chromosome 21 in fetal DS brain (Engidawork et al., this volume). Oxidative stress in fetal DS therefore may result from accumulation of hydrogen peroxide caused by amyloid beta (Behl et al., 1994), as soluble amyloid beta was detected in fetal DS brain (Teller et al., 1996) or direct changes in other hydrogen peroxide synthesizing or eliminating enzymes. Alternatively, it can be caused by increased activity of SOD-1, as activity was not measured in this study.

In line with published reports (Anneren and Epstein, 1987; Brooksbank and Balazs, 1984; de Hann et al., 1997), we did not find any significant alteration in expression of catalase in fetal DS brain. Brain contains low amount of catalase and the enzyme is largely located in peroxisomes, where some of the cellular hydrogen peroxide-generating enzymes are located (Halliwell and Gutteridge, 1989). The unaltered expression of catalase shows that hydrogen peroxide produced in this way is properly disposed of in fetal DS brain. Similarly, unchanged expression of phospholipid hydroperoxide glutathione peroxidase, an enzyme that could play a major role in protecting mammals from the toxicity of ingested lipid hydroperoxides, suggests that these agents are also not a threat for DS fetuses.

In most tissues glutathione peroxidase is present in much lower concentrations compared with peroxiredoxins (Kang et al., 1998), which may explain why it was not possible to identify glutathione peroxidase on 2D map of fetal brain. The thioredoxin system is an important conserved system for protection against oxidative stress by reducing peroxides such as hydrogen peroxide. The system is composed of three proteins thioredoxin peroxidase, thioredoxin and thioredoxin reductase (Chae et al., 1994a). Thioredoxin peroxidase proteins have been found to have regulatory functions in addition to their antioxidant properties. In this regard, the thioredoxin peroxidase-I and II could possibly serve as components of the signaling cascades mediated by growth factors and TNF-α (tumor necrosis factor-alpha) via regulation of the intracellular concentrations of hydrogen peroxide (Kang et al., 1998b). Thioredoxin peroxidase-1 inhibits activation of AP-1 (activated protein-1) either by competing for reducing equivalents or by scavenging oxidants formed by TNF-α (Shau et al., 1998). Thioredoxin-1 and antioxidant enzyme AOE372 also inhibit activation of NF-κB (nuclear factor-κB) induced by hydrogen peroxide or TNF-α (Jin et al., 1997; Kang et al., 1998b). Moreover, expression of thioredoxin peroxidase-II is correlated with proliferation (Prospèri et al., 1994) and the protein is shown to inhibit lipid peroxidation and regulate apoptosis (Zhang et al., 1997). Collectively these findings indicate that thioredoxin peroxidase I & II exert a negative modulatory effect on phenomena that have relevance to DS neuropathology. Hence, their downregulation may lead to alteration in signaling processes and oxidative stress in DS. It is interesting to note that expression of all thioredoxin peroxidases is not affected. For example, expression of antioxidant enzyme AOE 372, a thioredoxin peroxidase, was unchanged. One possible explanation is different thioredoxin peroxidases may serve restricted functions in a tissue specific manner (Jin et al., 1997). Alternatively, although expression of antioxidant enzyme AOE 372 was unchanged, its activity can be altered due to lack of dimerizing partner as a result of reduction in expression of other members of the subfamily. This seems likely, as heterodimer of two different peroxiredoxins was shown to be more active than counterpart homodimers (Jin et al., 1997). Expression of AOP2, a protein suggested to be involved in growth factor and cytokines mediated signal transduction by reducing levels of peroxides (Kang et al., 1998b; Phelan, 1999), was also reduced in fetal DS. This concomitant reduction of AOP2 strengthens the notion that failure to remove hydrogen peroxide resulting from signal transduction may lead to impaired signaling and oxidative stress in DS.

The glutathione transferases are a family of proteins that inactivate the toxic products of oxygen metabolism by conjugating with glutathione or by direct binding (Jakoby, 1978). These products include 4-hydroxy-alkenals such as 4-hydroxynonenal (4-HNE), a highly reactive cytotoxic substance capable of inhibiting glycolysis, nucleic acid and protein synthesis, and degrading proteins (Markesbery, 1999). The decrease in glutathione transferase in fetal DS may indicate the vulnerability of fetal DS neurons to the toxic action of 4-HNE. Glutathione, apart from being a physiological electron donor for the glutathione peroxidases, is a scavenger of hydroxyl radicals and singlet oxygen. Although there was no difference in glutathione reductase, the enzyme that regenerates reduced glutathione, the decrease in glutathione synthetase in fetal DS is bound to lower concentration of glutathione and this can lead to inefficient removal of oxidant species. On the other hand the unchanged levels of thioredoxin-like protein may indicate that there is no alteration in levels of the physiological electron donor per se, rather the alteration is associated with the electron acceptors, thioredoxin peroxidases. It appears therefore that it is the peroxiredoxins that are mostly affected in fetal DS among the antioxidant defense system. The observation of unchanged expression of selenium binding protein in DS is in agreement with unaltered expression of enzymes with redox-sensitive moiety such as glutathione peroxidases.

In conclusion, the view that hydrogen peroxide is an integral component of membrane receptor signaling is now widely accepted (Bae et al., 1997; Castagne et al., 1999; Sundaresan et al., 1995). As this signaling requires rapid removal of hydrogen peroxide for normal functioning, the decrease in AOP, thioredoxin peroxidase-I and II may lead to alterations in signal transduction mechanisms in fetal DS brain. Overexpression of SOD-1 does not appear to contribute to oxidative stress in fetal DS. Rather, reduced synthesis of glutathione and inefficient removal of hydrogen peroxide produced during signaling processes seem to be responsible. In addition, the findings have particular significance for the fetus, as it undergoes an abrupt transition from a hypoxic uterine environment to a relatively hyperoxic extrauterine environment.

Acknowledgements

We are highly indebted to Red Bull Company, Salzburg, Austria for the generous financial support. E.E is a scholarship holder of the Austrian Academic Exchange Program, ÖAD.

References

Anneren KG, Epstein CJ (1987) Lipid peroxidation and superoxide dismutase-1 and glutathione peroxidase activities in trisomy 16 fetal mice and human trisomy 21 fibroblast. Pediatr Res 21: 88–92

Bae SY, Kang SW, Seo MS, Bainess IC, Tekle E, Chock PB, Rhee SG (1997) Epidermal growth factor (EGF) induced generation of hydrogen peroxide. Role in EGF receptor mediated tyrosine phosphorylation. J Biol Chem 272: 217–221

Banks RE, Dunn MJ, Hochstrasser DF, Sanchez JC, Blackstock W, Pappin DJ, Selby PJ (2000) Proteomics: new perspectives, new biomedical opportunities. Lancet 356: 1749–1756

Behl C, Davis J, Lesley R, Shubert D (1994) Hydrogen peroxide mediates amyloid beta protein toxicity. Cell 77: 817–827

Berndt, P, Hobohm U, Langen H (1999) Reliable automatic protein identification from matrix-assisted laser desorption/ionization mass spectrometric peptide fingerprints. Electrophoresis 20: 3521–3526

Bradford M (1976) A rapid and sensitive method for the quantitation of microgram quantities of protein utilizing the principle of protein-dye binding. Anal Biochem 72: 248–254

Brooksbank BW, Balazs R (1984) Superoxide dismutase, glutathione peroxidase and lipoperoxidation in Down's syndrome fetal brain. Brain Res 318: 37–44

Busciglio J, Yankner BA (1995) Apoptosis and increased generation of active oxygen species in Down syndrome neurons in vitro. Nature 378: 776–779

Castagne V, Gautschi M, Lefevre K, Posada A, Clarke PGH (1999) Relationships between neuronal death and the cellular redox status. Focus on the developing nervous system. Prog Neurobiol 59: 397–423

Chae HZ, Chung SJ, Rhee SG (1994a) Thioredoxin dependent peroxide reductase from yeast. J Biol Chem 269: 27670–27678

Chae HZ, Robinson K, Poole LB, Church G, Storz G, Rhee SG (1994b) Cloning and sequencing of thiol-specific antioxidants from mammalian brain: alkyl hydroperoxide reductase and thiol-specific antioxidants define a large family of antioxidant enzymes. Proc Natl Acad Sci USA 91: 7017–7021

Chambers G, Lawrie L, Cash P, Murray GI (2000) Proteomics: a new approach to the study of disease. J Pathol 192: 280–288

de Haan JB, Wolvetang EJ, Cristiano F, Iannello R, Bladier C, Kelner MJ, Kola I (1997) Reactive oxygen species and their contribution to pathology in Down syndrome. Adv Pharmacol 38: 379–402

Engidawork E, Balic N, Fountoulakis M, Dierssen M, Lubec G (2001) β-Amyloid precursor protein, ETS-2 and collagen alpha 1 (VI) chain precursor, encoded on chromosome 21, are not overexpressed in fetal Down syndrome: further evidence against gene dosage effect. J Neural Transm (this volume)

Feaster WW, Kwok LW, Epstein CJ (1977) Dosage effects for superoxide dismutase-1 on nucleated cells aneuploid for chromosome 21. Am J Hum Genet 29: 563–570

Fountoulakis M, Langen H (1997) Identification of proteins by matrix assisted laser desorption ionization mass spectrometry following in-gel digestion in low salt, non-volatile buffer and simplified peptide recovery. Anal Biochem 250: 153–156

Fridovich I (1978) The biology of oxygen radicals. Science 201: 875–888

Gulesserian T, Seidl R, Hardmeier R, Cairns N, Lubec G (2001) Superoxide dismutase SOD1, encoded on chromosome 21, but not SOD2 is overexpressed in brains of patient with Down syndrome. J Invest Med 49: 41–46

Halliwell B, Gutterridge JMC (1989) Protection against oxidants in biological systems: the superoxide theory of oxygen toxicity. In: Halliwell B, Gutterridge JMC (eds) Free radicals in biology and medicine, 2nd edn. Oxford University Press, New York, pp 86–187

Hayn M, Kresmer K, Singewald N, Cairns N, Nemathova M, Lubec B, Lubec G (1996) Evidence against the involvement of reactive oxygen species in the pathogenesis of neuronal death in Down syndrome and Alzheimer's disease. Life Sci 59: 537–544

Iannello RC, Crack PJ, de Haan JW, Kola I (1999) Oxidative stress and neural dysfunction in Down syndrome. J Neural Transm 57: 257–267

Jakoby WB (1978) The glutathione-S-transferases: a group of multifunctional detoxification proteins. Adv Enzymol Relat Areas Mol Biol 46: 383–314

Jin D, Chae HZ, Rhee SG, Jeang KT (1997) Regulator role for a novel human thioredoxin peroxidase in NFK-B-activation. J Biol Chem 272: 30952–30961

Jin LW, Masliah E, Deteresa IR, Mallory M, Sundsmo M, Mori N, Sobel A, Saitoh T (1996) Neurofibrillary tangle associated alteration of stathmin in Alzheimer's disease. Neurobiol Aging 17: 331–341

Kang SW, Bainess IC, Rhee SG (1998a) Characterization of mammalian peroxiredoxin that contain one conserved cysteine. J Biol Chem 273: 6303–6311

Kang SW, Chae HZ, Seo MS, Kim K, Bainess IC, Rhee SG (1998a) Mammalian peroxiredoxin isoforms can reduce hydrogen peroxide generated in response to growth factors and tumor necrosis factor-alpha. J Biol Chem 273: 6297–6302

Langen H, Röder D, Juranville JF, Fountoulakis M (1997) Effect of the protein application mode and the acrylamide concentration on the resolution of protein spots separated by two-dimensional gel electrophoresis. Electrophoresis 18: 2085–2090

Markesbery WR (1999) The role of oxidative stress in Alzheimer's disease. Arch Neurol 56: 1449–1452

Odetti P, Angelini G, Dapino D, Zaccheo D, Garibaldi S, Dagna-Bricarelli F, Piombo G, Perri G, Smith M, Traverso N, Tabaton M (1998) Early glyco-oxidation damage in brains from Down's syndrome. Biochem Biophys Res Commun 243: 849–851

Phelan SA (1999) AOP2 (antioxidant protein 2): structure and function of a unique thiol-specific antioxidant. Antioxd Redox Signal 1: 571–584

Prospèri M-T, Apiou F, Dutrillaux B, Goubin G (1994) Organizational and chromosomal assignment of two human PAG gene loci: PAGA encoding a functional gene and PAGB a processed pseudogene. Genomics 19: 236–241

Seidl R, Greber S, Schuller E, Bernert G, Cairns N, Lubec G (1997) Evidence against increased oxidative DNA damage in Down syndrome brain. Neurosci Lett 235: 137

Sen CK, Packer L (1996) Antioxidant and redox regulation of gene expression. FASEB J 10: 709–720

Shau H, Huang ACJ, Faris M, Nazarian R, de Vellis J, Chen W (1998) Thioredoxin peroxidase (natural killer enhancing factor) regulation of activator protein-1 function in endothelial cells. Biochem Biophys Res Commun 249: 683–686

Shimbara N, Orino E, Sone S, Ogura T, Takashima M, Shono M, Tamura T, Yasuda H, Tanaka K, Ichihara A (1992) Regulation of gene expression of proteasomes (multiprotease complex) during growth and differentiation of human hematopoietic cells. J Biol Chem 267: 18100–181009

Sies H (1991) Oxidative stress. In: Sies H (ed) Oxidants and antioxidants. Academic Press, London, pp 650

Sies H (1994) Oxidative stress: from basic research to clinical medicine. In: Favier AE, Neve J, Faure P (ed) Trace elements and free radicals in oxidative diseases. AOCS PRESS, Champaign, Illinois, pp 1–7

Sundaresan M, Yu E-X, Ferrans VJ, Irani K, Finkel T (1995) Requirement for generation of hydrogen peroxide for platelet-derived growth factor signal transduction. Science 270: 296–299

Teller JK, Russo C, De Busk LM, Angelini G, Zaccheo D, Dagna-Bricarelli F, Scartezzini P, Bertolini S, Mann DMA, Tabaton M, Gambetti P (1996) Presence of soluble

amyloid β-peptide precedes amyloid plaque formation in Down's syndrome. Nature Med 2: 93–95
Zhang P, Liu B, Kang SW, Seo MS, Rhee SG, Obeid LM (1997) Thioredoxin peroxidase is a novel inhabitor of apoptosis with a mechanism distinct from that of Bcl-2. J Biol Chem 272: 30615–30618

Authors' address: Prof. Dr. G. Lubec, Department of Pediatrics, University of Vienna, Währinger Gürtel 18-20, A-1090 Vienna, Austria,
e-mail: Gert.Lubec@akh-wien.ac.at

Glial-neurotrophic mechanisms in Down syndrome

P. G. Nelson[1], **S. K. McCune**[2], **A. M. Ades**[2], and **K. B. Nelson**[3]

[1] Laboratory of Developmental Neurobiology, National Institute of Child Health and Human Development, Bethesda, MD,
[2] Department of Neonatology, Children's National Medical Center, Washington, DC, and [3] Neuroepidemiology Branch, National Institute of Neurological Disorders and Stroke, Bethesda, MD, U.S.A.

Summary. Complex interactions and interconnectivity between neurons are hallmarks of normal neuronal differentiation and development. Neurons also interact with other cell types, notably glia, and rely on substances released by glia for their normal function. A deficit in glial response may disturb this critical neuronal-glial-neuronal interaction in Down syndrome (DS), leading to loss of neurons and other defects of development, and contribute to cognitive limitation and early onset of Alzheimer disease.

The hypothesis this paper will discuss is that normal neural development involves an activity-dependent release of substances from neurons, and that these substances act upon glia cells which in turn release substances that influence neurons to promote their survival and development. This glial influence affects cortical neurons and also the subcortical cholinergic neurons that project to the cerebral and hippocampal cortices to maintain cortical neuronal excitability and activity. The neuronal activity stimulates glial secretion of sustaining substances, in a reciprocally interactive cycle. Some aspect of this "virtuous cycle" is deficient in Down syndrome. The result is a small but slowly increasing deficit in activity-dependent support by glia cells which produces a gradually increasing abnormality of cortical and subcortical, perhaps especially cholinergic, function.

Cholinergic neurons as supporters of cortical activity

Cholinergic input to cortex increases neuronal excitability and activity. Figure 1 represents a conceptual model in which cholinergic input to cortex from the basal forebrain, as well as catecholaminergic and serotonergic inputs to other cortical structures, serve to maintain cortical excitatory tone. Muscarinic cholinergic activation of cortical neurons decreases $K+$ conductances to increase excitability (Krnjevic et al., 1971), and such inputs are critical for expression of activity-dependent synaptic plasticity (Bear and Singer, 1986; Gu and Singer, 1989; Hohmann and Berger-Sweeney, 1998). A key feature of this model is

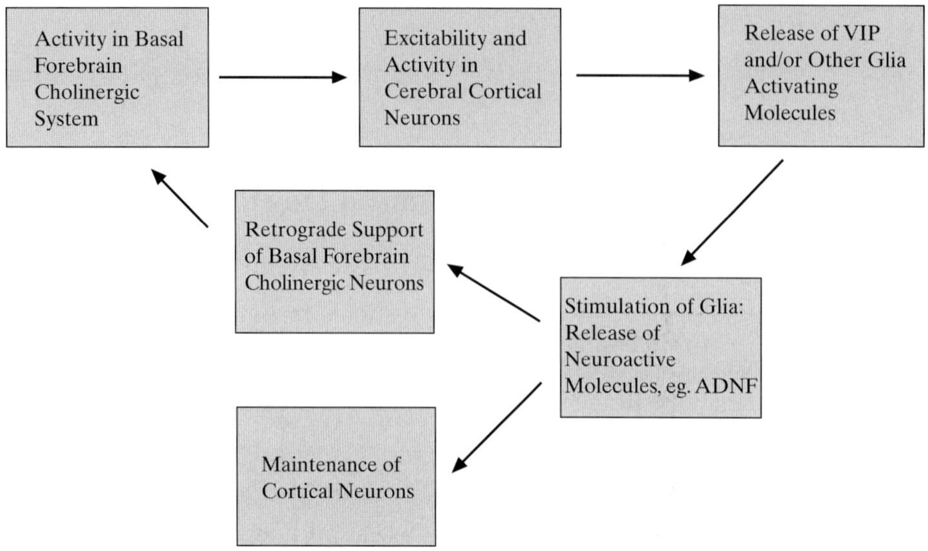

The Virtuous Cycle

Fig. 1. Schematic diagram of the relationships between cortex and basal forebrain, and the role of neural activity and glial function in maintaining normal structure and function of the system

that neuronal activity affects glial function so that glia secrete a number of neuroactive molecules including nerve growth factor (NGF) (Schwartz and Mishler, 1990) and activity-dependent neurotrophic factor (ADNF) (Brenneman, 1986). A prominent mediator of this glia-neuron interaction is vasoactive intestinal peptide (VIP), secreted in an activity dependent manner from neurons that extend across the entire thickness of the cerebral cortex. The function and survival of both cortical neurons and the projecting basal forebrain cholinergic neurons may be increased by the compounds released from the glia, in part because the neurons under the influence of VIP become more resistant to apoptotic influences including toxicity related to glutamate or beta-amyloid precursor protein (Brenneman et al., 1999). Secretion from glia of tumor necrosis factor-alpha, a neurotoxic cytokine, is also decreased under the action of VIP. A representation of these neuron-glia-neuron interactions is shown in Fig. 2.

The neuron-glia interaction and its mediators have been extensively studied in vitro. Those interactions and mediators, and the importance of these for brain development, are outlined next.

Electrical activity and neuronal survival

Electrical activity plays an important role in the regulation of neuronal survival during development (as well as thereafter), and thus affects the structure of nervous tissue (Brenneman et al., 1999). In primary cell cultures, the effect

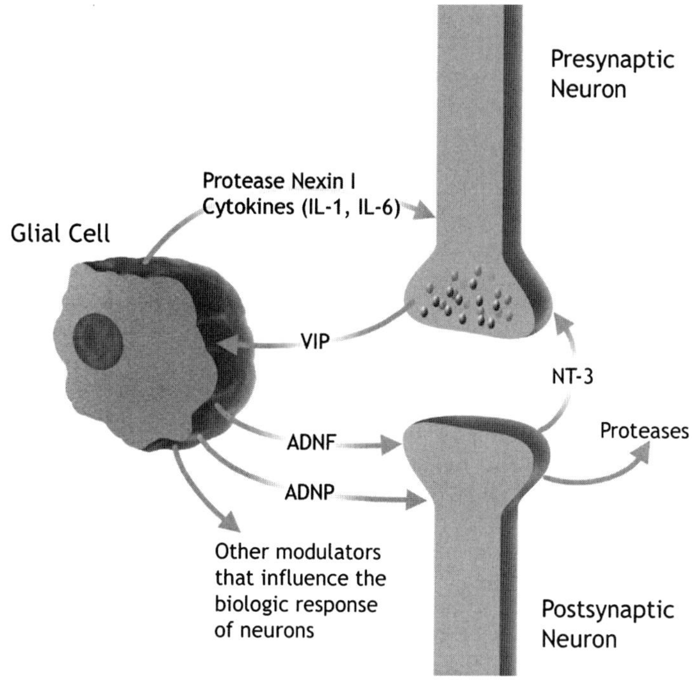

Fig. 2. Cartoon showing some aspects of the interaction between glia and neurons and the importance of electrical activity for this interaction. The release of glia-activating substances from the neurons (e.g. VIP) is dependent on electrical activity in the neurons

of electrical blockade on neuronal survival depends upon substances present in the medium surrounding the neurons. Removing the culture medium into which material was released by previous neuronal action, and then blocking new nerve activity with tetrodotoxin (TTX), results in increased neuronal death. If the media conditioned by previous nerve activity remained, however, neuronal cell death was minimized, whether that death was naturally occurring or was produced by TTX treatment (Brenneman et al., 1999). Studies seeking to identify the substance or substances released by nerve activity that preserved neurons from death found that VIP, a 28 amino acid peptide normally released from storage vesicles in neurons during neural firing, promoted neuronal survival in the face of electrical blockade with TTX (Brenneman, 1986). VIP-induced neuronal survival has been documented in spinal cord (Brenneman, 1986), and hippocampal (Blondel et al., 2000) and cerebral cortical cultures (Gressens et al., 1997). VIP affects a variety of neuronal subtypes including cholinergic neurons (Brenneman, 1986) and those that are NMDA responsive (Brenneman et al., 1990). The effects of VIP are regionally specific and may involve multiple signal transduction pathways.

Other actions of VIP, in addition to its role in neuronal survival, include promotion of cell proliferation, differentiation, and neurite outgrowth; protection against neurotoxic agents including gp120, the envelope protein of the immunodeficiency virus (Brenneman et al., 1998), the beta amyloid protein (Gozes et al., 1996), ethanol (Spong et al., 2001), and glutamate (Brown,

2000), the last-mentioned neuroprotective action being mediated by the potentiation of glutamate uptake by astrocytes and the release of certain cytokines (Brenneman et al., 1999). VIP has potent anti-inflammatory effects mediated through the decreased expression of pro-inflammatory cytokines and nitric oxide, and increase in the anti-inflammatory cytokine IL-10 (Gomariz et al., 2001). In addition, VIP improves the effectiveness of scavenging of reactive oxygen species (Kalfin et al., 1994). These neuroprotective effects are altered by substances that regulate VIP secretion. There appear to be marked changes in expression of VIP at different developmental stages, necessitating consideration of age of the material harvested and perhaps also time in culture.

VIP is important early in embryonic growth, brain and systemic, and brief early blockade of VIP results in permanent microcephaly (Glozner et al., 1999). Thus VIP appears to play a critical role in development by promoting cell proliferation, differentiation, and neurite outgrowth, in addition to its effects upon neuronal survival.

Regulators of VIP action and mediators of its effects

There are several regulators of VIP synthesis including activin, ciliary neurotrophic factor, and transforming growth factor-beta (Symes et al., 2000; Pitts et al., 2001), while VIP release is related to glutamate and to nerve activation. There are cAMP and cytokine response elements located in the promoter region of the VIP gene. Cytokines, including interleukins 1 and 6 and tumor necrosis factor-alpha, strongly induce VIP secretion from lymphocytes (Martinez et al., 1999).

The action of VIP released from neurons is not direct, but operates through neuroactive substances that VIP treatment causes astroglia to secrete (Fig. 2), including activity-dependent neurotrophic factor (ADNF), the related but not identical peptide activity-dependent neuroprotective protein (ADNP), cytokines, the chemokine RANTES, and protease nexin-1 (Brenneman et al., 1999). In all these actions, the powerful effects of VIP are developmentally and dose-regulated. For example IL-1, which promotes neuronal survival at one stage of development, is injurious to neuronal survival at another stage (Brenneman et al., 1995). Of note, the response curve for VIP is bell shaped, so that while the optimal concentration of VIP for sustaining neuronal survival is 30 pM, the survival promoting activity is diminished at concentrations greater than 10-9.

As noted, the neuroprotective activity of VIP is mediated by two peptides that VIP causes astrocytes to release, ADNF and ADNP, both of which are active in femtomolar concentrations (and are inactive at micromolar levels) (Gozes et al., 1999). Actions of these peptides and their even more active fragments include promotion of neuronal survival in cultures from spinal cord, hippocampus, and cerebral cortex.

ADNF causes the release of neurotrophin-3 which along with ADNF regulates NMDA receptor subunits and influences neuronal survival, dendritic length, and synaptogenesis in hippocampus, indicating that the VIP/ADNF/NT-3 pathway regulates early development of glutamatergic synapses (Blondel et al., 2000). In vivo, an ADNP-derived peptide prevents short-term memory deficits and aberrations in developmental reflexes in apolipoprotein E-deficient mice (Gozes et al., 1997). These effects of ADNF and ADNP, derived from astrocytes upon stimulation by VIP, may play important roles in the neuropathology observed in models of Down syndrome.

Abnormalities in developmentally and regionally specific neuronal growth and differentiation may be mediated through abnormal glial responses to VIP, which reduce the effective dose of ADNF and/or ADNP. These processes may be particularly important in the basal forebrain cholinergic system.

Neurons, glia, and cholinergic function in mouse models of DS

A segmental trisomy model, the Ts65Dn mouse, contains many of the genes found in the DS critical region (Davisson et al., 1993), survives into adulthood as the Ts 16 mice do not, and manifests many age-related neuropathologic and cognitive impairments similar to those of human DS. Phenotypically these animals have analogous craniofacial anomalies (Richtsmeier et al., 2000) to human DS and have decreased performance on memory and learning tasks as they age that are reminiscent of the cognitive deficits seen in human DS and AD (Holtzman et al., 1995; Granholm et al., 2000). This deterioration is temporally related to loss of basal forebrain cholinergic neurons, in human DS, AD and in the Ts65Dn mice (Holtzman et al., 1996). These mice also exhibit other abnormalities seen in human DS and AD such as astrocytic hypertrophy (Holtzman et al., 1995, 1996) and hippocampal cell loss (Insausti et al., 1998).

The loss of basal forebrain cholinergic neurons may be related to the presence of extra copies of the amyloid precursor protein, superoxide dismutase, and S100-B genes. In addition DS patients have increased expression of interleukin-1 and the pro-apoptotic genes, p53 and Bax, and downregulation of Bcl-2 and Gap 43 (de la Monte, 1999), factors that interact through glial and neuronal effects to enhance neuronal damage and apoptotic cell death in DS. Increased amyloid precursor protein expression leads to increased formation of B-amyloid which is directly toxic to neurons and involved in plaque formation. In addition amyloid precursor protein gene expression and amyloid precursor protein processing are enhanced by interactions of interleukin-1 and S100-B, both of which are made in astrocytes (Griffin et al., 1998). IL-1 and S100-B are also thought to be toxic to neurons in high concentrations, toxicity of the latter perhaps operating through Bcl-2 (Wang et al., 1999). S100-b through autocrine effects stimulates astrocytic proliferation, increases in free intracellular calcium, and changes in astrocytic morphology. II-1 also has autocrine effects that stimulate microglial expression of itself as well as IL-6 (Griffin et al., 1998). The positive feedback system that results from the DS associated alterations in IL-1 and S100-B expression is compounded by the increased generation of free radicals from b-amyloid deposition and increased oxidative stress from the imbalance of increased SOD1 in the absence of increases in glutathione peroxidase activity in the brain.

As already discussed, in the healthy non-DS brain many toxic effects, such as those of beta-amyloid or excitotoxicity due to excess glutamate, are mitigated by substances that VIP causes to be produced by astrocytes under stimulation by VIP. In the DS brain these protective mechanisms may be blunted. Astrocytes cultured from middle-aged Ts65Dn mice are hyporesponsive to VIP, and the supernatant of such cultures lack the neuroprotective effect ordinarily resulting from VIP stimulation of astrocytes. Binding of VIP was found to be upregulated in cerebral cortex of the Ts65Dn animals, while the VIP mRNA was unchanged, perhaps consistent with a receptor malfunction or with a defect in signalling pathways between receptor and nucleus (Ades et al., 2000).

The concentration of NGF in the hippocampal formation and in basal forebrain cholinergic neurons is decreased (Ford et al., 2000) and axonal transport of NGF impaired in Ts65Dn mice (Delcroix et al., 2000). Both NGF and VIP have roles in protecting neurons through substances produced by glia, so that abnormalities in these substances may contribute to increased neuronal apoptosis. The failure of VIP to induce the normal production from astrocytes of factors (ADNF, ADNP) that decrease vulnerability to beta amyloid toxicity and to oxidative or exicitoxic stress, and the decreasing production of VIP with age may thus be factors in the age-related neuropathologic changes in DS.

Glial-neurotrophic mechanisms in mouse models of DS

We postulate that in DS there is a primary defect in the ability of cortical glia to provide the normal support suggested in Fig. 1, due to inadequate functioning of receptors for VIP on the glia, or to deficiency in the mechanisms coupling VIP action to secretion of trophic materials. The basal forebrain is dependent upon cortical activity-related trophic effects, which may involve a retrograde mechanism. Such retrograde transport of trophins from cortical areas to basal forebrain neurons has been demonstrated (Seiler and Schwab, 1984) and a similar effect has been documented in sympathetic gangion neurons in vitro (Ure and Campenot, 1994) and in vivo (Korsching and Thoenen, 1983). Five sets of experiments in trisomic models of DS in vitro support this the hypothesis of a glial defect.

1. Caserta (1994) found that development of NPY neurons was abnormal in cultures from TS16 animals, with an excessive number of these cells that were morphologically maldeveloped. The effect was reversed by co culture of the TS16-derived neurons with euploid glia cells. Evidence of abnormality was manifested by neurons from nontrisomic animals that were grown on TS16 glia.

2. Nelson et al. (1997) measured cholinergic markers (cholineacetyl transferase [ChAT] and numbers of ChAT immunostained neurons) in cultures

from normal and TS16 mice. ChAT activity and ChAT neuron number were decreased in cultures from TS16 animals. Similar results were obtained when normal neurons were grown with TS16 glia. Furthermore, cholinergic function of TS-Derived neurons could be largely "rescued" by co-culture with glia from normal animals. Thus cholinergic function in these neurons seemed to be largely determined by the genotope of the glia with which they were grown.

3 and 4. Two groups [Muller et al., 1997; Bambrick et al., 1997) have shown abnormal Ca^{++} levels in TS16 glia. The resting intracellular Ca^{++} levels were higher than normal in these glia in culture, and the responses of intracellular calcium to various stimuli were reduced.

5. Perhaps most dramatically, glia cultures from TS65dn mice fail to secrete survival-promoting subtances when stimulated by VIP, in contrast to similar cultures from normal animals (Ades et al., 2000). Secretion of some trophic substances, including nerve growth factor, was paradoxically high, with a somewhat blunted response to VIP. Glia from the trisomic animals were strikingly deficient in this neurosupportive function.

Human DS

A recent study of neuropeptides and neurotrophins in archived neonatal blood of children with autism or mental retardation included four infants with DS (Nelson et al., 2001). Measurement by recycling immunoaffinity chromatography indicated that in all four infants with DS, neonatal concentrations of VIP and of brain derived neurotrophic factor exceeded the concentrations observed in any of 54 control infants. This small initial experience, obviously in need of replication and extension, suggests that further investigation of VIP and of substances it releases from astrocytes, including consideration of developmental stage, may be warranted in clinical DS and in experimental models.

Conclusion

We suggest that failure of DS astrocytes to respond to secretagogues released by activated nerve terminals may lead to diminished availability of sustances that normally sustain neurons, thus interrupting the normal "virtuous cycle" of glial-neurotrophic reciprocal interaction. The result may be a slowly increasing abnormality of cortical and subcortical function, perhaps especially cholinergic function, with resulting aberrations in neurologic development and precocious neurodegenerative change.

References

Ades AM, McCune SK, Crnic LS, Ring M, Brenneman DE, HiLL JM (2000) Vasoactive intestinal peptide dysfunction associated with a segmental trisomy model of Down syndrome [Abstract]. Soc Ped Res 79: 472

Bambrick LL, Golovina VA, Blaustein MP, Yarowsky PJ, Krueger BK (1997) Abnormal calcium homeostasis in astrocytes from the trisomy 16 mouse. Glia 19: 352–358

Bear MF, Singer W (1986) Modulation of visual cortical plasticity by acetylcholine and noradrenaline. Nature 320: 172–176

Blondel O, Collin C, McCarran B, Zhu X, Zamostiano R, Gozes I, Brenneman DE, McKay R (2000) A glia-derived signal regulating neuronal differentiation. J Neurosci 20: 8012–8020

Brenneman DE (1986) Role of electrical activity and trophic factors during cholinergic development in dissociated cultures. Can J Physiol Pharmacol 64: 356–362

Brenneman DE, Yu C, Nelson PG (1990) Multi-determinate regulation of neuronal survival: neuropeptides, excitatory amino acids and bioelectric activity. Int J Dev Neurosci 8: 371–378

Brenneman DE, Hill JM, Glazner GW, Gozes I, Phillips TM (1995) Interleukin-1 alpha and vasoactive intestinal peptide: enigmatic regulation of neuronal survival. Int J Dew Neurosci 13: 187–200

Brenneman DE, Westbrook GL, Fitzgerald SP, Ennist DL, Elkins KL, Ruff MR, Pert CB (1998) Neuronal cell killing by the envelope protein of HIV and its prevention by vasoactive intestinal peptide. Nature 335: 639–642

Brenneman DE, Hauser J, Phillips TM, Davidson A, Bassan M, Gozes I (1999) Vasoactive intestinal peptide: link between electrical activity and glia mediated neurotrophism. Ann NY Acad Sci 897: 17–26

Brown R (2000) Neuronal release of vasoactive intestinal peptide is important to astrocyte protection of neurons from glutamate toxicity. Mol Cell Neurosci 15: 465–475

Caserta MT (1994) Neuropeptide Y immunoreactive neurons in murine trisomy 16 cortical cultures. Plasticity of expression and differentiation Mo1. Chem Neuropathol 22: 197–210

Davisson MT, Schmidt C, Reeves RH, Irving NG, Akeson EC, Harris BS, Bronson RT (1993) Segmental trisomy as a mouse model for Down syndrome. Prog Clin Biol Res 384: 117–133

de la Monte SM (1999) Molecular abnormalities of the brain in Down syndrome, relevance to Alzheimer's neurodegeneration. J Neural Transm [Suppl] 57: 1–19

Delcroix JD, Cooper JD, Howe CL, Belichenko PV, Valleta JS, Lee AAT, Mobley WC (1999) Failure of axonal retrograde 1251-NGF transport by septal neurons in a mouse model of Down syndrome. J Neurosci 19: 523–527

Ford KA, Granholm ARE, Crnic LS, Hyde LA, Seo H, Ferree AW, Isacson O (2000) Alterations in APP and NGF levels in an animal model of Down syndrome. Neurosci Abstr 76: 421

Glazner G, Gressens P, Lee SJ, Gibney G, Gozes I, Gozes Y, Brenneman DE, Hill JM (1999) Activity-dependent neurotrophic factor: a potent regulator of embryonic growth and development. Anat Embryol 200: 65–71

Gomariz RP, Martinez C, Abad C, Leceta J, Delgado M (2001) Immunology of VIP: a review and theapeutical perspectives. Curr Pharm Des 7: 89–111

Gozes I, Bardea A, Reshef A, Zamostiano R, Zhukovsky S, Rubinraut S, Fridkin M, Brenneman DE (1996) Neuroprotective strategy for Alzheimer disease: intranasal administration of a fatty neuropeptide. Proc Natl Acad Sci 93: 427–432

Gozes I, Bachar M, Bardea A, Davidson A, Rubinraut S, Fridkin M, Giladi E (1997) Protection against developmental retardation in apolipoprotein E deficient mice by a fatty neuropeptide: implications for early treatment of Alzheimer's disease. J Neurobiol 33: 329–342

Gozes I, Bassan M, Zamostiano R, Pinhasov A, Davidson A, Giladi E et al. (1999) A novel signaling molecule for neuropeptide action: activity-dependent neuroprotective protein. Ann NY Acad Sci 897: 125–135

Granholm ARE, Sanders LA, Crnic LS (2000) Loss of cholinergic phenotype in basal forebrain coincides with cognitive decline in a mouse model of Down syndrome. Exp Neurol 161: 647–663

Gressens P, Marret S, Hill JM, Brenneman DE, Gozes I, Fridkin M, Evrard P (1997) Vasoactive intestinal peptide prevents excitotoxic cell death in the murine developing brain. J Clin Invest 100: 390–397

Griffin WS, Sheng JG, Royston MC, Gentleman SM, McKenzie JE, Graham DI, Roberts GW, Mark RE (1998) Glial-neuronl Interactions in Alzheimer's disease: the potential role of a 'cytokine cycle' in disease progression. Brain Pathol 8: 65–72

Gu Q, Singer W (1989) The role of muscarinic receptors in ocular dominance plasticity. EXS 57: 305–314

Hohmann CF, Berger-Sweeney J (1998) Cholinergic regulation of cortical development and plasticity, new twists to an old story. Persp Dev Neurobiol 5: 401–425

Holtzman DM, Kilbridge J, Chen K, Rabin J, Luche R, Carlson E, Epstein CJ, Mobley WC (1995) Preliminary characterization of the central nervous system in partial trisomy 16 mice. Prog Clin Biol Res 393: 227–240

Holtzman DM, Santucci D, Kilbridge J, Chua-Couzens J, Fontana DJ, Daniels SE, Johnson RM, Chen K, Sun Y, Carlson E, Alleva E, Epstein CJ, Mobley WC (1996) Developmental abnormalities and age-related neurodegeneration in a mouse model of Down syndrome. Proc Natl Acad Sci 93: 13333–13338

Insausti AM, Megias M, Crespo D, Cruz-Orive LM, Dierssen M, Vallina TF, Insausti R, Florez J (1998) Hippocampal volume and neuronal number in mice: a murine model of Down syndrome. Neurosci Lett 253: 175–178

Kalfin R, Maulik N, Engelman RM, Cordis GA, Milenov K, Kasakov L, Das DK (1994) Protective role of intracoronary vasoactive intestinal peptide in ischemic and reperfused myocardium. J Pharmacol Exp Ther 268: 953–958

Korsching S, Thoenen H (1983) Quantitative demonstration of the retrograde axonal transport of endogenous nerve growth factor, Neurosci Lett 39: 1–4

Krnjevic K, Pumain R, Renaud L (1971) The mechanism of exccitation by acetylcholine in the cerebral cortex. J Physiol 215: 247–268

Martinez C, Delgado M, Abad C, Gomariz RP, Ganea D, Leceta J (1999) Regulation of VIP production and secretion by murine lymphocytes. J Neuroimmunol 93: 126–138

Muller W, Heinemann U, Schuchmann S (1997) Impaired Ca-signaling in astrocytes from the Ts16 mouse model of Down syndrome. Neurosci Lett 223: 81–84

Nelson KB, Grether JK, Croen LA, Dambrosia JM, Dickens BF, Jelliffe LL, Hansen RL, Phillips TM (2001) Neuropeptides and neurotrophins in neonatal blood of children with autism or mental retardation. Ann Neurol 49: 597–606

Nelson PG, Fitzgerald S, Rapoport SI, Neale EA, Galdzicki Z, Dunlap V, Bowers L, van Agoston (1997) Cerrebral cortical astroglia from the trisomy 16 mouse, a model for Down syndrome, produce neuronal cholinergic deficits in cell culture. Proc Natl Acad Sci 94: 12644–12648

Pitts RL, Wang S, Jones EA, Symes AJ (2001) Transforming growth factor beta and ciliary neurotrophic factor synergisticly induce vasoactive intestinal peptide gene expression through the cooperation of Smad, Stat and AP-1 sites. J Biol Chem (Epublication ahead of print)

Richtsmeier JT, Baxter LL, Reeves RH (2000) Parallels of craniofacial caldevelopment in Down syndrome and Ts65Dn mice. Dev Dynam 17: 137–145

Schwartz JP, Mishler K (1990) Beta-adrenergic receptor regulation, through cyclic AMP, of nerve growth factor expression in rat cortical and cerebellar strocytes. Cell Mol Neurobiol 10: 447–457

Seiler M, Schwab ME (1984) Specific retrograde transport of nerve growth factor (NGF) from neocortex to nucleus basalis in the rat. Brain Res 300: 33–39

Spong C, Abebe DT, Gozes I, Brenneman DE, Hill JM (2001) Prevention of fetal demise and growth restriction in a mouse model of fetal alcohol syndrome. J Pharmacol Exp Ther 297: 774–779

Symes AJ, Pitts RL, Conover J, Kos K, Coulombre J (2000) Synergy of activin and ciliary neurotrophic factor signaling pathways in the induction of vasoactive intestinal peptide gene expression. Mol Endocrinol 14: 429–439

Ure DR, Campenot RB (1994) Leukemia inhibitory factor and nerve growth factor are retrogratedly transported and processed by cultured rat sympathetic neurons. Dev Biol 162: 339–347

Wang S, Rosengren LE, Franlund M, Hamberger A, Haglid KG (1999) Bcl 2 expression regulates cell sensitivity to S100-b mediated apoptosis. Brain Res Mol Brain Res 0: 167–176

Authors' address: P. G. Nelson, MD, PhD, Laboratory of Developmental Neurobiology, NICHD, National Institutes of Health, Building 49, Room 5A08, Bethesda, MD 20892-4480, U.S.A., e-mail: pgnelson@codon.nih.gov

Aberrant expression of dihydropyrimidinase related proteins-2,-3 and -4 in fetal Down Syndrome brain

R. Weitzdoerfer[1], **M. Fountoulakis**[2], and **G. Lubec**[1]

[1] Department of Pediatrics, University of Vienna, Vienna, Austria
[2] Gene Technologies, F. Hoffmann-La Roche, Basel, Switzerland

Summary. Pathfinding of growing axons to reach their target during brain development is a subtle process needed to build up contacts between neurons. Abnormalities in brain development in Down Syndrome (DS) are described in a couple of morphological reports but the molecular mechanisms underlying abnormal wiring in fetal DS brain are not yet elucidated. We therefore performed a study using the proteomic approach to show differences in protein levels involved in the guidance of axons between control and DS brain in early prenatal life. Proteins obtained from autopsy of human fetal abortus were applied on 2-dimensional gel, identified and quantified. We quantified 5 members of the semaphorin/collapsin family, the dihydropyrimidinase related proteins 1–4 and the collapsin response mediator protein-5 (CRMP-5) in 8 DS and 7 control cortex samples.

DRP-1 and CRMP-5 levels were comparable in the control and DS samples.

Evaluation of DRP-2, DRP-3 and DRP-4 revealed significantly decreased levels of 2 of the 15 spots assigned to DRP-2 and increased levels of one spot assigned to DRP-3 and increased DRP-4 in DS brain.

We conclude that as early as from the 19th week of gestation pathfinding cues of the outgrowing axons are impaired in DS. These findings may help to elucidate mechanisms leading to abnormalities in neural migration of DS brain.

Introduction

Information on the mechanisms involved in abnormal development of fetal brain in Down Syndrome (DS) is limited.

Abnormal synaptogenesis, dendritic spines, neuronal differentiation and migration in developing DS brain have been reported in morphological studies (Becker et al., 1991; Epstein, 1995; Marin-Padilla, 1976; Wisniewski and Bobinski, 1991; Wisniewski and Kida, 1994).

One of the underlying mechanisms necessary to build the network of connections between neurons during brain development is the guidance of

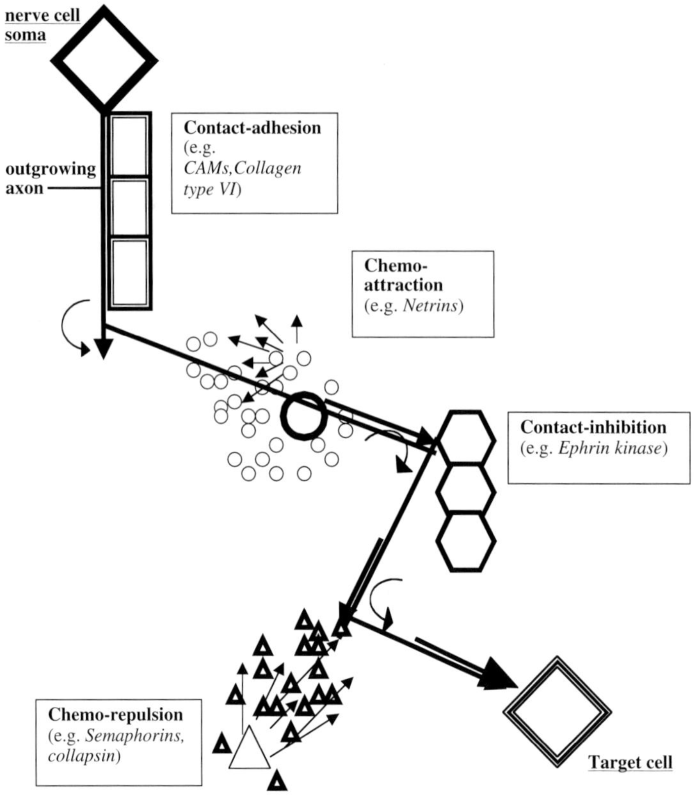

Fig. 1. Signals guiding the growing axon to its target can be separated into cell-adhesion signals, chemo-attraction, contact-inhibition and chemo-repulsion. Cell-adhesion is provided by e.g. cell-adhesion molecules (CAMs) and collagen type VI, suggested to play a role in cell-cell-adhesion and cell migration (Knupp and Squire, 2001; Trüeb and Bornstein, 1984) and revealed to be decreased in fetal DS brain (Engidawork et al., 2001) enlarging our findings of impaired levels of proteins involved in axon guidance. Chemo-attractants such as netrins build up a concentration gradient attracting the growth cone in direction of the highest gradient. Activation of e.g. ephrin kinase by contact of the growth cone with a cell leads to a turn-away of the axon. Semaphorins and collapsin belong to a group of proteins using chemo-gradients to repel the outgrowing axon by inducing growth cone collapse and turn of the recovered growth cone

axons to their appropriate target. This process is based on the function of a large number of molecular guidance cues signalling the correct path to the growing axon.

Different types of guidance signals such as cell-adhesion, chemo-attraction, contact-inhibition and chemo-repulsion are provided by these cues (see Fig. 1), received by the growth cone on tip of the growing axon and therefore directing the axon to the target cell. E.g. in presence of cell-adhesion molecules (CAMs) the axon attaches to a cell surface and extends along this border (Tessier-Lavigne and Goodman, 1996). Chemo-attractant factors, such as netrins, build up a concentration gradient along which the axon grows (Sanes and Jessell, 2000). In contrast to these guidance cues there exist molecules repelling the growth cone. This is provided either by contact-inhibition, e.g. activation of ephrin kinase, present on the growth cone is answered by a turn away of the growth cone, or collapse of the growth cone is induced followed by recovery and turn, a mechanism mediated for instance by collapsin a member of the semaphorin family (Nakamura et al., 2000).

We decided to study the levels of guidance molecules of the semaphorin/collapsin family in fetal control and DS brains using 2-dimensional (2-D) gel electrophoresis with subsequent matrix-associated laser desorption ionisation mass spectrometry (MALDI-MS), unambiguously identifying 5 members of the semaphorin/collapsin family, the dihydropyrimidinase related proteins 1–4 (DRP 1–4) (Hamajima et al., 1996; Wang and Strittmatter, 1997; Kato et al., 1998) and collapsin response mediator protein-5 (CRMP-5) (Fukada et al., 2000).

The aim of our study was to show whether proteins involved in the pathfinding mechanisms of the outgrowing axon were impaired in fetal DS brains at about the 19th week of gestation. We observed decreased levels of 2 spots assigned to DRP-2 and increased levels of one spot assigned to DRP-3 and increased levels of DRP-4 in DS brain as early as the 19th week of gestation indicating impaired axonal outgrowth and pathfinding in the development of DS brain.

Materials and methods

Fetal brain tissue and brain preparation

Abortus fetal brain sample (cerebral cortex) of controls (n = 7; 1 female and 6 males with mean age of 18.8 ± 2.2 weeks) and DS (n = 8; 2 females and 6 males with mean age of 19.8 ± 2.0 weeks) were obtained from Drs. M. Dierssen and J. C. Ferreres the Medical and Molecular Genetics Center-IRO, Hospital Duran i Reynals, Barcelona, Spain. All samples were stored at −70°C and the freezing chain was never interrupted. Brain tissue was suspended in 0.5 ml of sample buffer consisting of 40 mM Tris, 5 M urea (Merck, Darmstadt, Germany), 2 M thiourea (Sigma, St. Louis, MO, USA), 4% CHAPS (Sigma), 10 mM 1,4-dithioerythriol (Merck), 1 mM EDTA (Merck), 1 mM PMSF (Sigma) and 1 µl/mg of each pepstatin A, chymostatin, leupeptin and antipain. The suspension was sonicated for approximately 30 seconds and centrifuged further at 150,000 × g for 45 minutes to sediment undissolved material. The protein concentration of the supernatant was determined by the Coomassie blue method (Bradford, 1976).

Two-dimensional gel electrophoresis

Two-dimensional gel electrophoresis was performed essentially as reported (Langen et al., 1997). Samples of 2 mg protein were applied on immobilized pH 3–10 nonlinear gradient strips in sample cups at their basic and acidic ends. Focusing started at 200 V and the voltage was gradually increased to 5,000 V at 3 V/min and kept constant for a further 24 h (approximately 1,800,000 kVh totally). The second-dimensional separation was performed on 9–16% SDS gradient polyacrylamide gels. The gels (180 × 200 × 1.5 mm) were run at 40 mA per gel. After protein fixation for 12 h in 40% methanol, containing 5% phosphoric acid, the gels were stained with colloidal Coomassie blue (Novex, San Diego, CA, USA) for 48 h. Molecular masses were determined by running standard protein markers (Gibco, Basel, Switzerland), covering the range 10–200 kDa. pI values were used as given by the supplier of the IPG strips. Excess of dye was washed out from the gels with H_2O and the gels were scanned in an Agfa DUOSCAN densitometer (resolution 200). Electronic images of the gels were recorded using Photoshop (Adobe) and PowerPoint (Microsoft) software.

Matrix-assisted laser desorption ionization mass spectroscopy (MALDI-MS)

MALDI-MS analysis was performed as described elsewhere (Fountoulakis and Langen, 1997) with some modifications. The spots were excised with a spot picker and placed into 96-well microtiter plates. Each spot was destained with 100 μl of 30% acetonitrile in 50 mM ammonium bicarbonate and dried in a speedvac evaporator. Each dried gel piece was rehydrated with 4 μl of 3 mM Tris-HCl, pH 9.0, containing 50 ng trypsin (Promega, Madison, WI, USA). After 16 h at room temperature, 7 μl of H_2O were added to each gel piece and the samples were shaken in for 10 min. Four μl of 50% acetonitrile, containing 0.3% trifluoroacetic acid and the standard peptides des-Arg-bradykinin (Sigma, 904.4681 Da) and adrenocorticotropic hormone fragment 18–39 (Sigma, 2,465.1989 Da), in water were added to each gel piece and shaken for 10 min. The application of the samples was performed with a SymBiot I sample processor (PE Biosystems, Framingham, MA, USA). 1.5 μl of the peptide mixture was simultaneously applied with 1 μl of matrix, consisting of a saturated solution of α-cyano-4-hydroxycinnamic acid (Sigma) in 50% acetonitrile, containing 0.1% trifluoroacetic acid. Samples were analyzed in a time-of-flight mass spectrometer (Reflex 3, Bruker Analytics, Bremen, Germany). An accelerating voltage of 20 kV was used. Peptide matching and protein searches were performed automatically. The peptide masses were compared to the theoretical peptide masses of all available proteins from all species. Monoisotopic masses were used and a mass tolerance of 0.0025% was allowed. The algorithm used for determining the probability of a false positive match with a given MS-spectrum is described elsewhere (Berndt et al., 1999).

Quantification of proteins

Protein spots were outlined (first automatically and then manually) and quantified using the ImageMaster 2D Elite software (Amersham Pharmacia Biotechnology). The percentage of the volume of the spots representing a certain protein was determined in comparison with the total proteins present in the 2-D gel.

Statistical analysis

Software calculated the percentage of spot volume in each particular gel after subtraction of background values. Between-group differences were calculated by non-parametric Mann-Whitney U-test. Level of significance was set at $P < 0.05$. If the spot of a protein was not detectable a value of 0.01 was inserted to make statistical analysis possible.

Results

Identification and quantification of DRP-1,-2,-3 and -4 and CRMP-5

Protein spots were analyzed by MALDI-MS, following in-gel digestion. Peptide masses were matched with the theoretical peptide masses of all proteins of the SWISS-PROT database. We used internal standards to correct the

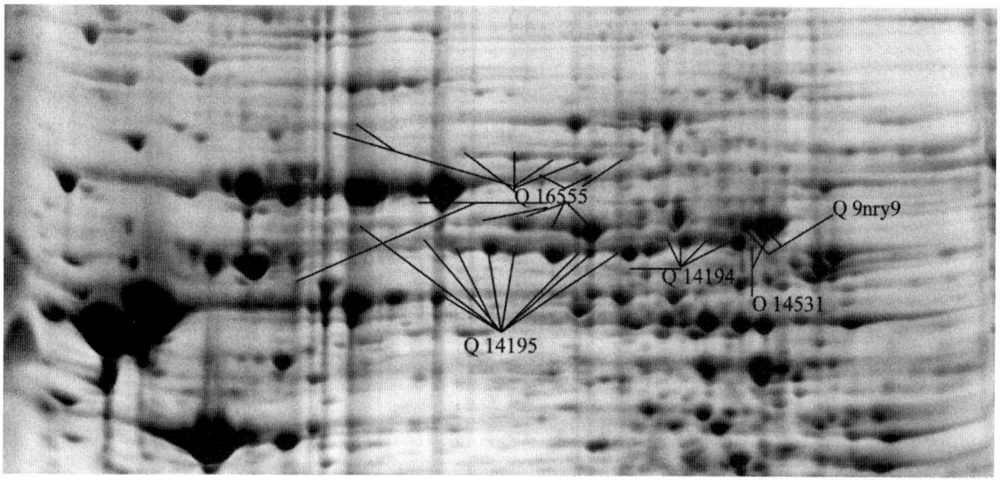

A Fetal sample

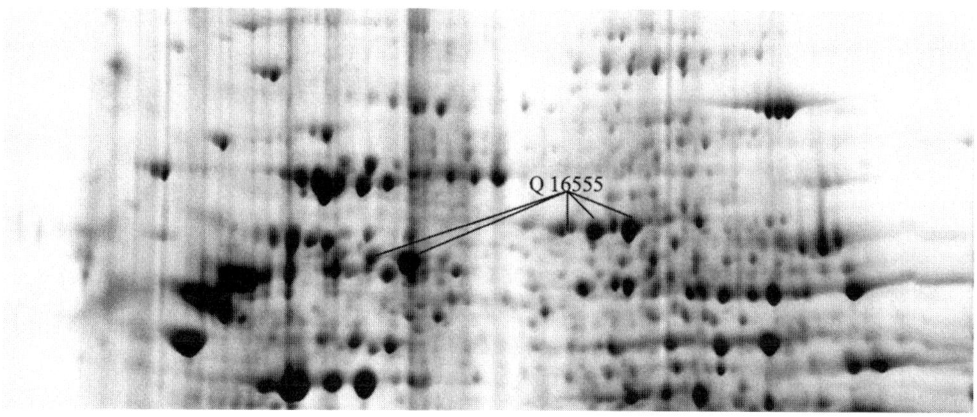

B Adult sample

Fig. 2. A Identification of proteins measured in a fetal control brain: *Q 14194*: Dihydropyrimidinase related protein-1, *Q 16555*: Dihydropyrimidinase related protein-2, *Q 14195*: Dihydropyrimidinase related protein-3, *Q 14531*: Dihydropyrimidinase related protein-4, *Q 9nry9*: Collapsin response mediator protein-5. **B** Dihydropyrimidinase related protein-2 identification in an adult control brain. In contrast to Fig. 2A only few spots represent this protein in adult samples

measured peptide mass thus reducing the windows of mass tolerance and increasing the confidence of identification. In this study 5 proteins were quantified using the Image Master 2D Elite software (see Fig. 2A). Expression levels of proteins were determined as a percentage of spot volume in the area of interest. SWISS-PROT accession numbers and names of the quantified proteins with MALDI information are given in Table 1.

The levels of 5 spots assigned to DRP 1 were comparable in DS and control brains.

DRP 2 was represented by 15 spots. 2 of these spots (detected in 7 control and 7 DS samples) were significantly reduced in DS brain as compared to controls (P = 0.018; P = 0.011).

Table 1. Identification (SWISS PROT), synonyms and homologues of proteins

Protein name	Accession number	pI	kDa	Matches	Probability	Synonyme	Homologues
Dihydropyrimidinase related protein-1 (DRP 1)	Q 14194	7.02	62.486	12	1.13E-20	Collapsin response mediator protein-1 (CRMP-1)	
Dihydropyrimidinase related protein-2 (DRP 2)	Q 16555	6.34	62.710	11	8.52E-18	Collapsin response mediator protein-2 (CRMP-2)	99% homology to rat TOAD-64 (turned on after division, 64 kDa), 98% to mouse UNC-33 and 98% to chicken CRMP-62
Dihydropyrimidinase related protein-3 (DRP 3)	Q 14195	6.46	62.323	13	1.80E-22	Unc-33 like phosphoprotein (ulip protein)	98% homology to mouse Ulip
Dihydropyrimidinase related protein-4 (DRP-4)	O 14531	7.21	62.493	9	1.66E-12	Ulip4 protein	
Collapsin response mediator protein-5	Q 9nry9	7.31	61.990	6	3.82E-08	—	

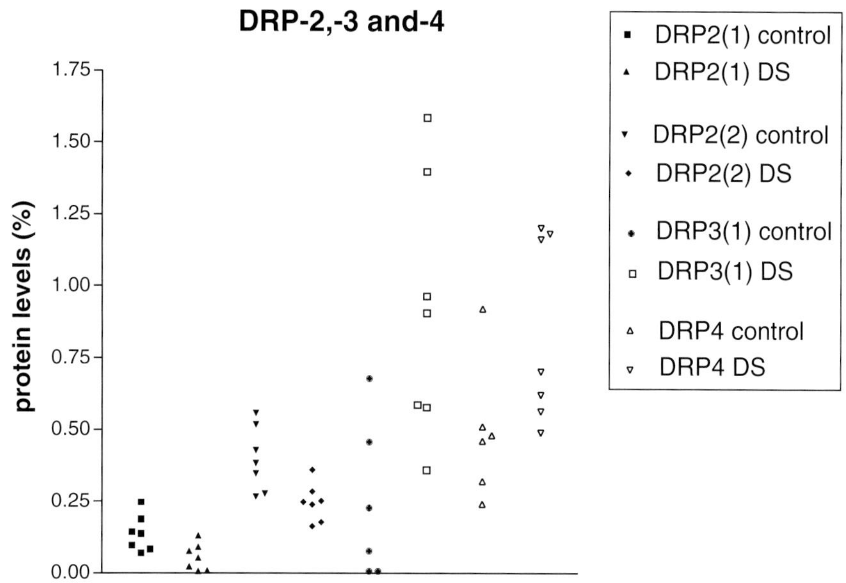

Fig. 3

9 spots represented the protein DRP 3 but only 7 of them could be quantified. One of the spots (detected in 6 control and 7 DS samples) showed significantly increased levels in DS brain compared to controls (P = 0.014).

DRP 4 (detected in 6 control and 7 DS brains) was represented by two spots but only one single spot could be quantified as the other one could not be clearly separated. Quantification revealed increased levels of DRP 4 in DS brains as compared to controls (P = 0.022).

Levels of CRMP-5 were comparable in DS and control samples.

Means and SD of all protein levels are given in Table 2.

Scattergrams of data of the proteins with significantly decreased or increased levels in DS samples as compared to controls are given in Fig. 3.

Discussion

For the sake of clarifying the everchanging nomenclature, synonyms and homologues, the reader is referred to Table 1.

The growth cone which is a sensory-motor structure at the tip of the growing axon plays a central role in axonal pathfinding receiving various guidance information from its surrounding and dynamically responds to them to decide growth direction to find the appropriate target (Sanes and Jessel, 2000). Molecules of different families controlling this process have been identified attracting or repelling the growth cone. Repelling signals which help to guide the growth cone to its destination are provided by e.g. ephrins and members of the semaphorin/collapsin family. Collapsin was identified as a

Table 2. Means ± SD of protein levels

Spot	1	2	3	4	5	6	7
DRP 1							
Control	0.66 ± 0.56	0.40 ± 0.24	9.61 ± 6.78	3.85 ± 3.19	5.51 ± 3.26		
DS	0.67 ± 0.53	0.77 ± 0.46	5.83 ± 2.35	3.03 ± 1.05	9.33 ± 2.27		
DRP 2							
Control	0.13 ± 0.09	0.13 ± 0.08	0.23 ± 0.17	0.62 ± 0.52	0.33 ± 0.22	0.65 ± 0.49	0.14 ± 0.06
DS	0.11 ± 0.03	0.13 ± 0.06	0.16 ± 0.12	0.41 ± 0.32	0.19 ± 0.07	0.28 ± 0.03	0.06 ± 0.05*
DRP 3							
Control	0.40 ± 0.29	0.12 ± 0.11	1.55 ± 1.44	0.84 ± 0.37	0.62 ± 0.22	2.42 ± 1.00	0.25 ± 0.27
DS	0.83 ± 0.77	0.10 ± 0.06	1.08 ± 0.72	0.62 ± 0.45	0.55 ± 0.40	1.81 ± 0.78	0.91 ± 0.45*
DRP 4							
Control	0.49 ± 0.24						
DS	0.84 ± 0.32*						
CRMP 5							
Control	3.90 ± 1.2						
DS	4.23 ± 1.00						

* $P < 0.05$

protein in brain with repulsive function to axons, inducing collapse of their growth cones (Luo et al., 1993) and there is now an increasing number of collapsin response mediator proteins: Goshima et al. identified a protein mediating collapsin induced growth cone collapse in developing chick nervous system. This intracellular protein, named collapsin response mediator protein (CRMP)-62, is related to UNC-33 (Goshima et al., 1995), a neural protein in nematodes which, if mutated leads to abnormalities in axonal outgrowth and to uncoordinated movements. Fukada et al. (2000) cloned CRMP5 as a novel member of this family. In the mouse it is evenly distributed predominantly in the developing CNS and localized to chromosome 5, which is not syntenic with human chromosome 21. It was not described so far in the human brain and no difference in the expression of CRMP5 in fetal control or DS brain was observed. No specific function can be assigned to this molecule and only interaction between CRMP family members has been described.

1996 Hamajima et al. introduced a novel gene family defined by dihydropyrimidinase and three related proteins with differential tissue distribution and function and called them DRP-family and synonyms for this group of proteins are (TOAD/Ulip/CRMP) TUC-family (Qinn et al., 1999). Members of the DRP family are involved in the axonal outgrowth and pathfinding through the transmission and modulation of extracellular signals (Goshima et al., 1995; Minturn et al., 1995; Byk et al., 1996). One of these signals is mediated by proteins of the collapsin/semaphorin family which is necessary to create the complex pattern of neuronal connectivity in collaboration with their receptor, neuropilin (He and Rosenfeld, 1991; Tessier-Lavigne and Goodmann, 1996).

DRP-1 is encoded on chromosome 4 and ubiquitously expressed in the human brain. No function, however, could be assigned to this protein yet and the tentative function is that of the other family members, i.e. involvement in neuronal migration and pathfinding. We unambiguously identified it in the

8	9	10	11	12	13	14	15	sum
								17.58 ± 13.16
								19.53 ± 2.56
0.40 ± 0.11	0.26 ± 0.13	3.04 ± 2.06	0.93 ± 0.79	0.97 ± 0.71	0.34 ± 0.26	0.12 ± 0.07	0.19 ± 0.11	8.46 ± 4.87
0.25 ± 0.07*	0.29 ± 0.27	1.74 ± 0.61	0.39 ± 0.12	0.41 ± 0.17	0.10 ± 0.04	0.05 ± 0.04	0.25 ± 0.10	4.73 ± 1.35
								6.11 ± 2.03
								5.90 ± 1.36

human fetal cortex represented by five spots. Quantification revealed that DRP-1 was comparable between controls and DS cortex.

Another member of this family, the dihydropyrimidinase related protein 2 (DRP-2), also referred to as collapsin response mediator protein 2 (CRMP-2), is a cytosolic protein with 99% homology to rat TOAD-64 (turned on after division, 64 kDa), 98% to mouse UNC-33 and 98% to chicken CRMP-62 (Hamajima et al., 1996), proteins known to take part in pathfinding and migration of neurons (Minturn et al., 1995; Goshima et al., 1995).

In our study we found DRP-2 represented by 15 spots and two spots (see Fig. 4A and 4B) were significantly reduced in DS brain. These two spots may reflect post-translationally modified DRP-2 proteins and the cause for the decrease in levels remains unclear. Differences in post-mortem time, storage time or gestation weeks are highly unlikely (data not shown). Dysregulated DRP-2 in adult DS brain has been described by our group previously (Lubec et al., 1999). As the study group of adult DS patients all showed Alzheimer's disease neuropathology and the DRP-2 — homologue CRMP-2 has been described to be associated with neurofibrillary tangles in Alzheimer's disease (Yoshida et al., 1997) we could not rule out that the finding was associated with neurodegeneration rather than with DS. It should also be mentioned that in our study on DRP-2 in fetal brain we found 15 proteins whereas in our observation on adult brain, only 5 DRP-2 proteins were observed, thus assigning a strong developmental role to these proteins (see Fig. 2A and 2B).

Our current finding of deteriorated DRP-2, however, provides evidence for a primary pathogenetic role of DRP-2 in DS. The altered DRP-2 proteins may well be responsible for the afore mentioned neuronal migration impairment observed in DS. A more precise deranged function could be postulated by the findings of Arimura and co-workers who identified the DRP-2 homologue CRMP-2 as a substrate of the GTPase Rho-kinase in brain and showed that phosphorylation of CRMP-2 by Rho-kinase plays partially a role in

CONTROL DS

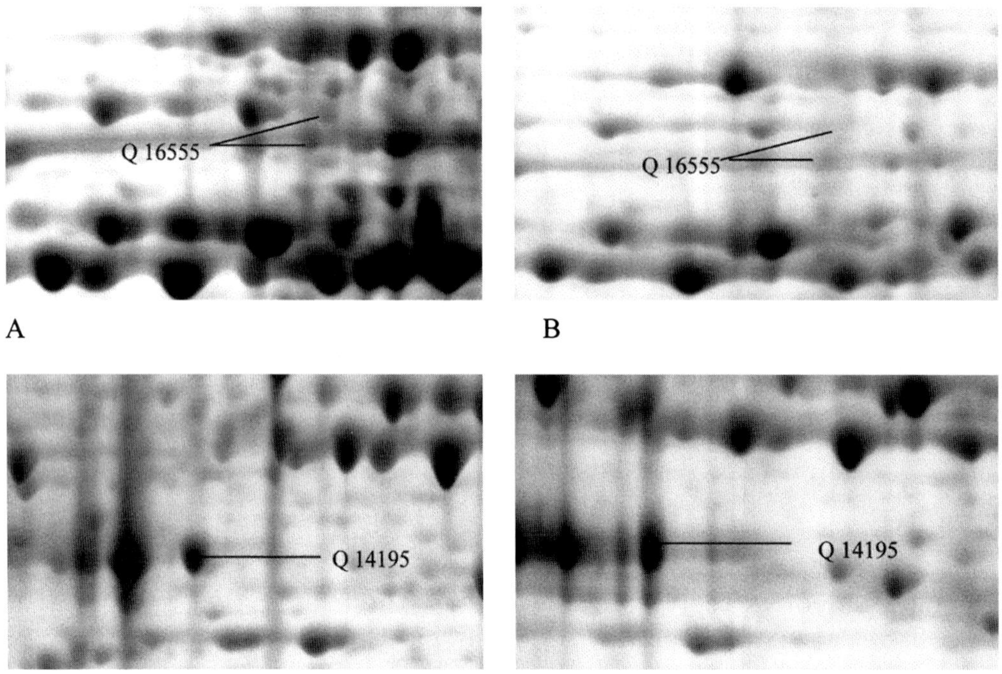

A B

C D

Fig. 4. **A** Two spots of the dihydropyrimidinase related protein-2 in fetal control brain. **B** Levels of these two spots of the dihydropyrimidinase related protein-2 in fetal DS brain were significantly decreased. **C** One spot of the dihydropyrimidinase related protein-3 in fetal control brain. **D** Protein level of this spot of the dihydropyrimidinase related protein-3 was significantly increased in fetal DS brain

morphology of the growth cone (Arimura et al., 2000). In a recent publication evidence for the involvement of microtubule dynamics of DRP-2 homologue CRMP-2 protein was provided (Gu and Ihara, 2000). This finding would be in agreement with the (unpublished results) observation of altered cytoskeleton structures in fetal DS.

Koyama localized a portion of the coding region of DRP-2 to human chromosome 8p21 (Koyama et al., 1995) while another portion of the coding DRP-2 sequence was aligned to chromosome 21. We have already challenged the gene dosage hypothesis (Greber et al., 1999) showing that several proteins encoded on chromosome 21 were not increased in trisomy 21 and decrease of two DRP-2 structures may well be compatible with chromosomal imbalance. Furthermore, deranged DRP-2 expression may affect the neuropilin — collapsin — receptor system as previous studies have shown the coordinated expression of DRP-2 and neuropilin (Pasterkamp et al., 1998).

Seven spots for DRP-3 were observed and one was significantly — more than threefold-increased (Fig. 4C and 4D). Matsuo et al. (2000) mapped the Ulip gene to chromosome 5q32. As nothing is described in terms of specific function we only can suggest that the calculated increase would interfere with the concerted action of other DRP's and cause and/or reflect abnormal neuronal migration.

DRP-4 showing several inconsistently expressed spots left only a single spot on 2-D gel electrophoresis for evaluation. DRP-4 also can be considered an orphan as no information on it is available and again, we simply can speculate on a tentative function as a guidance cue in neuronal migration. It was significantly elevated in fetal cortex of DS patients and thus may be contributing to aberrant axonal outgrowth and pathfinding in DS brain.

Taken together, we show that DRPs are dysregulated early in human life with the biological meaning of aberrant migration, pathfinding and signalling in fetal Down Syndrome, although we cannot entirely exclude that deranged DRPs are only representing the abnormal migrational processes and are not leading to abnormal pathfinding and wiring of the brain.

Acknowledgment

We are highly indebted to the Red Bull Company (Salzburg, Austria) for generous financial support.

References

Arimura N, Inagaki N, Chirara K, Menager C, Nakamura N, Amanot M, Iwamatsu A, Goshima Y, Kaibuchi K (2000) Phosphorylation of collapsin response mediator protein-2 by Rho-kinase. J Biol Chem 275: 23973–23980

Becker L, Mito T, Takashima S, Onodera K (1991) Growth and development of the brain in Down syndrome. In: Epstein CJ (ed) The morphogenesis of Down syndrome. Wiley-Liss, New York, p 133

Berndt P, Hobolm U, Langen H (1999) Reliable automatic protein identification from matrix-assisted laser desorption/ionisation mass spectrometric peptide fingerprints. Electrophoresis 20: 3521–3526

Bradford M (1976) A rapid and sensitive method for the quantitation of microgram quantities of protein utilizing the principle of protein-dye binding. Anal Biochem 72: 248–254

Byk T, Dobransky T, Cifuentes-Diaz C, Sobel A (1996) Identification and molecular characterization of unc-33-like phosphoprotein (Ulip), a putative mammalian homolog of the axonal guidance-associated unc-33 gene product. J Neurosci 16: 688–701

Engidawork E, Balic N, Fountoulakis M, Dierssen M, Lubec G (2001) β-Amyloid precursor protein, ETS-2 and collagen alpha 1 (VI) precursor, encoded on chromosome 21, are not overexpressed in fetal Down Syndrome: further evidence against gene dosage effect. J Neural Transm (this volume)

Epstein CJ (1995) Down syndrome (Trisomy 21). In: Scriver CR, Beaudet AL, Sly WS, Valle D (eds) The metabolic and molecular basis of inherited disease. McGraw Hill, New York, pp 749–794

Fountoulakis M, Langen H (1997) Identification of proteins by matrix-assisted laser desorption ionisation-mass spectrometry following in-gel digestion in low-salt, nonvolatile buffer and simplified peptide recovery. Anal Biochem 250: 153–156

Fukada M, Watakabe I, Yuasa-Kawada J, Kawachi H, Kuriowa A, Matsuda Y, Noda M (2000) Molecular characterization of CRMP5, a novel member of the collapsin response mediator protein family. J Biol Chem 275: 37957–37965

Goshima Y, Nakamura F, Strittmatter P, Strittmatter S (1995) Collapsin-induced growth cone collapse by an intracellular protein related to unc-33. Nature 376: 509–514
Greber-Platzer S, Shatzmann-Turhani D, Wollenek G, Lubec G (1999) Evidence against the current hypothesis of "gene dosage effect" of trisomy 21: ets-2, encoded on chromosome 21" is not overexpressed in hearts of patients with Down Syndrome. Biochem Biophys Res Commun 254: 395–399
Gu Y, Ihara Y (2000) Evidence that collapsin response mediator protein-2 is involved in the dynamics of microtubules. J Biol Chem 275: 17917–17920
Hamajima N, Matsuda K, Sakata S, Tamaki N, Sasaki M, Nonaka M (1996) A novel gene family defined by human dihydropyrimididnase and the three related proteins with differential tissue distribution. Gene 180: 157–163
He X, Rosenfeld MG (1991) Mechanisms of complex transcriptional regulation: implications for brain development. Neuron 17: 183–196
Kato Y, Hamajima N, Inagaki H, Okamura T, Kojin T, Sasaki M, Nonaka M (1998) Postmeiotic expression of the mouse dihydrpyrimidinase-related protein 3 (DRP3) gene during spermiogenesis. Mol Reprod Dev 51: 105–111
Knupp C, Squire JM (2001) A new twist in the collagen story-the type VI segmented super coil. EMBO J 20: 372–376
Koyama K, Sudo K, Nakamura Y (1995) Isolation of 115 human chromosome 8-specific expressed-sequence tags by exon amplification. Genomics 26: 245–253
Langen H, Roder D, Juranville J, Fountoulakis M (1997) Effect of protein application mode and acrylamide concentration on the resolution of protein spots separated by two-dimensional gel electrophoresis. Electrophoresis 18: 2085–2090
Lubec G, Nonaka M, Krapfenbauer K, Gratzer M, Cairns N, Fountoulakis M (1999) Expression of the dihydropyrimidinase related protein 2 (DRP-2) in Down syndrome and Alzheimer's disease brain is downregulated at the mRNA and dysregulated at the protein level. J Neural Transm [Suppl] 57: 161–177
Luo Y, Raible D, Raper J (1993) Collapsin: a protein in brain that induces the collapse and paralysis of neuronal growth cone. Cell 75: 217–227
Marin-Padilla M (1976) Pyamidal cell abnormalities in the motor cortex of a child with Down syndrome. J Comp Neurol 167: 63–75
Matsuo T, Staffer JK, Walker RL, Meltzer P, Thiele CJ (2000) Structure and promotor analysis of the human unc-33-like phosphoprotein gene: E-box required for maximal expression in neuroblastoma and myeloblasts. J Biol Chem 275: 16560–16568
Minturn JE, Fryer HJ, Geschwind DH, Hockfield S (1995) TOAD-64, a gene expressed early in neuronal differentiation in the rat, is related to unc-33, a C. elegans gene involved in axon outgrowth. J Neurosci 15: 6757–6766
Nakamura F, Kalb RG, Strittmatter SM (2000) Molecular basis of semaphorin-mediated axon guidance. J Neurobiol 44: 219–229
Pasterkamp RJ, De Winter F, Holtmaat AJ, Verhagen J (1998) Evidence for a role of the chemorepellent semaphoring III and ist receptor neuropilin-1 in the regeneration of primary olfactory axons. J Neurosci 18: 9962–9976
Quinn C, Gray G, Hockfield S (1999) A family of proteins implicated in axon guidance and outgrowth. J Neurobiol 41: 158–164
Sanes JR, Jessell TM (2000) The guidance of axons to their targets. In: Kandel ER, Schwartz JH, Jessell TM (eds) Principles of neural science. McGraw Hill, New York, pp 1063–1086
Tessier-Lavigne M, Goodman CS (1996) The molecular biology of axon guidance. Science 274: 1123–1133
Trüeb B, Bornstein P (1984) Characterization of the precursor form of type VI collagen. J Biol Chem 259: 8597–8604
Wang LH, Strittmatter SM (1997) Brain CRMP forms heterotetramers similar to liver dihydropyrimidinase. J Neurochem 69: 2261–2269
Wisniewski KE, Bobinski M (1991) Hypothalamic abnormalities in Down syndrome. In: Epstein CJ (ed) The morphogenesis of Down syndrome. Wiley-Liss, New York, p 133

Wisniewski KE, Kida E (1994) Abnormal neurogenesis and synaptogenesis in Down syndrome brain. Dev Brain Dysfunct 7: 289–301
Yoshida H, Watanabe A, Ihara Y (1998) Collapsin response mediator protein-2 is associated with neurofibillary tangles in Alzheimer's disease. J Biol Chem 273: 9761–9768

Authors' address: Prof. Dr. G. Lubec, Department of Pediatrics, University of Vienna, Währinger Gürtel 18, A-1090 Vienna, Austria, e-mail: gert.lubec@akh-wien.ac.at

Decreased protein levels of complex I 30-kDa subunit in fetal Down syndrome brains

S. H. Kim[1], **M. Fountoulakis**[2], **M. Dierssen**[3], and **G. Lubec**[1]

[1] Department of Pediatrics, University of Vienna, Austria
[2] F. Hoffmann — La Roche, Basel, Switzerland
[3] Medical and Molecular Genetics Center-IRO, Hospital Duran i Reynals, Barcelona, Spain

Summary. Defects of mitochondrial electron transport enzymes have been implicated in the pathogenesis of several neurodegenerative diseases. In previous work, we reported decreased protein levels of mitochondrial electron transport enzyme subunits in adult brain with Down syndrome (DS). However it is not clear whether cellular damage due to mitochondrial defects in brain of DS fetus begins in utero. Here we investigated the protein levels of mitochondrial electron transport enzymes in fetal DS brain using the proteomic technologies. Two-dimensional (2-D) gel electrophoresis, matrix-assisted laser desorption ionization mass spectroscopy (MALDI-MS) and specific software for quantification were used. The protein levels of complex I 30-kDa subunit were significantly decreased in cerebral cortex of fetal DS brain. We conclude that decreased mitochondrial electron transport enzyme subunits in fetal DS brains could contribute to the impaired energy and free radical metabolism affecting brain development in DS fetus. Furthermore, the defects of mitochondrial electron enzymes shown in adult DS brains could begin in utero and continue during the life span of the individual with DS.

Introduction

The mitochondrial electron transport for energy metabolism is composed of five protein-lipid complexes: NADH: ubiquinone oxidoreductase (complex I), succinate: ubiquinone oxidoreductase (complex II), ubiquinol: ferricytochrome c oxidoreductase (complex III), ferrocytochrome c: oxygen oxidoreductase (complex IV) and ATP synthase (complex V) (Hatefi, 1985). Each of these five major mitochondrial electron transport complexes is composed of numerous subunits and a proton gradient generated by complex I, III and IV is used by complex V to generate ATP from ADP.

Recently the defects of mitochondrial electron transport chain enzymes have been reported in brains with neurodegenerative diseases. The generalized depression of activity of electron transport complexes (complex I, II, III and IV) was demonstrated in brain from patients with Alzheimer's disease (AD) (Parker et al., 1994). The reduction of complex V was observed in the hippocampus of brain from AD patients using blue native polyacrylamide gel electrophoresis (Schagger and Ohm, 1995). In Parkinson disease (PD), a significant reduction of complex I activity was observed in the substantia nigra (Schapira et al., 1990) and deficiencies in complex I subunits (30-, 25- and 24-kDa subunits) were reported in the striata (Mizuno et al., 1989). In addition, mitochondrial enzyme deficiencies have been demonstrated in Down syndrome (DS) (Prince et al., 1994) and in previous work, we observed decreased protein levels of complex I 24- and 75-kDa subunit, complex III core protein 1 and complex V beta chain in brains from patients with AD and DS (Kim et al., 2000). However it is not clear whether brain damage due to the mitochondrial defects begin already in utero. These reasons made us study mitochondrial electron transport enzymes in fetal DS brain compared to controls. We evaluated the protein levels of complex I (24-, 30- and 51-kDa subunit), complex III (core protein 1 precursor and iron-sulfur protein) and complex V (alpha chain precursor and d chain) in fetal brain with DS. Two-dimensional (2-D) gel electrophoresis with subsequent matrix-assisted laser desorption ionization mass spectroscopy (MALDI-MS) identification and specific software for quantification of proteins was used. We also determined the protein levels of mitochondria-located proteins (mitochondrial subunit alpha of isocitrate dehydrogenase and fumarate hydratase) to show the specificity of the reduction rather than simple mitochondrial loss.

Methods

Brain samples preparation

Brain tissues of fetus with DS and controls were obtained from Dr. M. Dierssen (Medical and Molecular Genetics Center-IRO, Hospital Duran i Reynals, Barcelona, Spain). Cerebral cortex of controls (n = 7, male/female = 6/1, 18.79 ± 2.23 weeks) and fetuses with DS (n = 8, male/female = 6/2, 19.81 ± 2.00 weeks) were used. Brain tissues were suspended in 0.5 ml of sample buffer consisting of 40 mM Tris, 5 M urea (Merck, Darmstadt, Germany), 2 M thiourea (Sigma, St. Louis, MO, USA), 4% CHAPS (3-[(3-cholamidopropyl)dimethylammonio]-1-propanesulfonate, Sigma), 10 mM 1,4-dithioerythritol (Merck), 1 mM EDTA (ethylenediaminetetraacetic acid, Merck), 1 mM PMSF (phenylmethanesulfonyl fluoride, Sigma) and 1 μg/ml of each pepstatin A, chymostatin, leupeptin and antipain. The suspension was sonicated for approximately 30 seconds and centrifuged at 10,000 × g for 10 minutes and the supernatant was centrifuged further at 150,000 × g for 45 minutes to sediment undissolved material. The protein concentration of the supernatant was determined by the Coomassie blue method (Bradford, 1976).

Two-dimensional (2-D) gel electrophoresis

The 2-D gel electrophoresis was performed essentially as reported (Langen et al., 1997). Samples of 2 mg were applied on immobilized pH 3–10 nonlinear gradient strips (IPG, immobilized pH-gradient strips, Pharmacia Biotechnology, Uppsala, Sweden) in sample cups at their basic and acidic ends. Focusing stated at 200 V and the voltage was gradually increased to 5,000 V at 3 V/min and kept constant for a further 24 hours (approximately 180,000 kVh totally). The second-dimensional separation was performed on 9–16% SDS gradient polyacrylamide gels. The gels (180 × 200 × 1.5 mm) were run at 40 mA per gel. After protein fixation with 40% methanol containing 5% phosphoric acid for 12 hours, the gels were stained with colloidal Coomassie blue (Novex, San Diego, CA, USA) for 48 hours. Molecular masses were determined by running standard protein markers (Gibco, Basel, Switzerland), covering the range 10–200 kDa. Isoelectric point (pI) values were used as given by the supplier of the IPG strips. Excess of dye was washed out from the gels with H_2O and the gels were scanned in an Agfa DUOSCAN densitometer (resolution 200). Electronic images of the gels were recorded using PhotoShop (Adobe) and PowerPoint (Microsoft) software.

Matrix-assisted laser desorption ionization mass spectroscopy (MALDI-MS)

MALDI-MS analysis was performed as described elsewhere (Fountoulakis and Langen, 1997) with some modifications. The spots were excised with a spot picker and placed into 96-well microtiter plates. Each spot was destained with 100 µl of 30% acetonitrile in 50 mM ammonium bicarbonate and dried in a speedvac evaporator. Each dried gel pieces was rehydrated with 4 µl of 3 mM Tris-HCl, pH 9.0, containing 50 ng trypsin (Promega, Madison, WI, USA). After 16 hours at room temperature, 7 µl of H_2O were added to each gel piece and the samples were shaken in for 10 minutes. Four µl of 50% acetonitrile, containing 0.3% trifluoroacetic acid, the standard peptides des-Arg-bradykinin (Sigma, 904.4681 Da) and adrenocorticotropic hormone fragment 18–39 (Sigma, 2465.1989 Da), in H_2O were added to each gel piece and shaken for 10 minutes. The application of the samples was performed with a SymBiot I sample processor (PE Biosystems, Framingham, MA, USA). 1.5 µl of the peptide mixture was simultaneously applied with 1 µl of matrix, consisting of a saturated solution of α-cyano-4-hydroxycinnamic acid (Sigma) in 50% acetonitrile, containing 0.1% trifluoroacetic acid Samples were analyzed in a time-of-flight mass spectrometer (Reflex 3, Bruker Analytics, Bremen, Germany). An accelerating voltage of 20 kV was used. Peptide matching and protein searches were performed automatically. The peptide masses were compared to the theoretical peptide masses of all available proteins from all species. Monoisotopic masses were used and a mass tolerance of 0.0025% was allowed. The algorithm used for determining the probability of a false positive match with a given MS-spectrum is described elsewhere (Berndt et al., 1999).

Quantification of proteins

In partial 2D-gel images including each protein corresponding to mitochondrial electron transport enzyme subunits and the neighbouring proteins, the percentage of the volume of each protein was quantified as compared to the total proteins present using the ImageMaster 2D Elite software (Amersham Pharmacia Biotechnology).

Statistics

Results were expressed as means ± standard deviation (SD). Between-group differences were investigated by non-parametric Mann-Whitney U test. The level of significance was set at $P < 0.05$. All statistical analyses were performed by GraphPad Instat2 software, version 2.05.

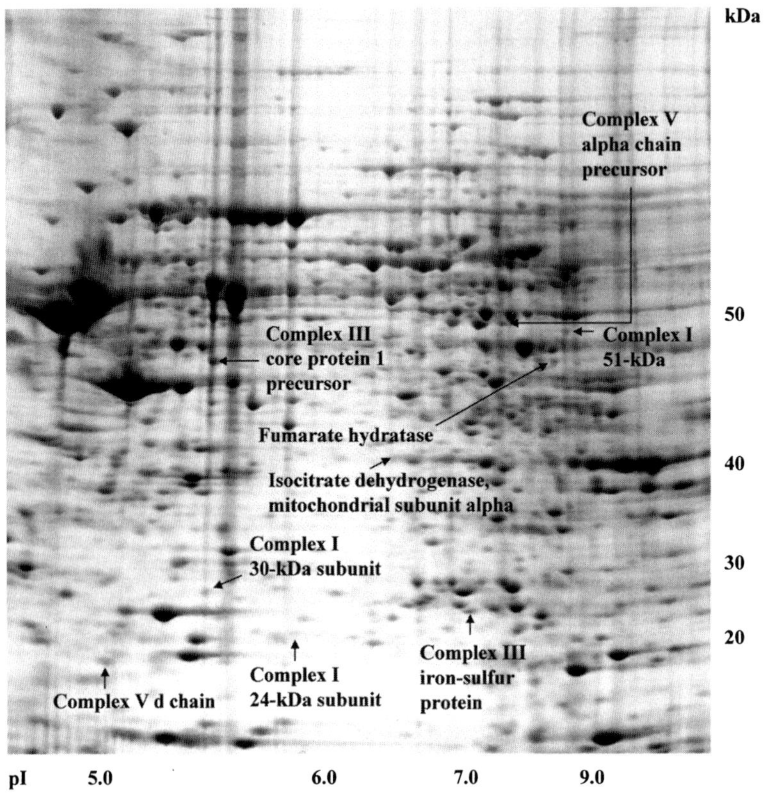

Fig. 1. The 2-D gel images of fetal brain. The brain proteins were extracted and separated on an immobilized pH 3–10 nonlinear gradient strip, followed by a 9–16% linear gradient polyacrylamide gel, as stated under experimental procedures. The gel was stained with Coomassie blue and the spots were analyzed by MALDI-MS

Results

Identification of human brain proteins

The fetal brain proteins were identified based upon molecular weight, pI value and peptide matches using 2-D gel electrophoresis and MALDI-MS analysis. Protein spots used in this study were presented in Fig. 1 and the data generated were summarized in Table 1.

Protein levels of mitochondrial electron transport enzymes subunits

As shown in Table 2, the protein levels of complex I 30-kDa subunit were significantly decreased in cerebral cortex of fetal DS brain compared to controls, but there was no significant difference in other proteins. The 2-D gel images including complex I 30-kDa subunit were presented in Fig. 2.

Table 1. Identification of mitochondrial proteins in 2-D gel

Components	pI	MW	Matches	Probability
Complex I				
24-kDa subunit	8.08	27,631	7	4.22E-06
30-kDa subunit	7.57	30,336	9	5.32E-14
51-kDa subunit	9.29	33,764	6	3.40E-08
Complex III				
core protein 1 precursor	6.33	53,269	6	1.35E-07
iron-sulfur subunit	8.35	29,918	4	1.22E-05
Complex V				
alpha chain precursor	9.93	59,827	6	8.66E-09
d chain	5.05	18,405	5	8.32E-07
Isocitrate dehydrogenase, mitochondrial subunit alpha	6.91	40,022	7	4.81E-08
Fumarate hydratase	9.36	54,773	5	7.03E-07

Protein separated by 2-D electrophoresis was identified by MALDI-MS, following digestion with trypsin. The peptide masses were matched with the theoretical peptide masses of all proteins from all species of the SWISS-PROT database

Protein levels of mitochondria-located proteins

As shown in Table 3, mitochondria-located proteins, mitochondrial subunit alpha of isocitrate dehydrogenase and fumarate hydratase, were not significantly changed in fetal DS compared to controls.

Table 2. The protein levels of mitochondrial electron transport enzymes in fetal brain with DS

Components	Control (n)	DS (n)
Complex I		
24-kDa subunit	0.35 ± 0.19 (7)	0.38 ± 0.35 (7)
30-kDa subunit	1.18 ± 0.23 (7)	0.61 ± 0.34** (8)
51-kDa subunit	0.35 ± 0.24 (7)	0.35 ± 0.20 (5)
Complex III		
core protein 1 precursor	2.00 ± 0.54 (7)	1.84 ± 0.70 (4)
iron-sulfur subunit	1.13 ± 0.66 (7)	1.02 ± 0.28 (7)
Complex V		
alpha chain precursor	6.25 ± 1.26 (7)	5.52 ± 1.45 (5)
d chain	1.94 ± 0.47 (7)	2.15 ± 0.37 (8)

The proteins from cerebral cortex from fetal DS and controls were separated on 2-D gels and visualized following stain with colloidal Coomassie blue. In partial gel images, including each protein and the neighboring proteins, the intensities of the spots representing each protein were quantified with ImageMaster 2D software, compared to the total proteins present. The statistical analyses were performed by Graphpad Instat2 software. Significant differences are indicated as follow: $^{**}P < 0.01$

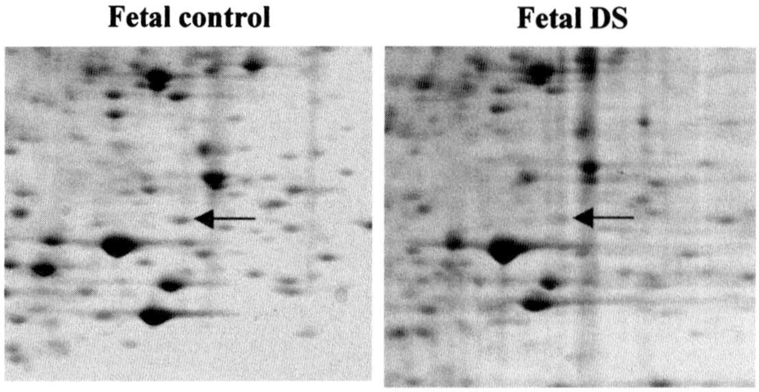

Fig. 2. The 2-D gel images of complex I 30-kDa subunit. This protein spot was indicated by arrows

Discussion

The brain is acutely dependent on energy supplies for normal functioning and mitochondria are the intracellular parts for the brain's energy supplies. Deficiencies of mitochondrial electron transport enzymes were proposed to contribute to the progression of the pathology of neurodegenerative diseases. Mitochondrial enzyme deficiencies in DS were reported (Prince et al., 1994) and in previous work we observed the decreased protein levels of mitochondrial electron transport enzymes in brains from patients with DS (Kim et al., 2000). Decreased levels of subunit protein in mitochondrial electron transport enzymes could lead to deficient mitochondrial function and may contribute to derangement of energy metabolism in DS brain. In addition, a consequence of defective mitochondrial energy production is the enhanced sensitivity to excitotoxicity, the alteration of Ca^{2+} buffering and the increased generation of free radicals such as superoxide and hydroxyl radicals, which are normally produced as by-products of oxidative metabolism. These free radicals oxidizing lipids, proteins, nucleic acids and other molecules could alter their structure and function. Mitochondria are themselves susceptible to damage by reactive oxygen species (ROS) (Schapira, 1996) and oxidative stress due to increased free radicals is the characteristic features in DS (de Haan et al., 1997). Impairment in the ability to repair oxidative damage to mitochondrial DNA was reported in DS (Druzhyna et al., 1998) and DS neurons showed

Table 3. The levels of proteins located in mitochondria in fetal brain with DS

Components	Control (n)	DS (n)
Isocitrate dehydrogenase, mitochondrial subunit alpha	1.07 ± 0.38 (6)	1.29 ± 0.84 (5)
Fumarate hydratase	0.52 ± 0.17 (7)	0.64 ± 0.28 (5)

The gels were prepared and the intensities of the spots representing each protein were quantified and analysed as described under legend to Table 2 increased free radical production preceding neuronal death compared with controls (Busciglio and Yankner, 1995). Additional evidence for increased oxidative stress in DS was provided by the findings of significantly elevated levels of 8-hydroxy-2′-deoxyguanosine (8-OHdG), a biomarker of oxidative damage to DNA, and malondialdehyde, a biomarker of lipid peroxidation evaluated by the chemical analyses of urine samples (Jovanovic et al., 1998; Iannello et al., 1999).

In fetal DS brain the perturbation of oxygen radical metabolism was demonstrated to show the increase in the activity of Cu/Zn-dependent superoxide dismutase without an adaptive rise in glutathione peroxidase (Brooksbank and Balazs, 1984). In addition, there is an increase in the lipoperoxidation potential, as indicated by enhanced thiobarbituric acid chromogen formation in vitro, associated with abnormalities of fatty acid composition of phospholipids.

Here we observed the decreased protein levels of complex I 30-kDa subunit in fetal DS brain using proteomic technologies. These results suggest that decreased subunits levels in mitochondrial electron transport enzymes could result in the impaired energy and free radical metabolism and also affect the development of fetal DS brain. Furthermore, brain damage due to impaired energy and free radical metabolism may begin early in life in the DS fetus and continue during the life span of the individual with DS.

References

Bendt P, Hobohm U, Langen H (1999) Reliable automatic protein identification from matrix-associated laser desorption/ionization mass spectrometric peptide fingerprints. Electrophoresis 20: 3521–3526

Bradford MM (1976) A rapid and sensitive method for the quantitation of microgram quantities of protein utilizing the principle of protein-dye binding. Anal Biochem 72: 248–254

Brooksbank BWL, Balazs R (1984) Superoxide dismutase, glutathion peroxidase and lipoperoxidation in Down's syndrome fetal brain. Brain Res 318: 37–44

Busciglio J, Yankner BA (1995) Apoptosis and increased generation of reactive oxygen species in Down's syndrome neurons in vitro. Nature 378: 776–779

de Haan JB, Wolvetang EJ, Cristiano F, Iannello R, Bladier C, Kelner MJ, Kola I (1997) Reactive oxygen species and their contribution to pathology in Down syndrome. Adv Pharmacol 38: 379–402

Druzhyna N, Nair RG, LeDoux SP, Wilson GL (1998) Defective repair of oxidative damage in mitochondrial DNA in Down's syndrome. Mutat Res 409: 81–89

Fountoulakis M, Langen H (1997) Identification of protein by matrix-assisted laser desorption ionization-mass spectroscopy following in-gel digestion in low-salt nonvolatile buffer and simplified peptide recovery. Anal Biochem 250: 153–156

Hatefi Y (1985) The mitochondrial electron transport and oxidative phosphorylation system. Ann Rev Biochem 54: 1015–1069

Iannello RC, Crack PJ, de Haan JB, Kola I (1999) Oxidative stress and neural dysfunction in Down syndrome. J Neural Transm [Suppl] 57: 257–267

Jovanovic SV, Clements D, MacLeod K (1998) Biomarkers of oxidative stress are significantly elevated in Down syndrome. Free Radic Biol Med 25: 1044–1048

Kim SH, Vlkolinsky R, Cairns N, Lubec G (2000) Decreased levels of complex III core protein 1 and complex V β chain in brains from patients with Alzheimer's disease and Down syndrome. Cell Mol Life Sci 57: 1810–1816

Langen H, Roeder D, Juranville J-F, Fountoulakis M (1997) Effect of the protein application mode and the acrylamide concentration on the resolution of protein spots separated by two-dimensional gel electrophoresis. Electrophoresis 18: 2085–2090

Mizuno Y, Ohta S, Tanaka M, Takamiya S, Suzuki K, Sato T, Oya H, Ozawa T, Kagawa Y (1989) Deficiencies in complex I subunits of the respiratory chain in Parkinson's disease. Biochem Biophys Res Commun 163: 1450–1455

Parker WD, Parks J, Filley CM, Kleinschmidt-DeMasters BK (1994) Electron transport chain defects in Alzheimer's disease brain. Neurology 44: 1090–1096

Prince J, Jia S, Bave U, Anneren G, Oreland L (1994) Mitochondrial enzyme deficiencies in Down syndrome. J Neural Transm [P-D Sect] 8: 171–181

Schapira AHV (1996) Oxidative stress and mitochondrial dysfunction in neurodegeneration. Curr Opin Neurol 9: 260–264

Schapira AH, Cooper JM, Dexter D, Clark JB, Jenner P, Marsden CD (1990) Mitochondrial complex I deficiency in Parkinson's disease. J Neurochem 54: 823–827

Schagger H, Ohm TG (1995) Human diseases with defects in oxidative phosphorylation. 2. F1F0 ATP-synthase defects in Alzheimer disease revealed by blue native polyacrylamide gel electrophoresis. Eur J Biochem 227: 916–921

Authors' address: Prof. Dr. G. Lubec, Department of Pediatrics, University of Vienna, Währinger Gürtel 18, A-1090 Vienna, Austria, e-mail: gert.lubec@akh-wien.ac.at

Selective upregulation of the ubiquitin-proteasome proteolytic pathway proteins, proteasome zeta chain and isopeptidase T in fetal Down syndrome

E. Engidawork[1], **J.-F. Juranville**[2], **M. Fountoulakis**[2], **M. Dierssen**[3], and **G. Lubec**[1]

[1] Department of Pediatrics, University of Vienna, Vienna, Austria
[2] F. Hoffman-La Roche, Basel, Switzerland
[3] Medical and Molecular Genetics Center, IRO Hospital Duran i Reynals, Barcelona, Spain

Summary. The ubiquitin-proteasome proteolytic pathway is involved in an important non-lysosomal proteolytic pathway that is responsible for the highly selective turnover of cellular proteins both under basal metabolic conditions as well as stress. Protein degradation by this pathway is attributed to the 20S proteasome that forms the catalytic core of the complex. Recently there has been increasing interest in the proteasome because of its possible role in neuron degeneration and death. Fetal Down syndrome (DS) neurons were demonstrated to degenerate and undergo apoptosis in vitro. We therefore investigated the expression of different proteins involved in this degradative pathway, including subunits of the 20S proteasome, ubiquitinating and deubiquitinating enzymes, and regulatory subunits of the 26S proteasome in control and DS fetal brains by two-dimensional electrophoresis (2-DE). After 2-DE, approximately 389 protein spots were successfully identified by matrix-associated laser desorption ionization mass spectroscopy (MALDI-MS) and this was followed by quantification of twenty three proteins of the pathway. The results indicate that all but two proteins exhibited no apparent alterations in their pattern of expression. Proteasome zeta chain, an alpha subunit of the 20S proteasome ($P < 0.05$) and ubiquitin carboxy-terminal hydrolase T (Isopeptidase T), a deubiquitinating enzyme ($P < 0.001$) were significantly increased in fetal DS compared to controls. Whilst the expression of proteasome iota ($n = 9$, $r = -0.9489$, $P = 0.0004$) and proteasome epsilon ($n = 9$, $r = -0.7227$, $P = 0.0311$) chains was decreased with age in fetal DS brain, no significant correlation was obtained in the other proteins with age. The data suggest that such selective upregulation may have relevance to the developmental abnormalities that characterize this disorder.

Introduction

The selective degradation of proteins via the ubiquitin-proteasome pathway involves the activation of a signaling cascade that generates the covalent attachment of a polyubiquitin chain to target proteins (Ciechanover and Schwartz, 1994; Ciechanover, 1994; Hochstrasser, 1995). This marking occurs through the formation of isopeptide bonds between the C-terminus of ubiquitin and the ε-amino group lysine residue on the target protein and effected by three enzymes: E1 (ubiquitin activating enzyme), E2 (ubiquitin conjugating enzyme, UBC) and E3 (ubiquitin protein ligase). E1 catalyzes activation of the C-terminal Gly by ATP. Following activation, E2 either transfers ubiquitin from E1 to the substrate that is bound to E3 or directly to the substrate. In successive reactions, a polyubiquitin chain is formed that serves as a signal for degradation by the 26S proteasome in a process that requires ATP, and releases free and utilizable ubiquitin (Ciechanover and Schwartz, 1994; Ciechanover, 1994).

Recycling of ubiquitin is essential to the function of the proteolytic machinery and is achieved by ubiquitin C-terminal hydrolases, ubiquitin isopeptidases or deubiquitinating enzymes. Hydrolases are required for dismantling the polyubiquitin chain prior, during, or following the degradative process and for proofreading, trimming, or processing activities (Ciechanover, 1994). They can be subdivided into two groups: ubiquitin specific proteases (UBPs) and ubiquitin C-terminal hydrolases (UCHs) (Hochstrasser, 1995; Wilkinson et al., 1995). UBPs are extremely diverse (Hochstrasser, 1995), and are implicated in growth regulation (Zhu et al., 1996; Naviglio et al., 1998) and modulation of transcription silencing (Moazed and Johnson, 1996) while UCHs are reported to have a role in Parkinson's disease (Leroy et al., 1998).

The proteasome (20S proteasome, macropain or multicatalytic protease) is a highly conserved large multicatalytic protease complex found in all eukaryotic and certain archaebacterial cells and is localized in both the cytoplasm and the nucleus (DeMartino et al., 1991; Peters, 1994; Coux et al., 1996; Rivett, 1998). In addition, it has been detected associated with endoplasmic reticulum (Rivett, 1998). Fourteen gene products have been identified in yeast that are classified as alpha and beta subunits (Groll et al., 1997). In mammalian proteasome, the gene products exceed fourteen owing to the inclusion of subunits induced by γ-interferon and involved in antigen presentation, such as LMP (low molecular mass protein)-2, LMP-7 and MECL1 that displace the housekeeping proteasome delta (Y), epsilon (X), and Z, respectively (Früh et al., 1994; Coux et al., 1995; Hilt and Wolf, 1996). The proteasome complex is a hollow cylindrical particle made up of four heptameric rings (Hilt and Wolf, 1996). The two outer rings are composed of seven alpha subunits that play a role in complex assembly and interaction with regulatory proteins, whereas the inner rings contain seven beta subunits that possess the catalytic active sites (Jentsch and Schlenker, 1995).

Proteasome constitutes the major non-lysosomal proteolytic pathway responsible for the highly selective turnover of intracellular proteins. It is involved in a host of cellular activities including DNA repair, cell cycle progression, metabolic adaptation, cell differentiation, signal transduction, stress response, antigen presentation, and removal of abnormal proteins (Ciechanover, 1994; Hochstrasser, 1995; Jentsch and Schlenker, 1995; Coux et al., 1996; Hilt and Wolf, 1996). These cellular functions are linked predominantly if not exclusively, to an ubiquitin and/or ATP dependent protein degradation pathway involving the 26S proteasome, whose catalytic core and proteolytic chamber is formed by the 20S proteasome (Murakami et al., 1992; Ciechanover, 1994; Peters, 1994; Hill and Wolf, 1996). The 26S proteasome contains two additional cap like structures (19S) on either end that are composed of ATPase and non-ATPase subunits. The capping proteins play a regulatory role and are thought to be required for substrate recognition, unfolding and subsequent funneling to the proteolytic core (Peters, 1994; Hochstrasser, 1995; Jentsch and Schlenker, 1995).

Down syndrome (DS) is the clinical expression of the most common postnatally viable human chromosomal abnormality resulting from triplication of all (trisomy) or portion (translocation) of chromosome 21 (Shapiro, 1999). DS is characterized by its distinct dysmorphic features, mental retardation, behavioral and cognitive impairments (Epstein, 1995). Limited information is available on gene expression in fetal DS. The ubiquitin-proteasome proteolytic pathway is essential to life and was shown to be influenced by oxidative stress (Reinheckel et al., 1998, 2000) and apoptosis (An et al., 2000), phenomena that have also been described in fetal DS (Busciligio and Yanker, 1995). Thus, we undertook this study to determine the expression of ubiquitinating and deubiquitinating enzymes as well as different subunits of the 20S proteasome and regulatory complex of the 26S proteasome in control and DS fetal brains by two-dimensional electrophoresis (2-DE). The data revealed an altered expression of only proteasome zeta chain, an alpha subunit of the 20S proteasome, and UBP T, a deubiquitinating enzyme, in fetal DS which may result in mishandling of proteins that leads to developmental defects seen in this disorder.

Material and methods

Fetal brain tissue

Brain tissue (cerebral cortex) of aborted control fetuses with no obvious abnormalities (n = 7; 1 female and 6 males with mean gestational age of 18.8 ± 2.2 weeks) and DS (n = 9; 2 females and 7 males with mean gestational age of 19.6 ± 2.0 weeks) was used for the experiment. The brain samples were obtained from the Medical and Molecular Genetics Center-IRO, Hospital Duran i Reynals, Barcelona, Spain. All samples were stored at −70°C and the freezing chain was never interrupted.

Two dimensional gel electrophoresis (2-DE)

Brain tissue was suspended in 0.5 ml of sample buffer consisting of 40 mM Tris, 5 M urea (Merck, Darmstadt, Germany), 2 M thiourea (Sigma, St. Louis, MO, USA), 4% CHAPS (Sigma), 10 mM 1,4-dithioerythritol (Merck), 1 mM EDTA (Merck) and a mixture of protease inhibitors, 1 mM PMSF and 1 µg of each pepstatin A, chymostatin, leupeptin and antipain. The suspension was sonicated for approximately 30 sec and centrifuged at

150,000g for 45min. The protein content in the supernatant was determined by the Coomassie blue method (Bradford, 1976).

2-DE was performed essentially as reported (Langen et al., 1997). Samples of 2mg protein were applied on immobilized pH 3–10 nonlinear gradient strips in sample cups at their basic and acidic ends. Focusing started at 200V and the voltage was gradually increased to 5,000V at 3V/min and kept constant for a further 24h (approximately 180,000kVh totally). The second-dimensional separation was performed on 9–16% SDS gradient polyacrylamide gels. The gels (180 × 200 × 1.5mm) were run at 40mA per gel. After protein fixation for 12h in 40% methanol, containing 5% phosphoric acid, the gels were stained with colloidal Coomassie blue (Novex, San Diego, CA, USA) for 48h. Molecular masses were determined by running standard protein markers (Gibco, Basel, Switzerland), covering the range 10–200kDa. pI values were used as given by the supplier of the IPG strips. Excess of dye was washed out from the gels with H_2O and the gels were scanned in an Agfa DUOSCAN densitometer (resolution 200). Electronic images of the gels were recorded using Photoshop (Adobe) and PowerPoint (Microsoft) software.

Matrix-associated laser desorption ionization mass spectroscopy (MALDI-MS)

MALDI-MS analysis was performed as described elsewhere (Fountoulakis and Langen, 1997) with some modifications. The spots were excised with a spot picker and placed into 96-well microtiter plates. Each spot was destained with 100μl of 30% acetonitrile in 50mM ammonium bicarbonate and dried in a speedvac evaporator. Each dried gel piece was rehydrated with 4μl of 3mM Tris-HCl, pH 9.0, containing 50ng trypsin (Promega, Madison, WI, USA). After 16h at room temperature, 7μl of H_2O was added to each gel piece and the samples were shaken in for 10min. Four μl of 50% acetonitrile, containing 0.3% trifluoroacetic acid and the standard peptides des-Arg-bradykinin (Sigma, 904.4681Da) and adrenocorticotropic hormone fragment 18–39 (Sigma, 2,465.1989Da), in water were added to each gel piece and shaken for 10min. The application of the samples was performed with a SymBiot I sample processor (PE Biosystems, Framingham, MA, USA). 1.5μl of the peptide mixture was simultaneously applied with 1μl of matrix, consisting of a saturated solution of α-cyano-4-hydroxycinnamic acid (Sigma) in 50% acetonitrile, containing 0.1% trifluoroacetic acid. Samples were analyzed in a time-of-flight mass spectrometer (Reflex 3, Bruker Analytics, Bremen, Germany). An accelerating voltage of 20kV was used. Peptide matching and protein searches were performed automatically. The peptide masses were compared to the theoretical peptide masses of all available proteins from all species. Monoisotopic masses were used and a mass tolerance of 0.0025% was allowed. The algorithm used for determining the probability of a false positive match with a given MS-spectrum is described elsewhere (Berndt et al., 1999).

Quantification of protein spots

Protein spots were outlined (first automatically and then manually) and quantified using the ImageMaster 2D Elite software (Amersham Pharmacia Biotechnology). The percentage of the volume of the spots representing a certain protein was determined in comparison with the total proteins present in the 2-D gel.

Statistical analysis

The software calculated the percentage of the spot volume in each particular gel after subtraction of the background values. Between-group differences were calculated using non-parametric Mann-Whitney U-test and within-group correlation were done using the Spearman rank-coefficient analysis. The level of significance was set at $P < 0.05$.

Results

Identification and quantification of ubiquitin-proteasome proteolytic machinery proteins on 2-DE map of fetal brain

Human proteins obtained from the cerebral cortex of fetal brain were applied on 2-D gel. The protein spots were visualized by staining with colloidal Coomassie blue. The protein spots were analyzed by MALDI-MS, following in-gel digestion. The peptide masses were matched with the theoretical peptide masses of all proteins of the SWISS-PROT database. Internal standards were used to correct the measured peptide mass thus reducing the windows of mass tolerance and increasing the confidence of identification. Approximately, 389 proteins were successfully identified and amongst them 23 proteins involved in the proteolytic pathway were quantified using the Image Master 2D Elite software. Figure 1 shows a representative gel from control brain containing fetal proteins separated by 2-DE, with accession number given only for the investigated proteins. The expression levels of the proteins were determined as a percentage of spot volume in the area of interest. The SWISS-PROT accession number and names of the quantified proteins along with the MALDI data and percent protein level in control and DS are given in Table 1.

Expression pattern of ubiquitin-proteasome proteolytic pathway proteins in control and DS fetal brain

Ubiquitinating and deubiquitinating enzymes

Four spots, one major and three minor, representing the ubiquitinating enzyme, E1 were identified on 2-DE gel (Fig. 1) that appear to have more or less the same molecular size but different pI values. As a single enzyme seems to catalyze activation of ubiquitin (Ciechanover, 1994), these multiple spots most likely represent different posttranslationally modified forms of the enzyme. However, the alternative that they may represent different isoforms can not be totally excluded. The expression of the major spot (third from left to right in Fig. 1) was comparable between control and DS (2.37 ± 0.90 vs. 2.45 ± 1.13). Likewise, there was no apparent difference in expression of the minor spots (data not shown) as well as the sum of the individual spots (Table 1) between the two groups. E2, another ubiquitinating enzyme, represented by one spot also did not show any significant change between the study populations, although the values in DS appear to be higher (~27.3%) than controls. With respect to the deubiquitinating enzymes, two isozymes of the UCH family; UCH L1 and UCH L3, and UBP T that belongs to UBP family of

Table 1. Quantitative analysis of proteins involved in the ubiquitin-proteasome proteolytic pathway in fetal human brain. The samples were analyzed and the proteins were identified and quantified as stated under Material and methods. The SWISS-PROT accession number of the proteins of interest is indicated in Fig. 1. The observed Mr and pI values, the matching peptides along with the probability as well as the relative protein levels (%) are given here. Within group analysis was performed by non-parametric Mann-Whitney U-test and the significance was set at $P < 0.05$, *$P < 0.05$, ***$P < 0.001$. (n = 7 for controls and 9 for DS)

Protein name	Accession number	pI	MW (Da)	Matches	Probability	Control	DS
1. *Ubiquitinating and deubiquitinating proteins*							
Ubiquitin activating enzyme E1	P22314	5.71	118,798	13	7.78E-18	4.06 ± 1.06	4.08 ± 1.43
Ubiquitin conjugating enzyme E2	Q16781	6.55	17,183	5	1.47E-05	0.77 ± 0.26	0.98 ± 0.30
Ubiquitin carboxy terminal hydrolase L1 (UCH L1)	P09936	5.28	25,150	5	1.75E-06	8.63 ± 1.64	9.14 ± 1.22
Ubiquitin carboxy terminal hydrolase L3 (UCH L3)	P15374	4.69	26,337	5	1.16E-05	0.27 ± 0.10	0.20 ± 0.08
Ubiquitin specific processing protease T (UBP T)	P45974	4.76	96,637	8	3.63E-06	0.26 ± 0.10	0.69 ± 0.21***
2. *20S proteasome subunits*							
Alpha subunits							
Proteasome zeta chain	P28066	4.52	26,623	5	1.00E-07	0.64 ± 0.23	1.02 ± 0.24*
Proteasome iota chain	P34062	6.72	27,837	5	8.10E-07	0.93 ± 0.24	1.01 ± 0.18
Proteasome component C8	P25788	5.07	28,512	5	8.03E-06	0.36 ± 0.14	0.24 ± 0.14
Proteasome component C2	P25786	6.60	29,821	5	5.74E-06	1.10 ± 0.32	1.49 ± 0.49
Proteasome component C3	P25787	7.51	25,865	5	1.72E-06	0.28 ± 0.14	0.23 ± 0.07
Proteasome component C9	P25789	7.74	29,750	4	5.96E-05	0.74 ± 0.17	0.74 ± 0.14
Proteasome component C7-L	P49721	7.03	22,992	5	1.02E-04	0.65 ± 0.23	0.73 ± 0.18
Proteasome subunit HSPC	AA99402	8.73	28,056	6	1.00E-05	0.27 ± 0.11	0.19 ± 0.04
Beta subunits							
Proteasome component C5	P20618	8.20	26,700	7	4.01E-09	0.63 ± 0.33	0.61 ± 0.32
Proteasome beta chain	P28070	5.83	29,230	5	1.95E-05	0.14 ± 0.06	0.12 ± 0.04
Proteasome epsilon (X) chain	P28074	8.76	23,053	5	3.85E-05	0.17 ± 0.06	0.18 ± 0.07
Regulatory complex of 26S proteasome							
26S protease regulatory subunit 6b (TBP-7)	P43686	4.95	47,420	6	2.01E-07	1.26 ± 0.85	1.61 ± 0.74
26S protease regulatory subunit 7 (mss1 protein)	P35998	5.73	49,002	10	7.54E-16	0.65 ± 0.23	0.76 ± 0.37
26S protease regulatory subunit 8 (p45)	P47210	8.36	45,795	10	7.33E-12	0.93 ± 0.37	0.65 ± 0.22
26S protease regulatory subunit s10b (p42)	Q92524	7.49	44,418	10	1.60E-16	1.33 ± 0.52	0.92 ± 0.26
26S protease regulatory subunit s6a (TBP-1)	P17980	4.92	49,372	10	3.74E-16	0.45 ± 0.18	0.50 ± 0.22
26S protease regulatory subunit s12 (p40, mov 34)	P51665	6.54	37,094	5	4.95E-06	0.94 ± 0.15	0.80 ± 0.26
26S protease regulatory subunit p28	O75832	6.05	24,697	5	3.39E-05	0.15 ± 0.07	0.15 ± 0.06

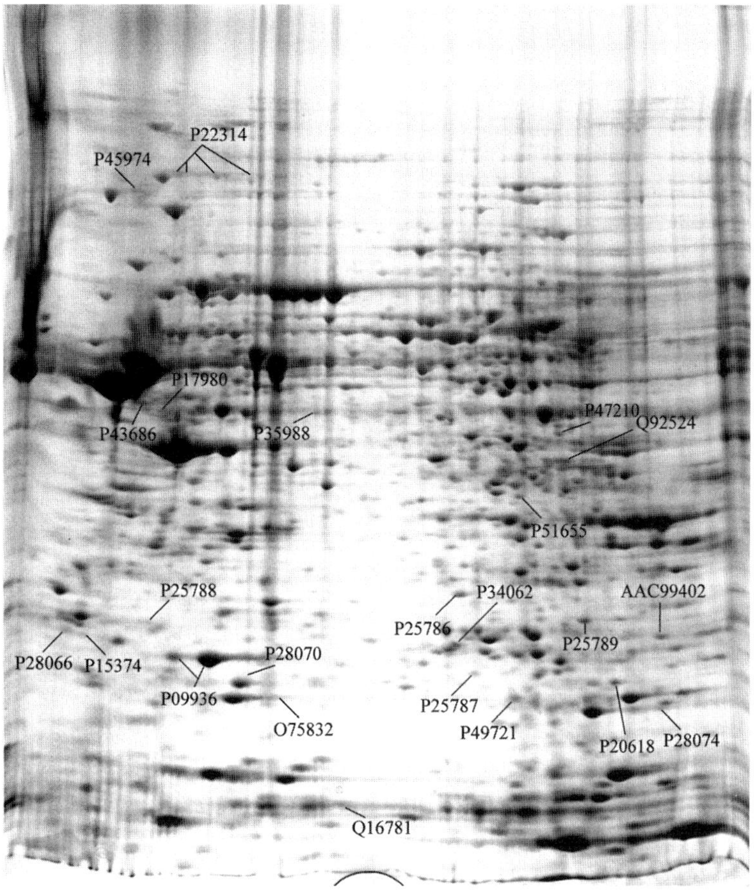

Fig. 1. Two-dimensional map of human fetal brain proteins. The proteins from cerebral cortex of fetal brain were extracted and analyzed on pH 3–10 nonlinear IPG strips, followed by a 9–16% sodiumdodecyl sulphate polyacrylamide gels. The gels were stained with Coomassie blue and identified by MALDI-MS. The spots representing proteins involved in ubiquitin-proteasome proteolytic pathway are designated with their SWISS-PROT number. Name of the proteins is given in Table 1

enzymes were identified. UCH L1 was represented by two spots, whereas UCH L3 and UBP T were represented each by one spot (Fig. 1). No significant difference was seen between the two groups in expression of UCH L3. So does, the individual (data not shown) as well as the sum (Table 1) of both spots of UCH L1, although it tended to increase in DS than controls. However, a significant elevation was observed in expression of UBP T in fetal DS brain ($P < 0.001$) (Fig. 2).

20S proteasome

2-DE identified eleven spots that stand for eleven different subunits of the 20S proteasome. Eight of them were alpha subunits and the other three beta

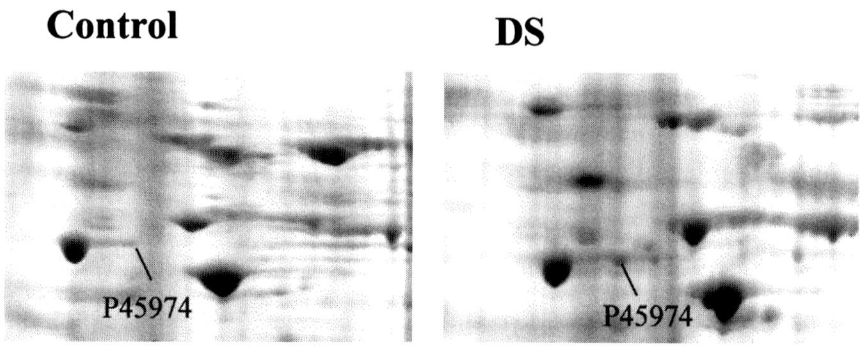

Fig. 2. Partial two-dimensional gel images of human fetal brain proteins. The gels were prepared and the protein was analyzed as described under legend to Fig. 1. The spot corresponding to ubiquitin carboxy-terminal hydrolase T (Isopeptidase T) is indicated with the accession number

subunits. The alpha subunits include proteasome component C2, C3, C7-L, C8 and C9; proteasome iota, zeta and HSPC (PSMA7) chains. The beta subunits comprise proteasome beta, epsilon and gamma chains. Quantitation did not yield a consistent trend of either increase or decrease in both alpha and beta subunits. With respect to alpha subunits, amongst the nine subunits identified expression of C9 was almost the same in both groups and that of C3, C8, and PSMA7 were decreased in DS by ~18%, 33% and 30%, respectively. By contrast, levels of C2, iota and zeta chains were increased by 35%, 9% and 59%, respectively with the increase being significant only in zeta chain ($P < 0.05$) (Fig. 3). The beta subunits exhibited somewhat a different pattern. Expression of the three identified beta chains; proteasome beta, epsilon and gamma chains was relatively the same in both controls and DS brain.

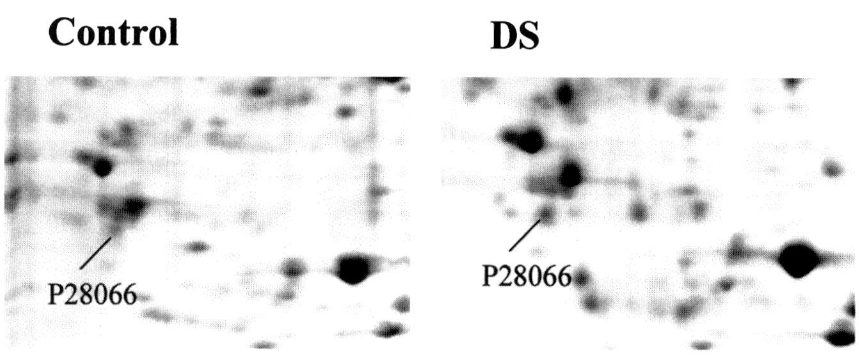

Fig. 3. Partial two-dimensional gel images of human fetal brain proteins. The gels were prepared and the protein was analyzed as described under legend to Fig. 1. The spot corresponding to proteasome zeta chain is indicated with the accession number

Regulatory subunits of 26S proteasome (19S)

The capping protein (19S) complex is composed of at least 18 proteins (Coux et al., 1996) that have a regulatory role and together with the 20S proteasome constitute the 26S proteasome. Seven such proteins, each represented by one spot, were identified by 2-DE. These include four ATPase subunits: tat binding protein (TBP)-1 (subunit 6a), TBP-7 (subunit 6b), subunit 7 (mss1) and p45 (subunit 8), and three non-ATPase subunits: p42 (subunit 10b), p40 (subunit 12) and subunit p28. Quantitation of the regulatory subunits of the 26S complex did not reveal any major alteration in expression of any of the subunits identified. Nevertheless, a decreasing or increasing trend was noted in both the ATPase as well as non-ATPase subunits. In fetal DS, levels of subunit 6a, 6b and 7 were increased insignificantly by ~10%, 33% and 17%, respectively, compared to controls, but that of subunit p28 was identical. Levels of subunit 8, s10 and s12 however, showed an opposite trend and decreased by ~30%, 31% and 15%, respectively.

Spearman rank-correlation analysis revealed significant correlation only between age and expression of proteasome iota ($n = 9$, $r = -0.9489$, $P = 0.0004$) and proteasome epsilon ($n = 9$, $r = -0.7227$, $P = 0.0311$) chains in fetal DS brain.

Discussion

This paper attempts to describe the expression pattern of different proteins involved in the ubiquitin-proteasome proteolytic pathway using 2DE. And data from the study indicate that no detectable alterations were apparent in fetal DS compared to controls in the majority of the investigated proteins except for isopeptidase T and proteasome zeta chain.

The ubiquitin/proteasome system stands out from the non-lysosomal proteolytic systems because it has the capacity to degrade almost any protein, but remains a harmless neighbor to intracellular proteins (Jentsch and Schlenker, 1995). This pathway identifies target proteins to be destroyed by making use of the ubiquitin conjugation system that involves E1, E2 and E3. The lack of significant change in expression of E1 and E2 may reflect that there is no conspicuous flaw in protein targeting in fetal DS. The 19S complex is composed of several proteins and serve as the mouth for the 20S digestive machinery, and provides components necessary for promoting selective degradation of ubiquitinated proteins, including ubiquitin recognition and binding site, isopeptidase and ATPase activities (Coux et al., 1996). Once again here no major alterations, except for a tendency to change, were observed in expression of subunits of the 19S complex that are identified by 2-DE. This may imply that in fetal DS, the substrate recognition as well as the unfoldase activities are not different to that of their age matched controls as far as expression is concerned.

The proteasome complex exists in both an ATP-independent 20S (~700 kDa) and ATP-dependent 26S (~1,700 kDa) forms in mammalian cells and the former is the proteolytic core that is responsible for the selective degradation of proteins destined to be removed by this system (Coux et al., 1996). More than fourteen gene products have been reported for 20S proteasome in mammalian cells (Rivett, 1998), however, only eleven of them were identified by 2-DE in fetal brain, several of which were alpha subunits. This may indicate that the unidentified subunits, which most of them were beta subunits, could be low abundance proteins that fell below the detection limit of 2-DE. Indeed, in chick embryonic muscle proteasome subunits expression is shown to be developmentally regulated and most of the less abundant proteins were of the beta types (Ahn et al., 1991). Previous studies on proteasome expression demonstrated varying results. It was reported to be unchanged in Alzheimer's disease (AD) (Keller et al., 2000a), decreased during aging associated with oxidative stress (Keller et al., 2000b), and increased in malignant tumor cells (Kumatori et al., 1990). All these studies measured the immunoreactivity of either the complex as a whole or the subunits as a group (alpha or beta), so it is not clear whether a specific alteration in the individual subunits is apparent or not. In the present study, however, it is shown that except for proteasome zeta chain, all the subunits identified were unaltered in fetal DS. This subunit displayed a significant elevation in fetal DS compared to controls (Fig. 3). The reason that caused the selective upregulation of proteasome zeta chain is presently unknown. It may include signaling molecules, such as reactive oxygen species (ROS), since they are known to increase expression of certain proteins (Castagne et al., 1999) and shown to be increased in fetal DS (Busciglio and Yanker, 1995). Loss of proteasome activity has been reported in AD (Keller et al., 2000a) and Parkinson's disease (McNaught and Jenner, 2001). Oxidative stress is implicated in this inhibition, as it was shown to impair proteasome activity in different cell lines and the inhibition is ascribed largely to oxidative modification of the subunits side-chain amino acids rather than protein loss (Reinheckel et al., 1998; Okada et al., 1999; Keller et al., 2000b). The proteasome complex is assembled from multiple alpha and beta subunit isoforms that associate in various combinations (Früh et al., 1994). Variation in either alpha and beta subunit levels may therefore affect proteasome activity by limiting either the rate of the complex formation or altering the stoichiometry of alpha and beta subunits. Alpha subunits were shown to be less susceptible to oxidative insults than beta subunits (Keller et al., 2000b). ROS may therefore use other means, for e.g., altering stoichiometry by selective upregulation, to further inhibit proteasome activity. This is not unlikely after all, considering the fact that concentration driven displacement can be used as a general mechanism to vary the subunit composition of the proteasome during assembly without changing its quaternary structure (Früh et al., 1994). Hence, the upregulated proteasome zeta chain may displace some key but less abundant subunits from the complex leading to formation of an inefficient enzyme complex which could accentuate the structural and functional abnormalities in DS brain. The decrease in expression of proteasome iota and epsilon chains with age in fetal DS brain points to the fact that these chains would not take part in proteasome assembly as age progresses and need to be replaced, supporting the existence of such a displacement mechanism. The possibility that the selective upregulation might have been caused to promote the proteasome assembly thereby maintaining activity against the eerie of oxidative stress in fetal DS is also can not be dismissed completely. Further studies need to be done at activity level to substantiate this result.

Multiple forms of deubiquitinating enzymes are identified (Zhu et al., 1996) and their precise function is unclear. Some deubiquitinating enzymes can apparently remove ubiquitin from ubiquitin conjugated substrates prior to proteasome-substrate binding, thereby impeding the turnover of such proteins. Others remove ubiquitin following degradation allowing recycling of ubiquitin and promote degradation (Hochstrasser, 1995; Coux et al., 1996). The UCH family of enzyme are suggested to hydrolyze small adducts of ubiquitin and generate free monomeric ubiquitins, but not to deubiquitinate ubiquitin-protein conjugates or disassemble polyubiquitin chains (Larsen et al., 1998). It appears to be no overt alteration in ubiquitin recycling mediated by this family of enzymes, as the expression of UCH L1 and UCH L3 in fetal DS was unchanged. This finding suggests that these enzymes may not have any relevance to DS pathology. In contrast to the UCH family, the UBP family is extremely divergent and represent majority of the deubiquitinating enzymes (Hochstrasser, 1996). UBP T primarily acts on unanchored ubiquitin chains, oligomers with a free ubiquitin carboxy-terminus (Wilkinson et al., 1995). These chains can be peptide remnants generated during substrate degradation by the proteasome (Hadari et al., 1992) or synthesized by a number of enzymes from free ubiquitins (Hochstrasser, 1995; Amerik et al., 1997). If such chains are left unattended will clog the proteasome and negatively affect the proteolytic process. UBP T by disassembling the unanchored chains, thus stimulates ubiquitin dependent proteolysis by the proteasome (Hadari et al., 1992; Wilkinson et al., 1995). However, overexpression of UBP T, as observed in the present study, may inhibit the proteolysis of proteins by rapidly dismantling ubiquitinated intermediates. Indeed, UBP14 a yeast isopeptidase most closely related and functionally homologous to UBP T, was found to inhibit proteolysis of a subset of proteins when overexpressed, and to deubiquitinate a significant fraction of proteasome bound polyubiquitinated substrates before degradation when proteasome activity was reduced by mutation (Amerik et al., 1997). For normal functioning of ubiquitin-proteasome pathway, optimum concentration of free ubiquitin chains must be maintained (Amerik et al., 1997). UBPs play crucial roles in the growth and development of eukaryotic organisms by controlling cellular ubiquitin chains via regulation of the ubiquitin chain assembly and/or disassembly rates. Failure to keep this balance will ultimately result in developmental defects (Amerik et al., 1997). The altered expression of UBP T in fetal DS can upregulate proteins eliminated by ubiquitination and have a negative modulatory role in growth and development, such as p53.

The present study describes, for the first time, the expression pattern of proteins involved in one of the major protein turnover pathways in fetal brain. The study revealed overexpression of two proteins of the pathway in fetal DS.

This upregulation may alter the delicate balance of the system and probably account for the developmental abnormalities observed in DS brain.

Acknowledgements

We are highly indebted to Red Bull Company, Salzburg, Austria for the generous financial support. E.E is a recipient of scholarship from the Austrian Academic Exchange Program, ÖAD.

References

Ahn JY, Hong SO, Kwak KB, Kang SS, Tanaka K, Ichihara A, Ha DB, Chung CH (1991) Developmental regulation of proteolytic activities and subunit pattern of 20S proteasome in chick embryo muscle. J Biol Chem 266: 15746–15749

Amerik A Yu, Swaminathan S, Krantz BA, Wilkinson KD, Hochstrasser M (1997) In vivo disassembly of free polyubiquitin chains by yeast Ubp 14 modulates rates of protein degradation by the proteasome. EMBO J 16: 4826–4838

An WG, Hwang SG, Trepel JB, Blagosklonny MV (2000) Protease inhibitor induced apoptosis: accumulation of p53, p21/WAF-1/CIP1 and induction of apoptosis are independent marker of proteasome inhibition. Leukemia 14: 1276–1283

Berndt P, Hobohm U, Langen H (1999) Reliable automatic protein identification from matrix-assisted laser desorption/ionization mass spectrometric peptide fingerprints. Electrophoresis 20: 3521–3526

Bradford M (1976) A rapid and sensitive method for the quantitation of microgram quantities of protein utilizing the principle of protein-dye binding. Anal Biochem 72: 248–254

Busciglio J, Yankner BA (1995) Apoptosis and increased generation of reactive oxygen species in Down's syndrome neurons in vitro. Nature 378: 776–779

Castagne V, Gautschi M, Lefevre K, Posada A, Clarke PG (1999) Relationships between neuronal death and the cellular redox status. Focus on the developing nervous system. Prog Neurobiol 59: 397–423

Ciechanover A (1994) The ubiquitin-proteasome proteolytic pathway. Cell 79: 13–21

Ciechanover A, Schwartz AL (1994) The ubiquitin mediated proteolytic pathway: mechanisms of recognition of the proteolytic substrate and involvement in the degradation of native cellular proteins. FASEB J 8: 182–191

Coux O, Tanaka K, Goldberg AL (1996) Structure and functions of the 20S and 26S proteasomes. Annu Rev Biochem 65: 801–847

DeMartino GN, Orth K, McCullough ML, Lee LW, Munn TZ, Moomaw CR, Dawson PA, Slaughter CA (1991) The primary structures of four subunits of the human high molecular weight proteinase, macropain (proteasome) are distinct but homologous. Biochim Biophys Acta 1079: 29–38

Epstein CJ (1995) Down Syndrome. In: Scriver CR, Beaudet AL, Sly WS, Valle D (eds) The metabolic and molecular bases of inherited disease, 7th edn, vol 1. McGraw Hill, New York, pp 749–794

Fountoulakis M, Langen H (1997) Identification of proteins by matrix assisted laser desorption ionization mass spectrometry following in-gel digestion in low salt, non-volatile buffer and simplified peptide recovery. Anal Biochem 250: 153–156

Früh K, Gossen M, Wang K, Bujard H, Peterson PA, Yang Y (1994) Displacement of housekeeping proteasome subunits by MHC-encoded LMPs: a newly discovered

mechanism for modulating the multicatalytic proteinase complex. EMBO J 13: 3236–3244
Groll M, Ditzel L, Lowe J, Stock D, Bochtler M, Bartunik HD, Huber R (1997) Structure of 20S proteasome from yeast at 2.4 Å resolution. Nature 386: 463–471
Hadari T, Warms JV, Rose IA, Hershko A (1992) A ubiquitin C-terminal isopeptidase that acts on polyubiquitin chains. Role in protein degradation. J Biol Chem 267: 719–727
Hilt W, Wolf DH (1996) Proteasomes: destruction as a program. Trends Biochem Sci 2: 96–102
Hochstrasser M (1995) Ubiquitin, proteasomes and the regulation of intracellular protein degradation. Curr Opin Cell Biol 7: 215–223
Hochstrasser M (1996) Ubiquitin dependent protein degradation. Ann Rev Genet 30: 405–439
Jentsch S, Schlenker S (1995) Selective protein degradation: a journey's end within the proteasome. Cell 82: 881–884
Keller JN, Hanni KB, Markesbery WR (2000a) Impaired proteasome function in Alzheimer's disease. J Neurochem 75: 436–439
Keller JN, Huang FF, Markesbery WR (2000b) Decreased levels of proteasome activity and proteasome expression in aging spinal cord. Neurosci 98: 149–156
Kumatori A, Tanaka K, Inamura N, Sone S, Ogura T, Matsumoto T, Tachikawa T, Shin S, Ichihara A (1990) Abnormally high expression of proteasomes in human leukemic cells. Proc Natl Acad Sci USA 87: 7071–7075
Langen H, Röder D, Juranville JF, Fountoulakis M (1997) Effect of the protein application mode and the acrylamide concentration on the resolution of protein spots separated by two dimensional gel electrophoresis. Electrophoresis 18: 2085–2090
Larsen CN, Krantz BA, Wilkinson KD (1998) Substrate specificity of deubiquitinating enzymes: ubiquitin C-terminal hydrolases. Biochemistry 37: 3358–3368
Leroy E, Boyer R, Auberger G, Leube B, Ulm G, Mezey E, et al (1998) The ubiquitin pathway in Parkinson's disease. Nature 395: 451–452
McNaught KStP, Jenner P (2001) Proteasomal function is impaired in substantia nigra in Parkinson's disease. Neurosci Lett 297: 191–194
Moazed D, Johnson AD (1996) A deubiquitinating enzyme interacts with SIR4 and regulates silencing in S. cerevisiae. Cell 86: 667–677
Murakami Y, Matsufuji S, Kameji T, Hayashi S, Igarashi K, Tamura T, Tanaka K, Ichihara A (1992) Ornithine decarboxylase is degraded by the 26S proteasome without ubiquitination. Nature 360: 597–599
Naviglio S, Matteucci C, Matoskova B, Nagase T, Nomura N, Di Fiore PP, Draetta GF (1998) UBP Y: a growth regulated human ubiquitin isopeptidase. EMBO J 17: 3241–3250
Okada K, Wangpoengtrakul C, Osawa T, Toyokuni S, Tanaka K, Uchida K (1999) 4-OH-2-nonenal mediated impairment of intracellular proteolysis during oxidative stress. J Biol Chem 274: 23787–23793
Peters JM (1994) Proteasomes: protein degradation machines of the cell. Trends Biochem Sci 19: 377–382
Reinheckel T, Sitte N, Ullrich O, Kuckelkorn U, Davies KJA, Grune T (1998) Comparative resistance of the 20S and 26S proteasome to oxidative stress. Biochem J 335: 637–642
Reinheckel T, Ullrich O, Sitte N, Grune T (2000) Differential impairment of 20S and 26S proteasome activities in human hematopoietic K562 cells during oxidative stress. Arch Biochem Biophys 377: 65–68
Rivett AJ (1998) Intracellular distribution of proteasomes. Curr Opin Immunol 10: 110–114
Shapiro BL (1999) The Down syndrome critical region. J Neural Transm 57: 41–60

Wilkinson KD, Tashayev VL, O'Connor LB, Larsen CL, Kasperek E, Pickart CM (1995) Metabolism of the polyubiquitin degradation signal: structure, mechanism, and role of isopeptidase T. Biochemistry 34: 14535–14546

Zhu Y, Carrol M, Papa FR, Hochstrasser M, D'Andrea AD (1996) DUB-1, a deubiquitinating enzyme with growth suppressing activity. Proc Natl Acad Sci USA 93: 3275–3279

Authors' address: Prof. Dr. G. Lubec, Department of Pediatrics, University of Vienna, Währinger Gürtel 18–20, A-1090 Vienna, Austria, e-mail: Gert.Lubec@akh-wien.ac.at

Functional genomics of Down syndrome: a multidisciplinary approach

M. Dierssen, E. Martí, C. Pucharcós, V. Fotaki, X. Altafaj, K. Casas, A. Solans, M. L. Arbonés, C. Fillat, and **X. Estivill**

Down Syndrome Research Group, Medical and Molecular Genetics Center-IRO, Hospital Duran i Reynals, L'Hospitalet de Llobregat, Barcelona, Spain

Summary. The availability of the DNA sequence of human chromosome 21 (HSA21) is a landmark contribution that will have an immediate impact on the study of the role of specific genes to Down syndrome (DS). Trisomy 21, full or partial, is a major cause of mental retardation and other phenotypic abnormalities, collectively known as Down syndrome (DS), a disorder affecting 1 in 700 births. The identification of genes on HSA21 and the elucidation of the function of the proteins encoded by these genes have been a major challenge for the human genome project and for research in DS. Over 100 of the estimated 300-500 genes of HSA21 have been identified, but the function of most remains largely unknown. It is believed that the overexpression of an unknown number of HSA21 genes is directly or indirectly responsible for the mental retardation and the other clinical features of DS. For this reason, HSA21 genes that are expressed in tissues affected in DS patients are of special interest.

Introduction

Trisomy of HSA21 produces a variety of developmental anomalies recognized as DS. The disorder affects one in about 700 newborns and is the most significant genetic cause of mild to moderate mental retardation, due to alterations in neural development. Individuals with trisomy 21 may show several abnormalities to various extents, including craniofacial dismorphology, short stature, congenital defects of the heart and gastrointestinal tract, infertility (mostly in men), immune system alterations, thyroid dysfunction or hematopoietic disorders that include a higher incidence of childhood leukemias. At a gross morphological level, DS brains are smaller than normal, brain weight is reduced disproportionately, mostly affecting the cerebellum and brainstem, and the depth and number of sulci are reduced. Neuronal number is reduced in distinct regions, including the cochlear nuclei, cerebellum, hippocampus, the cholinergic neurons of the basal forebrain, the granular layers of the cerebral cortex and areas of the brain stem, and abnormal neuronal morphology is observed, especially in the cerebral cortex. In addition, over the age of 35, DS individuals exhibit neuropathology alterations typical of Alzheimer's disease (AD) and frequently show premature aging (see Pueschel, 1992; Epstein, 1995; Flórez, 1992 for review). There are many other disorders involving HSA21 genes, such as familial amiotrophic lateral sclerosis, familial AD, Ewing's sarcoma and other primitive neuroectodermal tumors and acute myeloid leukemia.

Defining how an extra copy of all or part of HSA21 results in the phenotype of DS is a specific case of the more general problem of explaining how chromosomal imbalance produces abnormalities in morphology and function. No single mechanism can explain the deleterious consequences of aneuploidy, and therefore, there is no simple solution to counteract its phenotypic impact. While some loci may have a greater phenotypic effect, it is the cumulative effect of imbalance of many genes that determines the overall phenotype. Two hypothesis have been proposed to explain DS phenotype: i) the amplified developmental instability hypothesis (Shapiro, 1999) suggests that the DS phenotype is the result of a non-specific disturbance of chromosome balance, resulting in a disruption of homeostasis; ii) in contrast, the gene dosage hypothesis proposes that DS phenotypes may stem directly from the cumulative effect of overexpression of specific chromosome 21 gene products or indirectly through the interaction of these HSA21 gene products with the whole genome, transcriptome or proteome. Evidence from different murine models points to specific genes affecting phenotypes rather than the unspecific effect of the amount of extra genetic material (see Pritchard and Kola, 1999). However, we still cannot answer the question of how three copies of normal genes contribute to the abnormal phenotype of DS. Most commonly, DS is due to the presence of an extra copy of a complete chromosome 21 and it is assumed that the DS phenotypic features are a direct consequence of the overexpression of some number of genes contained within 21q. Based on the analysis of human individuals with segmental trisomy 21, it has been proposed that genetic loci situated in the segment of 21q22.2 that is referred to as the DS critical region (DSCR) could harbor genes with major effect. Data using a larger number of partial trisomy cases showed that only the most centromeric region of 21q could be excluded from containing relevant genes, in particular for mental retardation (Davisson et al., 1990). However, the resolution of this approach is limited by the high phenotypic variation among DS individuals and to date no convincing evidence exists of association between any particular phenotypic trait and overexpression of a specific HSA21 gene. It is, however, worth considering questions such as if all chromosome 21 genes are overexpressed, if overexpression of some genes could not drive any phenotypic effect or how many genes are relevant to DS.

The availability of the essentially complete sequence of 21q will have an immediate impact on the study of the genetic aspects of DS by providing a comprehensive catalogue of the genes on HSA21 (Becker and Joost, 1999). However, the biological function of most of the 225 identified genes the expression patterns of the novel genes or the evolutionary conservation of, in particular, those genes lacking functional associations remain largely unknown, as does their contribution, if any, to the DS phenotype. Even knowing the molecular defect, it is difficult to decipher the complex pathophysiology of the disease. Thus, we are now facing a new era, the postgenomics in which the goal will be to identify protein function and physiological role of gene products.

It should be born in mind that DS is a complex disorder because it is caused by an extra copy of a whole set of genes. The impact of gene overdosage on the transcription level may vary because some genes are highly regulated. Thus, the presence of three copies of many genes may result either in overexpression or repression of few transcripts. In addition, they can interact with each other and/or affect expression of genes located on other chromosomes. The effects of three copies of a gene can be even more dramatic at the protein expression level.

Two hundred and twenty-five is a small number for the complete gene content of approximately 1% of the human genome. It is significantly less than the number previously predicted. Previous data have consistently suggested that chromosome 21 was relatively gene-poor overall, and extremely so in some regions (Gardiner, 1997; Xu et al., 1995). This may be a factor that justifies that trisomy 21 is compatible with life (Korenberg et al., 1991).

Promising candidate genes and the encoded proteins are currently identified on the basis of the previous knowledge of the function of the protein encoded, their expression pattern, or from functional studies in models organisms.

Transcription factors are attractive candidates because imbalance of one component of a transcription factor complex may alter the effectiveness of the activation or repression of transcription of target genes. Genes within the ubiquitin pathway may alter rates of target protein degradation. Cell adhesion was long ago postulated, with intriguing preliminary data (Kurnit et al., 1985) to play a role in altering rates and extents of cell migration during development. Overexpression of one potassium channel gene has been shown to disregulate expression of other channel genes, affecting neuronal network excitability (Sutherland et al., 1999). If mental retardation and cognitive deficits are the primary focus of study, any of these genes could be relevant, such is the extent of our current understanding of the complex developmental processes leading to these conditions.

A relatively small number of HSA21 genes have been ascribed to a particular functional association, although in some cases this association is not very definitive as occurs with the presence of a transmembrane domain. In the cases where there is a lack of protein or functional domain data may be due to the lack of complete coding sequence information, expression patterns may help in prioritizing genes for further study. For relevance to mental retardation and cognitive deficits, genes with brain-specific expression such as PCP4 (Ziai et al., 1988) are of interest. Equally interesting are examples of brain-specific alternative processing, as is seen with Intersectin and DSCAM (Guipponi et al., 1998; Yamakawa et al., 1998). In an analysis of a number of novel gene models, alternative processing, some of it brain-specific, was observed in the majority of cases (Slavov et al., 2000).

Model organisms will provide the basis for functional studies of the known and novel chromosome 21 genes. The genomes of *Saccharomyces cerevisiae*, *Caenorhabditis elegans* and *Drosophila* (Goffeau et al., 1996; Adams et al., 2000) have been completely sequenced, and thus the complete set of proteins of each of these organisms is known. Comparison of the translations of all annotated HSA21 genes with the *Drosophila* set identified 23 HSA21 gene products with similarity to a *Drosophila* protein over the complete length. These similarities involve basic biochemical/biological functions and include SOD1 (superoxide dismutase), GART (a purine biosynthesis enzyme), CBS (cystathionine beta-synthetase), and those involved in RNA splicing and the ubiquitin pathway. A further set of 31 genes showed informative matches but only over a domain or subregion of the human protein. Previously known homologs include DYRK1A (MNBH) and SIM2 (*Drosophila* single-minded). Perhaps most interesting in both sets are those genes for which there is little or no functional data. Determining the phenotypes of mutants in the *Drosophila* genes is likely to shed light on the function of the homologous human genes. Genetic approaches available in the mouse make this model organism particularly powerful in the functional analysis of candidate genes and in defining molecular pathways underlying the pathogenesis of human disease.

Regions of HSA21 are conserved within segments of three mouse chromosomes. The centromere-proximal region of HSA21 through the MX genes is homologous with the telomeric region of MMU16. The next approximately 2 Mb segment of HSA21 is homologous with the centromere-proximal region of MMU17, and the telomeric 2 Mb of HSA21 is homologous with an internal segment of MMU10. On the basis of current data, the order of HSA21 homologues in the MMU16 and 10 segments appears to be completely conserved, although the boundaries of these regions are still approximate (Wilsthire et al., 1999; Mouse Genome Database: http://www.informatics.jax.org). Defining the endpoints of these homologous regions is critical for evaluating genotype-phenotype correlations within existing mouse models.

Currently, the best mouse models of DS are the MMU16 segmental trisomies, Ts65Dn and Ts1Cje. Ts65Dn is trisomic for the region spanning an undefined distance proximal to App through Mx to presumably the telomere of chromosome 16. The phenotype of Ts65Dn includes working memory impairment and long term memory deficits; delayed development and lower body weight; motor dysfunction; decreased responsiveness to pain; hyperactivity; and decreased ability to inhibit behavior (reviewed in Davisson and Costa, 1999; Crnic and Pennington, 2000; see also Costa et al., 1999; Martínez-Cué et al., 1999). Particularly interesting are observations of age-related loss of cholinergic neurons, decreased numbers of asymmetric synapses in the temporal cortex, abnormalities in neuron number in hippocampal regions, and deficiencies of beta-noradrenergic transmission within the hippocampus and cerebral cortex (Holtzman et al., 1996; Granholm et al., 2000; Kurt et al., 2000; Insausti et al., 1998; Dierssen et al., 1997). On the basis of current comparative mapping data, some of the known genes are present in only two copies in Ts65Dn and therefore would not contribute to its phenotypic features. The phenotypic consequences of these genes must be assessed in some fashion, because the Ts65Dn lacks some features of DS. Constructing single-gene transgenic mice expressing each of these and then combining each with the Ts65Dn by breeding would be laborious and probably of limited success. An alternative is to generate additional segmental trisomies using the Cre-lox system (Zheng et al., 2000).

The Ts1Cje mouse is a more recent model that is trisomic for the region of MMU16 from Sod1 through Mx and present phenotypic differences with Ts65Dn such as hypoactivity, no loss of cholinergic neurons, and no deficits in the visible platform part of the water maze tests (Sago et al., 1998). There are some differences in the genes trisomic in both models and it is tempting to conclude that these may be responsible for the phenotypes.

Multiple genes can be "added" to the Ts65Dn using transgenics carrying bacterial chromosomes (BACs), to look for enhanced DS-relevant phenotypes. The human sequence will be useful here in ensuring that clones are extensive enough to contain appropriate regulatory regions. Single-gene knockouts can also be "subtracted" from the Ts65Dn mouse model, to search for amelioration of phenotype. However, the most direct assessment of gene function will require mutation or overexpression of individual genes or sets of genes to identify promising candidates and design potential therapeutics can begin.

Dyrk1A/mnbh

The efforts of different laboratories have allowed the isolation of a number of candidate genes for DS neuropathology. One of the principal candidates is the mammalian homologue of *Drosophila minibrain*, named *Dyrk1A*. Due to its mapping position on HSA21, expression pattern in mouse and human, overexpression in DS fetuses and putative *mnb* analogous role during neurogenesis, *DYRK1A* is an outstanding candidate for implication in some DS phenotypes. The *Drosophila mnb* gene encodes a serine-threonine protein kinase, which is expressed in distinct neuroblast proliferation centers during postembryonic neurogenesis (Tejedor et al., 1995). Due to its relationship to the *Drosophila* gene *mnb* whose mutation results in specific defects in neurogenesis, and based on functional experiments on transgenic mice, *DYRK1A* has been proposed as a good candidate gene for mental retardation in DS. The kinase is characterized by its ability to catalyze tyrosine-directed autophosphorylation as well as phosphorylation of serine/threonine residues in substrates (Tejedor et al., 1995; Kentrup et al., 1996). Its cellular function is yet unknown. DYRK1A is, however, known to translocate into the nucleus and supposed to be involved in the control of cell growth and development. RNA in situ hybridization in mice detected expression in several regions of the mouse brain, including the olfactory bulb, the cerebellum, the cerebral cortex, the pyramidal cell layer of the hippocampus and several hypothalamic nuclei (Guimerà et al., 1996; Guimerà et al., 1999). This expression pattern corresponds to some of the regions of the brain that are abnormal in individuals with DS and suggests that altered expression of *DYRK1A* could have detrimental consequences in DS patients (Guimerà et al., 1996). Overexpression of DYRK1A has been found in DS fetal brain and its murine homologue *Dyrk1A* is also overexpressed in the trisomic murine model Ts65Dn. The pathogenetic impact of overexpression of *DYRK1A* on Down syndrome needs further elucidation. Data from murine models suggests its involvement in development and in brain function. Trangenic mice containing YACs encompassing a 2 Mb region of HSA21 (YAC152F7tel) displayed behavioral and learning defects (Smith et al., 1997). The only gene subsequently found to map to this genomic fragment was *DYRK1A*. Moreover *Dyrk1A* is in triplicate in a mouse model with segmental trisomy of MMU16, the Ts65Dn (see before), thus enabling to compare both models. Recently, a transgenic mouse has been generated with doxycycline regulated *Dyrk1A* transgene expression in the forebrain, in which phosphorylation of mouse Dyrk1A has been demonstrated in primary hippocampal cultures in the course of NGF induced differentiation.

Transgenic mice overexpressing specifically the full-length cDNA of *Dyrk1A* under the control of the inducible sheep metallothionein-I promoter (Altafaj et al., 2001) exhibit a delay in neuromotor preweaning development, an altered motor coordination and mild hyperactivity accompanied with a reduced emotionality.

TgDyrk1A are hyperactive with respect to controls in situations that usually provoke caution in normal mice, a result that can be interpreted as a failure of these mice to inhibit activity. Similar to our findings, the partial MMU16 trisomic model, the Ts65Dn mouse, is also unable to inhibit exploration of the open arms in the plus maze (Escorihuela et al., 1998). This inability to inhibit behavior has been ascribed to dysfunction of prefrontal cortex, specifically related with hyperactivity in rodents (Kolb and Gib, 1990). Patients with DS are more likely to be hyperactive than normal children (Pueschel et al., 1991) but the prefrontal deficit is more markedly expressed as a deficient behavioral inhibiting function (Kopp et al., 1983). Thus, specific overexpression of *Dyrk1A* leads to alterations in motor behavior, which could explain to some extent the behavioral alteration of DS patients. Those mice On the contrary, the transgenic TgYAC157Ftel were described as hypoactive (Smith et al., 1997), but this hypoactivity was very slight and only observed in non-stressful conditions (dimly illuminated arena), in which no significant hyperactivity was observed in TgDyrk1A. Thus overexpression of *Dyrk1A* may have fairly specific, although subtle, long term consequences on motor performance.

There are also data suggesting that overexpression of *Dyrk1A* also leads to specific learning impairment. In most of the existing models the paradigm used to assess these alterations is the Morris water maze task. In this task all the existing models with overexpression of Dyrk1A showed different degrees of impairment. Interestingly, similar results in the reverse platform test where the mice are required to learn a novel position of the hidden platform, moved to the quadrant opposite to the orginal location have been obtained in

YAC152F7tel transgenic mice (Smith et al., 1997), TgDyrk1A and in the partial MMU16 trisomy models (Escorihuela et al., 1995).

All these models are poor in learning the new location of the platform and they persist in searching in the wrong quadrant, indicating a poor cognitive flexibility. This flexibility requires that the animal does not persevere in choosing the wrong solution, and has been ascribed to the correct functioning of the prefrontal cortex. The fact that reversal and probe learning are more impaired than initial discrimination learning suggest that forebrain structures, underlying behavioral flexibility, but also probably hippocampal structures, are likely to be impaired. Thus imbalance of the *DYRK1A* region could be important in determining the alteration in forebrain regions with an impact in cognitive function. In the probe test, all murine models that overexpress *Dyrk1A* have shown different degrees of impairment. YAC152F7tel showed mild impairment in the hidden platform and probe tests (Smith et al., 1997) whereas Ts65Dn and Ts1Cje are more importantly impaired (Reeves et al., 1995; Escorihuela et al., 1995, 1998; Sago et al., 1998).

The specific contribution of *Dyrk1A* to the partial MMU16 model phenotype's and probably to DS is highlighted by a model of specific overexpression of this genes, the TgDyrk1A, showing a specific deficit in the development of learning strategies. Moreover, this impairment is specifically related to an alteration in reference memory but not in working memory, an alteration comparable with those detected in the Ts65Dn mice (Escorihuela et al., 1998) and clearly extend the findings described for the YAC152F7tel mice (Smith et al., 1997).

Thus specific *Dyrk1A* overexpression in mice leads to developmental, motor and cognitive deficits. These findings clearly indicate that DYRK1A overexpression should be sufficient to cause learning deficits and neurodevelopmental alterations in DS patients.

Dscr1

Gene products that are involved in the control of intracellular signaling pathways are good candidates to participate in the pathogenesis of DS. We have identified a new family of calcineurin binding proteins that are able to regulate calcineurin signaling pathways in vivo (Fuentes et al., 2000). The human family consists of three members, DSCR1 (Fuentes et al., 1995), ZAKI-4 (Mizayaski et al., 1996) and DSCR1L2 (GenBank accession no. AAF01684), all of which have mouse orthologs. DSCR1 maps on HSA21, is trisomic in DS individuals and the presence of an extra copy of the gene leads to its overexpression in fetal brain. Homologous proteins exist in lower eukaryotes, invertebrates and non-mammals vertebrates.

Taking into account that several members of the family including yeast Rcn1p and both DSCR1 and ZAKI-4 are able to bind CaNA and inhibit its signaling functions, the family has been named "calcipressins" and the DSCR1 gene product is now named as human calcipressin 1 (Fuentes et al., 2000).

DSCR1 and ZAKI-4, although showing distinct expression patterns in other tissues, share expression in heart and skeletal muscle: two organs where fine regulation of calcineurin activity is relevant (Molkentin et al., 1998; Semsarian et al., 1999). In fact, calcineurin deregulation has been suggested to be involved in both cardiac and skeletal muscle hypertrophy (Molkentin et al., 1998; Musaro et al., 1999; Semsarian et al., 1999).

DSCR1 is expressed as four protein isoforms through the use of four alternative first exons (Fuentes et al., 1997). Although all the different DSCR1 spliced transcripts are expressed in heart and skeletal muscle, only the exon 1-containing DSCR1 transcript can be detected in fetal and adult brain, and only the exon 4-containing mRNA is detected in placenta and kidney. These two DSCR1 transcripts encode predicted polypeptides of 197 amino acids of unknown function, which differ in their N-terminal 29 amino acids.

DSCR1 is a calcineurin catalytic A subunit binding protein which has its interaction domain in CaNA to the linker region between the catalytic domain and also presents a CaNB binding domain (Fuentes et al., 2000). Further, overexpression of DSCR1 can repress the transcriptional activation of a NF-AT-dependent promoter in response to phorbol 12-myristate 13-acetate (PMA) and ionophore by inhibiting the nuclear translocation of NF-AT. In particular, DSCR1 and ZAKI-4 inhibit NF-AT nuclear accumulation in response to calcium signaling, and consequently modulate NF-AT-dependent transcription (Fuentes et al., 2000). Although current data do not address the mechanism of DSCR1-mediated inhibition, the effects could be explained through the inhibition of calcineurin phosphatase activity by DSCR1 binding. Alternatively, NF-AT and DSCR1 might compete for calcineurin binding if their interaction sites overlap in the calcineurin molecule. Therefore, DSCR1 and the other related proteins in humans, may represent physiological endogenous inhibitors for calcineurin and define a new negative regulatory pathway in calcineurin signaling in mammals.

Calcium is used by the cell to modulate gene expression programs by various means such as direct binding of calcium by the transcriptional repressor DREAM, which leads to a reduction in its DNA activity (Carrión et al., 1999), or Ca2+/calmodulin kinase IV phosphorylation of the transcription factor CREB allowing its binding to the co-activator p300 (Chawla et al., 1998). The best known example of calcium regulation through phosphatases is the NF-AT transcription factor (reviewed in Crabtree, 1999). In this case, an increase in intracellular Ca2+ leads to the activation of calcineurin, which then dephosphorylates cytosolic NF-AT family members. This modification unmasks their nuclear localization signals and promotes translocation into the nucleus where they bind cooperatively to DNA with other transcription factors such as AP-1, c-MAF or GATA4. This signaling pathway is important for the immune response (reviewed Rao et al., 1997), cardiac and skeletal muscle hypertrophy (Molkentin et al., 1998; Musaro et al., 1999; Semsarian et al., 1999), slow fiber differentiation in skeletal muscle (Chim et al., 1998), cardiac valve development (de la Pompa et al., 1998; Ranger et al., 1998) and the differentiation of a pre-adipocyte cell line to adipocytes in culture (Ho et al., 1998).

In spite of the wide spectrum of calcineurin physiological roles, the list of its modulators is short. Calmodulin, the main regulator, acts as an enhancer of the phosphatase activity in response to increases in cytosolic Ca2+ concentrations. FKBP12 targets the phosphatase to intracellular calcium channels such as the ryanodine receptor (Sun et al., 1998; Aramburu et al., 1998) and the inositol 1,4,5-triphosphate receptor (Cameron et al., 1995). AKAP79 is a scaffold protein that also binds cAMP-dependent protein kinase and protein kinase C (Klauck et al., 1996) and inhibits the phosphatase activity upon binding. A238L acts as an inhibitor and is thought to be used by the virus to evade the host defense systems (Miskin et al., 1998). CHP is a CaNB homologous protein and inhibits calcineurin activity by impairing the assembly of the heterodimer (Lin et al., 1999). And finally, cain/cabin, is a recently described non-competitive inhibitor (Lai et al., 1998; Sun et al., 1998). The calcipressin family illustrates a distinct type of calcineurin regulator whose members can affect calcineurin-mediated signaling.

Calcineurin is the only serine/threonine protein phosphatase under the control of Ca2+/calmodulin (Klee et al., 1998). It functions as a heterodimer composed of a catalytic A subunit (CaNA) and a calcium binding regulatory B subunit (CaNB). DSCR1 interacts with the calcineurin molecule through the catalytic A subunit in both non-stimulated and calcium-stimulated cells (Fuentes et al., 2000), suggesting that the interaction can occur independently of the activation state of calcineurin, and therefore independently of CaM binding. This region is well conserved in all the CaNAs sequenced to date, including the three human isoforms α, β and γ, suggesting that DSCR1 may interact with all of the human isoforms.

DSCR1 expression is induced by calcium, through a calcineurin-dependent mechanism. These results support a model in which DSCR1 functions as a feedback inhibitor of calcineurin. An inhibitory feedback loop may be necessary to avoid sustained calcineurin activity in situations of prolonged Ca2+ stimulus. Each calcipressin family member might respond differentially to various stimuli providing an additional level of complexity to regulate calcineurin in vivo.

Although DS is a multifactorial disease, the role of the HSA21 gene DSCR1 as an inhibitor of calcineurin signaling could provide new insights towards the possible involvement of this signaling pathway in some of the clinical abnormalities observed in patients with DS. These aspects should be dissected further by analysis of mouse models of overexpression and disruption of Dscr1.

ITNS

Intersection (ITSN) maps to chromosome 21q22.1, between markers D21S320 and D21S325. *ITSN* exists with a high degree of similarity in flies, frogs and mammals, suggesting a conserved role in higher eukaryotes. A new human gene encoding for intersection 2 (ITSN2), a putative paralogue gene of intersection (which we herein will name intersection 1, ITSN1) has been isolated. *ITSN1* is a binding partner of dynamin that has been shown to participate in clathrin-mediated endocytosis. Human ITSN1 is expressed as three main transcripts of approximately 15, 6 and 4.5 kb (Guipponi et al., 1998). In distinction to ITSN2, the longest one corresponds to the long isoform and it is expressed almost exclusively in brain a fact that establishes a clear difference between both genes. Although widely expressed, ITSN2 expression is tissue-dependent and likely regulated during development.

ITSN1 is expressed as two major alternative spliced transcripts that differ in their 3′-ends (Guipponi et al., 1998; Pucharcós et al., 1999). The intersection 2 multimodular composition, the high degree of similarity with intersection 1 and its subcellular localization suggest a role in clathrin-mediated endocytosis. This fact has been confirmed by showing that its overexpression of intersection's SH3 domains inhibits transferrin uptake, posibly by disrupting targeting of accessory proteins of clathrin-coated pit formation (Pucharcós et al., 2000; Tong et al., 2000). Alternative splicing of *ITSN2* generates a short isoform with two EH domains, a coiled-coil region and five SH3 domains, and a longer isoform containing extra carboxy domains (DH, PH and C2 domains), suggesting that it could act as a guanine-nucleotide exchange factor for Rho-like GTPases. Intersection-2 isoforms show a subcellular distribution similar to other components of the endocytic machinery and co-localize with Eps15. Moreover, its overexpression inhibits transferrin internalization, supporting a functional role for *ITSN2* in endocytosis (Pucharcós et al., 2000).

The clathrin-mediated endocytosis pathway occurring at the plasma membrane of eukaryotic cells plays a crucial role in synaptic vesicle recycling in neurons, as well as in the internalization of receptor-ligand complexes and the maintenance of membrane identity in all cells. The uptake process mediated by clathrin-coated vesicles (CCVs) seems to require the coordinated interaction of many proteins to form large macromolecular complexes, and even though a growing list of proteins involved in CCV formation has emerged in recent years, the precise mechanism of this process remains unknown (reviewed in Cremona et al., 1997; Schmid, 1997; Marsh and McMahon, 1999).

The main structural component of CCVs is the 190-kDa protein clathrin heavy chain, which when associated with a 25-kDa clathrin light chain forms complexes called triskelions. The triskelions can organize themselves into a caged lattice forming the scaffold of the nascent CCVs (Pishvaee and Payne, 1998). Clathrin is recruited to the plasma membrane by the adapter protein AP-2, a heterotetramer that recognizes the signals present at the cytoplasmic tails of membrane proteins (Kirchhausen et al., 1997). AP-2 also binds to Eps15 (Fazioli et al., 1993), epsin (Chen et al., 1998), amphiphysin heterodimers (David et al., 1996; Wigge et al., 1997), auxilin (Owen et al., 1999) and AP180 (Wang et al., 1995).

The interaction occurs through the AP-2-adaptin appendage domain and a conserved DPF/W (aspartate-proline-phenylalanine/tryptophan) motif present in the ligands (Owen et al., 1999). All these molecules participate not only in the rearrangement of the clathrin coat, but also act as antennae to attract other proteins to the growing CCVs. The fission of the clathrin-coated invaginated pit into free vesicles is thought to be mediated by the GTPase dynamin, which oligomerizes itself forming a helical collar around the neck of the endocytic pits (Hinshaw and Schmid, 1995). The carboxy-end of dynamin contains a proline/arginine-rich domain able to interact with several SH3 (Src3 homology) domain-containing proteins including amphiphysin I and II, endophilin, syndapin and intersection (reviewed in Schmid et al., 1998; Marsh and McMahon, 1999).

The number of proteins involved in clathrin-mediated endocytosis from the plasma membrane has grown dramatically in the past few years revealing a complex picture of protein-protein interactions. Intersection 1 and its paralogous intersection 2 described here belong to this group of proteins and likely act as scaffolding on providing a great variety of interacting domains.

Human intersection is a multimodular protein containing two EH (Eps15 homology) domains, a central coiled-coil region and five consecutive SH3 domains. It exists also as a longer isoform, mainly expressed in brain, with an extended carboxy-end region composed of a DH (Dbl homology), a PH (pleckstrin homology) and a C2 domain (Giupponi et al., 1998; Pucharcós et al., 1999). Intersection orthologues have been identified in rat [named EHSH1 (Okamoto et al., 1999)], mouse [named Ese1 (Sengar et al., 1999)], Xenopus laevis (Yamabhai et al., 1998) and *Drosophila melanogaster* [named Dap160 (Roos et al., 1998)]. The short isoform of intersection binds to several molecules involved in clathrin-mediated endocytosis through its different domains. Thus, the EH domains bind to epsins (Sengar et al., 1999; Yamahbai et al., 1998) the coiled-coil region interacts with the central region of Eps15 and Eps15R (Sengar et al., 1999) and the SH3 domains interact with dynamin (Okamoto et al., 1999; Sengar et al., 1999; Yamahbai et al., 1998) and synaptojanin (Okamoto et al., 1999; Yamahbai et al., 1998). Intersection has also been proposed to be a connection between endocytosis and exocytosis since its coiled-coil region binds to SNAP25 and SNAP23 (Okamoto et al., 1999). The ability of intersection to interact with multiple components of the endocytic and exocytic machinery leads to the suggestion that it acts as a scaffolding protein necessary for the formation of CCVs.

Genes expressing during development, and more particularly in the developing brain may be of special importance in DS. Moreover, processes involving proteins with the capacity to interact with a myriad of other gene products may be more susceptible to the effects of dosage imbalance. Abnormal synaptic transmission has been reported in DS brains that determine a reduced efficiency of synaptic transmission, in which it is feasible that ITSN might participate.

BACE2

One of the most common features of DS is the presence in young individuals of neuropathological changes of Alzheimer's disease (AD). AD is characterized by the progressive accumulation of neuritic plaques in the brain limbic and association cortices. The neuritic plaques contain extracellular deposits of amyloid-β protein, an endoproteolytic product of APP. APP is anchored to internal membranes through its C-terminal region, and undergoes specific endoproteolytic cleavages during its transport to the cell surface. The enzymes responsible for these cleavages are membrane-anchored proteases known as α, β and γ secretases. A combination of β and γ secretase cleavages leads to the production of Aβ (Seubert et al., 1993). None of the APP secretases have been yet clearly identified. Metalloprotease disintegrins may be responsible for the α secretase cleavage (Buxbaum et al., 1998). Presenilin 1 (PS1) could be the γ secretase, since mice deficient in PS1 show decreased Abeta accumulation. (De Strooper et al., 1998). Recently, the transmembrane aspartic protease BACE cDNA with characteristics of a β secretase has been described (Hussain et al., 1999; Sinha et al., 1999; Vassar et al., 1999; Yan et al., 1999).

BACE2 (Solans et al., 2000) maps to chromosome 21 and encodes for a 518 amino acid polypeptide with the signature of the aspartyl proteases (reviewed by Rawlings and Barret, 1995), a widely distributed family of proteolytic enzymes whose human members include the secretory proteins pepsin, gastriscin and renin; the intracellular catepsins D and E; BACE; and napsin. All of these enzymes have in common an active site cleft containing two catalytic aspartic residues. Despite their amino acid similarity, there are considerable differences regarding their catalytic properties, tissue distribution, cellular location, and subsequent physiological roles. Although BACE2 shares a degree of amino acid identity with the other members of the family, several special features are distinguishable in both BACEs.

BACE2 is probably glycosilated and thus, two potential N-linked oligosacharide attachment sites are present. The most distinct feature of BACE2 is the C-terminus. This end, though similar to BACE, is very different from that of other human APs in both its extension and the presence of a transmembrane segment and several cysteine residues able to form disulfide bonds. With a predicted N-terminal signal sequence and a C-terminal membrane-anchoring sequence, BACE2 has the topology of a type 1 integral membrane protein and, therefore, may represent a membrane secretase (Hooper et al., 1997).

The gene BACE2 consists of nine exons extending over 108 kb of genomic DNA around marker D21S53 at chromosome 21q22.3. BACE2 is longer than the other human AP genes, but it maintains its genomic organization, although with the two aspartic residues in exons 2 and 6 and the two lobes connecting the peptide encoded by the first half of exon 5. Similarly BACE exons 2 to 9 appear to be located in exactly the same positions they occupy in BACE2, supporting the idea of both BACEs being paralogous genes and establishing a new subfamily within the AP family of proteins. Human BACE2 is expressed in a variety of adult and fetal tissues, suggesting a possible involvement of BACE2 in essential cellular functions during fetal and adult life. This expression pattern differs from the pattern of mRNA expression of BACE, which shows a relatively specific expression in pancreas

(Vassar et al., 1999; Yan et al., 1999). BACE2 is expressed as multiple transcripts of approximately 2, 3 and 9 kb, suggesting that its expression is regulated by either different promoters or by post-transcriptional processes, such as alternative splicing or the use of polyadenylation sites. Several alternative splicing variants involving exons 7 and 8 have been described for BACE2 (Solans et al., 2000) whose biological significance is unclear, although it has been speculated that the loss of exon 8 would generate a protein product lacking the transmembrane segment and, therefore, provides the possibility of having both a membrane-associated and a secreted enzyme.

The association of BACE and BACE2 with the membrane via the carboxy-terminal transmembrane domain may increase the efficiency or specificity of cleavage of membrane-bond substrates. This suggestion is consistent with the identification of APP as a substrate of BACE, given that APP is a type I.

Analysis of the complete sequence of chromosome 21 has provided the first look at all candidate DS genes. The next steps require verifying and refining the predicted, incomplete gene models, defining new models as necessary, and isolating complete cDNAs for each gene. With complete coding sequences, protein sequences can be examined for motifs, domains, and biochemical characteristics that may suggest function. The most challenging problem will then be determining the functions of these genes and the other "known" genes. While it is tempting to focus on genes whose protein characteristics suggest a hypothesis for relevance to some aspect of DS, the more than 100 genes distributed throughout the chromosome that have no functional association are too large a dataset to ignore. For these and other genes on 21q, detailed expression analysis may be informative. Demonstration that a gene shows increased expression in the trisomic state by northern blot or RT-PCR analysis, followed by RNA tissue in situ hybridization to define specific cell types, brain regions and developmental stages of expression, may help in selecting genes of greater or lesser interest.

References

Adams MD, Celniker SE, Holt RA, Evans CA, Gocayne JD, Amanatides PG, Scherer SE, Li PW, Hoskins RA, Galle RF, et al (2000) The genome sequence of Drosophila melanogaster. Science 287: 2185–2195

Aramburu J, García-Cózar F, Raghavan A, Okamura H, Rao A, Hogan PG (1998) Selective inhibition of NFAT activation by a peptide spanning the calcineurin targeting site of NFAT. Mol Cal 1: 627–637

Barrett AJ, Rawlings ND (1995) Families and clans of serine peptidases. Arch Biochem Biophys 318: 247–250

Benmerah A, Begue B, Dautry-Varsat A, Cerf-Bensussan N (1996) The ear of alpha-adaption interacts with the COOH-terminal domain of the Eps 15 protein. J Biol Chem 271: 12111–12116

Becker W, Joost HG (1999) Structural and functional characteristics of Dyrk, a novel subfamily of protein kinases with dual specificity. Prog Nucl Acid Res Mol Biol 62: 1–17

Buxbaum JD, Liu KN, Luo Y, Slack JL, Stocking KL, Peschon JJ, Johnson RS, Castner BJ, Cerretti DP, Black RA (1998) Evidence that tumor necrosis factor alpha converting enzyme is involved in regulated alpha-secretase cleavage of the Alzheimer amyloid protein precursor. J Biol Chem 273: 27765–27767

Cameron AM, Steiner JP, Roskams AJ, Ali SM, Ronnett GV, Synder SH (1995) Calcineurin associated with the inositol 1, 4, 5-triphosphate receptor-FKBP12 complex modulates Ca^{2+} flux. Cell 83: 463–472

Carrión AM, Link WA, Ledo F, Mellstrom B, Naranjo JR (1999) DREAM is a Ca^{2+}-regulated transcriptional repressor. Nature 398: 80–84

Chawla S, Hardingham GE, Quinn DR, Bading H (1998) CBP: a signal-regulated transcriptional coactivator controlled by nuclear calcium and CaM kinase IV. Science 281: 1505–1509

Chen H, Fre S, Slepnev VI, Capua MR, Takei K, Butler MH, Di Fiore PP, De Camilli P (1998) Epsin is an EH-domain-binding protein implicated in clathrin-mediated endocytosis. Nature 394: 793–797

Chin ER, Olson EN, Richardso JA, Yang Q, Humphries C, Shelton JM, Wu H, Zhu W, Bassel-Duby R, Williams RS (1998) A calcineurin-dependent transcriptional pathway controls skeletal muscle fiber type. Genes Dev 12: 2499–2509

Costa AC, Walsh K, Davisson MT (1999) Motor dysfunction in a mouse model for Down syndrome. Physiol Behav 68: 211–220

Crabtree GR (1999) Generic signals and specific outcomes: signalling through Ca^{2+}, calcineurin, and NF-AT. Cell 96: 611–614

Crnic LS, Pennington BF (2000) Down syndrome: neuropsychology and animal models. Progr Infancy Res 1: 69–111

Cremona O, de Camilli P (1997) Synaptic vessicle endocytosis. Curr Opin Neurobiol 7: 323–330

Davisson MT, Schmidt C, Akeson EC (1990) Segmental trisomy for murine chromosome 16: a new system for studying Down syndrome. In: Patterson D, Epstein CJ (eds) Molecular genetics of chromosome 21 and Down syndrome. Wiley-Liss, New York, pp 263–280

David C, McPherson PS, Mundigl O, de Camilli P (1996) A role of amphiphysin in synaptic vesicle endocytosis suggested by its binding to dynamin in nerve terminals. Proc Natl Acad Sci USA 93: 331–335

Davisson MT, Costa ACS (1999) Mouse models of Down syndrome. In: Popko B (ed) Mouse models in the study of genetic neurological disorders, vol 9. Plenum Press, New York, pp 297–327 (Adv Neurochem)

De la Pompa JL, Timmerman LA, Takimoto H, Yoshida H, Elia AJ, Samper E, Potter J, Wakeham A, Marengere L, Langille BL, et al (1998) Role of NF-Atc transcription factor in morphogenesis cardiac valvles and septum. Nature 392: 182–186

De Strooper B, Saftig P, Craessaerts K, Vanderstichele H, Guhde G, Annaert W, Von Figura K, Van Leuven F (1998) Deficiency of presenilin-1 inhibits the normal cleavage of amyloid precursor protein. Nature 391: 387–390

Dierssen M, Vallina IF, Baamonde C, Garcia-Calatayud S, Lumbreras MA, Florez J (1997) Alterations of central noradrenergic transmission in Ts65Dn mouse, a model for Down syndrome. Brain Res 749: 238–244

Epstein CJ (1995) The metabolic and molecular basis of inherited disease. In: Scriver CR, Beaudet AL, Sly WS, Valle D (eds) Down syndrome, vol 1, 7th edn. Mac Graw-Hill, New York

Escorihuela RM, Fernández-Teruel A, Vallina IF, Baamonde C, Lumbreras M, Diersen M, Tobena A, Florez J (1995) A behavioral assessment of Ts65Dn mice: a putative Down syndrome model. Neurosci Lett 199: 143–146

Escorihuela RM, Vallina IF, Martínez-Cué C, Baamonde C, Dierssen MM, Tobena A, Flórez J, Fernández-Teruel A (1998) Impaired short- and long-term memory in Ts65Dn mice, a model for Down syndrome. Neurosci Lett 247: 171–174

Fazioli F, Minichiello L, Matoskova B, Wong WT, Di Fiore PP (1993) eps15, a novel tyrosine kinase substrate, exhibits transforming activity. Mol Cell Biol 13: 5814–5828

Flórez J (1992) Neurologic abnormalities. In: Pueschel SM, Pueschel JK (eds) Biomedical concerns in persons with Down syndrome. Paul H Brookes, Baltimore, pp 159–173

Fuentes JJ, Pritchard MA, Planas AM, Bosch A, Ferrer I, Estivill X (1995) A new human gene from the Down syndrome critical region encodes a proline-rich protein highly expressed in fetal brain and heart. Hum Mol Gen 4: 1935–1944

Fuentes JJ, Pritchard MA, Estivill X (1997) Genomic organization, alternative splizing, and expression patterns of the DSCR1 (Down syndrome candidate region 1) gene. Genomics 44: 358–361

Fuentes JJ, Genescà L, Kingsbury TJ, Cunningham KW, Pérez-Riba M, Estivill X, de la Luna S (2000) DSCR1, overexpressed in Down syndrome, is an inhibitor of calcineurin-mediated signalling pathways. Hum Mol Gen 9: 1681–1690

Gardiner K (1997) Clonability and gene distribution on the human chromosome 21: reflections of junk DNA content? Gene 205: 39–46

Genome sequence of the nematode C. elegans: a platform for investigating biology (1998) The C. elegans sequencing. Science 282: 2012–2018

Goffeau A, Barrell BG, Bussey H, Davis RW, Dujon B, Feldmann H, Galibert F, Hoheisel JD, Jacq C, Johnston M, Louis EJ, Mewes HW, Murakami Y, Philippsen P, Tettelin H, Oliver SG (1996) Life with 6,000 genes. Science 274, 546: 563–567

Granholm AC, Sanders LA, Crnic LS (2000) Loss of cholinergic phenotype in basal forebrain coincides with cognitive decline in a mouse model of Down's syndrome. Exp Neurol 161: 647–663

Guimerà J, Casas C, Pucharcos C, Solans A, Doménech A, Planas AM, Ashley J, Lovett M, Estivill X, Pritchard MA (1996) A human homologue of human minibrain (MNB) is expressed in the neuronal regions affected in Down syndrome and maps to the critical region. Hum Mol Gen 5: 1305–1310

Guimerà J, Casas C, Estivill X, Pritchard M (1999) Human minibrain homologue (MNBH/DYRK1): characterization, alternative splizing, differential tissue expression, and overexpression in Down syndrome. Genomics 57: 407–418

Guipponi M, Scott HS, Chen H, Schebesta A, Rossier C, Antonorakis SE (1998) Two isoforms of a human intersection (ITSN) protein are produced by brain-specific alternative splizing in a stop codon. Genomics 53: 369–376

Hinshaw JE, Schmid SL (1995) Dynamin self-assembles into rings suggesting a mechanism for coated vesicle budding. Nature 374: 190–192

Ho IC, Kim JH, Eooney JW, Spiegelman BM, Glimcher LH (1998) A potential role for the nuclear factor activated T cell family of transcriptional regulatory proteins in adipogenesis. Proc Natl Acad Sci USA 95: 15537–15541

Holtzman DM, Santucci D, Kilbridge J, Chua Couzens J, Fontana DJ, Daniels SE, Johnson RM, Chen K, Sun Y, Carlson E, Alleva E, Epstein CJ, Mobley WC (1996) Developmental abnormalities and age-related neurodegeneration in a mouse model of Down syndrome. Proc Natl Acad Sci USA 93: 13333–13338

Hussain I, Powell D, Howlett DR, Tew DG, Meek TD, Chapman C, Gloger IS, Murphy KE, Southan CD, Ryan DM, Smith TS, Simmons DL, Walsh FS, Dingwall C, Christie G (1999) Identification of a novel aspartic protease (Asp 2) as beta-secretase. Mol Cell Neurosci 14: 419–427

Insausti AM, Megias M, Crespo D, Cruz-Orive LM, Dierssen M, Vallina IF, Insausti R, Florez J, Vallina TF (1998) Hippocampal volume and neuronal number in Ts65Dn mice: a murine model of Down syndrome. Neurosci Lett 253: 175–178

Kentrup H, Becker W, Heukelbach J, Wilmes A, Schurmann A, Huppertz C, Kaniulainen H, Joost HG (1996) Dyrk, a dual specificity protein kinaase with unic structural features whose activity is dependent on tyrosine residues between subdomains VII and VIII. J Biol Chem 271: 3488–3495

Kirchhausen T, Bonifacino JS, Riezman H (1997) Linking cargo to vesicle formation: receptor tail interactions with coat proteins. Curr Opin Cell Biol 4: 488–495

Klauck TM, Faux MC, Labudda K, Langeberg LK, Jaken S, Scott JD (1996) Coordination of three signaling enzimes by AKAP79, a mammalian scaffold protein. Science 271: 1589–1592

Klee CB, Ren H, Wang X (1998) Regulation of the calmodulin-stimulated protein phosphatase calcineurin. J Biol Chem 273: 13367–13370

Kolb B, Gibb R (1990) Anatomical correlates of behavioral change after neonatal prefrontal lesions in rats. Prog Brain Res 85: 241–255

Kopp CB, Krakow JB, Johnson KL (1983) Strategy production by Down syndrome children. Am J Ment Defic 88: 164–169

Kronenberg JR, Kalousek DK, Anneren G, Pulst S-M, Hall JG, Epstein CJ, Cox DR (1991) Delection of chromosome 21 and intelligence: molecular definition of the lesion. Hum Genet 87: 112–118

Kurnit DM, Aldrige JF, Matsuoka R, Mathysse S (1985) Increased adhesiveness of trisomy 21 cells and atrioventricular calnal malformations in Down syndrome: a stochastic model. Am J Med Genet 20: 385–399

Kurt MA, Davies DC, Kidd M, Dierssen M, Florez J (2000) Synaptic deficit in the temporal cortex of partial trisomy 16 (Ts65Dn) mice. Brain Res 858: 191–197

Lai MM, Burnett PE, Wolosker H, Blackshaw S, Synder SH (1998) Cain, a novel physiologic protein inhibitor of calcineurin. J Biol Chem 273: 18325–18331

Lin X, Sikkink RA, Rusnak F, Barber DL (1999) Inhibition of calcineurin phosphatase activity by calcineurin B homologous protein. J Biol Chem 274: 36125–36131

Marsh M, McMahon HT (1999) The structural era of endocytosis. Science 285: 215–220

Martinez-Cue C, Baamonde C, Lumbreras MA, Vallina IF, Dierssen M, Florez J (1999) A murine model for Down syndrome shows reduced responsiveness to pain. Neuroreport 10: 1119–1122

Miskin JE, Abrams CC, Goatley LC, Dixon LK (1998) A viral mechanism for inhibition of cellular phosphatase calcineurin. Science 281: 562–565

Mizayaki T, Kanou Y, Murata Y, Ohmori S, Niwa T, Maeda K, Yamamura H, Seo H (1996) Molecular cloning of a novel thyroid hormone-responsive gene, ZAKI-4, in human skin fibroblasts. J Biol Chem 271: 14567–14571

Molkentin JD, Lu JR, Antos CL, Markham B, Richardson J, Robbins J, Grant SR, Olson EN (1998) A calcineuri-dependent pathway for cardiac hypertrophy. Cell 93: 215–228

Musaro A, McCullagh JK, Naya FJ, Olson EN, Rosenthal N (1999) IGF-1 induces skeletal myocite hypertrophy through calcineurin in association with GATA-2 and NF-Atc1. Nature 400: 581–585

Okamoto M, Schoch S, Sudhof TC (1999) EHSH1/intersection, a protein that contains EH and SH3 domains and binds to dynamin and SNAP-25. A protein connection between exocytosis and endocytosis? J Biol Chem 274: 18446–18454

Owen DJ, Vallis Y, Noble ME, Hunter JB, Dafforn TR, Evans PR, McMahon HT (1999) A structural explanation for the binding of multiple ligands by the alpha-adaption appendage domain. Cell 97: 805–815

Pishvaee B, Payne GS (1998) Clathrin coats–threads laid bare. Cell 95: 443–446

Pritchard MA, Kola I (1999) The "gene dosage effect" hypothesis versus the "amplified developmental instability" hypothesis in Down syndrome. J Neural Transm [Suppl] 57: 293–303

Pucharcos C, Fuentes JJ, Casas C, de la Luna S, Alcántara S, Arbonés ML, Soriano E, Estivill X, Pitchard M (1999) Alu-Splice cloning of human Intersection (ITSN), a putative multivalent binding protein expressed in proliferating and differentiating neurons and overexpressed in Down syndrome. Eur J Hum Genet 7: 704–712

Pucharcos C, Estivill X, de la Luna S (2000) Intersection 2, a new multimodular protein involved in clathrin-mediated endocytosis. FEBS Lett 478: 43–51

Pueschel SM (1992) Phenotypic characteristics. In: Pueschel SM, Pueschel JK (eds) Biomedical concerns in persons with Down syndrome. Paul H. Brookes, Baltimore, pp 1–12

Pueschel SM, Bernier JC, Pezullo JC (1991) Behavioral observations in children with Down's syndrome. J Ment Defic Res 35: 502–511

Ranger AM, Grusby MJ, Hodge MR, Gravallese EM, De la Brousse FC, Hoey T, Mickanin C, Baldwin HS, Glimcher LH (1998) The transcription factor, NF-ATc is essential for cardiac valve formation. Nature 392: 186–189

Rao A, Luo C, Hogan PG (1997) Transcription factors of the NFAT family: regulation and function. Ann Rev Immunol 15: 707–747

Reeves RH, Irving NG, Moran TH, Wohn A, Kitt C, Sisodia SS, Schmidt C, Bronson RT, Avisson MT (1995) A mouse model for Down syndrome exhibits learning and behaviour deficits. Nat Genet 11: 177–184

Roos J, Kelly RB (1998) Dap160, a neural-specific Eps15 homology and multiple SH3 domain-containing protein that interacts with Drosophila dynamin. J Biol Chem 273: 19108–19119

Sago H, Carlson EJ, Smith DJ, Kilbridge J, Rubin EM, Mobley WC, Epstein CJ, Huang TT (1998) Ts1Cje, a partial trisomy 16 mouse model for Down syndrome, exhibits learning and behavioral abnormalities. Proc Natl Acad Sci USA 95: 6256–6261

Shapiro BL (1999) The Down syndrome critical region. J Neural Transm [Suppl] 57: 41–60

Semsarian C, Wu MJ, Ju YK, Marciniec T, Yeoh T, Allen DG, Harvey RP, Graham RM (1999) Skeletal muscle hypertrophy is mediated by a Ca^{2+}-dependent calcineurin signalling pathway. Nature 400: 576–581

Sengar AS, Wang W, Bishay J, Cohen S, Egan SE (1999) The EH and SH3 domain Ese proteins regulate endocytosis by linking to dynamin and Eps15. EMBO J 18: 1159–1171

Seubert P, Oltersdorf T, Lee MG, Barbour R, Blomquist C, Davis DL, Bryant K, Fritz LC, Galasko D, Thal LJ, Lieberburg I, Schenk DB (1993) Secretion of beta-amyloid precursor protein cleaved at the amino terminus of the beta-amyloid peptide. Nature 361: 260–263

Schmid SL (1997) Clathrin-coated vesicle formation and protein sorting: an integrated process. Annu Rev Biochem 66: 511–548

Schmid SL, McNiven MA, De Camilli P (1998) Dynamin and its partners: a progress report. Curr Opin Cell Biol 10: 504–512

Smith DJ, Stevens ME, Sudanagunta SP, Bronson RT, Makhinson M, Watabe AM, O'Dell TJ, Fung J, Weier HG, Cheng J, Rubin EM (1997) Functional screening of 2 Mb of human chromosome 21q22.2 in transgenic mice implicates minibrain in learning defects associated with Down syndrome. Nat Genet 16: 28

Sinha S, Anderson JP, Barbour R, Basi GS, Caccavello R, Davis D, Doan M, Dovey HF, Frigon N, Hong J, Jacobson-Croak K, Jewett N, Keim P, Knops J, Lieberburg I, Power M, Tan H, Tatsuno G, Tung J, Schenk D, Seubert P, Suomensaari SM, Wang S, Walker D, Zhao J, McConlogue L, John V (1999) Purification and cloning of amyloid precursor protein beta-secretase from human. Nature 402: 537–540

Slavov D, Hattori M, Sakaki Y, Rosenthal A, Shimizu N, Minoshima S, Kudoh J, Yaspo ML, Ramser J, Reinhardt R, Reimer C, Clancy K, Rynditch A, Gardiner K (2000) Criteria for gene identification and features of genome organization: analysis of 6.5 Mb of DNA sequence from human chromosome 21. Gene 247: 215–232

Smith DJ, Stevens ME, Sudanagunta SP, Bronson RT, Makhinson M, Watabe AM, O'Dell TJ, Fung J, Weier HU, Cheng JF, Rubin EM (1997) Functional screening of 2 Mb of human chromosome 21q22.2 in transgenic mice implicates minibrain in learning defects associated with Down syndrome. Nat Genet 16: 28–36

Solans A, Estivill X, de La Luna S (2000) A new aspartyl protease on 21q22.3, BACE2, is highly similar to Alzheimer's amyloid precursor protein beta-secretase. Cytogenet Cell Genet 89: 177–184

Stowell MH, Marks B, Wigge P, McMahon HT (1999) Nucleotide-dependent conformational changes in dynamin: evidence for a mechanochemical molecular spring. Nature Cell Biol 1: 27–32

Sun L, Woun HD, Loh C, Stolow M, He W, Liu JO (1998) Abin 1, a negative regulator for calcineurin signalling in T lymphocytes. Immunity 8: 703–711

Sutherland ML, Williams SH, Abedi R, Overbeek PA, Pfaffinger PJ, Noebels JL (1999) Overexpression of a Shaker-type potassium channel in mammalian central nervous system dysregulates native potassium channel gene expression. Proc Natl Acad Sci USA 96: 2451–2455

Tejedor F, Zhu XR, Kaltenbach E, Ackermann A, Baumann A, Canal I, Heisenberg M, Fischbach KF, Pongs O (1995) Minibrain: a new protein kinase family involved in postembryonic neurogenesis in Drosophila. Neuron 14: 287–301

Tong XK, Hussain NK, Heuvel E, Kurakin A, Abi-Jaude E, Quinn CC, Olson MF, Marais R, Baranes D, Kay BK, Mc Pherson PS (2000) The endocytic protein intersection is a major binding partner for the Ras exchange factor mSos1 in rat brain. EMBO J 19: 1263–1271

Vassar R, Bennett BD, Babu-Khan S, Kahn S, Mendiaz EA, Denis P, Teplow DB, Ross S, Amarante P, Loeloff R, Luo Y, Fisher S, Fuller J, Edenson S, Lile J, Jarosinski MA, Biere AL, Curran E, Burgess T, Louis JC, Collins F, Treanor J, Rogers G, Citron M (1999) Beta-secretase cleavage of Alzheimer's amyloid precursor protein by the transmembrane aspartic protease BACE. Science 286: 735–741

Wang LH, Sudhof TC, Anderson RG (1995) The appendage domain of alpha-adaption is a high affinity binding site for dynamin. J Biol Chem 270: 10079–10083

Wigge P, Kohler K, Vallis Y, Doyle CA, Owen D, Hunt SP, McMahon HT (1997) Amphiphysin heterodimers: potential role in clathrin-mediated endocytosis. Mol Biol Cell 8: 2003–2015

Wiltshire T, Pletcher M, Cole SE, Villanueva M, Birren B, Lehoczky J, Dewar K, Reeves RH (1999) Perfect conserved linkage across the entire mouse chromosome 10 region homologous to human chromosome 21. Genome Res 9: 1214–1222

Xu H, Wei H, Tassone F, Graw S, Gardiner K, Weissman SM (1995) A search for genes from the dark band regions of human chromosome 21. Genomics 27: 1–8

Yamabhai M, Hoffman NG, Hardison NL, McPherson PS, Castagnoli L, Cesareni G, Kay BK (1998) Intersection, a novel adaptor protein with two Eps15 homology and five Src homology 3 domains. J Biol Chem 273: 31401–31407

Yamakawa K, Huot YK, Haendelt MA, Hubert R, Chen XN, Lyons GE, Korenberg JR (1998) DSCAM: a novel member of the immunoglobulin superfamily maps in a Down syndrome region and is involved in the development of the nervous system. Hum Mol Genet 7: 227–237

Yan R, Bienkowski MJ, Shuck ME, Miao H, Tory MC, Pauley AM, Brashier JR, Stratman NC, Mathews WR, Buhl AE, Carter DB, Tomasselli AG, Parodi LA, Heinrikson RL, Gurney ME (1999) Membrane-anchored aspartyl protease with Alzheimer's disease beta-secretase activity. Nature 402: 533–537

Ziai MR, Sangameswaran L, Hempstead JL Danho W, Morgan JI (1988) An immunochemical analysis of the distribution of a brain-specific polypeptide, PEP-19. J Neurochem 51: 1771–1776

Zheng B, Sage M, Sheppeard EA, Jurecic V, Bradley A (2000) Engineering mouse chromosomes with Cre-loxP: range, efficiency, and somatic applications. Mol Cell Biol 20: 648–655

Authors' address: Dr. M. Dierssen, Down Syndrome Research Group, Medical and Molecular Genetics Center-IRO, Hospital Duran i Reynals, Autov. Castelldefels km 2.7, E-08907 L'Hospitalet de Llobregat, Barcelona, Spain.

Unaltered expression of Fas (CD95/APO-1), Caspase-3, Bcl-2 and Annexins in brains of fetal Down syndrome: evidence against increased apoptosis

E. Engidawork[1], **N. Balic**[2], **J.-F. Juranville**[3], **M. Fountoulakis**[3], **M. Dierssen**[4], and **G. Lubec**[1]

[1] Department of Pediatrics, and
[2] Institute of Medical and Chemical Laboratory Diagnostics, AKH, University of Vienna, Vienna, Austria
[3] F. Hoffmann-La Roche, Basel, Switzerland
[4] Medical and Molecular Genetics Center, IRO Hospital Duran i Reynals, Barcelona, Spain

Summary. Apoptosis is the mechanism by which cells are programmed to die under a wide range of physiological and developmental stimuli. Accumulating evidence indicates that enhanced apoptosis (programmed cell death) in Down syndrome (DS) may play a role in mental retardation and precocious neurodegeneration of the Alzheimer-type. In this regard, alteration of several apoptosis related proteins have been reported in adult DS brain. Fetal DS neurons exhibited increased reactive oxygen species leading to early apoptosis, however, expression of apoptosis related proteins in fetal DS, has never been considered. To address this issue, we investigated the expression of proteins involved in apoptosis including Fas (CD95, APO-1), caspase-3, Bcl-2 and annexins in the cerebral cortex of control and DS fetal brain by western blot and two dimensional electrophoresis. Here, we report that no detectable changes were obtained in fetal DS brain in the expression of Fas, caspase-3, Bcl-2 and Annexins (I, II, V, and VI) compared to controls. In parallel experiment, we also examined the expression of neuron specific enolase (NSE), a neuronal marker found to be decreased in adult DS brain, to see if there is any neuronal loss and no difference was observed between the two groups. Protein expression did not correlate with age. The unchanged levels of Fas, Bcl-2 and annexins together with unaltered caspase-3 expression, a predominant caspase that executes apoptosis in the developing nervous system, suggest that enhanced apoptosis may not be apparent in fetal DS brain as demonstrated for adult DS brain.

Introduction

Down syndrome (DS) is the most common chromosomal disorder with developmental brain abnormalities (Wisniewski and Kida, 1994) that results in early mental retardation and precocious, age dependent Alzheimer-type neurodegeneration (Epstein, 1995; Wisniewski et al., 1985). Brains from adult individuals with DS manifest abundant senile plaques, neurofibrillary tangles and extensive neuronal loss throughout the cortex (Mann, 1988). Although the mechanism leading to neuronal cell death is not clearly delineated, there is accumulating evidence that apoptosis may be involved (Anderson et al., 2000; Sawa et al., 1997; Seidl et al., 1999). Apoptosis, a genetically programmed cell death that is supposed to serve the process of neuronal selection during central nervous system development (Hutchins and Barger, 1998), can be maneuvered by pathological processes to cause amplified cell death. This seems the reason why apoptosis is now recognized as a common event in acute and chronic neurodegenerative disorders, such as Alzheimer's disease (Dragunow et al., 1997), a disorder closely related to DS.

The current knowledge about molecular mechanisms of apoptosis assumes that the balance between interacting pro- and antiapoptotic factors is the critical element to set a cell death program into motion or to hold it in check, called the "death-life rheostat" (Kroemer, 1997). Fas (CD95/APO-1) is one of the proapoptotic proteins that belongs to the tumour necrosis factor/nerve growth factor receptor superfamily of cell surface proteins and has been shown to mediate receptor dependent programmed cell death. When engaged by CD95 ligand, CD95 oligomers are formed on the target cell and a signal to enter the cell's intrinsic apoptotic pathway is transmitted via the cytoplasmic so called "death domain", which in turn triggers the caspase cascade executing apoptosis (Becher et al., 1998a). Caspases are cysteine proteases that play an important role as an effector arm of the cell death machinery (Cohen, 1997). Among the identified caspases, caspase-3 appears to play a significant role in neuronal apoptosis particularly, in the developing nervous system (Kuan et al., 1999; Srinivasan et al., 1998). Another family of proteins that regulate neuronal apoptosis in response to a wide variety of stimuli are encoded by the Bcl-2 (B cell leukemia-2 gene product) gene family. The interplay between the Bcl-2 and caspase gene families has a paramount importance in regulating cell death in the developing nervous system (Srinivasan et al., 1998). Bcl-2, an important antiapoptotic member of the family that is found inserted into the mitochondrial membrane and the endoplasmic reticulum, appears to be the primary factor responsible for keeping cells from choosing the apoptotic pathway by binding and neutralizing the proapoptotic factors (Bax). In addition, Bcl-2 attenuates apoptosis by decreasing generation of ROS (Reactive oxygen species) and maintaining the mitochondrial membrane potential (Kroemer, 1997).

Other proteins implicated in apoptosis are Annexins. Annexins form a multigene family of soluble calcium binding proteins that interact with phospholipids and cellular membranes in a calcium dependent manner. To date, 10 annexins have been identified in vertebrates (I–VII, XI, XIII) (Osterloh et al., 1998) with annexin V being the most widely distributed in mammalian organisms (Kaetzel et al., 1989). The genuine functions of annexins are not well established, however, there has been an accumulation of information on and speculations about the cellular activities of annexins. These include: membrane related activities such as, exocytosis, endocytosis and intracellular vesicle trafficking (Creutz, 1992; Francis et al., 1992; Raynal and Pollard, 1994); DNA synthesis and cell proliferation (Chiang et al., 1999); and apoptosis (Gidon-Jeangirard et al., 1999a, b; Sakamoto et al., 1996; Tzima and Walker, 2000). Annexin I, a putative second messenger of glucocorticoids that are known to induce apoptosis, is demonstrated to enhance apoptosis by altering the cellular sensitivity to ROS (Sakamoto et al., 1996). Targeted disruption of the annexin II gene has been describe to affect multiple physiological activities including apoptosis (Chiang et al., 1999). Annexin V, another member of the family, has been used as a marker for apoptosis owing to its binding to externalized phosphatidylserine, an event well documented to be a hallmark of cells undergoing apoptosis (Martin et al., 1995). Annexin VI message was also shown to be upregulated during programmed cell death in PC 12 cells and developing sympathetic neurons (Wang et al., 1997).

Fetal DS neurons in vitro exhibited increased ROS that resulted in early apoptosis (Busciglio and Yanker, 1995). Although alterations of the Bcl-2 gene family (Sawa et al., 1997) and Fas (Seidl et al., 1999) are demonstrated in adult DS brain, to our knowledge nothing has so far been described about the expression of Fas, Bcl-2 and annexins in fetal DS. In the present study, the expression of these proteins was investigated in the cerebral cortex of control and DS fetal brains by western blotting and two-dimensional electrophoresis (2-DE).

Material and methods

Fetal brain tissue

Brain tissue (cerebral cortex) of aborted control fetuses with no obvious abnormalities (n = 7; 1 female and 6 males with mean gestational age of 18.8 ± 2.2 weeks) and DS (n = 9; 2 females and 7 males with mean gestational age of 19.6 ± 2.0 weeks) was used for the experiment. The brain samples were obtained from the Medical and Molecular Genetics Center-IRO, Hospital Duran i Reynals, Barcelona, Spain. All samples were stored at −70°C and the freezing chain was never interrupted.

Western blot

Sample preparation

Frozen brain samples were thawed on ice and homogenized in 4 volumes of TBS buffer (100 mM Tris, 150 mM NaCl, pH 7.5) containing protease inhibitors, (1 μg/ml each of antipain, pepstatin A, and leupeptin and 10 mM of phenylmethylsulfonyl fluoride) and 0.05% Tween 20. Samples were homogenized on ice for 30 seconds (6–8 strokes) with a Potter-Elvehjem homogenizer. The brain homogenate then was centrifuged for 10 min at 10,000 × g and 4°C. To the supernatant, equal volume of sample buffer (125.5 mM Tris-HCl base, 70 mM sodium dodecyl sulfate, 0.001% Bromophenol blue, 20% glycerol, and 2% 2-mercaptoethanol, pH 6.8) was added and the mixture was heated at 95°C for 15 min. The concentration of protein was determined by the Bicinchoninic acid assay using bovine serum albumin as a standard (Pierce, USA).

Immunoblotting

Protein extracts containing 30μg of Fas, caspase-3 or Bcl-2 were subjected to sodium dodecyl sulfate-polyacrylamide gel electrophoresis (200V for 3hrs), and transferred to polyvinylidene fluoride membranes (Immobilon™-P, MilliPore, Bedford, USA) (192V for 1hr). The membranes were then washed two times for 5min in Tris buffer (100mM Tris-HCl, 150mM NaCl, pH 7.5) and later blocked with blotto (1.5% non-fat dry milk in TBST (100mM Tris-HCl, 150mM NaCl, pH 7.5, 1mM $MgCl_2$, 0.1% Tween-20)) overnight at 4°C followed by incubation with anti-Fas antibody (Cat # AM35, Oncogene Research Products, Boston, USA) (1:200), anti-caspase-3 antibody (Cat # H-277, Hybridomus, Copenhagen, Denmark) (1:1,000) and anti Bcl-2 antibody (Cat # AF827, R&D Systems, Inc., Minneapolis, USA) (1:1,000) at room temperature for 4 and 3 hours, respectively on a rocking shaker. The membranes were washed three times for 15 minutes in blocking solution followed by incubation with anti mouse IgG (H + L) (Cat # 1031-05) and anti-rabbit IgG (H + L) (Cat # 4050-05) (Southern Biotechnology Associates Inc Birmingham, USA) horseradish peroxidase conjugated secondary antibody in blotto (1:2,000) for one hour at room temperature on a rocking shaker. The membranes were washed again three times for 15 minutes followed by incubation with NEN chemiluminescence reagent for 1 minute and subsequent exposure to films (Kodak Blue XB-1, Rochester, USA). Developed films were scanned and densities of the immunoreactive bands were calculated by RFLP Scan 2.1 software program (Scanalytics, Billerica, USA). Protein size markers (Cat # 161-0372) were purchased from BIORAD.

Two dimensional gel electrophoresis (2-DE)

Brain tissue was suspended in 0.5ml of sample buffer consisting of 40mM Tris, 5M urea (Merck, Darmstadt, Germany), 2M thiourea (Sigma, St. Louis, MO, USA), 4% CHAPS (Sigma), 10mM 1,4-dithioerythritol (Merck), 1mM EDTA (Merck) and a mixture of protease inhibitors, 1mM PMSF and 1μg of each pepstatin A, chymostatin, leupeptin and antipain. The suspension was sonicated for approximately 30sec and centrifuged at 150,000g for 45min. The protein content in the supernatant was determined by the Coomassie blue method (Bradford, 1976).

2-DE was performed essentially as reported (Langen et al., 1997). Samples of 2mg protein were applied on immobilized pH 3–10 nonlinear gradient strips in sample cups at their basic and acidic ends. Focusing started at 200V and the voltage was gradually increased to 5,000V at 3V/min and kept constant for a further 24h (approximately 180,000kVh totally). The second-dimensional separation was performed on 9–16% SDS gradient polyacrylamide gels. The gels (180 × 200 × 1.5mm) were run at 40mA per gel. After protein fixation for 12h in 40% methanol, containing 5% phosphoric acid, the gels were stained with colloidal Coomassie blue (Novex, San Diego, CA, USA) for 48h. Molecular masses were determined by running standard protein markers (Gibco, Basel, Switzerland), covering the range 10–200kDa. pI values were used as given by the supplier of the IPG strips. Excess of dye was washed out from the gels with H_2O and the gels were scanned in an Agfa DUOSCAN densitometer (resolution 200). Electronic images of the gels were recorded using Photoshop (Adobe) and PowerPoint (Microsoft) software.

Matrix-associated laser desorption ionization mass spectroscopy (MALDI-MS)

MALDI-MS analysis was performed as described elsewhere (Fountoulakis and Langen, 1997) with some modifications. The spots were excised with a spot picker and placed into 96-well microtiter plates. Each spot was destained with 100μl of 30% acetonitrile in 50mM ammonium bicarbonate and dried in a speedvac evaporator. Each dried gel piece was rehydrated with 4 μl of 3 mM Tris-HCl, pH 9.0, containing 50 ng trypsin (Promega, Madison, WI, USA). After 16 h at room temperature, 7 μl of H_2O was added to each gel piece and the samples were shaken in for 10 min. Four μl of 50% acetonitrile, containing 0.3% trifluoroacetic acid and the standard peptides des-Arg-bradykinin (Sigma, 904.4681 Da) and adrenocorticotropic hormone fragment 18–39 (Sigma, 2,465.1989 Da), in water were added to each gel piece and shaken for 10 min. The application of the samples was performed with a SymBiot I sample processor (PE Biosystems, Framingham, MA, USA). 1.5 μl of the peptide mixture was simultaneously applied with 1 μl of matrix, consisting of a saturated solution of α-cyano-4-hydroxycinnamic acid (Sigma) in 50% acetonitrile, containing 0.1% trifluoroacetic acid. Samples were analyzed in a time-of-flight mass spectrometer (Reflex 3, Bruker Analytics, Bremen, Germany). An accelerating voltage of 20 kV was used. Peptide matching and protein searches were performed automatically. The peptide masses were compared to the theoretical peptide masses of all available proteins from all species. Monoisotopic masses were used and a mass tolerance of 0.0025% was allowed. The algorithm used for determining the probability of a false positive match with a given MS-spectrum is described elsewhere (Berndt et al., 1999).

Quantification of protein spots

Protein spots were outlined (first automatically and then manually) and quantified using the ImageMaster 2D Elite software (Amersham Pharmacia Biotechnology). The percentage of the volume of the spots representing a certain protein was determined in comparison with the total proteins present in the 2-D gel.

Statistical analysis

Values are expressed as mean ± standard deviation of optical density for western blotting and percentage of the spot volume in each particular gel after subtraction of the background values for 2-DE. Inter-group differences were analyzed using non-parametric Mann-Whitney U-test and within-group correlation was done using the Spearman rank-coefficient analysis. The level of significance was set at $P < 0.05$.

Results

Whereas Fas migrated on immunoblot as a single band at 36 kDa (Fig. 1A), two bands one minor and one major were detected for both caspase-3 (Fig. 1B) and Bcl-2 (Fig. 1C). The minor bands were at ~50 kDa (caspase-3) and ~70 kDa (Bcl-2), whereas that of the major ones at 27 kDa for both proteins. Cells synthesize caspases as zymogens that subsequently activated by proteolytic cleavage upon induction of apoptosis to generate large and small sububits, which together constitute the active enzyme. In fetal brain, the pattern of caspase-3 cleavage has not been described. However, available data obtained from studies on different cell lines indicate that the 32 kDa procaspase is cleaved into an active large (p18) and small (p12) fragments (Srinivasan et al., 1998). Hence, the 27 kDa caspase-3 immunoreactive band detected may stand for procaspase-3 and the 50 kDa for caspase complexes. We failed to detect the active caspases, although the antibody does recognize both the precursor as well as the active fragments. The major band that

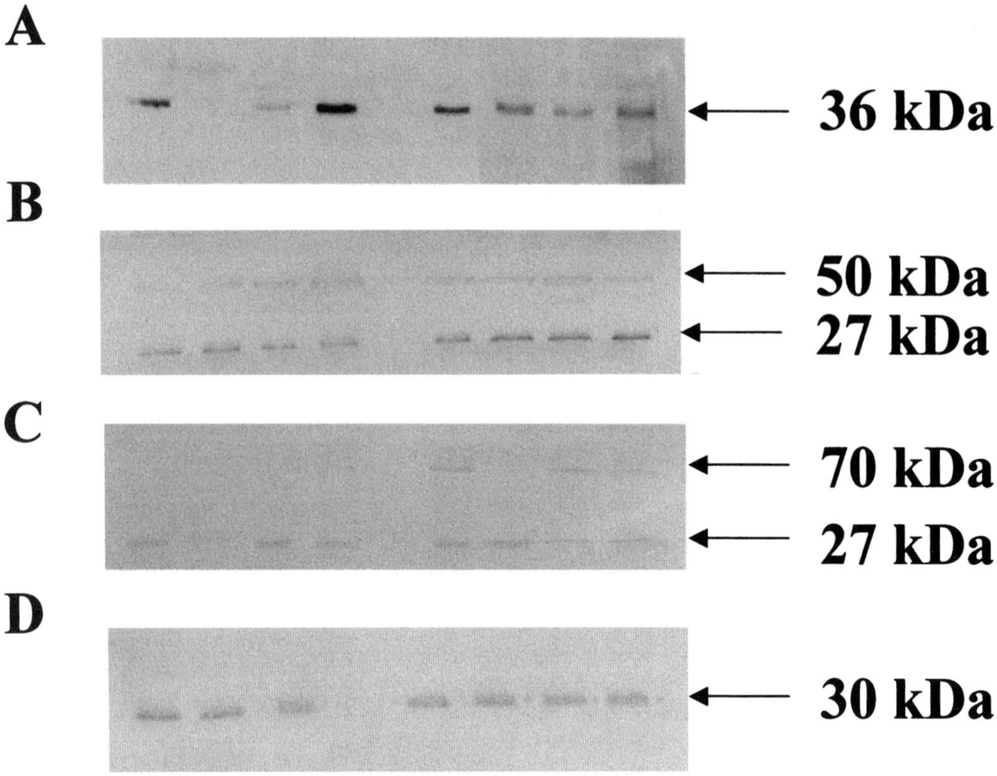

Fig. 1. Immunodetection of Fas (**A**), caspase-3 (**B**), Bcl-2 (**C**) and NSE (**D**) in cerebral cortex of fetal control and DS. Brain homogenate containing 5 (NSE) and 30 (the rest) μg protein/lane from the cerebral cortices of 7 controls and 8 DS brains were used for the immunodetection of Fas, caspase-3, Bcl-2 and NSE. The immunoblots of 4 controls (lanes 1–4) and 4 DS (lanes 5–8) cases were shown in the upper panels

migrated at 27 kDa denotes Bcl-2, whereas the minor band could be a protein complex containing Bcl-2.

Subsequent densitometric analysis of the immunoreactive bands did not reveal any significant change in Fas immunoreactivity (IR) between DS and controls (Table 1). A large variation had been noticed among the control group, which essentially contributed to increased mean value of controls over DS. The observation of unchanged expression did not exclude both caspase-3 and Bcl-2. Neither the IR of the individual bands nor the sum of the IR of all bands exhibited any detectable changes between DS and controls (Table 1, data for minor band not shown). This finding was surprising, as it is discordant with the current knowledge of apoptosis and DS. To support our finding, we performed analysis of neuron specific enolase (NSE), a neuronal marker, by western blotting, since DS is associated with neuronal loss and NSE was found to be decreased in adult DS brain (Gulesserian et al., 2000). To this end, 5 μg of NSE was applied and analyzed as described in Material and Methods, using monoclonal mouse anti-NSE antibody (MAB324, CHEMICON International, Inc., Temecula, USA) (1:3,000) and goat anti-mouse IgG (H + L) (Cat. No. 1031-05, Southern Biotechnology Associates, Inc., Birmingham,

Table 1. Levels of apoptosis related proteins and NSE in cerebral cortex of fetal control and DS brain. Following western blot, Fas, Bcl-2 and NSE immunoreactivity of 7 controls and 8 DS cases were quantified by scanning densitometry. Annexins were resolved and identified by 2-DE and MALDI-MS. This was followed by quantification using the 2D Image Master. Values are expressed as mean ± standard deviation of relative percent level for annexins and arbitrary optical density for the others. Inter-group differences were analyzed using non-parametric Mann-Whitney U-test and the significance was set at $P < 0.05$ for a two-tailed test

Protein	Control	DS
Fas	1.39 ± 1.16 (n = 7)	0.58 ± 0.39 (n = 8)
Caspase-3	0.74 ± 0.36 (n = 6)	0.93 ± 0.33 (n = 7)
Bcl-2	0.43 ± 0.28 (n = 7)	0.34 ± 0.06 (n = 8)
NSE	0.52 ± 0.21 (n = 7)	0.45 ± 0.16 (n = 8)
Annexin I	0.53 ± 0.28 (n = 7)	0.29 ± 0.15 (n = 9)
Annexin II	0.22 ± 0.07 (n = 7)	0.16 ± 0.04 (n = 9)
Annexin V	0.32 ± 0.17 (n = 7)	0.24 ± 0.12 (n = 9)
Annexin VI	0.31 ± 0.08 (n = 7)	0.39 ± 0.22 (n = 9)

USA) (1:2,000). The analysis revealed a single band of NSE that ran at 30kDa (Fig. 1D) and quantitative data obtained from this analysis did not show any apparent change in NSE IR (Table 1) among the groups confirming our finding.

2-DE

The protein spots were visualized by staining with colloidal Coomassie blue. The protein spots were analyzed by MALDI-MS, following in-gel digestion. The peptide masses were matched with the theoretical peptide masses of all proteins of the SWISS-PROT database. We used internal standards to correct the measured peptide mass thus reducing the windows of mass tolerance and increasing the confidence of identification. Approximately 389 proteins were resolved, including six annexins (I–VI). Figure 1 shows a representative gel of brain proteins separated by 2-DE from a control subject indicating only identified annexins by their name and SWISS-PROT number. Among the identified annexins, the annexins thought to be involved in apoptosis: annexin I, II, V and VI were quantified using the Image Master 2D Elite software and their expression levels were determined as a percentage volume of total proteins in the area of interest. The relative percent protein level and the MALDI data for the quantified annexins are given in Table 1 and 2, respectively. The data generated by 2-DE did not show any significant difference in the relative percentage level of proteins in any of the quantified annexins between controls and DS.

Neither of the proteins investigated both by western blotting as well as 2-DE did show any significant correlation between protein expression and age (data not shown).

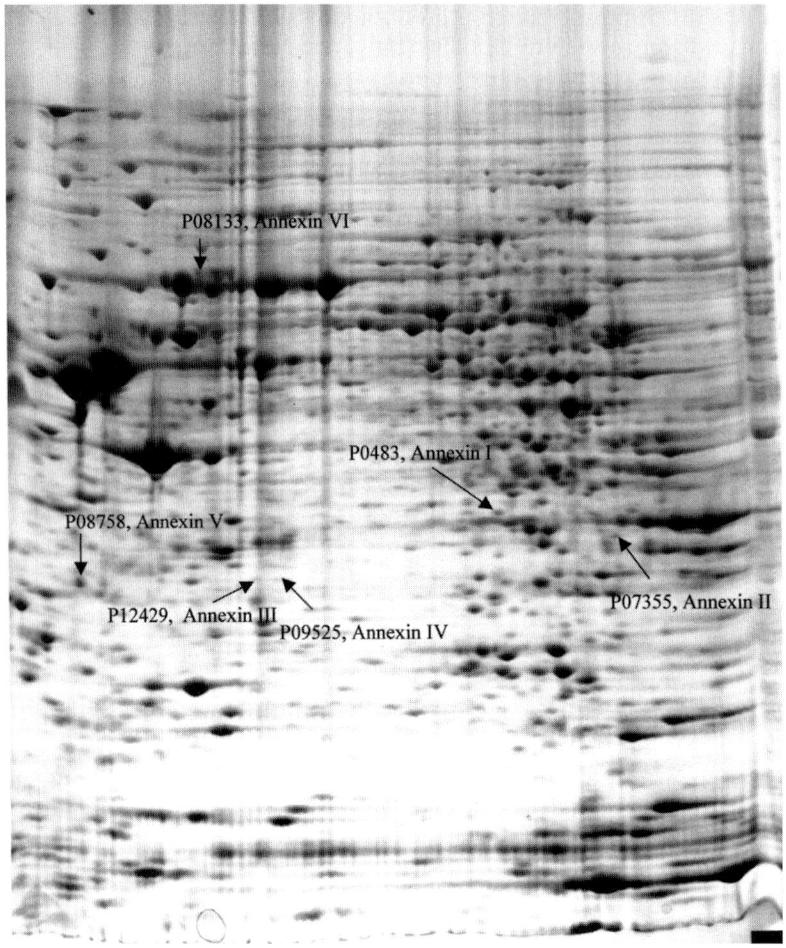

Fig. 2. Two-dimensional map of human fetal brain proteins. The proteins from cerebral cortex of fetal brain were extracted and analyzed on pH 3–10 nonlinear IPG strips, followed by a 9–16% sodiumdodecyl sulphate polyacrylamide gels. The gels were stained with Coomassie blue and identified by MALDI-MS. The spots representing annexins are designated with their SWISS-PROT number

Table 2. MALDI-MS data for quantified annexins in human fetal brain. The samples were analyzed and the proteins were identified and quantified as stated under Material and Methods. The SWISS-PROT accession number and name of the different annexins resolved are indicated in Fig. 2. The observed Mr and pI values, the matching peptides along with the probability as of the quantified annexins are given here

Protein name	pI	Mol. Size (Da)	Matches	Probability
Annexin I (Lipocortin I)	7.02	38,787	5	8.15E-05
Annexin II (Lipocortin II)	7.82	38,676	10	1.95E-16
Annexin V (Lipocortin V)	4.76	35,840	15	4.14E-29
Annexin VI (Lipocortin VI)	5.37	76,036	6	1.06E-07

Discussion

It was the aim of this study to gain insight whether the naturally occurring cell death mechanism can be manipulated by pathological processes during development using western blotting and 2-DE. Cell death occurs mainly to match the size of each neuronal population to the magnitude of its target fields and to eliminate erroneous or inadequate projections (Hutchins and Barger, 1997). In addition, it is suggested to have a role during morphogenesis of the developing brain (Kuan et al., 2000). Given the role of cell death in development, any alteration in the normal apoptotic process would cause disorganized neuronal circuitry that may lead to developmental disorder. DS being a neurodevelopmental disorder and altered apoptosis was reported in fetal DS neurons in vitro (Busciglio and Yanker, 1995) and fetal DS heart (Saphier and Yeh, 1998), it was expected to see alteration in expression of proteins involved in apoptosis. Interestingly, Fas, a suicide receptor that was shown to be increased in adult DS (Seidl et al., 1999) and animal models of fetal alcohol syndrome (Cheema et al., 2000) did not exhibit any detectable changes between controls and DS. This finding implies two possibilities. The first is unlike the adult DS brain, the death receptor pathway may be dispensable for apoptosis in fetal DS brain and the second is naturally occurring neuronal death may not be dysregulated. In human fetal brain, the observation that susceptibility to cell death via this pathway depends on state of maturity and expression is not necessarily coupled to susceptibility (Becher et al., 1998b) appears to favor the first possibility.

Caspases have emerged as catalysts of the execution phase of apoptosis resulting in the systematic and eventual disassembly of the cell. This final step in apoptotic cell death emanates from concerted cleavage of a set of substrates of the caspases (Rao and White, 1997). Caspase-3 has been identified as one of the major effector caspases responsible for the cellular dismantling in both the cytochrome c as well as death receptor mediated apoptotic pathways (Gulesserian et al., 2001; Kuan et al., 2000; Rao and White, 1997). Moreover, it was documented that caspase-3 is a key regulator of programmed cell death in neural development, as targeted disruption of caspase-3 results in marked hyperplasia of the embryonic nervous tissue and leads to either embryonic or neonatal death (Kuan et al., 2000). In adult DS brain, procaspase-3 was decreased and like here we were unable to detect the active caspase-3 (Gulesserian et al., 2001), a situation that could be attributed to rapid clearance of cells containing active caspases (Shimohama et al., 1999). In fetal DS brain, however, procaspase-3 was unchanged. As large amount of caspases are synthesized as zymogens in advance and activated on demand (Thornberry and Lazebnik, 1998), we could have noticed a difference in procaspase-3 levels between the two groups, like we did in adults, had there been enhanced apoptosis in fetal DS. Hence, this finding cast doubt on the presence of dysregulated apoptosis in fetal DS.

Bcl-2 is a protooncogene first discovered as a novel transcript associated with the t(14; 18) translocation breakpoint common in follicular B cell lymphomas. It is expressed in a variety of tissues, including the nervous system and is developmentally regulated (Merry et al., 1994). Bcl-2 gene family proteins are believed to control apoptosis, with Bcl-2 functioning as a repressor of cell death (Krosmeyer, 1995). Here, we found unchanged Bcl-2 levels in fetal DS compared to controls. Expression of Bcl-2 was reported to be decreased in adult DS brains (Sawa et al., 1997). This depletion of antiapoptotic mechanisms regulated by Bcl-2 was suggested to render neurons vulnerable to apoptotic stimuli (Wang et al., 1999). Considering the fact that Bcl-2 functions upstream of the caspases and thereby inhibit their activation (Rao and White, 1997), the unchanged Bcl-2 is concordant with unaltered caspase-3 levels indicating enhanced apoptosis may not be apparent in fetal DS brain. On the other hand, studies in fetal brain show that brain regions with high apoptotic activities are not always correlated with decreased Bcl-2 expression and high levels of Bcl-2 do not necessarily indicate suppression of apoptosis (Chan and Yew, 1998). This observation led to the assumption that Bcl-2 apart from being an apoptosis regulator might also have other roles to play during neural development (Chan and Yew, 1998). The lack of change in its expression thus indicate that Bcl-2 may not play a crucial role in apoptosis during development and its contribution perhaps is confined in complementing other antiapoptotic proteins to promote survival of specific neuronal subpopulations (Kuan et al., 2000).

The translocation of plasma membrane phosphotidylserine from the cytosolic portion to the external surface of the plasma membrane is one of the well documented hallmarks of cells undergoing apoptosis (Martin et al., 1995). This externalization functions as a cellular marker or tag for neurons that are to be identified for subsequent removal by microglia (Adayev et al., 1998; Maiese and Vincent, 2000). Annexin V is a calcium dependent phospholipid binding protein that has high affinity for phosphatidylserine. This binding enables annexin V to be used as a marker for apoptosis and inhibits the proinflammatory activity of the dying cell during apoptosis (Reutelingsperger and van Heerde, 1997). Recently, Annexin V has been demonstrated to delay apoptosis, an effect ascribed to increase in intracellular calcium (Gidon-Jeangirard et al., 1999a, b). Moreover, annexin V levels were shown to be reduced in Parkinson's disease owing to consumption of the protein during neuronal apoptosis in the midbrain (Vermes et al., 1999). Annexin I is an anti-inflammatory and stress protein induced by glucocorticoids and stresses, such as oxidative stress (Rhee et al., 2000). Annexin I promotes ROS mediated rat thymocyte apoptosis (Sakamoto et al., 1996) and its increased expression was found to be associated with enhanced apoptosis in U937 cells (Canaider et al., 2000). Furthermore, increased annexin I IR was observed during rat mammary regression, a process that heavily relies on apoptosis (McKanna et al., 1995). And yet we did not find any difference in both annexin I and V between the two subjects. Although annexin II is mainly involved in cellular DNA synthesis and proliferation, the reason why it is used as a marker for cancer, targeted disruption of its gene also affects apoptosis (Chiang et al., 1999). No difference had also been observed in its expression between the two groups. Annexin II expression in fetal brain is also associated with proliferation of radial glia, astrocytic progenitors suggested to be linked with glial tumors

(Reeves et al., 1992). The unchanged levels of annexin II apart from its implication in apoptosis, may also reflect normal proliferation and maturation of radial glia into adult astrocytes. Taken together, all the above findings strongly imply that among the annexins at least, annexin I and V are closely associated with apoptosis. The unchanged levels of the quantified annexins in general and annexin I & V in particular therefore may indicate that either they have a limited distribution and amounts are low in fetal brain (Johnson et al., 1989) to take significant part in apoptotic processes or there is no enhanced apoptosis in fetal DS. Failure to see any detectable changes in the other investigated apoptotic proteins (particularly caspase-3) and NSE, a neuronal marker, on top of annexins makes this later notion highly likely, however, further work is needed to substantiate it.

In conclusion, apoptosis related proteins shown to be altered in adult DS brain are unaffected in fetal DS brain. These findings suggest that either the same protein may assume different roles depending on brain development or degree of apoptosis in fetal DS may not be different from that of controls.

Acknowledgements

We are highly indebted to Red Bull Company, Salzburg, Austria for the generous financial support. E.E is a recipient of scholarship from the Austrian Academic Exchange Program, ÖAD.

References

Adayev T, Estephan R, Meserole S, Mazza B, Yurkow EJ, Banerjee P (1998) Externalization of phosphatidylserine may not be an early signal of apoptosis in neuronal cells, but only the phosphatidylserine-displaying apoptosis cells are phagocytosed by microglia. J Neurochem 71: 1854–1864

Anderson AJ, Stoltzner S, Lai Florence Su J, Nixon RA (2000) Morphological and biochemical assessment of DNA damage and apoptosis in Down syndrome and Alzheimer's disease, and effect of postmortem tissue archival on TUNEL. Neurobiol Aging 21: 511–524

Becher B, Barker PA, Owens T, Antel PJ (1998a) CD95-CD95L: can the brain learn from the immune system? Trends Neurosci 21: 114–117

Becher B, D'Souza SD, Troutt AB, Antel JP (1998b) Fas expression on human fetal astrocytes without susceptibility to fas mediated toxicity. Neuroscience 84: 627–634

Berndt P, Hobohm U, Langen H (1999) Reliable automatic protein identification from matrix-assisted laser desorption/ionization mass spectrometric peptide fingerprints. Electrophoresis 20: 3521–3526

Bradford M (1976) A rapid and sensitive method for the quantitation of microgram quantities of protein utilizing the principle of protein-dye binding. Anal Biochem 72: 248–254

Busciglio J, Yanker BA (1995) Apoptosis and increased generation of reactive oxygen species in Down's syndrome in vitro. Nature 378: 776–779

Canaider S, Solito E, de Coupade C, Flower RJ, Russo-Marie F, Goulding NJ, Perretti M (2000) Increased apoptosis in U937 cells over-expressing lipocortin 1 (annexin 1). Life Sci 66: PL265–270

Chan WY, Yew DT (1998) Apoptosis and Bcl-2 oncoprotein expression in the human fetal central nervous system. Anat Rec 252: 165–175

Cheema ZF, West JR, Miranda RC (2000) Ethanol induces Fas/APO mRNA and cell suicide in the developing cerebral cortex. Alcohol Clin Exp Res 24: 535–543

Chiang Y, Rizzino A, Sibenaller ZA, Wold MS, Vishwanatha JK (1999) Specific down-regulation of annexin II expression in human cells interferes with cell proliferation. Mol Cell Biochem 199: 139–147

Creutz CE (1992) The annexins and exocytosis. Science 258: 924–931

Dragunow M, MacGibbon GA, Lawlor P, Butterworth N, Connor B, Henderson C, Walton M, Woodgate A, Hughes P, Faull RL (1997) Apoptosis, neurotrophic factors and neurodegeneration. Rev Neurosci 8: 223–265

Epstein CJ (1995) Down Syndrome. In: Scriver CR, Beaudet AL, Sly WS, Valle D (eds) The metabolic and molecular bases of inherited disease, 7th edn, vol I. McGraw Hill, New York, pp 749–794

Fountoulakis M, Langen H (1997) Identification of proteins by matrix assisted laser desorption ionization mass spectrometry following in-gel digestion in low salt, non-volatile buffer and simplified peptide recovery. Anal Biochem 250: 153–156

Francis JW, Balazovich KJ, Smolen JE, Margolis DI, Boxer LA (1992) Human neutrophil annexin I promotes granule aggregation and modulates Ca2+ dependent membrane fusion. J Clin Invest 90: 537–544

Gidon-Jeangirard C, Hugel B, Holl V, Toti F, Laplanche JL, Meyer D, Freyssinet JM (1999a) Annexin V delays apoptosis while exerting an external constraint preventing the release of CD4+ and PrPc+ membrane particles in a human T lymphocyte model. J Immunol 162: 5712–5718

Gidon-Jeangirard C, Solito E, Hofmann A, Russo-Marie F, Freyssinet JM, Martinez MC (1999b) Annexin V counteracts apoptosis while inducing Ca2+ influx in human lymphocytic T cells. Biochem Biophys Res Commun 265: 709–715

Gulesserian T, Engidawork E, Cairns N, Lubec G (2000) Increased serotonin protein levels in frontal cortex of patients with Down syndrome. Neurosci Lett 296: 53–57

Gulesserian T, Engidawork E, Yoo BC, Cairns N, Lubec G (2001) Alteration of caspases and their regulatory proteins in brains of Down syndrome. J Neural Transm (this volume)

Hutchins JB, Barger SW (1998) Why neurons die: cell death in the nervous system. Anat Rec 253: 79–90

Johnson MD, Kamso-Pratt JM, Whetsell WO, Pepinsky RB (1989) Lipocortin-1 immunoreactivity in the normal human central nervous system and lesions with astrocytosis. Am J Clin Pathol 92: 424–429

Kaetzel MA, Hazarika P, Dedman JR (1989) Differential distribution of three 35-kDa annexin calcium dependent phospholipid binding proteins. J Biol Chem 264: 14463–14470

Kroemer G (1997) The proto-oncogene Bcl-2 and its role in regulating apoptosis. Nature Med 3: 614–620

Krosmeyer SJ (1995) Regulators of cell death. Trends Genet 11: 101–105

Kuan C-Yi, Roth KA, Flavell RA, Rakic P (2000) Mechanisms of programmed cell death in the developing brain. Trends Neurosci 23: 291–297

Langen H, Röder D, Juranville JF, Fountoulakis M (1997) Effect of the protein application mode and the acrylamide concentration on the resolution of protein spots separated by two dimensional gel electrophoresis. Electrophoresis 18: 2085–2090

Maiese K, Vincent AM (2000) Membrane asymmetry and DNA degradation: Functionally distinct determinants of neuronal programmed cell death. J Neurotics 59: 568–580

Mann DM (1988) The pathological association between Down syndrome and Alzheimer disease. Mech Aging Div 43: 99–136

Martin SJ, Reutelingsperger CP, McGahon AJ, Rader JA, van Schie RC, LaFace DM, Green DR (1995) Early redistribution of plasma membrane phosphatidylserine externalization is a general feature of apoptosis regardless of the initiating stimulus: inhibition by overexpression of Bcl-2 and Abl. J Exp Med 182: 1545–1556

McKanna JA (1995) Lipocortin I in apoptosis: mammary regression. Anat Rec 242: 1–10

Merry DE, Veis DJ, Hickey WF, Korsmeyer SJ (1994) Bcl-2 expression is widespread in the developing nervous system and retained in the adult PNS. Development 120: 301–311

Osterloh D, Wittbrodt J, Gerke V (1998) Characterization and developmentally regulated expression of four annexins in the killifish medaka: DANN. Cell Biol 17: 835–847

Rao L, White E (1997) Bcl-2 and the ICE family of apoptotic regulators: making a connection. Curr Opin Gene Dev 7: 52–58

Raynal P, Pollard MB (1994) Annexins: the problem of assessing the biological role for a gene family of multifunctional calcium and phospholipid binding proteins. Biochem Biophys Acta 1197: 63–93

Reeves SA, Chavez-Kappel C, Davis R, Rosenblum M, Israel MA (1992) Developmental regulation of annexin II (Lipocortin 2) in human brain and expression in high grade glioma. Cancer Res 52: 6871–6876

Reutelingsperger CP, van Heerde WL (1997) Annexin V, the regulator of phosphatidylserine catalyzed inflammation and coagulation during apoptosis. Cell Mol Life Sci 53: 527–532

Rhee HJ, Kim GY, Huh JW, Kim SW, Na DS (2000) Annexin I is a stress protein induced by heat, oxidative stress and a sulfhydryl-reactive agent. Eur J Biochem 267: 3220–3225

Sakamoto T, Repasky WT, Uchida K, Hirata A, Hirata F (1996) Modulation of cell death pathways to apoptosis and necrosis of H2O2 treated rat thymocytes by lipocortin I. Biochem Biophys Res Commun 220: 643–647

Saphier CJ, Yeh J (1998) Altered apoptosis levels in hearts of human fetuses with Down syndrome. Am J Obstet Gynecol 179: 962–965

Sawa A, Oyama F, Cairns N, Amano N, Matsushita M (1997) Aberrant expression of Bcl-2 gene family in Down's syndrome brains. Mol Brain Res 48: 53–59

Seidl R, Fang-Kircher S, Bidmon B, Cairns N, Lubec G (1999) Apoptosis-associated proteins p53 and APO-1/Fas (CD95) in brains of adult patients with Down syndrome. Neurosci Lett 260: 9–12

Shimohama S, Tanino H, Fujimoto S (1999) Changes in caspase expression in Alzheimer's disease: comparison with development and aging. Biochem Biophys Res Commun 256: 381–384

Srinivasan A, Roth KA, Sayers RO, Shindler KS, Wong AM, Fritz LC, Tomasselli KJ (1998) In situ immunodetection of activated caspase-3 in apoptotic neurons in the developing nervous system. Cell Death Diff 5: 1004–1016

Thornberry NA, Lazebnik Y (1998) Caspases: enemies within. Science 281: 1312–1322

Tzima E, Walker JH (2000) Platelet annexin V: the ins and outs. Platelets 11: 542–251

Vermes I, Steur EN, Reutelingsperger C, Haanen C (1999) Decreased concentration of annexin V in Parkinsonian cerebrospinal fluid: speculation on the underlying cause. Mov Disord 14: 1008–1010

Wang S, Dibenedetto AJ, Pittman RN (1997) Genes induced in programmed cell death of neuronal PC12 cells and developing sympathetic neurons in vivo. Dev Biol 188: 322–336

Wang S, Rosengren LE, Franlund M, Hamberger A, Haglid KG (1999) Bcl-2 expression regulates cell sensitivity to S100ß mediated apoptosis. Mol Brain Res 70: 167–176

Wisniewski KE, Kida E (1994) Abnormal neurogenesis and synaptogenesis in Down syndrome brain. Dev Brain Dysfunct 7: 289–301

Wisniewski KE, Wisniewski HM, Wen GY (1985) Occurrence of neuropathological changes and dementia of Alzheimer's disease in Down syndrome. Ann Neurol 17: 278–282

Authors' address: Prof. Dr. G. Lubec, Department of Pediatrics, University of Vienna, Währinger Gürtel 18–20, A-1090 Vienna, Austria, e-mail: Gert.Lubec@akh-wien.ac.at

Alteration of caspases and other apoptosis regulatory proteins in Down syndrome

T. Gulesserian[1], **E. Engidawork**[1], **B. C. Yoo**[1], **N. Cairns**[2], and **G. Lubec**[1]

[1] Department of Pediatrics, University of Vienna, Vienna, Austria
[2] Department of Neuropathology, Institute of Psychiatry, King's College, London, United Kingdom

Summary. Apoptosis has been implicated in the selective neuronal loss of Down syndrome (DS). Apoptosis activates a family of cysteine proteases with specificity for aspartic acid residues referred to as caspases that play a key role in dismantling a cell committed to die. Caspase activity is regulated by a variety of proteins that possess a domain resembling the prodomains of caspases. Little is known, however, about the changes of caspases and their regulatory proteins in DS. Here, we investigated levels of nine such different proteins by western blot technique in frontal cortex and cerebellum of control and DS subjects. The protein levels of DFF45 (DNA fragmentation factor 45), and FLIP (FADD like interleukin-1β-converting enzyme inhibitory proteins) were significantly decreased whereas that of RICK (RIP-like interacting CLARP kinase) increased in both regions of DS. In contrast, cytochrome c, Apaf-1 (apoptosis protease activating factor-1), procaspase-9 and ARC (apoptosis repressor with caspase recruitment domain) were unchanged. Procaspase-3 and -8 were significantly decreased in frontal cortex but no significant change was observed in cerebellum. Regression analysis revealed no correlation between postmortem interval and levels of the investigated proteins. However, inconsistent correlation was found between age and levels of proteins as well as amongst the density of individual proteins.

These findings demonstrate that dysregulation of apoptotic proteins does exist in DS brain and may underlie the neuropathology of DS. The study further suggests that apoptosis in DS may occur via the death receptor pathway independent of cytochrome c. Hence, therapeutic strategies that target caspase activation may prove useful in combating neuronal loss in this disorder.

Introduction

Apoptosis or cellular suicide is a morphologically and biochemically distinct form of programmed cell death that plays an essential role in development, homeostasis, and defense in multicellular organisms (Cryns and Yuan, 1998).

The cell death machinery is conserved throughout evolution and is composed of activators, inhibitors, and effectors (Chinnaiyan et al., 1996). The effector arm of the cell death pathway is composed of a novel family of cysteinyl aspartate specific proteases related to the *C. elegans* (*Caenorhabditis elegans*) cell death product CED-3, termed caspases (Alnemri et al., 1996). Thus far, at least 14 family members have been identified and can be divided into two groups based on the length of the prodomain (Srinivasula et al., 1997; Zhivotovsky et al., 1999). Several of these caspases, notably caspase-2, -3, -4, -6, -7, -8, -9, and -10 have been implicated in induction of apoptosis (Cohen, 1997). Caspases with large prodomains are known as upstream caspases, as they are thought to be involved in the initiation of the apoptotic response and appear to be activated through interaction with specific adaptor proteins. Whereas those with short prodomains are called downstream or effector caspases and are probably activated predominantly, if not exclusively, through cleavage by upstream caspases (Earnshaw et al., 1999; Reed, 1999).

Caspases are synthesized as inactive precursors with an N-terminal prodomain, a large subunit, and small subunit. Proteolytic cleavage at specific aspartate residues produces two subunits of approximate molecular masses of 20 and 10 kDa, which on association form an active heterotetramer. Caspase activation may proceed by autoactivation, transactivation, or proteolysis by distinct proteases (Nunez et al., 1998; Thornberry and Lazebnik, 1998). The proapoptotic function of the active enzymes is mediated by processing key intracellular substrates at a DXXD motif by the effector caspases (Cohen, 1997; Nicholson and Thornberry, 1997; Earnshaw et al., 1999).

Little is known about the regulation of caspase activity during apoptosis. In the nematode *C. elegans*, activation of the cell death protease CED-3 is positively regulated by CED-4 and inhibited by CED-9 through direct protein-protein interactions (Chinnaiyan et al., 1997; Wu et al., 1997). Likewise, Apaf-1 (apoptotic protease activating factor-1), a human protein that resembles *C. elegans* CED-4, interacts with caspase-9, a step that is required for the activation of the downstream protease caspase-3 (Zou et al., 1997). The cytochrome c: Apaf pathway in turn, is regulated by Bcl-2 family members (Earnshaw et al., 1999). The prodomains of several apical caspases contain a protein module termed CARD (caspase activation and recruitment domain) or DED (death effector domain) that is conserved in several apoptosis regulatory molecules, including Apaf-1 and cellular inhibitors of apoptosis proteins (Hofmann et al., 1997; Wellington and Hayden, 2000). The CARD has been proposed to play a regulatory role in apoptosis by allowing proteins such as Apaf-1 to associate with caspase-9 (Li et al., 1997). A number of novel CARD or DED containing proteins that modulate apoptosis have been identified, including ARC (apoptosis repressor with CARD) (Koseki et al., 1998), FLIP (FADD like interleukin-1β-converting enzyme inhibitory proteins) (Irmler et al., 1997), and RICK (RIP-like interacting CLARP kinase) (Inohara et al., 1998).

Down syndrome (DS) is a neurodevelopmental disorder with neurodegenerative component, resulting from trisomy of the distal segment of chromosome 21. Recently, biochemical evidence for an apoptotic mechanism of cell death in DS has been presented (Anderson et al., 2000) and neuronal apoptosis in mouse trisomy 16, an animal model of DS, was shown to be mediated by caspases (Bambrick and Krueger, 1999). Moreover, there is accumulating evidence that defects in apoptosis may lead to several pathologies including ischemic injury, cancer, and neurodegenerative disorders (Thompson, 1995; Wellington and Hayden, 2000). These reports formed the Rationale for us to examine the levels of different proteins involved in apoptosis including caspase-3, caspase-8, caspase-9, Apaf-1, cytochrome c, RICK, ARC, FLIP, and DFF45 (DNA fragmentation factor 45) in brain of patients with DS. The findings demonstrate that apoptosis in DS may be mediated via cytochrome c independent pathway.

Materials and methods

Materials

Antibodies against caspase-9 (210-759-C100), FLIP (210-736-C100), DFF45 (210-737-C100), ARC (210-212-C100), Apaf-1, and RICK (210-213-C100) were purchased from ALEXIS BIOCHEMICALS (San Diego, USA). Rabbit polyclonal anti-caspase-8 (PC335) and anti-caspase-3 (H-277) were purchased from ONCOGENE Research Products (Boston, USA) and HybriDomus (Copenhagen, Denmark), respectively. Monoclonal mouse anti-cytochrome c (804-122-C100) was from PharMingen (San Diego, USA). Horseradish peroxidase-conjugated anti-mouse IgG (H + L) (1031-05) and anti-rabbit IgG (H + L) (4050-05) were obtained from Southern Biotechnology Associates Inc (Birmingham, USA). Enhanced chemiluminescence reagents were from NEN™ Life Science Products, Inc (NEN™ Life Science Products, Inc., Boston, USA) and protein size markers (Rainbow) were from Amersham (Uppsala, Sweden).

Human brain postmortem tissue

Postmortem brain samples (frontal cortex, superior frontal gyrus and cerebellum) of patients with DS (n = 8; 2 females and 6 males; 55.9 ± 8.0 years old) and controls (n = 10; 3 females and 7 males; 62.6 ± 8.0 years old) were obtained from the Medical Research Council's, Brain Bank for Neurodegenerative Diseases, Department of Neuropathology, Institute of Psychiatry, London, UK. The major cause of death in DS was bronchopneumonia and heart disease in controls. The control brains were from individuals with no history of neurological or psychiatric illness whereas in all DS brains there was evidence of abundant beta A plaques and neurofibrillary tangles. Postmortem interval of brain dissection was 32.4 ± 20.7 for DS and 37.3 ± 18.4 hours for controls. All samples were stored at −70°C and the freezing chain was never interrupted.

Sample preparation

Frozen brain samples were thawed on ice and homogenized in 4 volumes of TBS buffer (100 mM Tris, 150 mM NaCl, pH 7.5) containing protease inhibitors, (1 μg/ml each of antipain, pepstatin A, and leupeptin and 10 mM of phenylmethylsulfonyl fluoride) and 0.05% Tween 20. Samples were homogenized on ice for 30 seconds (6–8 strokes) with a Potter-Elvehjem homogenizer. The brain homogenate then was centrifuged for 10 min at

10,000 × g and 4°C. To the supernatant, equal volume of sample buffer (125.5 mM Tris-HCl base, 70 mM sodium dodecyl sulfate, 0.001% Bromophenol blue, 20% glycerol, and 2% 2-mercaptoethanol, pH 6.8) was added and the mixture was heated at 95°C for 15 min. The concentration of protein was determined by the Bicinchoninic acid assay using bovine serum albumin as a standard (Pierce, USA).

Immunoblotting

Protein extracts containing either 100 µg (cytochrome c, RICK, DFF 45, ARC, FLIP), 50 µg (capase-3, -8, and -9) or 20 µg (Apaf-1) were subjected to sodium dodecyl sulfate-polyacrylamide gel electrophoresis (200 V for 3 hrs), and transferred to polyvinylidene fluoride membranes (Immobilon™-P, MilliPore, Bedford, USA) (192 V for 1 hr). The membranes were then washed two times for 5 min in Tris buffer (100 mM Tris-HCl, 150 mM NaCl, pH 7.5) and later blocked with blotto (1.5% non-fat dry milk in TBST (100 mM Tris-HCl, 150 mM NaCl, pH 7.5, 1 mM $MgCl_2$, 0.1 % Tween-20)) overnight at 4°C followed by incubation with the respective primary antibodies in blotto (1:500 dilution for caspase-3, 2.5 µg/ml for caspase-8, 1:2,000 for Apaf-1, and 1:1,000 for the others) at room temperature for 3 hours on a rocking shaker. The membranes were washed three times for 15 minutes with blotto prior to incubation with respective horse-radish peroxidase conjugated secondary antibody in blotto (1:1,000 for caspase-3 and 1:2,000 for the others) for one hour at room temperature on a rocking shaker. The membranes were washed again three times for 15 minutes followed by incubation with NEN chemiluminescence reagent for 1 minute and subsequent exposure to films (Kodak Blue XB-1, Rochester, USA). Developed films were scanned and densities of the immunoreactive bands were calculated by RFLP Scan 2.1 software program (Scanalytics, Billerica, USA).

Data analysis

Values are expressed as mean ± standard deviation of optical density. Inter-group differences were analyzed by non-parametric Mann-Whitney U-test and the significance was set at $P < 0.05$ for a two-tailed test. The relationship between age or postmortem interval and density of the proteins as well as the relation amongst the density of individual proteins was examined with linear regression analysis.

Results

Proapoptotic proteins

Caspases

Caspases are synthesized as inactive proforms which on receiving an apoptotic signal activated and subsequently cleaved into a prodomain and two subunits. The cleavage appears to occur in an ordered fashion with cleavage between the large and small subunits preceding removal of the prodomain. Caspase-3 is the major terminal effector or terminator caspase that acts on multiple death substrates to culminate in the classic morphologic features in cells undergoing apoptosis (Jänicke et al., 1998; Porter and Jänicke, 1999). Initial

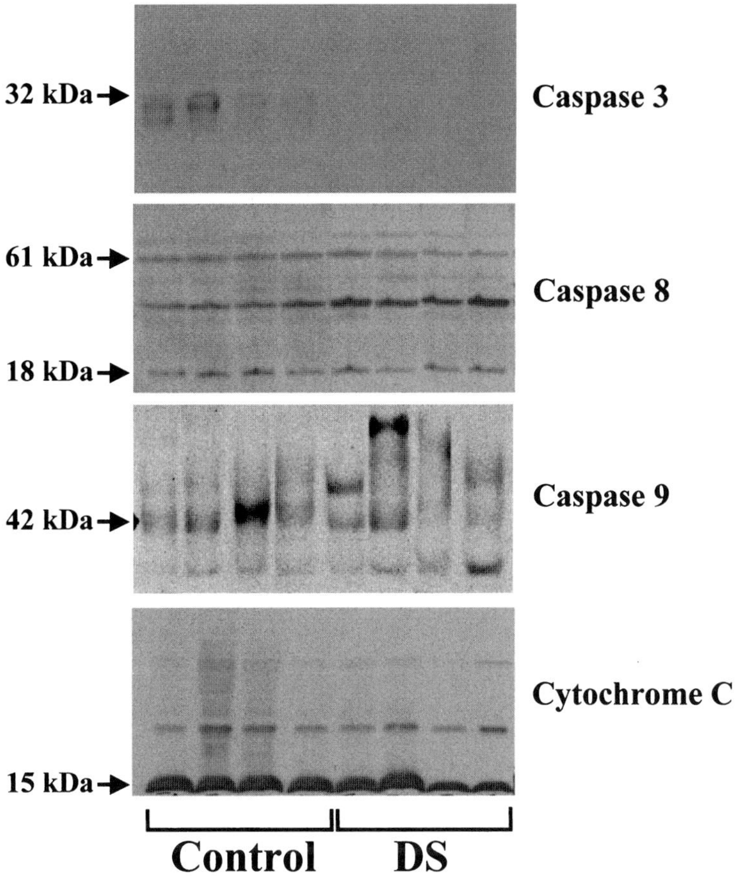

Fig. 1. Immunodetection of caspase-3, -8, -9, and cytochrome c in frontal cortex of control, and DS. Brain homogenate containing 50 (caspases) and 100 (cytochrome c) μg protein/lane from the frontal cortices of 10 control and 8 DS brains were used for the immunodetection of caspases and cytochrome c, respectively. The immunoblots of 4 controls (lanes 1–4) and 4 DS (lanes 5–8) cases were shown in the upper panels

cleavage of caspase-3 yields a P20 or P19 (depending on which aspartate site is cleaved) and P12 fragments and subsequent cleavage of P20 or P19 fragment give rises to the P17 protein (Cohen, 1997). On immunoblots, a 32 kDa procaspase-3 protein was detected with anticaspase-3 antibody (Fig. 1). The antibody also reacted with two other proteins with apparent molecular size of 30 and 27 kDa that may correspond to caspase-3 lacking a proarm and/or heterodimer of the active subunits. Quantification of caspase-3 precursor immunoreactivity (IR) revealed comparable DS cerebellar levels to that of controls (Table 1A) whereas levels were significantly reduced in DS frontal cortex ($P < 0.0001$) (Table 1B). The decrement in total caspase-3 IR showed more or less the same pattern except that DS cerebellum also displayed a significant change ($P < 0.05$).

Caspase-8 is considered as the apical caspase and mediates signal transduction downstream of death receptors located on the plasma membrane

Table 1. Levels of proapoptotic proteins in cerebellum (**A**) and frontal cortex (**B**) of controls and DS. Following western blot, immunoreactivity of 10 control and 6–8 DS cases were quantified by scanning densitometry. Values are expressed as mean ± standard deviation of arbitrary optical density and inter-group differences were analyzed by non-parametric Mann-Whitney U-test and the significance was set at $P < 0.05$ for a two-tailed test. *$P < 0.05$, **$P < 0.01$, ***$P < 0.001$, ****$P < 0.0001$ compared to controls. *TIR* total immunoreactivity

A. Cerebellum

Protein	Control	DS
a) Caspase 3		
TIR	2.88 ± 1.31 (n = 9)	1.32 ± 0.94* (n = 7)
Precurssor (32 kDa)	1.81 ± 1.08 (n = 9)	0.82 ± 0.61 (n = 7)
30 kDa	0.47 ± 0.28 (n = 9)	0.23 ± 0.11 (n = 7)
27 kDa	0.48 ± 0.18 (n = 9)	0.43 ± 0.35 (n = 7)
b) Caspase 8		
TIR	7.81 ± 2.29 (n = 9)	8.61 ± 1.62 (n = 6)
Precurssor (61 kDa)	2.81 ± 0.84 (n = 9)	2.41 ± 0.32 (n = 6)
P18	2.29 ± 1.31 (n = 9)	2.16 ± 1.11 (n = 6)
P43	2.72 ± 0.90 (n = 9)	4.04 ± 0.95* (n = 6)
c) Caspase 9		
TIR	3.31 ± 1.61 (n = 10)	3.36 ± 0.60 (n = 6)
Precurssor (42 kDa)	1.35 ± 1.04 (n = 10)	1.28 ± 1.08 (n = 6)
56 kDa	0.53 ± 0.36 (n = 10)	0.34 ± 0.09 (n = 6)
29 kDa	1.42 ± 0.96 (n = 10)	1.64 ± 0.98 (n = 6)
d) RICK	2.07 ± 1.07 (n = 9)	4.13 ± 2.55* (n = 6)
e) Cytochrome c		
TIR	16.37 ± 5.77 (n = 9)	15.41 ± 3.83 (n = 6)
60 kDa	2.06 ± 1.25 (n = 9)	2.44 ± 1.26 (n = 6)
25 kDa	3.72 ± 2.85 (n = 9)	3.15 ± 1.42 (n = 6)
15 kDa	10.59 ± 2.05 (n = 9)	9.83 ± 1.66 (n = 6)
f) Apaf-1		
TIR	9.88 ± 1.63 (n = 9)	8.25 ± 1.53 (n = 6)
130 kDa	0.51 ± 0.24 (n = 9)	0.42 ± 0.34 (n = 6)

(Peter and Krammer, 1998). A 43 and 11 kDa proteins are obtained on first cleavage followed by the 18 kDa. Procaspase-8 ran on SDS-PAGE with an apparent molecular mass of 61 kDa accompanied by two other proteins with molecular mass of 43 and 18 kDa (Fig. 1). The two bands may represent the partially processed large subunit that results from initial cleavage and the fully processed P18 subunit, respectively as the antibody reacts with the 18 kDa subunit. No significant difference was observed in the total caspase-8 as well as P18 IR between controls and DS in the regions examined. By contrast, a significant decrease in the IR of caspase-8 precursor in DS frontal cortex ($P < 0.05$) was observed. Interestingly, an extremely significant ($P < 0.001$) and significant ($P < 0.05$) increase in P43 IR in DS frontal cortex and cerebellum, respectively had also been noted (Table 1A, B).

Caspase-9 also known as Apaf-3 mediates apoptotic signals after mitochondrial damage (Li et al., 1997). A 37 kDa partially processed large subunit

Table 1. *Continued*

B. Frontal

Protein	Control	DS
a) Caspase 3		
TIR	2.44 ± 1.58 (n = 10)	0.10 ± 0.08**** (n = 8)
Precurssor (32 kDa)	1.29 ± 1.08 (n = 10)	0.03 ± 0.01**** (n = 8)
30 kDa	1.29 ± 1.08 (n = 10)	0.02 ± 0.01**** (n = 8)
27 kDa	0.35 ± 0.27 (n = 10)	0.05 ± 0.02** (n = 8)
b) Caspase 8		
TIR	4.81 ± 1.62 (n = 10)	4.09 ± 1.62 (n = 8)
Precurssor (61 kDa)	2.38 ± 0.86 (n = 10)	1.57 ± 0.46* (n = 8)
P18	2.22 ± 1.16 (n = 10)	1.45 ± 0.52 (n = 8)
P43	0.44 ± 0.17 (n = 10)	1.07 ± 0.41*** (n = 8)
c) Caspase 9		
TIR	3.82 ± 1.37 (n = 10)	5.08 ± 1.77 (n = 6)
Precurssor (42 kDa)	2.15 ± 1.21 (n = 10)	1.28 ± 1.17 (n = 6)
56 kDa	0.37 ± 0.12 (n = 10)	2.47 ± 2.04 (n = 6)
29 kDa	1.45 ± 1.13 (n = 10)	1.13 ± 1.05 (n = 6)
d) RICK	1.58 ± 0.73 (n = 10)	4.72 ± 2.93** (n = 6)
e) Cytochrome c		
TIR	20.97 ± 4.68 (n = 10)	19.00 ± 2.08 (n = 8)
60 kDa	0.45 ± 0.25 (n = 10)	0.46 ± 0.20 (n = 8)
25 kDa	5.90 ± 3.25 (n = 10)	3.73 ± 2.18 (n = 8)
15 kDa	14.44 ± 3.02 (n = 10)	14.80 ± 3.22 (n = 8)
f) Apaf-1		
TIR	0.59 ± 0.17 (n = 9)	0.65 ± 0.24 (n = 8)
130 kDa	0.12 ± 0.07 (n = 9)	0.16 ± 0.13 (n = 8)

that later fully processed to an 18 kDa, and a 12 kDa fragments are obtained following cleavage. Three bands with molecular size of 56, 42, and 29 kDa were detected by making use of anti-caspase-9 antibody (Fig. 1). The 42 kDa protein represents procaspase-9 and whereas the 27 kDa could be a split product of the 42 kDa that would not be processed to the active enzyme. The 56 kDa protein may be a splice variant, pathologically modified caspase or caspase complex. Both the 56 and 42 kDa proteins were unchanged in DS cerebellum as well as frontal cortex. Likewise, the total caspase-9-IR exhibited no significant change (Table 1A, B).

RICK, Apaf-1, and Cytochrome c

RICK represents a novel kinase that regulate apoptosis mediated by the CD95 receptor pathway (Inohara et al., 1998). It contains an N-terminal kinase catalytic domain and a C-terminal CARD domain. The protein migrated as a single band with an apparent molecular weight of ~40 kDa (Fig. 2). Densitometric analysis showed a significant increase ($P < 0.05$) of RICK-IR in DS cerebellum (Table 1A) and a highly significant increase ($P < 0.01$) in

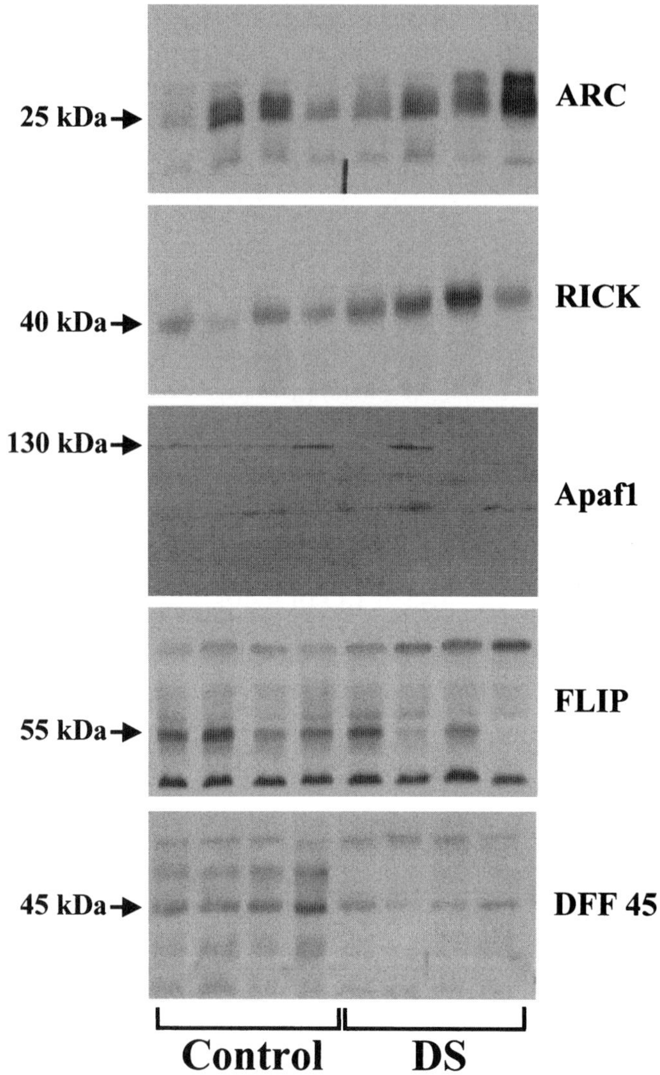

Fig. 2. Immunodetection of Apaf-1, RICK, ARC, FLIP, and DFF45 in frontal cortex of control and DS. Brain homogenate containing 20 μg (Apaf-1) and 100 μg (the others) protein/lane from the frontal cortices of 10 control and 8 DS brains were used for the immunodetection. The immunoblots of 4 controls (lanes 1–4) and 4 DS (lanes 5–8) cases were shown in the upper panels

frontal cortex (Table 1B). Apaf-1 is the mammalian homologue of the key cell death gene CED-4 in *C. elegans* (Zou et al., 1997). Although Apaf-1 is not a caspase, its N-terminal region contains a CARD, suggesting that it may recruit caspase-9 through their respective CARDs (Zou et al., 1997). However, this domain is not exposed unless dATP or ATP is present (Li et al., 1997). Anti-Apaf-1 antibody detected Apaf-1 that migrated as a 130 kDa and two other bands with molecular size of ~53 kDa and ~30 kDa (Fig. 2). The 53 kDa protein may correspond to a truncated form of Apaf-1 reported recently (Saleh et al., 1999) and the third band could be a split product. Apaf-1 did not show any significant change in both regions of DS so did the total IR (Table 1A, B). Cytochrome c is normally present exclusively in mitochondria and thought to be released to the cytosol in response to a variety of apoptosis inducing agents. Release of cytochrome c has a pivotal role in binding Apaf-1 and causing caspase-9 activation that executes the death program (Li et al., 1997). Based on its presumed role we analyzed its level and the western blot pattern showed one minor and two major bands with molecular size of 52, 25, and 15 kDa, respectively with the 15 kDa protein representing cytochrome c (Fig. 1). Neither the total nor the 15 kDa cytochrome c-IR did exhibit any significant change in both cerebellum and frontal cortex of DS patients (Table 1A, B).

Antiapoptotic proteins

Among the proteins that negatively modulate apoptosis, DFF 45, ARC, and FLIP can be cited as an example. A total of 5 bands with three major bands ranging in molecular size from 25 to 60 kDa were detected using anti-DFF 45 antibody with DFF 45 migrating as a 45 kDa protein (Fig. 2). The two other major bands that lie above and below the 45 kDa protein could be a heterodimer of DFF40/DFF45 and cleavage product of DFF45, respectively as inactive DFF40 complexed with DFF45 was separated by gel filtration (Enari et al., 1998) and DFF45 was shown to be cleaved into smaller fragments (Liu et al., 1997). When the total IR for DFF 45 was calculated, frontal cortex of DS ($P < 0.0001$) patients displayed a highly significant reduction (Table 2). Although the numerical mean in DS cerebellum appears to be lower than controls, a high degree of variation among the results precluded statistical significance. The intensity for the protein representing DFF 45, however, showed a highly significant reduction in both DS cerebellum ($P < 0.001$) and frontal cortex ($P < 0.0001$) (Table 2). FLIP also ran as a 55 kDa protein along with two other bands with apparent molecular mass of 37 and 80 kDa (Fig. 2). Since FLIP is a substrate for caspases (Irmler et al., 1997) the 80 kDa and 37 kDa protein may represent a complex of caspase and FLIP and a cleavage product of FLIP, respectively. The IR for the 55 kDa protein displayed an extremely significant decrease in both DS cerebellum ($P < 0.001$) as well as frontal cortex ($P < 0.0001$). Neither DS cerebellum nor frontal cortex showed any significant difference in the total as well as the 37 kDa protein IR ($P < 0.05$) (Table 2). One major band at 25 kDa that represents ARC and two minor bands at 15 and 80 kDa were detected using anti-ARC antibody (Fig. 2). The 25 kDa ARC-IR tended to increase both in cerebellum and frontal cortex of DS but did not reach significance level. Likewise, the total ARC-IR showed insignificant increase in the test subject (Table 2). Linear regression analysis revealed no significant correlation between postmortem interval and levels of all the investigated proteins. However, inconsistent correlation was found between age and levels of proteins as well as amongst the density of individual proteins (data not shown).

Table 2. Levels of antiapoptotic proteins in frontal cortex and cerebellum of control and DS cases. Following western blot, immunoreactivity of 10 controls and 6–8 DS cases were quantified by scanning densitometry. Values are expressed as mean ± standard deviation of arbitrary optical density and inter-group differences were analyzed by non-parametric Mann-Whitney U-test and the significance was set at $P < 0.05$ for a two-tailed test. ***$P < 0.001$, ****$P < 0.0001$ compared to controls. *TIR* total immunoreactivity

Protein	Control	DS
Cerebellum		
a) DFF 45		
TIR	11.75 ± 7.11 (n = 8)	5.77 ± 1.73 (n = 6)
DFF 45	2.65 ± 1.52 (n = 8)	0.31 ± 0.29*** (n = 6)
b) ARC		
TIR	5.57 ± 3.92 (n = 8)	12.47 ± 8.49 (n = 6)
25 kDa	2.78 ± 1.50 (n = 8)	4.87 ± 3.38 (n = 6)
c) FLIP		
TIR	5.81 ± 2.10 (n = 8)	6.76 ± 3.33 (n = 6)
55 kDa	2.03 ± 0.72 (n = 8)	0.46 ± 0.28*** (n = 6)
37 kDa	2.83 ± 1.11 (n = 8)	3.15 ± 0.79 (n = 6)
Frontal		
a) DFF 45		
TIR	6.18 ± 1.92 (n = 10)	2.32 ± 0.94**** (n = 8)
DFF 45	1.83 ± 1.06 (n = 10)	0.17 ± 0.13**** (n = 8)
b) ARC		
TIR	7.33 ± 4.91 (n = 10)	9.15 ± 6.73 (n = 8)
25 kDa	2.55 ± 1.19 (n = 10)	3.10 ± 2.85 (n = 8)
c) FLIP		
TIR	10.65 ± 3.00 (n = 10)	8.70 ± 2.52 (n = 8)
55 kDa	4.04 ± 1.88 (n = 10)	0.56 ± 0.32**** (n = 8)
37 kDa	4.04 ± 1.88 (n = 10)	5.17 ± 1.36 (n = 8)

Discussion

The results presented in this study demonstrate alteration in different proteins involved in the apoptotic process, lending further support for the hypothesis that apoptosis may account for the neuronal loss observed in DS. Levels of procaspases were decreased in DS compared to controls, the decrease being more severe in frontal cortex than cerebellum confirming the differential vulnerability of different brain regions. The decrease in procaspases level may partly be explained by activation of precursors, as similar decrease was also observed in ceramide induced apoptosis in human neuroblastoma cell lines (Ito et al., 1999). In addition, several of the known procaspases are found in multiple intracellular compartments and translocate from these compartments to others following apoptotic stimuli (Zhivotovsky et al., 1999) which could also contribute to the observed reduction. Although apoptosis may have a role in enhancing caspase expression, when we consider the fact that large amounts of caspases are synthesized as zymogens in advance and activated on demand (Thornberry and Lazebnik, 1998), activation seems quiet likely a logical explanation for the decrease in procaspases. The finding of increased P43 which represents the partially processed large subunit of caspase-8 in DS points to the rapid processing of the procaspases. Contrary to this notion, we did not find any change in the fully processed P18 large subunit of caspase-8 probably due to rapid clearance of cells containing activated caspases (Shimohama et al., 1999). It has recently been shown that caspases are substrates of calpains and cleavage by calpains yields inactive fragments (Chua et al., 2000). Thus, calpain cleavage of caspases may serve as an alternative explanation for the decrease in procaspases and production of different fragments.

At least two pathways of caspase activation have been described (Fig. 3). In the first, cytochrome c has a pivotal role in binding Apaf-1 and causing caspase-9 activation leading to the cascade of caspase activation that executes the death program (Li et al., 1997). In the second, apoptosis is activated by cell surface death receptors and caspase-8 plays a central role. It is activated upstream of mitochondria and directly processes the effector caspases like caspase-3 (Peter and Krammer, 1998). The lack of detectable translocation of cytochrome c from mitochondria to cytosol is suggestive that caspase activation in DS may be effected by the receptor mediated pathway. This is further supported by the unchanged levels of Apaf-1, a cytoplasmic protein that serves as a docking protein for cytochrome c and caspase-9 (Li et al., 1997), which may also explain the insignificant decrease of procaspase-9. Indeed, in many types of cells caspase activation and apoptosis were shown to be induced via cytochrome c independent pathway (Chauhan et al., 1997; Holinger et al., 1999). Hence, it appears likely that mitochondria participate in apoptosis in DS, not because of cytochrome c but rather due to oxidative stress as they are the major source of reactive oxygen species. The receptor pathway involves death receptors like CD95(APO-1/Fas). CD95 belongs to the tumor necrosis factor/nerve growth factor receptor superfamily of cell surface proteins and has been shown to mediate receptor dependent programmed cell death. When engaged by CD95 ligand, CD95 oligomers are formed on the target cell and a signal to enter the cell's intrinsic apoptotic pathway is transmitted via the cytoplasmic so called "death domain", which in turn triggers the caspase cascade executing apoptosis (Becher et al., 1998).

The unchanged levels of cytosolic cytochrome c leaves the receptor pathway to emerge as the likely pathway responsible for caspase activation. If this is the case, what initiates caspase activation via this pathway? There appear to be some candidates which could be considered as a triggering factor. The first one is Aβ. Aβ a major component of senile plaques in DS, is considered as key etiological factor for neuronal loss mediated by apoptosis in vitro (Lee et al., 2000). Caspases are activated in neurons exposed to Aβ and caspase inhibitors block Aβ induced cell death (Mattson et al., 1998). Thus, it is suggested that Aβ might initiate apoptosis by cross-linking death receptors (Ivins et al., 1999). In addition, APP is a substrate of caspases, with caspase-6, and -8 displaying higher efficiency than caspase-3 and -9 (Pellegrini et al., 1999) indicating a self-reinforcing mechanism that commits the cell to the death pathway. The second candidate is oxidative stress. Reactive oxygen species (ROS) play an important role in cell death induced by diverse stimuli. There is fair evidence indicating the association between ROS and CD95. ROS were shown to increase CD95 mRNA as well as protein expression in microglial cells (Vogt et al., 1998). Furthermore, 4-hydroxynonenal, a diffusible product of lipid peroxidation, was reported to cause Fas induced activation of caspase-8 (Liu et al., 2000). Oxidative stress has been considered a possible contributor to neurodegeneration associated with the formation of amyloid deposits and neurofibrillary tangles (Kitamura et al., 1998). In vitro studies of fetal cerebral cortical neurons isolated from cases of Down syndrome indicate increased generation of reactive oxygen species that results in induction of apoptosis which is ablated by free radical scavengers (Busciglio and Yanker, 1995). Hence, it seems that ROS either directly or through Aβ can activate the

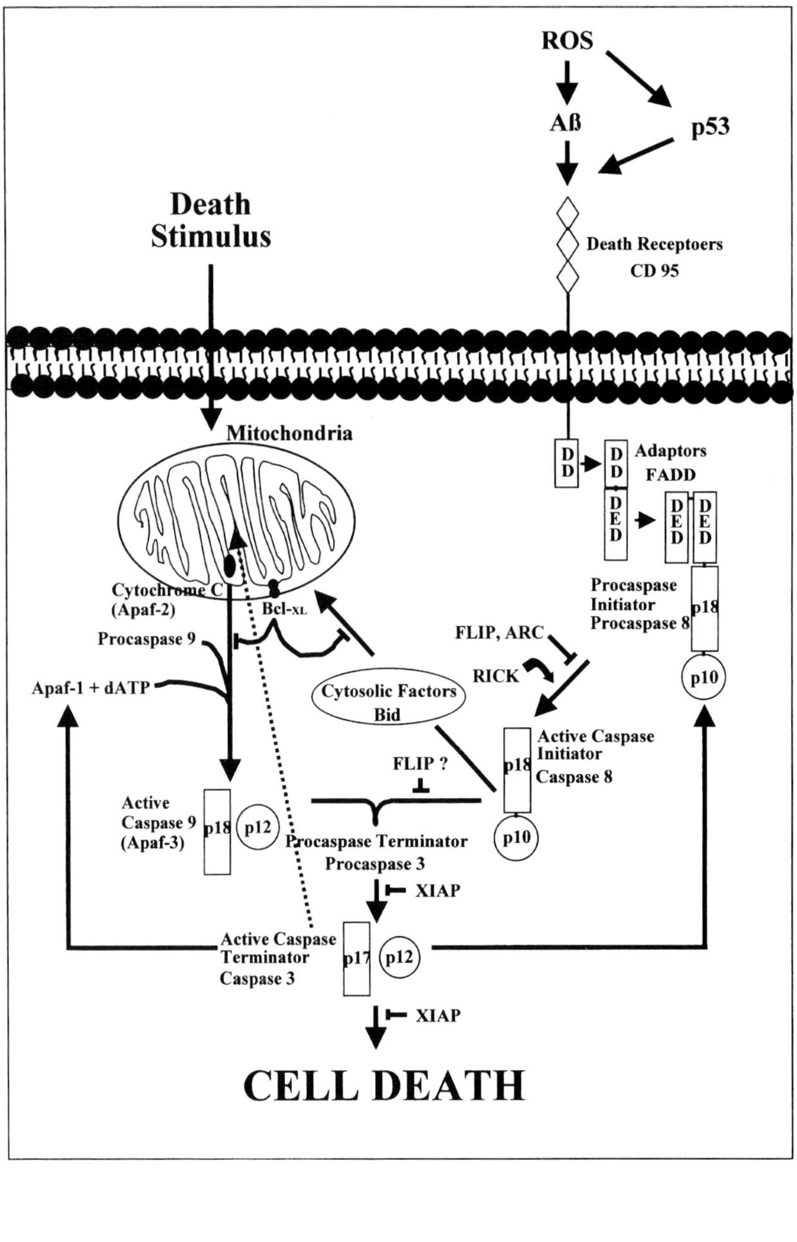

receptor pathway. The third one is p53. p53 is a stress induced transcription factor that can be activated by a number of adverse stimuli including DNA damage, hypoxia, and ROS. Increased expression of p53 protein under these conditions cause growth arrest or apoptosis (Lowe et al., 1993; Polyak et al., 1997). p53 and CD95 proteins may be inter-related because in non-neuronal cells, P53 activation was observed to induce CD95 expression (Müller et al., 1998; Bennet et al., 1998). A similar correlate was also reported in DS (de la Monte et al., 1998; Seidl et al., 1999). The upregulation of p53 (Castagne et al., 1999) and CD95 (Vogt et al., 1998) by ROS indicates the strong association between them and provides support for the involvement of p53 and ROS in activating the receptor pathway. Taken together, it is plausible that apoptosis in DS may be induced by factors, such as Aβ activation of caspase-8 in a cytochrome c independent manner.

The apoptotic regulatory proteins also showed a similar alteration as that of caspases. RICK contains an N-terminal serine-threonine kinase catalytic domain like RIP and a C-terminal that has significant similarity to the prodomain of caspases (Inohara et al., 1998). Although the precise mechanism by which RICK promotes apoptosis is not known, its expression augments apoptosis initiated by caspase-8 and -10 (Inohara et al., 1998). The increase in RICK observed in DS frontal cortex as well as cerebellum strengthen the proposed role of caspase-8 in mediating apoptotic cell death in this disorder. ARC is a novel CARD containing protein that inhibits CED-3,

Fig. 3. Proposed pathway of apoptotic death activated by factors like Aβ in DS (adapted from Cryans and Yuan, 1998; Ivins et al., 1999). Caspases can be activated by one of the two interacting pathways: a cytochrome c (left) and receptor (right) pathway. In the cytochrome c pathway, apoptotic stimuli such as DNA damaging agents (ROS ?) trigger the release of cytochrome c from mitochondria into the cytosol, a process that probably involves Bax redistribution in the reverse direction and is inhibitable by Bcl-2. Cytochrome c binds with Apaf-1 in the presence of dATP (ATP), a step that eventually exposes the CARD domain of Apaf-1 for binding and subsequent activation of procaspase-9. Active caspase-9 in turn, proteolytically activates the downstream terminator caspases such as caspase-3 that cause cell death by cleaving key intracellular proteins. In contrast, in the receptor pathway Aβ aggregation induced by ROS may either cross-link the death receptors or cause the release of death ligands that leads to the recruitment of a variety of DD bearing proteins such as FADD. FADD in turn recruits the initiator caspases such as caspase-8 to the complex through the DED module, an event that leads to the proteolytic cleavage of active caspase-8. This is further facilitated by p53 mediated upregulation of the death receptors. Active caspase-8 can initiate the caspase cascade by directly cleaving, bypassing mitochondria, the terminator caspases that results in the demise of the cell. Alternatively, active caspase-8 can activate the C-terminal fragment of the proapoptotic Bcl-2 family protein, Bid to cause the release of cytochrome c. This typically exemplifies the crosstalk between the two pathways. Activation of procaspase-8 and -9 and further release of cytochrome c by the active terminator caspases occur as a self reinforcing mechanism to ensure death of the cell. The receptor pathway can be inhibited by preventing procaspase-8 recruitment and/or activation by FLIP and ARC whereas stimulated by RICK. Moreover, FLIP may also act as a persistent pseudosubstrate reminiscent of p35, a viral protein that inhibits caspases. The mammalian homologues of the baculovirus IAPs serve as inhibitors of the terminator caspases capase-2, and -8 through direct binding to death proteases (Koseki et al., 1998). Though human ARC is mainly expressed in skeletal muscle and cardiac tissue (Koseki et al., 1998), the western blot pattern shows its expression in the brain as well (Fig. 2). Although ARC is not significantly increased in DS brains, the trend it showed is consistent with the hypothesis that antiapoptotic mechanisms that inhibit caspase activation may be upregulated with persistent apoptotic signal such as Aβ (Lee et al., 2000).

FLIP, another apoptotic regulating protein is also known by several names such as, Casper (Shu et al., 1997), FLAME-1 (Srinivasula et al., 1997), CASH (Goltsev et al., 1997), and CLARP (Inohara et al., 1997). It is expressed in mammalian cells as two alternatively spliced messages, as $FLIP_L$ and $FLIP_S$. The protein is composed of two N-terminal DEDs fused to a C-terminal caspase like domain devoid of protease activity as it lacks key residues involved in catalysis (Irmler et al., 1997). The role of FLIP in apoptosis is controversial, some studies reported its apoptosis inducing properties (Inohara et al., 1997; Shu et al., 1997) while others advocate for its anti-apoptotic property (Irmler et al., 1997; Srinivasula et al., 1997; Hennino et al., 2000; Kataoka et al., 2000). Cleavage of FLIP by caspases in vitro results into two fragments topologically equivalent to the large and small subunits of caspases, the same is thought to occur in vivo (Srinivasula et al., 1997). This cleavage may result in a strong FLIP-caspase-8 interaction, thereby blocking further processing and activation of caspase-8, reminiscent of p35 (Irmler et al., 1997). The decrease in the 55 kDa FLIP levels obtained in DS in the present study supports such a scenario. Indeed, the 37 kDa protein, which is thought to be a split product of FLIP, constitutes 90.2% of the total pool of FLIP in DS frontal cortex compared to controls which was 57.4%.

DFF consists of two subunits of 45 kDa and 40 kDa, from which the 45 kDa subunit can be cleaved by caspase-3 into smaller peptides in cells undergoing apoptosis (Liu et al., 1997; Enari et al., 1998). DFF45/ICAD is thought to serve as specific chaperone allowing the proper folding of the nascent DFF40/CAD polypeptide then it remains complexed with CAD to inhibit its DNAse activity (Enari et al., 1998). Moreover, DFF45 is also suggested to inhibit DFF40 translocation into the nucleus (Enari et al., 1998). The reduction in DFF45 in DS is in line with its degradation by caspase-3 and further confirms the contribution of apoptosis to DS neuronal death.

Our results described here provide evidence for the alteration in levels of different pro- and anti-apoptotic proteins in DS. The study suggests that apoptosis triggered by factors e.g., Aβ activation of caspases via cytochrome c independent pathway may account for the neuronal loss observed in this disorder. Therapeutic strategies aimed at caspase inhibition therefore might have some benefit in DS.

Acknowledgements

We are highly indebted to Red Bull Company, Salzburg, Austria for the generous financial support and Mrs. N. Balic for technical help. E.E. is a recipient of scholarship from the Austrian Academic Exchange Program, ÖAD.

References

Alnemri ES, Livingston DJ, Nicholson DW, Salvesen G, Thornberry NA, Wong WW, Yuan J (1996) Human ICE/CED-3 proteases nomenclature. Cell 87: 171

Anderson AJ, Stoltzner S, Lai F, Su J, Nixon RA (2000) Morphological and biochemical assessment of DNA damage and apoptosis in Down syndrome and Alzheimer's disease, and effect of postmortem tissue archival on tunnel. Neurobiol Aging 21: 511–524

Bambrick LL, Krueger BK (1999) Neuronal apoptosis in mouse trisomy 16: mediation by caspases. J Neurochem 72: 1769–1772

Becher B, Barker PA, Owens T, Antel PJ (1998) CD95-CD95L: can the brain learn from the immune system? Trends Neurosci 21: 114–117

Bennett M, Macdonald K, Chan SW, Luzio JP, Simari R, Weissberg P (1998) Cell surface trafficking of Fas: a rapid mechanism of p53 mediated apoptosis. Science 282: 290–293

Busciglio J, Yanker BA (1995) Apoptosis and increased generation of reactive oxygen species in Down's syndrome neurons in vitro. Nature 378: 776–779

Castagne V, Gautschi M, Lefevre K, Posada A, Clarke PG (1999) Relationships between neuronal death and the cellular redox status. Focus on the developing nervous system. Prog Neurobiol 59: 397–423

Chauhan D, Pandey P, Ogata A, Teoh G, Krett N, Halgren R, Rosen S, Kufe D, Kharbanda S, Anderson K (1997) Cytochrome c dependent and independent induction of apoptosis in multiple myeloma cells. J Biol Chem 272: 29995–29997

Chinnaiyan AM, Dixit VM (1996) The cell death machine. Curr Biol 6: 555–562

Chinnaiyan AM, O'Rourke K, Lane BR, Dixit VM (1997) Interaction of CED-4 with CED-3 and CED-9: a molecular framework for cell death. Science 275: 1122–1126

Chua BT, Guo K, Li P (2000) Direct cleavage by the calcium activated protease calpain can lead to inactivation of caspases. J Biol Chem 275: 5131–5135

Cohen GM (1997) Caspases: the executioners of apoptosis. Biochem J 326: 1–16

Cryns V, Yuan J (1998) Proteases to die for. Gene Dev 12: 1551–1570

de la Monte SM, Sohn YK, Ganju N, Wands JR (1998) p53 and CD95 associated apoptosis in neurodegenerative disorders. Lab Invest 78: 401–411

Earnshaw WC, Martins LM, Kaufmann SH (1999) Mammalian caspases: structure, activation, substrates, and functions during apoptosis. Annu Rev Biochem 68: 383–424

Enari M, Sakahira H, Yokoyama H, Okawa K, Iwamatsu A, Nagata S (1998) A caspase activated DNAse that degrades DNA during apoptosis, and its inhibitor ICAD. Nature 391: 43–50

Goltsev YV, Kovalenko AV, Arnold E, Varfolomeev EE, Brodianskii VM, Wallach D (1997) CASH, a novel caspase homologue with death effector domains. J Biol Chem 272: 19641–19644

Hennino A, Berard M, Casmayor-Palleja M, Krammer PH, Defrance T (2000) Regulation of the Fas death pathway by FLICE-inhibitory protein in primary human B cells. J Immunol 165: 3023–3030

Hofman K, Bucher P, Tschopp J (1997) The CARD domain: a new apoptotic signaling motif. Trends Biochem Sci 22: 155–156

Holinger EP, Chittenden T, Lutz RJ (1999) Bak BH3 peptides antagonize Bcl-xL function and induce apoptosis through cytochrome c-independent activation of caspases. J Biol Chem 274: 13298–13304

Inohara N, del Peso L, Koseki T, Chen S, Núñez G (1998) RICK, a novel protein kinase containing a caspase recruitment domain, interacts with CLARP and regulates CD95-mediated apoptosis. J Biol Chem 273: 12296–12300

Inohara N, Koseki T, Hu Y, Chen S, Nunez G (1997) CLARP, a death effector domain containing protein interacts with caspase-8 and regulates apoptosis. Proc Natl Acad Sci USA 94: 10717–10722

Irmler M, Thome M, Hahne M, Schneider P, Hofmann K, Steiner V, Bodmer JL, Schroter M, Burns K, Mattmann C, Rimoldi D, French LE, Tschopp J (1997) Inhibition of death receptor signals by cellular FLIP. Nature 388: 190–195

Ito A, Uehara T, Tokumitsu AI, Okuma Y, Nomura Y (1999) Possible involvement of cytochrome c release and sequential activation of caspases in ceramide induced apoptosis in SK-N-MC cells. Biochim Biophys Acta 1452: 263–274

Ivins KJ, Thornton PL, Rohn TT, Cotman CW (1999) Neuronal apoptosis induced by β-amyloid is mediated by caspase-8. Neurobiol Dis 6: 440–449

Jänicke RU, Sprengart ML, Wati MR, Porter AG (1998) Caspase-3 is required for DNA fragmentation and morphological changes associated with apoptosis. J Biol Chem 273: 9357–9360

Katoka T, Budd RC, Holler N, Thome M, Martinon F, Irmler M, Burns K, Hahne M, Kennedy N, Kvacsovics M, Tschopp J (2000) The caspase-8 inhibitor FLIP promotes activation of NF-kB and Erk signaling pathways. Curr Biol 10: 640–648

Kitamura Y, Shimohama S, Kamoshima W, Ota T, Matsuoka Y, Nomura Y, Smith MA, Perry G, Whithouse PJ, Taniguchi T (1998) Alterations of proteins regulating apoptosis. Bcl-2, Bcl-x, Bax, Bak, Bad, ICH-1, and CPP32, in Alzheimer's disease. Brain Res 780: 260–269

Koseki T, Inohara N, Chen S, Nunez G (1998) ARC, an inhibitor of apoptosis expressed in skeletal muscle and heart that interacts selectively with caspases. Proc Natl Acad Sci USA 95: 5156–5160

Lee SA, Zheng TS, Flavell RA, Rakic P, Roth KA (2000) Amyloid beta induced neuronal death is Bax dependent but caspase independent. J Neuropathol Exp Neurol 4: 271–279

Li P, Nijhawan D, Budihardjo I, Srinivasula SM, Ahmed M, Alnemri ES, Wang X (1997) Cytochrome c and dATP dependent formation of Apaf-1/caspase-9 complex initiates an apoptosis cascade. Cell 91: 479–489

Liu W, kato M, Akhand AA, Hayakawa A, Suzuki H, Miyata T, Kurokawa K, Hotta Y, Ishikawa N, Nakashima I (2000) 4-Hydroxynonenal induces a cellular redox status related activation of the caspase cascade for apoptotic cell death. J Cell Sci 113: 635–641

Liu X, Zou H, Slaughter C, Wang X (1997) DFF, a heterodimeric protein that functions downstream of caspase-3 to trigger DNA fragmentation during apoptosis. Cell 89: 175–184

Lowe SW, Schmitt EM, Smith SW, Osborne BA, Jacks T (1993) p53 is required for radiation induced apoptosis in mouse thymocytes. Nature 362: 847–849

Mattson MP, Partin J, Begley JG (1998) Amyloid β peptide induces apoptosis related events in synapses and dendrites. Brain Res 807: 167–176

Müller M, Wilder S, Bannasch D, Israeli D, Lehlbach K, Li-Wiber M, Friedman SL, Galle PR, Stremmel W, Oren M, Krammer P (1998) p53 activates the CD95 gene in response to DNA damage by anticancer drugs. J Exp Med 188: 2033–2045

Nicholson DW, Thornberry NA (1998) Caspases: killer proteases. Trends Biochem Sci 22: 299–306

Nunez G, Benedict MA, Hu Y, Inohara N (1998) Caspases: the proteases of the apoptotic pathway. Oncogene 17: 3237–3245

Pellegrini L, Passer BJ, Tabaton M, Ganjei JK, D'Adamio L (1999) Alternative, non-secretase processing of Alzheimer's β-amyloid precursor protein during apoptosis by caspase-6 and -8. J Biol Chem 274: 21011–21016

Peter ME, Krammer PH (1998) Mechanism of CD95 (APO-1/Fas) mediated apoptosis. Curr Opin Immunol 10: 545–551

Polyak K, Xia Y, Zweier JL, Kinzler KW, Vogelstein B (1997) A model for p53 induced apoptosis. Nature 389: 300–305

Porter AG, Jänicke RU (1999) Emerging roles of caspase-3 in apoptosis. Cell Death Differ 6: 99–104

Reed JC (1999) Dysregulation of apoptosis in cancer. J Clin Oncol 17: 2941–2953

Saleh A, Srinivasula SM, Acharya S, Fishel R, Alnemri ES (1999) Cytochrome c and dATP mediated oligomerization of Apaf-1 is a prerequisite for proscaspase-9 activation. J Biol Chem 274: 17941–17945
Seidl R, Fang-Kircher S, Bidmon B, Cairns N, Lubec G (1999) Apoptosis associated proteins p53 and APO-1/FAS (CD95) in brains of adult patients with Down syndrome. Neurosci Lett 260: 9–12
Shimohama S, Tanino H, Fujimoto S (1999) Changes in caspase expression in Alzheimer's disease: comparison with development and aging. Biochem Biophys Res Commun 256: 381–384
Shu HB, Halpin DR, Goeddel DV (1997) Casper is a FADD and caspase related inducer of apoptosis. Immunity 6: 751–763
Srinivasula SM, Ahmad M, Ottilie S et al. (1997) FLAME-1, a novel FADD like antiapoptotic molecule that regulates Fas/TNFR-1 induced apoptosis. J Biol Chem 272: 18542–18545
Thompson CB (1995) Apoptosis in the pathogenesis and treatment of disease. Science 267: 1456–1462
Thornberry NA, Lazebnik Y (1998) Caspases: enemies within. Science 281: 1312–1322
Vogt M, Bauer MK, Ferrari D, Schulze-Osthoff K (1998) Oxidative stress and hypoxia/reoxygenation triggers CD95 (APO-1/Fas) ligand expression in microglial cells. FEBS Lett 429: 67–72
Wellington CL, Hayden MR (2000) Caspases and neurodegeneration: on the cutting edge of new therapeutic approaches. Clin Genet 57: 1–10
Wu D, Wallen HD, Inohara N, Núñez G (1997) Interaction and regulation of Caenorhabditis elegans death protease CED-3 by CED-4 and CED-9. J Biol Chem 272: 21449–21454
Zhivotovsky B, Samali A, Gahm A, Orrenius S (1999) Caspases: their intracellular localization and translocation during apoptosis. Cell Death Differ 6: 644–651
Zou H, Henzel WJ, Liu X, Lutschg A, Wang X (1997) Apaf-1, a human protein homologous to C. elegans CED-4, participates in cytochrome c dependent activation of caspase-3. Cell 90: 405–413

Authors' address: Prof. Dr. G. Lubec, Department of Pediatrics, University of Vienna, Währinger Gürtel 18-20 A-1090 Vienna, Austria, e-mail: Gert.Lubec@akh-wien.ac.at

Expression of apoptosis related proteins: RAIDD, ZIP kinase, Bim/BOD, p21, Bax, Bcl-2 and NF-κB in brains of patients with Down syndrome

E. Engidawork[1], **T. Gulesserian**[1], **R. Seidl**[1], **N. Cairns**[2], and **G. Lubec**[1]

[1] University of Vienna, Department of Pediatrics, Austria
[2] Department of Neuropathology, Institute of Psychiatry, King's College, London, United Kingdom

Summary. Down syndrome (DS) is a genetic disease that exhibits significant neuropathological parallels with Alzheimer's disease (AD). One of the features of DS, neuronal loss, has been hypothesized to occur as a result of apoptosis. An increasing number of proteins are implicated in apoptosis and several of them were shown to be altered in AD, however, the knowledge in DS is far from complete. To further substantiate the hypothesis that apoptosis is the underlying mechanism for neuronal loss and contribute towards the current knowledge of apoptosis in DS, we analyzed the expression of apoptosis related proteins in frontal cortex and cerebellum of DS by western blot and ELISA techniques. Quantitative analysis revealed a significant increase in DS frontal ($P < 0.0001$) and cerebellar ($P < 0.05$) Bim/BOD (Bcl-2 interacting mediator of cell death/Bcl-2 related ovarian death gene), cerebellar Bcl-2 ($P < 0.01$) as well as p21 ($P < 0.05$) levels compared to controls. No significant change was detected in Bax, RAIDD (receptor interacting protein (RIP)-associated ICH-1/CED-3-homologus protein with death domain), ZIP (Zipper interacting protein) kinase and NF-κB p65 levels in both regions, although frontal cortex levels of RAIDD, Bcl-2 and p21 levels tended to increase. In addition, a 45 kDa truncated form of NF-κB p65 displayed a significant elevation ($P < 0.05$) in DS cerebellum. No significant correlation had been obtained between postmortem interval and level of the proteins analyzed. With regard to age, it was only NF-κB p65 that showed significant correlation ($r = -0.8964$, $P = 0.0155$, $n = 9$) in frontal cortex of controls. These findings provide further evidence that apoptosis indeed accounts for the neuronal loss in DS but Bax and RAIDD do not appear to take part in this process.

Introduction

Apoptosis is a genetically programmed mode of cell death that exists in all multicellular species (Wolf and Green, 1999). Its major function is to facilitate the survival of the organism by the controlled elimination of damaged or superfluous cells. Small protein-protein interaction domains that are present in apoptotic proteins, including receptors, adaptors and effectors, are important components of the cell death pathways. The adaptor domains serve to transduce the death signal and include the DD (death domain), DED (death effector domain) and CARD (caspase activation and recruitment domain). RAIDD (receptor interacting protein (RIP)-associated ICH-1/CED-3-homologus protein with DD) (Duan and Dixit, 1997) also known as CRADD (caspase and RIP adaptor with DD) (Ahmad et al., 1997) is a recently identified such molecule with DD and CARD to interact with RIP and caspases, respectively (Duan and Dixit, 1997; Hofmann et al., 1997). Death receptors mediated apoptosis not only triggers a DD cascade that communicates with the downstream effectors, such as the caspases but also a protein kinase cascade (Aravind et al., 1999). Several Ser/Thr protein kinases are specifically involved in apoptosis. ZIP (Zipper interacting protein) kinase is a recently identified kinase that belongs to a novel Ser/Thr kinase subfamily. It contains a C-terminal leucine zipper that mediates activation via homodimerization and an N-terminal catalytic domain (Kawai et al., 1998). It binds to ATF4, member of ATF/CREB family, and the catalytic domain is closely related to DAP (death associated protein) kinase, hence the name Dlk (DAP like) or ZIP kinase (Kawai et al., 1998; Page et al., 1999). NF-κB (nuclear factor Kappa B), designates a family of redox sensitive transcription factors that participate in the activation of a wide range of genes crucially involved in immune and inflammatory functions. NF-κB is typically present in the cytoplasm bound to the inhibitory IκB proteins. Activation involves the signal induced degradation of inhibitory proteins, allowing NF-κB to translocate into the nucleus to modulate the expression of pro- or anti-apoptotic genes (Castagne et al., 1999). NF-κB has been shown to be activated in human brains of patients affected by several neurological diseases, although whether this activation contributes to cell survival or cell death pathways is an intriguing and debated issue (Castagne et al., 1999; Grilli and Memo, 1999; Mattson et al., 2000).

Because of the opposing options faced by a cell in development: proliferation versus death, apoptosis is connected to cell cycle regulation. p21 (WAF-1, Cip1, Sid1) a predominantly nuclear protein (El-Diery et al., 1994) and inhibitor of cyclin dependent kinase, is involved in both cell cycle control and apoptosis (Aravind et al., 1999). Furthermore, products of many genes induced by p21 are suggested to play a causal role or to be associated with different age related diseases (Chang et al., 2000). Bcl-2 gene family proteins are believed to control apoptosis (Krosmeyer, 1995). Several members of the gene family have been identified with each having several splice variants (Tsujimoto et al., 1985; Oltvazi et al., 1993). Members of the Bcl-2 family regulate apoptosis by either inhibiting (e.g., Bcl-2) or promoting (e.g., Bax) cell death (Aravind et al., 1999). Bim/BOD (Bcl-2 interacting mediator of cell death/Bcl-2 related ovarian death gene) is a new BH3 domain containing protein (Hsu et al., 1998; O'Connor et al., 1998) that provokes apoptosis by interacting with diverse members of the pro-survival Bcl-2 subfamily.

DS (Down syndrome) is a neurodevelopmental disorder with neurodegenerative component, resulting from trisomy of the distal segment of chromosome 21. DS is characterized clinically by somatic anomalies, mental retardation and precocious dementia. Brains from adult individuals with DS manifest abundant senile plaques, neurofibrillary tangles and extensive neuronal loss, hallmarks of AD (Alzheimer's disease), throughout the cortex (Mann, 1988). The mechanism leading to neuronal cell death in DS is not yet clearly established, however, an increasing bulk of evidence indicates that apoptosis may be involved (de la Monte, 1998; Seidl et al., 1999; Anderson et al., 2000). The present study was initiated to further substantiate this notion. To this effect, levels of seven different proteins involved in apoptosis (RAIDD, ZIP kinase, Bim/BOD, Bax, Bcl-2, NF-κB and p21) were determined by western blot and ELISA techniques in brains of patients from DS. The results demonstrate that apoptosis indeed has a role in DS neuronal loss, however, it does seem not to involve Bax and RAIDD.

Material and methods

Brain samples

Postmortem brain samples (frontal cortex, superior frontal gyrus, and cerebellum) of nine DS patients with Alzheimer' disease (AD)-like neuropathology (3 females and 6 males; mean age of 55.9 ± 8.0 years and postmortem interval of 32.4 ± 20.7 hours) and 10 controls with no history of neurological or psychiatric illness (3 females and 7 males; mean age 62.6 ± 8.0 years and postmortem interval of 37.3 ± 18.4 hours) were obtained from the Medical Research Council's London Brain Bank for Neurodegenerative Diseases, Department of Neuropathology, Institute of Psychiatry, London, UK. The major cause of death in DS was bronchopneumonia and heart disease in controls. All samples were stored at −70°C and the freezing chain was never interrupted.

Immunoblotting

Frozen brain samples were thawed on ice and homogenized in 4 volumes of TBS buffer (100 mM Tris, 150 mM NaCl, pH 7.5) containing protease inhibitors, (1 μg/ml each of antipain, pepstatin A and leupeptin, and 10 mM of phenylmethylsulfonyl fluoride) and 0.05% Tween 20 for 30 seconds (6–8 strokes) with a Potter-Elvehjem homogenizer. The brain homogenate then was centrifuged for 10 min at 4°C and 10,000 g. To the supernatant, equal volume of sample buffer (125.5 mM Tris-HCl base, 70 mM sodium dodecyl sulfate, 0.001% Bromophenol blue, 20% glycerol, and 2% 2-mercaptoethanol, pH 6.8) was added and the mixture was heated at 95°C for 15 min. The concentration of protein was determined by the Bicinchoninic acid assay using bovine serum albumin as a standard (Pierce, USA). Protein extracts containing either 20 μg (NF-κB), 25 μg (Bax), 50 μg (ZIP kinase), or 100 μg (RAIDD, Bim/BOD) were subjected to sodium dodecyl sulfate-polyacrylamide gel electrophoresis (200 V for 3 hrs), and transferred to polyvinylidene fluoride membranes (Immobilon™-P, MilliPore, Bedford, USA) (192 V for 1 hr). Following transfer, membranes were washed 2 × 5 min in Tris buffer (100 mM Tris-HCl, 150 mM NaCl, pH 7.5) and later blocked with blotto (1.5% non-fat dry milk in TBST (100 mM Tris-HCl, 150 mM NaCl, pH 7.5, 1 mM $MgCl_2$, 0.1% Tween-20)) overnight at 4°C. This was followed by incubation with primary polyclonal rabbit anti-NF-κB p65 CT (#06-418, UPSTATE biotechnology, Lake Placid, NY, USA) (1:1,000), anti-human Bax (AB2916, CHEMICON, Temecula, CA, USA) (1:250); anti-human ZIP kinase (Cat. #210-216-C100, ALEXIS BIOCHEMICALS, San Diego, USA) (1:1,000); anti-human RAIDD (Cat. #210-733-C100, ALEXIS BIOCHEMICALS, San Diego, USA) (1:1,000) and anti-human Bim/BOD (Cat. #210-214-C100, ALEXIS BIOCHEMICALS, San Diego, USA) (1:1,000) antibody in blotto at 4°C overnight on a rocking shaker. The membranes were then washed 3 × 15 minutes with blotto prior to incubation with respective secondary horseradish peroxidase conjugated goat anti-rabbit IgG (H + L) (#4050-05, Southern Biotechnology Associates Inc., Birmingham, USA) in blotto for one hour at room temperature on a rocking shaker. The membranes were washed again 3 × 15 minutes followed by incubation with NEN chemiluminescence reagent for 1 minute and subsequent exposure to films (Kodak Blue XB-1, Rochester, USA). Developed films were scanned and densities of the immunoreactive bands were calculated by RFLP Scan 2.1 software program (Scanalytics, Billerica, USA).

ELISA

Frozen brain samples were thawed on ice and homogenized in a buffer (50 mM Tris, containing 5 mM EDTA, 0.2 mM PMSF, 1 μg/ml pepstatin and 0.5 μg/ml leupeptin, adjusted to pH 7.4) for 30 seconds (six strokes) with a Potter-Elvehjem homogenizer. Antigen extraction reagent (provided in the kits) was added to the brain homogenate and incubated for 30 minutes with occasional vortexing. The mixture was then centrifuged at 4°C and 3,000 g for 5 minutes. The supernatant was used for ELISA and determination of total protein by the Lowry method (Lowry et al., 1951). ELISA quantitation was performed with mouse monoclonal antibodies specific for human p21/WAF-1 and Bcl-2 protein (Oncogene Research Products, Cambridge, USA; ELISA kits Cat #QIA28 and QIA23, respectively). Antibody binding was detected with horseradish peroxidase conjugated secondary antibody. The color reaction was quantified using a spectrophotometer with dual wavelengths of 450/540 nm.

Data analysis

Values are expressed as mean ± standard deviation of optical density and U/mg protein for western blotting and ELISA, respectively. Inter-group differences were analyzed by non-parametric Mann-Whitney U-test and intra-group correlation with linear regression analysis. The significance was set at $P < 0.05$ for a two-tailed test.

Results

Western blot revealed RAIDD, ZIP kinase, Bim/BOD and Bax immunoreactive (IR) proteins with molecular size of 22, 52, 23, and 21 kDa, respectively (Fig. 1A–C & E). NF-κB also migrated as a 65 kDa band with two other bands at 45 and 32 kDa (Fig. 1D). NF-κB consists of homo- and heterodimeric proteins that belong to the Rel family of transcription factors, including p50, p52, p65 (Rel A), c-Rel, and Rel B all of which contain a so called Rel homology region. In mammals, the most abundant form of NF-κB consists of the p50/p65 heterodimer associated with member of IκB family (Simeonids et al., 1997). Accordingly, the 65 kDa corresponds to the p65 subunit of NF-κB

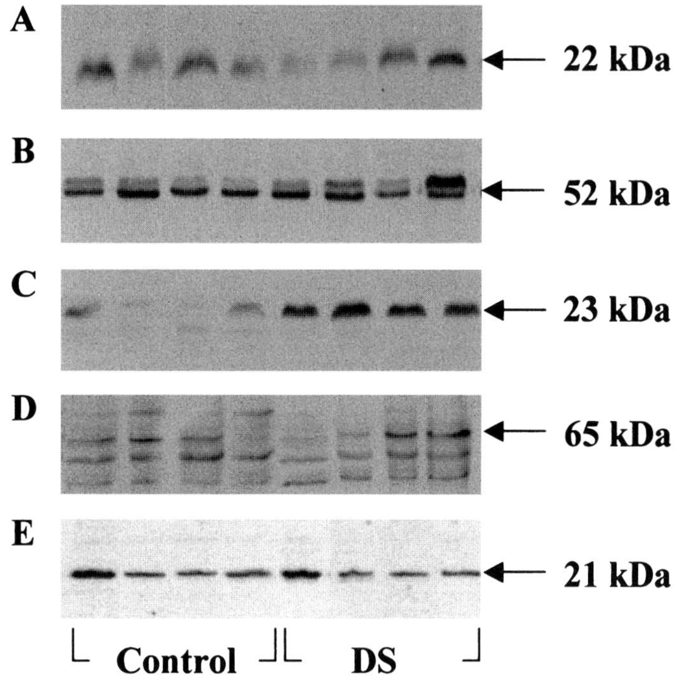

Fig. 1. Immunodetection of apoptosis related proteins in frontal cortex; **A** RAIDD, **B** ZIP kinase, **C** Bim/BOD, **D** NF-κB and **E** Bax. Brain homogenate containing 25 (Bax), 50 (ZIP kinase), 100 (RAIDD, Bim/BOD), and 20 (NF-κB) μg protein/lane from the frontal cortices of 10 controls and 8 DS brains were used for the immunodetection. The immunoblots of 4 controls (lanes 1–4), and 4 DS (lanes 5–8) cases were shown in the upper panels

and the remaining two bands may be truncated products generated via proteolytic processing of the p65 protein.

Subsequent densitometric analysis showed that levels of RAIDD an adaptor molecule with DD, were found to be quite similar to controls in cerebellum but insignificantly increased in frontal cortex (Table 1). ZIP kinase member of the novel subfamily of Ser/Thr kinases, likewise did not exhibit any detectable changes in DS frontal cortex as well as cerebellum compared to controls (Table 1). The cortical as well as cerebellar levels of the death agonist, Bax was also comparable to controls. By contrast, its BH3 containing homologue, Bim/BOD levels were significantly elevated in both DS frontal cortex ($P < 0.0001$) and cerebellum ($P < 0.05$) (Table 1). With regard to NF-κB, p65 levels were comparable between controls and DS in both regions examined. Similarly, neither the total NF-κB IR nor the 32 kD protein did show any significant alterations. The 45 kDa protein, however, exhibited a significant increase by ~106% ($P < 0.05$) in DS cerebellum but unaltered in the frontal cortex.

Quantitative analysis by making use of ELISA produced more or less a similar pattern of alterations in the proteins analyzed as that observed with western blot. No significant difference was detected in DS frontal cortex p21 levels, although levels appeared to be relatively higher (33%) compared to

Table 1. Levels of apoptosis related proteins in two brain regions. Following western blot, immunoreactivity of 5–10 controls and 6–8 DS cases were quantified by scanning densitometry. Values are expressed as mean ± standard deviation of arbitrary optical density and inter-group differences were analyzed by non-parametric Mann-Whitney U-test and the significance was set at $P < 0.05$ for a two-tailed test. *$P < 0.05$, ****$P < 0.0001$ compared to controls

Region/Protein	Control	DS
Frontal		
Bim (BOD)	0.46 ± 0.24 (n = 10)	1.90 ± 0.34**** (n = 8)
RAIDD	0.92 ± 0.38 (n = 10)	1.44 ± 0.75 (n = 8)
Zip Kinase (52 kDa)	2.60 ± 0.56 (n = 9)	2.44 ± 1.16 (n = 7)
Zip Kinase (26 kDa)	0.80 ± 0.24 (n = 9)	0.90 ± 0.24 (n = 7)
Bax	2.95 ± 2.97 (n = 9)	3.54 ± 4.83 (n = 9)
NF-κB total IR	4.33 ± 1.31 (n = 6)	4.10 ± 2.19 (n = 6)
NF-κB 65 kDa	1.60 ± 0.73 (n = 6)	1.73 ± 1.62 (n = 6)
NF-κB 45 kDa	1.21 ± 0.41 (n = 6)	1.11 ± 0.54 (n = 6)
NF-κB 32 kDa	0.63 ± 0.33 (n = 6)	0.84 ± 0.36 (n = 6)
Cerebellum		
Bim (BOD)	0.47 ± 0.18 (n = 9)	0.74 ± 0.14* (n = 6)
RAIDD	0.88 ± 0.27 (n = 8)	0.84 ± 0.48 (n = 6)
Zip Kinase (52 kDa)	0.14 ± 0.07 (n = 5)	0.13 ± 0.05 (n = 6)
Zip Kinase (26 kDa)	0.46 ± 0.11 (n = 5)	0.47 ± 0.12 (n = 6)
Bax	2.82 ± 1.69 (n = 9)	2.44 ± 1.08 (n = 9)
NF-κB total IR	0.74 ± 0.43 (n = 8)	1.02 ± 0.25 (n = 6)
NF-κB 65 kDa	0.21 ± 0.17 (n = 8)	0.21 ± 0.14 (n = 6)
NF-κB 45 kDa	0.15 ± 0.08 (n = 8)	0.31 ± 0.09* (n = 6)
NF-κB 32 kDa	0.18 ± 0.06 (n = 8)	0.24 ± 0.07 (n = 6)

controls. A significant increase, however, was observed in DS cerebellum by ~82% ($P < 0.05$) (Fig. 2A). Bcl-2 a pro-survival protein, likewise showed an insignificant increase by ~18% in frontal cortex of DS whereas it was significantly increased by ~66% in DS cerebellum ($P < 0.01$) (Fig. 2B). Linear regression analysis was performed to test the influence of age and postmortem interval on levels of the investigated proteins. No significant correlation was obtained with age and postmortem interval in all proteins except NF-κB p65. NF-κB p65 showed a significant correlation with age in frontal cortex of controls ($r = -0.8964$, $P = 0.0155$, $n = 6$).

Discussion

This paper describes levels of apoptosis related proteins in DS using western blot and ELISA. Alterations were noted both in frontal cortex and cerebellum, the later being relatively more affected than the former consistent with its regular involvement in DS in morphological/neuropathological terms.

RAIDD is an adaptor protein similar to FADD. Both of these proteins contain an N-terminal domain, which interacts with homologous prodomain

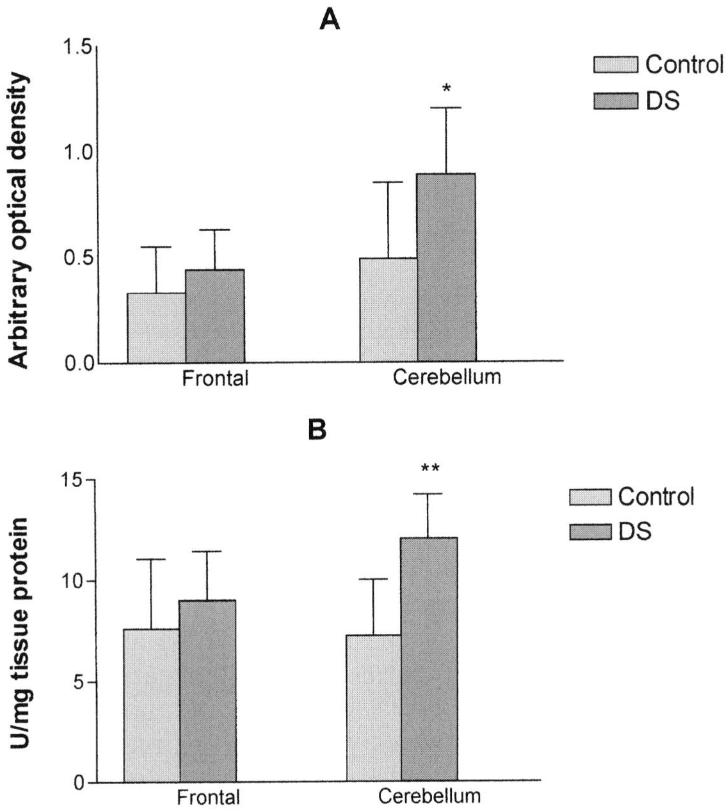

Fig. 2. Concentration (U/mg tissue protein) of **A** WAF-1 and **B** Bcl-2 in two brain regions. Concentration of 9 controls and 9 DS cases were quantified by ELISA. Values are expressed as mean ± standard deviation of protein concentration and inter-group differences were analyzed by non-parametric Mann-Whitney U-test and the significance was set at $P < 0.05$ for a two-tailed test. $^*P < 0.05$ and $^{**}P < 0.01$ compared to controls

of the recruited caspases, and a C-terminal death domain. The presence of a CARD region and the ability to interact with procaspase-2 suggest that RAIDD may play a role in apoptosis (Duan and Dixit, 1997). At present, the signaling pathway by which RAIDD induces apoptosis is unknown. We have recently proposed that apoptosis in DS may occur via the death receptor pathway (Gulesserian et al., this volume) and here we did not find any detectable changes in RAIDD levels between control and test groups. Moreover, RAIDD was shown to be dispensable for Fas ligand induced apoptosis in non-neuronal cells (Villunger et al., 2000). Taken together these data suggest that RAIDD may not be a required factor for apoptosis in DS. Thus, the exact role of RAIDD in apoptosis remains to be determined.

In addition to activation of caspases, signaling events such as activation of protein kinases are involved in apoptosis (Andersson, 1997). ZIP kinase is one of the kinases shown to have a role in apoptosis and the cell death inducing activity depends on its intrinsic kinase activity (Kawai et al., 1998). The subcellular localization of ZIP kinase determines its apoptotic activity (Kogel et al., 1999). As the enzyme is tightly associated with nuclear structures, translocation to or retention in the cytoplasm is mandatory to exert an apoptotic activity (Kogel et al., 1999; Page et al., 1999). This was shown to be effected by co-expression of ZIP kinase and PAR-4 (prostate apoptosis response-4) (Page et al., 1999). Indeed, in AD increase in ZIP kinase (Engidawork et al., 2001) is ascribed to elevated expression of PAR-4 (Mattson et al., 1999). This suggests that although all DS brains employed in the present study display AD neuropathology, the two disorders may not totally share the same apoptotic mechanisms.

p21 potentially regulates cell cycle progression by inhibiting the activity of cyclin and cyclin dependent kinases, and proliferating cell nuclear antigen, a processivity factor for DNA polymerase δ (Chen et al., 1995; Sherr and Roberts, 1995). In addition, p21 has a complex role in apoptosis (Chang et al., 2000). A positive correlation of p21 and p53 dependent apoptosis was reported (EL-Diery et al., 1994) and later p21 was shown to function in p53 dependent cell cycle arrest but not cell death (Wu et al., 1996). p21 is induced by p53 dependent and independent mechanisms (Steinman et al., 1994). Increased expression of p53 was reported in DS (Seidl et al., 1999) favoring the p53 dependent mechanism to account for the upregulation of p21 in the present study. High levels of p21 expression in p53 negative cells do not result in apoptosis (Steinman et al., 1994) indicating that upregulation per se is neither necessary nor sufficient for apoptosis. Hence, it seems that p53 dependent apoptosis in DS probably mediated not by upregulating p21 but CD95 (Seidl et al., 1999). Indeed, recent studies uncovered a new antiapoptotic role for p21 (Polyak et al., 1996) that appears to involve inhibition of proapoptotic signals mediated by stress activated MAP kinase cascade (Shim et al., 1996; Asada et al., 1999) as stress activated protein kinase/c-Jun N-terminal kinase (SAPK/JNK) a stress activated MAP kinase, is involved in transmitting proapoptotic signals (Verheij et al., 1996). Hence, it appears likely that the upregulated p21 may mediate growth arrest in the nucleus and protect cells from apoptosis in the cytoplasm. Given that monocytes can protect themselves from oxidative stress by modulating the nuclear import or export signal of p21, since it is the cytoplasmic but not the nuclear protein that exhibits antiapoptotic property (Asada et al., 1999), neurons may do likewise when they face conditions associated with oxidative stress, such as DS. With ongoing apoptosis a compensatory response may be initiated to protect the remaining neurons from death. This could include modulation of the nuclear export or import signal of p21 that results in translocation or retention of the protein in the cytoplasm.

The Bcl-2 family member proteins critically regulate apoptosis by either inhibiting or promoting cell death. The ability of the pro-survival and anti-survival Bcl-2 family members to form heterodimers makes it possible that each type might titrate the other, potentially accounting for their opposing action (O'Connor et al., 1998). No significant difference in cortical and cerebellar Bax levels were found between control and DS subjects. This finding is in agreement with previous reports in DS (Sawa et al., 1997). The reason for unaltered Bax levels could be one of the following. If Bax expression is increased as a prelude to cell death in neurodegenerative conditions, the comparable levels to controls observed in this study could be attributed to loss of Bax containing neurons with disease progression (Nagy and Esiri, 1997) or, alternatively, apoptosis in DS may be mediated by a Bax related anti-survival Bcl-2 protein rather than Bax itself. The later seems likely because we did find a significant increase in Bim/BOD levels, a protein with BH3 motif that provokes apoptosis (Hsu et al., 1998; O'Connor et al., 1998). Pro-apoptotic Bcl-2 proteins can be classified into two subgroups: those with BH1, BH2 and BH3 domains (e.g., Bax) and those with the BH3 domain only (e.g., Bim/BOD). Bax regulates apoptosis not only by dimerization with antiapoptotic Bcl-2 proteins but by regulating cytochrome c release and subsequent caspase activation. In contrast, Bim/BOD probably initiates apoptosis mainly through dimerization to antagonize the function of anti-apoptotic Bcl-2 proteins (Hsu et al., 1998) or alternatively it can directly induce cell death unless complexed by the pro-survival proteins (O'Connor et al., 1998). The unchanged Bax levels obtained in this study are consistent with our previous finding of unaltered cytochrome c levels in DS, and strongly argue for our hypothesis of cytochrome c independent apoptosis in DS (Gulesserian et al., 2001). Unlike Bax, a significant increase in Bcl-2 levels were observed in DS cerebellum. The frontal cortex of DS also showed an increased Bcl-2 concentration although it did not reach level of significance. Reports published so far on Bcl-2 expression in DS range from none (Nagy and Esiri, 1997) through decreased (Sawa et al., 1997) to increased (Yoshioka et al., 2000) expression. The disparity may be ascribed to different cell type or brain regions used and/or neuronal versus glial contribution. The upregulation of Bcl-2 may represent a compensatory response tending to protect surviving neurons from apoptosis (Satou et al., 1995). Such an upregulation may facilitate DNA repair after oxidative stress (Deng et al., 1999) as oxidative stress is a major factor affecting the brain during aging and neurodegeneration.

NF-κB is related to apoptosis in a complex manner. The presence of multiple independent transcription activating domains (Moore et al., 1993) enables NF-κB to differentially regulate a number of genes containing κB sites. Consequently, it could initiate a neuroprotective or neurodegenerative gene expression depending on the cellular context (Kaltschmidt et al., 1997). Nevertheless, an increasing body of evidence indicates that activation of NF-κB can prevent apoptosis of the cell in which it is activated at least, during acute neurodegenerative conditions (Kaltschmidt et al., 1997, 1999; Mattson et al., 2000). In light of these observations, the comparable levels of DS NF-κB p65 levels to controls may be explained by: i) reduction as a result of progressive loss of cells due to amyloid β (Aβ) and/or reactive oxygen species (ROS) mediated apoptosis. High Aβ and ROS levels were shown to inhibit NF-κB expression and promote apoptosis (Castagne et al., 1999; Kaltschmidt et al., 1999). This seems plausible, as Aβ is the major component of senile plaques in DS patients and DS is associated with increased ROS generation (de Haan et al., 1995). Moreover, unlike controls in, DS NF-κB p65 levels did not decrease with age indicating the potential contribution of disease progression; ii) an alternative explanation is that a protease may be activated that cleaves p65 into a less active protein. This later possibility can not be excluded by virtue of the fact that in vitro, p65 can be converted to a truncated form via proteolytic processing that removes the transactivation function (Franzoso et al., 1994). Consistent with this finding, increased levels of the 45 kDa protein which is assumed to be a truncated form of p65 was found in DS cerebellum, a region involved in DS neuropathology. Given a widespread antiapoptotic role for NF-κB (Mattson et al., 2000), such proteolytic inactivation may also contribute to inhibition of NF-κB as the later represents an important mechanism for inducing or enhancing apoptosis (Kaltschmidt et al., 1999; Mattson et al., 2000).

In conclusion, we generated evidence for dysregulation of apoptosis related proteins in brain regions affected by the neuropathology of DS. The current data reinforce our previous findings that apoptosis mediated via the death receptor pathway, may account for the neuronal loss observed in DS. The results also suggest that therapies directed at upregulation of Bcl-2 and downregulation of Bim/BOD may ameliorate the degenerative process in DS.

Acknowledgements

We are highly indebted to the Red Bull Company, Salzburg, Austria, for generous financial support. E.E. is a recipient of scholarship from the Austrian Academic Exchange Service (ÖAD).

References

Ahmad M, Srinivasula SM, Wang L, Talanian RV, Litwack G, Fernandes-Alnemri T, Alnemri ES (1997) CRADD, a novel human apoptotic adaptor molecule for caspase-2, and FASL/tumor necrosis factor receptor interacting protein, RIP. Cancer Res 57: 615–619

Anderson AJ, Stoltzner S, Lai Florence Su J, Nixon RA (2000) Morphological and biochemical assessment of DNA damage and apoptosis in Down syndrome and Alzheimer's disease, and effect of postmortem tissue archival on TUNEL. Neurobiol Aging 21: 511–524

Andersson P (1997) Kinase cascades regulating entry into apoptosis. Microbiol Mol Biol Rev 66: 33–46

Aravind L, Dixit VM, Koonin EV (1999) The domains of death: evolution of the apoptosis machinery. TIBS 24: 47–53

Asada M, Yamada T, Ichijo H, Delia D, Miyazono K, Fukumoro K, Mizutani S (1999) Apoptosis inhibitory activity of cytoplasmic p21 (Cip1/WaF1) in monocyte differentiation. EMBO J 18: 1223–1234

Castagne V, Gautschi M, Lefevre K, Posada A, Clarke PGH (1999) Relationships between neuronal death and the cellular redox status. Focus on the developing nervous system. Prog Neurobiol 59: 397–423

Chang BD, Watanabe K, Broude EV, Fang J, Poole JC, Kalinichenko TV, Roninson IB (2000) Effects of p21 (WAF-1, Cip-1, Sdi-1) on cellular gene expression: implication for carcinogenesis, senescence and age related disease. Proc Natl Acad Sci USA 94: 11357–11362

Chen J, Jackson PK, Kirschner MW, Duzta A (1995) Separate domains of p21 involved in the inhibition of CDK kinase and PCNA. Nature 374: 386–388

de Hann JB, Cristiano F, Iannello RC, Kola I (1995) Cu/Zn-superoxide dismutase and gluthatione peroxidase during aging. Biochem Mol Biol Int 35: 1281–1297
de la Monte SM, Sohn YK, Ganju YK, Wands JR (1998) p53 and CD95 associated apoptosis in neurodegenerative diseases. Lab Invest 78: 401–411
Deng G, Su JH, Ivins KJ, Van Houten B, Cotman CW (1999) Bcl-2 facilitates recovery from DANN damage after oxidative stress. Exp Neurol 159: 309–318
Duan H, Dixit VM (1997) RAIDD is a new death adaptor molecule. Nature 385: 86–89
El-Diery WS, Harper JW, O'Connor PM, Velculescu VE, Canman CE, Jackmann J, Pietenpol JA (1994) WAF1/CIP1 is induced in p53 mediated G1 arrest and apoptosis. Cancer Res 54: 1169–1174
Engidawork E, Gulesserian T, Seidl R, Cairns N, Lubec G (2001) Expression of apoptosis related proteins in brains of patients with Alzheimer's disease. Neurosci Lett 303: 79–82
Franzoso G, Biswas P, Poli G, Carlson LM, Brown KD, Tomita-Yamaguchi M, Fauci AS, Siebelist UK (1994) A family of serine proteases expressed exclusively in myleo-monocytic cells specifically process the NF-κB subunit p65 in vitro and may impair HIV replication in these cells. J Exp Med 180: 1445–1456
Grilli M, Memo M (1999) Nuclear factor-kappa B/Rel proteins: a point of convergence of signaling pathways relevant in neuronal function and dysfunction. Biochem Pharmacol 57: 1–7
Gulesserian T, Engidawork E, Yoo BC, Cairns N, Lubec G (2001) Comparable apoptotic pathways in frontal cortex and cerebellum of patients with Down syndrome (this volume)
Hofmann K, Bucher P, Tschopp J (1997) The CARD domain: a new apoptotic signaling motif. Trends Biochem Sci 22: 155
Hsu SY, Lin P, Hsueh AJ BOD (Bcl-2 related ovarian death gene) is an ovarian BH3 domain containing proapoptotic Bcl-2 protein capable of dimerization with diverse antiapoptotic Bcl-2 members. Mol Endocrinol 12: 1432–1440
Kaltschmidt B, Uherek M, Volk B, Baeuerle PA, Kaltschmidt C (1997) Transcription factor NF-κB is activated in primary neurons by amyloid β peptides in neurons surrounding early plaques from patients with Alzheimer's disease. Proc Natl Acad Sci USA 94: 2642–2647
Kaltschmidt B, Uherek M, Volk B, Wellmann H, Kaltschmidt C (1999) Inhibition of NF-κB potentiates amyloid β peptide mediated neuronal apoptosis. Proc Natl Acad Sci USA 96: 9409–9414
Kawai T, Matsumoto M, Takeda K, Sanjo H, Akira S (1998) ZIP kinase, a novel serine/threonine kinase which mediates apoptosis. Mol Cell Biol 18: 1642–1651
Kogel D, Bierbaum H, Preuss U, Scheidtmann KH (1999) C-terminal truncation of Dlk/ZIP kinase leads to abrogation of nuclear transport and high apoptotic activity. Oncogene 18: 7212–7218
Krosmeyer SJ (1995) Regulators of cell death. Trends Genet 11: 101–105
Lowry OH, Rosebrough NJ, Farr AL, Randal RJ (1951) Protein measurement with the folin phenol reagent. J Biol Chem 193: 265–275
Mann DMA (1988) The pathological association between Down Syndrome and Alzheimer's disease. Mech Aging Dev 43: 99–136
Mattson MP, Culmsee C, Yu Z, Camandola S (2000) Roles of nuclear factor kappa B in neuronal survival and plasticity. J Neurochem 74: 443–456
Mattson MP, Duan W, Chan SL, Camandola S (1999) Par-4: an emerging pivotal player in neuronal apoptosis and neurodegenerative disorders. J Mol Neurosci 13: 17–30
Moore PA, Ruben SM, Rosen CA (1993) Conservation of transcriptional activation functions of the NF-κB p50 and p65 subunits in mammalian cells and Sacchromyces cerevisiae. Mol Cell Biol 13: 1666–1674
Nagy ZS, Esiri MM (1997) Apoptosis related protein expression in the hippocampus in Alzheimer's disease. Neurobiol Aging 18: 565–571

O'Connor L, Strasser A, O'Reilly LA, Hausmann G, Adams JM, Cory S, Huang DC (1998) Bim: a novel member of the Bcl-2 family that promotes apoptosis. EMBO J 17: 384–395

Oltvazi ZN, Milliman CL, Krosmeyer SJ (1993) Bcl-2 hetrodimerizes in vivo with a conserved homolog Bax that accelerates programmed cell death. Cell 74: 609–619

Page G, Kogel D, Rangnekar V, Scheidtmann KH (1999) Interaction partners of Dlk/ZIP kinase: co-expression of Dlk/ZIP kinase and Par-4 results in cytoplasmic retention and apoptosis. Oncogene 18: 7265–7273

Polyak K, Waldman T, He TC, Kinzler KW, Vogelstein B (1996) Genetic determinants of p53 induced apoptosis and growth arrest. Genes Dev 10: 1945–1952

Satou T, Cummings BJ, Cotman CW (1995) Immunoreactivity for Bcl-2 protein within neurons in the Alzheimer's disease brain increases with disease severity. Brain Res 697: 35–43

Sawa A, Oyama F, Cairns NJ, Amano N, Matsushita M (1997) Aberrant expression of Bcl-2 gene family in Down syndrome brains. Mol Brain Res 48: 53–59

Seidl R, Fang-Kircher S, Bidmon B, Cairns N, Lubec G (1999) Apoptosis associated protein p53 and APO-1/Fas (CD95) in brains of adult patients with Down syndrome. Neurosci Lett 260: 9–12

Sherr CJ, Roberts JM (1995) Inhibitors of mammalian G1 cyclin dependent kinases. Genes Dev 9: 1149–1163

Shim J, Lee H, Park J, Kim H, Choi EJ (1996) A nonenzymatic p21 protein inhibitor of stress activated protein kinases. Nature 381: 804–806

Simeonidis S, Liang S, Chen G, Thanos D (1997) Cloning and functional characterization of mouse IκB epsilon. Proc Natl Acad Sci USA 94: 14372–14377

Steinman RA, Hoffman B, Iro A, Guillouf C, Liebermann DA, el-Houseini ME (1994) Induction of p21 (Waf-1/CIP1) during differentiation. Oncogene 9: 3389–3396

Tsujimoto Y, Cossman J, Jette E, Croce CM (1985) Involvement of the Bcl-2 gene in human follicular lymphoma. Science 228: 1440–1443

Verheij M, Bose R, Lin XH, Yao B, Jarvis WD, Grant S, Birrer MJ, Szabo E, Zon LI, Kyriakis JM, Haimovitz-Friedman A, Fuks Z, Kolesnick RN (1996) Requirement for ceramide initiated SAPK/JNK signaling in stress induced apoptosis. Nature 380: 75–79

Villunger A, Huang DCS, Holler N, Tschopp J, Strasser A (2000) Fas ligand induced c-Jun kinase activation in lymphoid cells requires extensive receptor aggregation but is independent of DAXX, and Fas-mediated cell death does not involve DAXX, RIP or RAIDD. J Immunol 165: 1337–1343

Wolf BB, Green DR (1999) Suicidal tendencies: apoptotic cell death by caspase family proteinases. J Biol Chem 274: 20049–20052

Wu H, Wade M, Krall L, Grisham J, Xiong Y, Van Dyke T (1996) Targeted in vivo expression of the cyclin dependent kinase inhibitor p21 halts hepatocytes cell cycle progression, postnatal liver development and regeneration. Genes Dev 10: 245–260

Yoshioka K, Yamamoto S, Moriguchi N, Miyata H, Tsukiyama K, Isokawa S, Horiuchi F, Takemura T (2000) Overexpression of Bcl-2 in transient abnormal myleopoiesis associated with Down syndrome. Ann Hematol 79: 319–321

Authors' address: Prof. Dr. G. Lubec, Department of Pediatrics, University of Vienna, Währinger Gürtel 18-20, A-1090 Vienna, Austria, e-mail: Gert.Lubec@akh-wien.ac.at

Increased brain protein levels of carbonyl reductase and alcohol dehydrogenase in Down Syndrome and Alzheimer's disease

B. Balcz[1], **L. Kirchner**[1], **N. Cairns**[2], **M. Fountoulakis**[3], and **G. Lubec**[1]

[1] Department of Pediatrics, University of Vienna, Vienna, Austria
[2] Department of Neuropathology, Brain Bank, King's College, London, United Kingdom
[3] F. Hoffmann-La Roche, Basel, Switzerland

Summary. Oxidative stress is considered to be crucial in the pathogenesis of Alzheimer's disease-like neurodegeneration. An elevation of carbonyl compounds that are biomarkers of and leading to oxidative stress has been demonstrated in Down Syndrome (DS) and Alzheimer's Disease (AD) brains and seems to be the result of a multifactorial process. Carbonyls, which are cytotoxic metabolic intermediates, are detoxified by either oxidation catalyzed by aldehyde dehydrogenase (ALDH), or by reduction to their corresponding alcohols by carbonyl reductase (CBR) and/or alcohol dehydrogenase (ADH). We quantified protein levels of CBR and its agonist/antagonist ADH using 2D gel electrophoresis with subsequent MALDI-identification and specific software in several brain regions of DS as well as AD patients and compared them to levels of CBR and ADH in brains of controls. Protein levels of both enzymes were increased in several brain regions because of enzyme induction by elevated carbonyls in DS and AD. Increased CBR in DS (trisomy 21) may be due to a gene dosage effect as the gene encoding CBR has been mapped to chromosome 21. These findings may confirm the proposed increase of reactive carbonyls in the brain thus supporting the involvement of oxidative stress and contribute to the understanding of carbonyl handling in brain of both neurodegenerative disorders.

Introduction

Carbonyl compounds — aldehydes and ketones — are widely distributed in living matter. They are encountered as metabolic intermediates, as mediators of highly specific biological effects and many of them are cytotoxic. They may react with primary amine groups to form Schiff bases, which in term subsequently rearrange into more chemically stable structures. In biological systems, such reactions may disrupt normal oligonucleotide structure, are thus mutagenic and may interfere with the biological activity of numerous structural or enzymatic polypeptides. They can react covalently with secondary

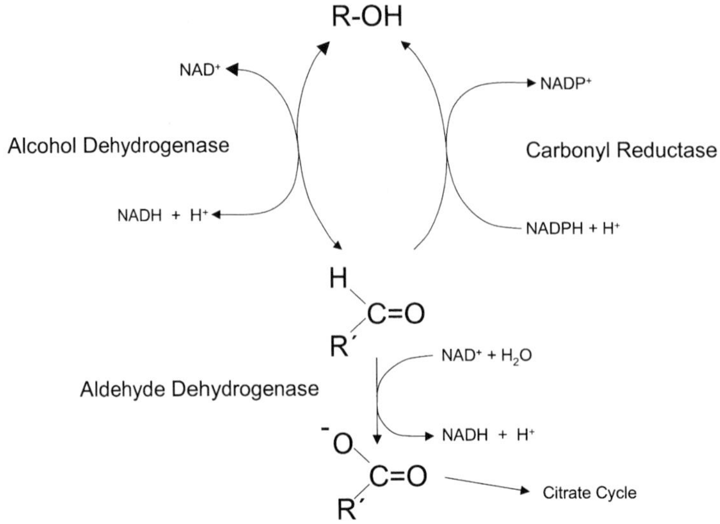

Fig. 1. Carbonyl metabolism. Carbonyls, which are cytotoxic metabolic intermediates, are detoxified by either oxidation catalyzed by aldehyde dehydrogenase (ALDH), or by reduction to their corresponding alcohols by carbonyl reductase (CBR) and/or alcohol dehydrogenase (ADH)

amine groups and thiol groups in proteins and also with lipids (Shapiro, 1998). Therefore, detoxification of these reactive carbonyls is of high importance and is under the control of a series of enzymes. For this purpose, carbonyls are either oxidized (generally catalyzed by the NAD+ dependent aldehyde dehydrogenase) or reduced (Fig. 1). Carbonyl reductase (CBR; SWISSPROT accession number P16152) catalyzes the reduction of a large number of biologically and pharmacologically active carbonyl compounds to their corresponding alcohols and has been mapped to 21q22.1 (Lemieux et al., 1993). It is localized in the cytoplasm of the cell and consists of a single polypeptide chain with a molecular weight of 30 kD (Wermuth, 1981). During catalysis the pro 4S hydrogen atom of the nicotinamide ring of NADPH is transferred to the substrate. CBR can be found in many human tissues with high concentrations in the nervous system (Wirth and Wermuth, 1992).

Alcohol dehydrogenase (ADH; SWISS PROT accession number P14550) catalyzes the reversible oxidation of many primary and secondary alcohols to aldehydes and ketones, and also the reduction of aldehydes and ketones to the corresponding alcohols. Human ADH is a dimeric protein consisting of two subunits with a molecular weight of 40 kD each (Agarwal, 1997) and three classes of isozymes with various subclasses have been identified (Deetz et al., 1984; Testa, 1994).

An increase of reactive carbonyls leads to oxidative stress, which is considered to play an important role in the pathogenesis of Alzheimer's disease (AD) and in the Alzheimer's disease like neurodegeneration found in brains of individuals with Down Syndrome (DS) from the fourth decade (Lyras et al., 1997; Markesberry, 1997; Smith et al., 1997). Brain pathology in both disorders is characterized by neuritic plaques and neurofibrillary tangles that allow a definite diagnosis after the patient's death (Selkoe, 1999). Immunohistochemical studies show the presence of oxidative stress products in neuritic plaques and neurofibrillary tangles (Markesberry and Carney, 1999). In postmortem brain of AD patients lipid peroxidation (evaluated by determination of lipid derived aldehydes) is increased in various brain regions, mainly in parietal and temporal cortex (Palmer and Bums, 1994; Lyras et al., 1997; Smith et al., 1997). In both, DS (Jovanovic et al., 1998) and AD (Smith et al., 1998) brains, an increase of carbonyls has been reported that may be the result of reduced metabolization due to a deficiency in aldehyde dehydrogenase (Lubec et al., 1999; Kamino et al., 2000). There is also evidence for carbonyl-related posttranslational modification of neurofilament protein in AD. This may contribute to cytoskeletal abnormalities, critical for the pathogenesis of neurofibrillary pathology (Smith et al., 1995).

The pathology of neurodegeneration in AD and DS is morphologically nearly identical, showing neuritic plaques and neurofibrillary tangles. A decrease of ALDH has been shown in DS brains (Lubec et al., 1999) and an increased risk of AD in individuals with ALDH deficiency has also been reported (Kamino et al., 2000). In both disorders the major metabolic pathway for detoxification of carbonyl compounds by ALDH was proposed to be blocked leading to increased brain aldehydes. We therefore decided to study carbonyl handling enzymes in DS (trisomy 21) and AD brains. We selected CBR as it is located on chromosome 21 and representing a major pathway of aldehyde metabolization (Forrest et al., 1990), and ADH which catalyzes the oxidation of alcohols to aldehydes but, depending on substrate availability, can also act as an aldehyde reducing enzyme (Deetz, 1984).

Materials and methods (Langen et al., 1999; Lubec et al., 1999)

Brain samples for the determination of protein levels

Seven brain regions (temporal, frontal, occipital, parietal cortex, cerebellum, thalamus and caudate nucleus) of karyotyped patients with DS (n = 9; 2 females, 7 males; 58.1 ± 7.9 years old), patients with AD (n = 13; 4 females, 9 males; 58.5 ± 7.6 years old) and controls (n = 18; 5 females, 13 males; 50.0 ± 16.9 years old) with no history of neurologic or psychiatric abnormality were used for studies at the protein expression level. Post mortem brain samples were obtained from the MRC London Brain Bank for Neurodegenerative Diseases, Institute of Psychiatry. In all DS and AD brains there was evidence of abundant neuritic plaques and neurofibrillary tangles. The major cause of death was bronchopneumonia in DS and AD patients and heart disease in controls. Post mortem interval of brain dissection in DS was 31.4 ± 19.6 h, in AD 30.3 ± 26.2 h, and in controls 38.8 ± 18.8 h. Tissue samples were stored at −70°C and the freezing chain was never interrupted.

Two dimensional gel electrophoresis

Brain tissue was suspended in 0.5 ml of sample buffer consisting of 40 mM Tris, 5 M urea (Merck, Darmstadt, Germany), 2 M thiourea (Sigma, St. Louis, MO, USA), 4% CHAPS

(Sigma), 10 mM 1,4 dithioerythriol (Merck), 1 mM EDTA (Merck) and a mixture of protease inhibitors, 1 mM PMSF and 1 mcg of each pepstatin A, chymostatin, leupeptin and antipain. The suspension was sonicated for approximately 30s and centrifuged at 10.000 × g for 10 min and the supernatant was centrifuged further at 150.000 × g for 45 min. The brain content in the supernatant was determined by the Coomassie blue method (Bradford, 1976).

The 2-D gel electrophoresis was performed essentially as reported (Langen et al., 1997). Samples of approximately 1.5 mg were applied on immobilized pH 3–10 nonlinear gradient strips (IPG, Pharmacia Biotechnology, Uppsala, Sweden), at both, the basic and acidic ends of the strips. The proteins were focused at 300 V for 1 h, after which the voltage was gradually increased to 3,500 V within 6 h. Focusing was continued at 5,000 V for 48 h. The second-dimensional separation was performed on 9–16% linear gradient polyacrylamide gels (chemicals from Serva, Heidelberg, Germany and Bio-Rad, Hercules, CA, USA). The gels were stained with colloidal Coomassie blue (Novex, San Diego, CA, USA) for 48 h, destained with water and scanned in a Molecular Dynamics Personal densitometer. The images were processed using Photoshop (Adobe) and Powerpoint (Microsoft) software. Protein spots were quantified using the ImageMaster 2D Elite software (Amersham Pharmacia Biotechnology).

Results

Human brain extracts from seven regions were separated by 2-D gels and visualized following stain with colloidal Coomassie blue. The protein spots were analyzed by MALDI-MS, following in-gel digestion. The peptide masses were matched with the theoretical masses of all species of the SWISS-PROT database (Figs. 2 and 3). Internal standards were used to correct the measured peptide determined by MALDI-MS (Fountoulakis and Langen, 1997).

The spots representing ADH and CBR were quantified in the 2-D gels of samples from the same brain region in DS, AD and controls using the ImageMaster 2D Elite software. Protein levels were determined as a percentage of total proteins present in the gel part considered.

A significant increase ($P < 0.05$) of CBR could be shown in the cerebellum, thalamus, caudate nucleus, parietal and temporal cortex in AD brains and in thalamus, caudate nucleus and temporal cortex in DS (Fig. 4).

A significant increase ($P < 0.05$) of ADH could be shown in the cerebellum, temporal and occipital cortex in AD brains and in the temporal cortex in DS (Fig. 5).

Discussion

As shown in the results we detected increased protein levels of both carbonyl handling enzymes in brains of both neurodegenerative disorders. This is in agreement with previous findings of increased oxidative stress and, in particular with accumulation of reactive carbonyls in the brain in AD and DS (Ianello et al., 1999). Methodologically, we unambiguously identified CBR and ADH using 2 dimensional electrophoresis with subsequent MALDI identification of spots.

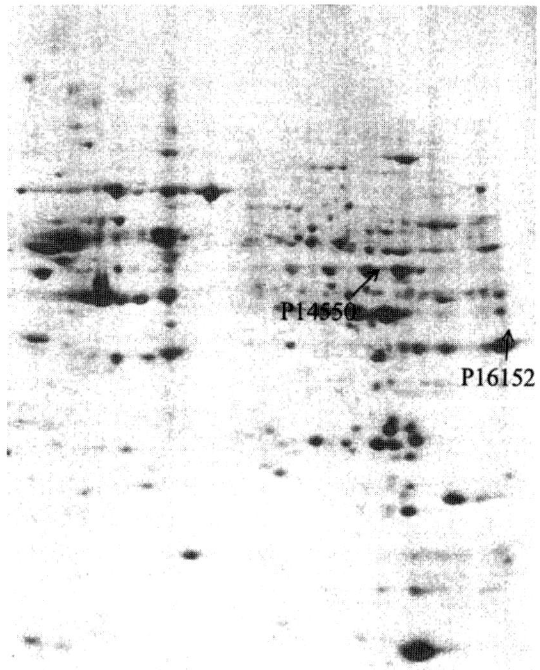

Fig. 2. CBR (P16152) and ADH (P14550) identified using 2 dimensional electrophoresis with subsequent MALDI identification of spots. The complete gel is shown

In DS brain elevated CBR protein levels may well be seen as a consequence of a gene dosage effect, as the gene encoding CBR has been mapped to chromosome 21 in band 21q22.1 (Lemieux et al., 1993). The presence of a third copy of the CBR gene may have been responsible for elevated CBR protein resembling findings of increased chromosome 21 gene products SOD1 and amyloid precursor protein in DS (Hensley et al., 1994). SOD1 plays a key role in the metabolism of reactive oxygen species by catalysing the dismutation of superoxide anions to hydrogen peroxide which is accumulating in DS brain (Ianello et al., 1999). This in turn may be responsible for the

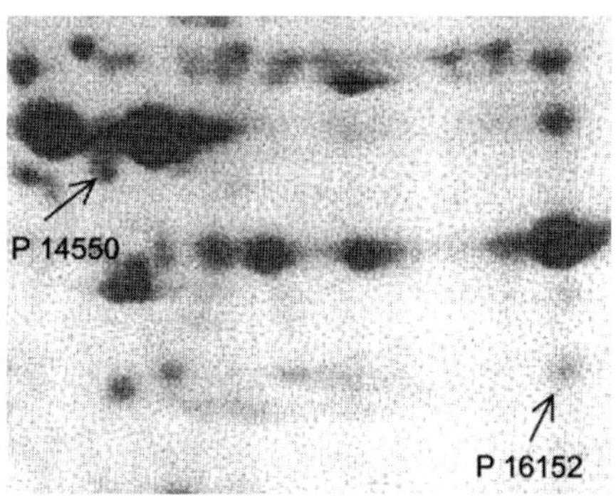

Fig. 3. Magnification and detail of CBR and ADH spots from the 2 dimensional electro phoresis shown in Fig. 2

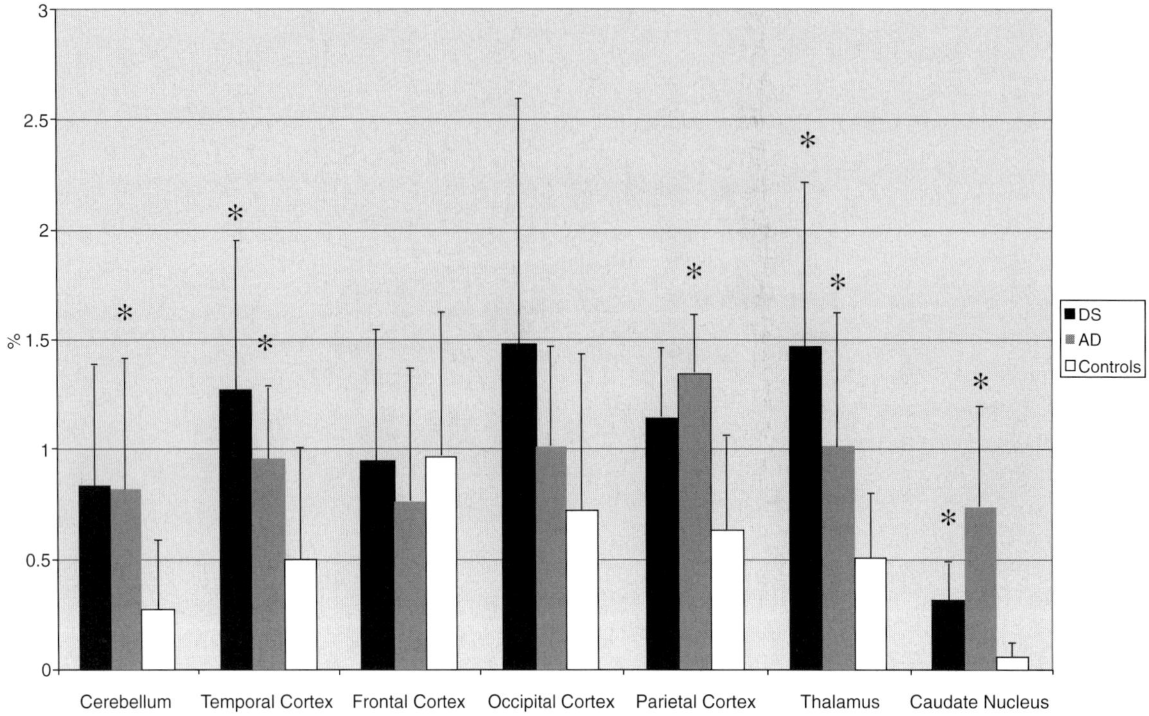

Fig. 4. Levels of carbonyl reductase in brains of DS and AD patients and in brains of controls. Protein levels are determined as a percentage of total proteins present in the gel part considered. An asterisks represents statistical difference from controls at the $p < 0.05$ level

generation of active oxygen species and reactive carbonyls. Increased carbonyls would then lead to the induction of CBR.

The precursor of Amyloid β, the β-amyloid precursor protein (APP) encoded on chromosome 21 and overexpressed in DS and AD brains, is also of particular interest for understanding the pathogenetic mechanisms of both DS and AD (Selkoe, 1999). In DS patients the gene product may be overexpressed due to a third copy of chromosome 21; in AD the reason for the accumulation is not known. Amyloid β (Aβ) deposits are involved in the generation of reactive oxygen species and several studies have clearly demostrated a relationship between high concentrations of Aβ and the production of ROS (Behl et al., 1994; Hensley et al., 1994). A fair association has been demonstrated between oxidative stress on neurons and the formation of plaques and tangles in AD (Yan et al., 1996). Oxidative stress impairs mitochondrial function resulting in energy depletion. Impaired cellular energy metabolism in turn increases the amyloidogenic processing of APP leading again to increased Aβ production and subsequent oxidative stress thus initiating a vicious cycle (Gabuzda et al., 1994). On the other hand, primary mitochondrial damage per se may have contributed to oxidative stress in AD.

But what would explain increased ADH protein in brains of DS and AD patients? It is important to state that ADH isozymes do not only act as

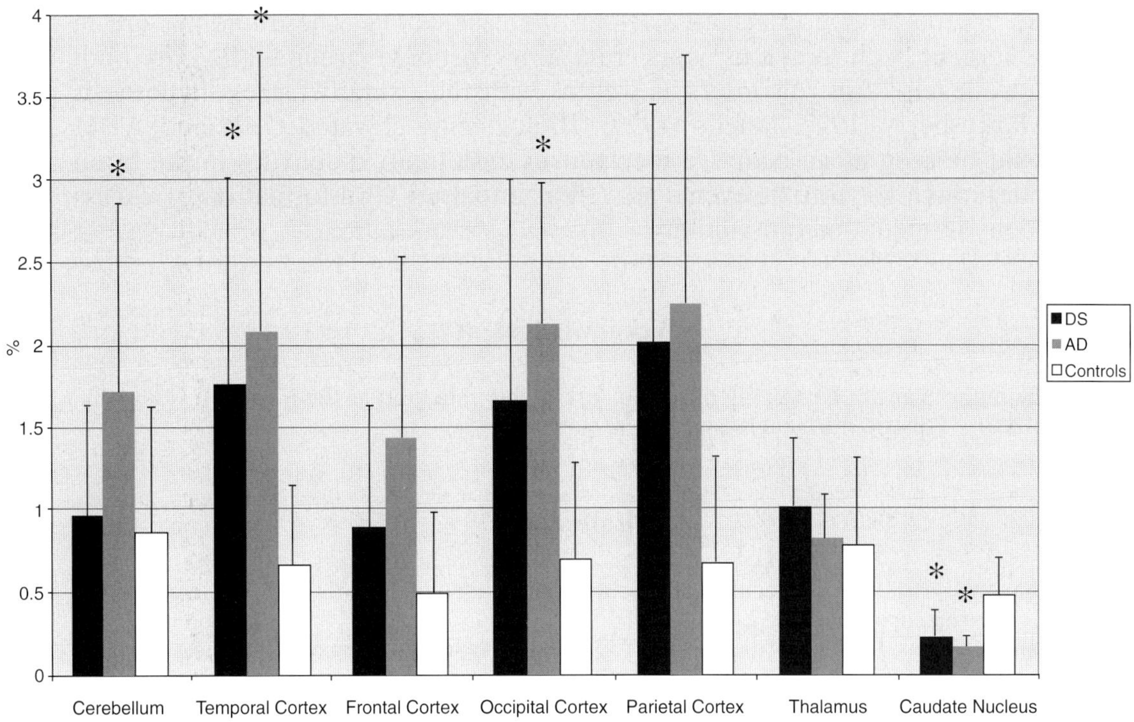

Fig. 5. Levels of alcohol dehydrogenase in brains of DS and AD patients and in brains of controls. Protein levels are determined as a percentage of total proteins present in the gel part considered. An asterisks represents statistical difference from controls at the $p < 0.05$ level

antagonists to carbonyl reductase by catalysing the oxidation of a wide variety of alcohols but that they can also catalyse the reduction of aldehydes and ketones to their corresponding alcohols. Physiologic conditions and the equilibrium favour in fact the latter process (Deetz et al., 1984; Testa, 1994). The optimal activity of ethanol oxidation is found between pH 9 and pH 11 whereas the optimal activity for the reduction of acetaldehyde to ethanol is between pH 6 and pH 7. Thus, the turnover rate for the reduction of aldehydes is much higher at physiological pH than that for the oxidation of the corresponding alcohols. Broad substrate specificity is characteristic of all mammalian alcohol dehydrogenases and the multiplicity of ADH isozymes argue for a much more diverse role for ADH in the metabolism of other alcohols and/or aldehydes. Elevated levels of carbonyls, present in DS as well as AD patients' brains could therefore account for elevated ADH by enzyme induction.

The reason for the uneven regional distribution of increased carbonyl handling enzymes remains unclear. Differences in the accumulation of carbonyls in the individual brain regions may be responsible as well as metabolic factors and enzymes including ALDH (Lubec et al., 1999).

In conclusion, we demonstrated increased protein levels of CBR and ADH in DS and AD brains which may be explained by enzyme induction and a gene dosage effect for CBR in the case of DS. These findings are in agreement with previous work revealing carbonyl accumulation in both neurodegenerative disorders as well as with the oxidative stress hypothesis (Markesberry, 1997; Ianello, 1999). Biologically, elevated CBR and ADH could be seen as an adaptive mechanism to detoxify aldehydes in the brain counteracting cytotoxic events as well as impaired fibril formation by excess aldehyde-mediated cross-linking.

Acknowledgment

The authors at the University of Vienna are highly indebted to the Red Bull Company, Salzburg, Austria, for generous financial support.

References

Agarwal DP (1997) Molecular genetic aspects of alcohol metabolism and alcoholism. Pharmacopsychiat 30: 79–84

Behl C, Davis JB, Lesley R, Schubert D (1994) Hydrogen peroxide mediates amyloid beta protein toxicity. Cell 77: 817–827

Bradford M (1976) A rapid and sensitive method for the quantitation of microgram quantities of protein utilizing the principle of protein-dye binding. Anal Biochem 72: 248–254

Deetz JS, Luehr CA, Vallee BL (1984) Human liver alcohol dehydrogenase isozymes: reduction of aldehydes and ketones. Biochemistry 23: 6822–6828

Epstein CJ (1992) Down Syndrome (Trisomy 21). In: Scriver CR, Beaudet AL, Sly WS, Valle D (eds) The metabolic and molecular basis of inherited disease. McGraw Hill, New York, pp 749–794

Forrest L, Akman S, Krutzik S, Paxton RJ, Sparkes RS, Doroshow J, Felsted RL, Glover CJ, Mohandas T, Bachur NR (1990) Induction of a human carbonyl reductase gene located on chromosome 21. Biochim Biophys Acta 1048: 149–155

Fountoulakis M, Langen H (1997) Identification of proteins by matrix-assisted laser desorption ionization-mass spectroscopy following in-gel digestion in low-salt, non-volatile buffer and simplified peptide recovery. Anal Biochem 250: 153–156

Gabuzda D, Busciglio J, Chen LB, Matsudaira P, Yankner BA (1994) Inhibition of energy metabolism alters the processing of amyloid precursor protein and induces a potentially amyloidogenic derivative. J Biol Chem 269: 13623–13628

Hensley K, Carney JM, Mattson MP, Aksenova M, Harris M, Wu JF, Floyd RA, Butterfield DA (1994) A model for beta-amyloid aggregation and neurotoxicity based on free radical generation by the peptide — relevance to Alzheimer's disease. Proc Natl Acad Sci USA 9: 3270–3274

Ianello RC, Crack PJ, de Haan JB, Kola I (1999) Oxidative stress and neural dysfunction in Down syndrome. J Neural Transm [Suppl] 57: 257–267

Jovanovic SV, Clements D, MacLeod K (1998) Biomarkers of oxidative stress are significantly elevated in Down syndrome. Free Radic Biol Med 25: 1044–1048

Kamino K, Nagasaka K, Imagawa M, Yamamoto H, Yoneda H, Ueki A, Kitamura S, Namekata K, Miki T, Ohta S (2000) Deficiency in mitochondrial aldehyde dehydrogenase increases the risk for late-onset Alzheimer's disease in the Japanese population. Biochem Biophys Res Commun 273: 192–196

Langen H, Roeder D, Juranville J-F, Fountoulakis M (1997) Effect of the protein application mode and the acrylamide concentration on the resolution of protein spots separated by two-dimensional gel electrophoresis. Electrophoresis 18: 2085–2090

Langen H, Berndt P, Roder D, Cairns N, Lubec G, Fountoulakis M (1999) Two-dimensional map of human brain proteins. Electrophoresis 20: 907–916
Lemieux N, Malfoy B, Forrest GL (1993) Human carbonyl reductase (CBR) localized to band 21q22.1 by high-resolution fluorescence in situ hybridization displays gene dosage effects in trisomy 21 cells. Genomics 15: 169–172
Lubec G, Labudova O, Cairns N, Berndt P, Langen H, Fountoulakis M (1999) Reduced aldehyde dehydrogenase levels in the brain of patients with Down syndrome. J Neural Transm [Suppl] 57: 21–40
Lyras L, Cairns NJ, Jenner A, Jenner P, Halliwell B (1997) An assessment of oxidative damage to proteins, lipids and DNA in brain from patients with Alzheimer's disease. J Neurochem 68: 2061–2069
Markesberry WR (1997) Oxidative stress hypothesis in Alzheimer's disease. Free Radic Biol Med 23: 134–147
Markesberry WR, Carney JM (1999) Oxidative alterations in Alzheimer's disease. Brain Pathol 9: 133–146
Palmer AM, Bums MA (1994) Selective increase in lipid peroxidation in the inferior temporal cortex in Alzheimer's disease. Brain Res 645: 338–342
Selkoe DJ (1999) Translating cell biology into therapeutic advances in Alzheimer's disease. Nature 399 [Suppl]: A23–A31
Shapiro HK (1998) Carbonyl-trapping therapeutic strategies. Am J Ther 5: 323–353
Smith MA, Rudnicka-Nawrot M, Richey PL, Praprotnik D, Mulvihill P, Miller CA, Sayre LM, Perry G (1995) Carbonyl-related posttranslational modification of neurofilament protein in the neurofibrillary pathology of Alzheimer's disease. J Neurochem 64: 2660–2666
Smith MA, Richey Harris PL, Sayre LM, Beckman JS, Perry G (1997) Widespread peroxynitrite-mediated damage in Alzheimer's disease. J Neurosci 17: 2653–2657
Smith MA, Sayre LM, Anderson VE, Harris PL, Beal MF, Kowall N, Perry G (1998) Cytochemical demonstration of oxidative damage in Alzheimer's disease by immunochemical enhancement of the carbonyl reaction with 2,4-dinitrophenylhydrazine. J Histochem Cytochem 46: 731–735
Testa B (1994) Dehydrogenation of alcohols and aldehydes, carbonyl reduction. In: Testa B, Caldwell J (eds) The metabolism of drugs and other xenobiotics. Biochemistry of redox reactions. Academic Press, San Diego, CA, pp 41–69
Wermuth B (1981) Purification and properties of an NADPH-dependent carbonyl reductase from human brain. J Biol Chem 256: 1206–1213
Wirth H, Wermuth B (1992) Immunohistochemical localization of carbonyl reductase in human tissues. J Histochem Cytochem 40: 1857–1863
Yan SD, Chen X, Fu J, Chen M, Zhu H, Roher A, Slattery T, Zhao L, Nagashima M, Morser J, Migheli A, Nawroth P, Stern D, Schmidt AM (1996) RAGE and amyloid-beta peptide neurotoxicity in Alzheimer's disease. Nature 382: 685–691

Authors' address: Prof. Dr. G. Lubec, Department of Pediatrics, University of Vienna, Währinger Gürtel 18, A-1090 Vienna, Austria, e mail: gert.lubec@akh-wien.ac.at

Carbohydrate handling enzymes in fetal Down Syndrome brain

E. Kitzmueller[1], **S. Greber**[1], **M. Fountoulakis**[2], and **G. Lubec**[1]

[1] Department of Pediatrics, University of Vienna, Austria
[2] F. Hoffmann La Roche, CNS Preclinical Research, Gene Technologies, Basle, Switzerland

Summary. Impaired glucose metabolism in adult Down Syndrome (DS) has been well-documented in vivo and information on the underlying biochemical defect i.e. aberrant glucose handling enzymes is already available. Nothing is known on carbohydrate handling, however, in early life of DS patients, when no secondary phenomena as e.g. Alzheimer-like neuropathology occur in the brain yet.

We therefore determined a series of key enzymes of carbohydrate metabolism in fetal control and DS brain during the early second trimenon. We used two-dimensional electrophoresis with subsequent MALDI characterization and specific software for quantification of protein spots.

We observed comparable levels of phosphoglycerate mutase, phosphoglycerate kinase 1; fructose-biphosphate aldolase A, fructose bisphosphate aldolase C; ribose-phosphate pyrophosphokinase 1; D-phosphoglycerate dehydrogenase, 6-phosphogluconolactonase; aflatoxin B1 aldehyde reductase 1, aldose reductase; inosine-5′-monophosphate dehydrogenase 2; galactokinase, in brain of fetal controls and DS.

We conclude that our biochemical findings point to the fact that DS patients start early life with unchanged glucose handling, pentose phosphate shunt, glycolysis, sugar aldehyde, guanine nucleotide- and ribonucleoside formation and galactose metabolism.

Introduction

Down Syndrome (DS) is the most frequent genetic cause of dementia and although the trisomic state can be directly associated with the phenotype, the underlying pathomechanisms are far from being understood. Nearly 100% of subjects with DS over age 40 years show neuropathological and neurochemical abnormalities on postmortem brain examinations that are indistinguishable from those seen in Alzheimer's disease (AD; Burger and Vogel, 1973; Wisniewski et al., 1985). A series of impaired metabolic functions in DS has been reported (Epstein, 1992) including deteriorated glucose metabolism. Specifically, Hsia and coworkers reported increased glucose-6-phosphate dehydrogenase (G-6-P-D) and 6-phosphogluconate-dehydrogenase (6-PG-D) in DS as early as in (1971); neither enzyme is encoded on chromosome 21.

Anneren and coworkers have shown that phosphofructokinase (PFK, liver type) activity was elevated in fibroblasts aneuploid for chromosome 21 (1987). As this isoenzyme is encoded on chromosome 21q22.3 (Van Keuren et al., 1985; Vora and Franke, 1981) the authors suggested a gene dosage effect for the explanation of the PFK increase. The increase of G-6-P-D and 6-PG-D described by Hsia could therefore be seen as compensatory/adaptive to the elevated PFK. Glucose metabolism has been evaluated in brain of adult DS patients in vivo using positron emission tomography (PET) with [^{18}F] fluorodeoxyglucose, a well-established method for the measurement of regional cerebral glucose metabolism rates in humans. The DS patients showed lower than normal rates of absolute glucose metabolism in temporal and parietal association neocortical areas (Schapiro et al., 1992), resembling a pattern seen in AD (Haxby et al., 1985). Dani and coworkers (1996) showed in a longitudinal study that older subjects with DS without cognitive impairment maintain regional cerebral glucose metabolism within the normal range for many years, although repeated measurements showed a progressive decline of glucose metabolism over time and after the onset of dementia.

In a recent publication Pietrini and coworkers (1997) demonstrated by a PET study that abnormalities in cerebral glucose metabolism of patients with DS at risk for AD prior to dementia, appeared during stimulation in cortical regions typically affected in AD. Their data indicate that a stress paradigm can detect metabolic abnormalities in the preclinical stages of AD despite normal cerebral metabolism at rest.

In a previous report we detected aberrant enzyme activities in adult patients with DS and AD (Labudova et al., 1999). In this publication we show that during early life of DS individuals no derangement at the protein level occurred for a series of carbohydrate metabolizing enzymes.

Materials and methods

Fetal brain tissue and brain preparation

Abortus fetal brain sample (cerebral cortex) of controls (n = 7; 1 female and 6 males with mean age of 18.8 ± 2.2 weeks) and DS (n = 8; 2 females and 6 males with mean age of 19.8 ± 2.0 weeks) were obtained from Drs. M. Dierssen and J. C. Ferreres the Medical and Molecular Genetics Center-IRO, Hospital Duran i Reynals, Barcelona, Spain. All samples were stored at −70°C and the freezing chain was never interrupted. Brain tissue was suspended in 0.5 ml of sample buffer consisting of 40 mM Tris, 5 M urea (Merck, Darmstadt, Germany), 2 M thiourea (Sigma, St. Louis, MO, USA), 4% CHAPS (Sigma), 10 mM 1,4-dithioerythriol (Merck), 1 mM EDTA (Merck), 1 mM PMSF (Sigma) and 1 μl/mg of each pepstatin A, chymostatin, leupeptin and antipain. The suspension was sonicated for approximately 30 seconds and centrifuged further at 150,000 × g for 45 minutes to sediment undissolved material. The protein concentration of the supernatant was determined by the Coomassie blue method (Bradford, 1976).

Two-dimensional gel electrophoresis

Two-dimensional gel electrophoresis was performed essentially as reported (Langen et al., 1997). Samples of 2 mg protein were applied on immobilized pH 3–10 non-linear gradient strips in sample cups at their basic and acidic ends. Focusing started at 200 V and the voltage was gradually increased to 5,000 V at 3 V/min and kept constant for a further 24 h (approximately 180,000 kVh totally). The second-dimensional separation was performed on 9–16% SDS gradient polyacrylamide gels. The gels (180 × 200 × 1.5 mm) were run at 40 mA per gel. After protein fixation for 12 h in 40% methanol, containing 5% phosphoric acid, the gels were stained with colloidal Coomassie blue (Novex, San Diego, CA, USA) for 48 h. Molecular masses were determined by running standard protein markers (Gibco, Basel, Switzerland), covering the range 10–200 kDa. pI values were used as given by the supplier of the IPG strips. Excess of dye was washed out from the gels with H_2O and the gels were scanned in an Agfa DUOSCAN densitometer (resolution 200). Electronic images of the gels were recorded using Photoshop (Adobe) and PowerPoint (Microsoft) software.

Matrix-assisted laser desorption ionization mass spectroscopy (MALDI-MS)

MALDI-MS analysis was performed as described elsewhere (Fountoulakis and Langen, 1997) with some modifications. The spots were excised with a spot picker and placed into 96-well microtiter plates. Each spot was destained with 100 µl of 30% acetonitrile in 50 mM ammonium bicarbonate and dried in a speedvac evaporator. Each dried gel piece was rehydrated with 4 µl of 3 mM Tris-HCl, pH 9.0, containing 50 ng trypsin (Promega, Madison, WI, USA). After 16 h at room temperature, 7 µl of H_2O were added to each gel piece and the samples were shaken in for 10 min. Four µl of 50% acetonitrile, containing 0.3% trifluoroacetic acid and the standard peptides des-Arg-bradykinin (Sigma, 0044681 Da) and adrenocorticotropic hormone fragment 18–39 (Sigma, 2465.1989 Da), in water were added to each gel piece and shaken for 10 min. The application of the samples was performed with a SymBiotI sample processor (PE Biosystems, Framingham, MA, USA). 1.5 µl of the peptide mixture was simultaneously applied with 1 µl of matrix, consisting of a saturated solution of α-cyano-4-hydroxycinnamic acid (Sigma) in 50% acetonitrile, containing 0.1% trifluoroacetic acid. Samples were analyzed in a time-of-flight mass spectrometer (Reflex 3, Bruker Analytics, Bremen, Germany). An accelerating voltage of 20 kV was used. Peptide matching and protein searches were performed automatically. The peptide masses were compared to the theoretical peptide masses of all available proteins from all species. Monoisotopic masses were used and a mass tolerance of 0.0025% was allowed. The algorithm used for determining the probability of a false positive match with a given MS-spectrum is described elsewhere (Berndt et al., 1999).

Quantification of proteins

Protein spots were outlined (first automatically and then manually) and quantified using the ImageMaster 2D Elite software (Amersham Pharmacia Biotechnology). The percentage of the volume of the spots representing a certain protein was determined in comparison with the total proteins present in the 2-D gel.

Results

As shown in Fig. 1 we clearly separated a series of spots (SWISSPROT numbers indicated) which were unambiguously identified (Table 1 providing molecular masses, pI, matches).

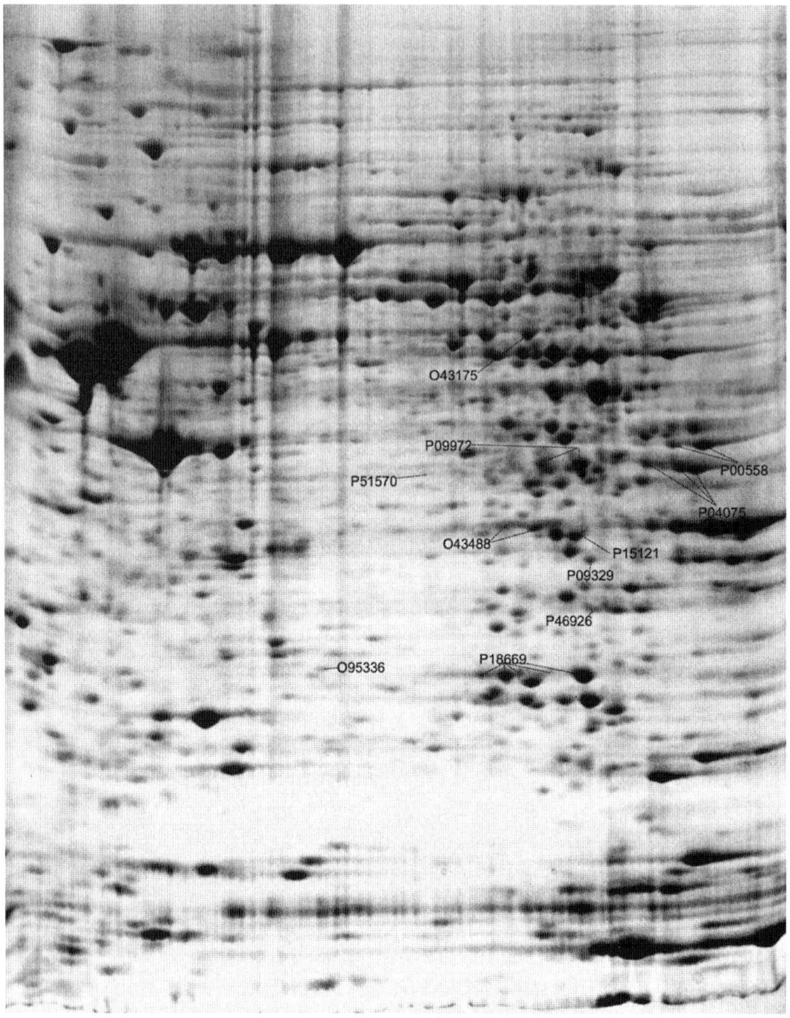

Fig. 1. Two dimensional gel electrophoresis showing the separation of protein spots with the corresponding SWISSPROT accession number. Only spots with fair separation were used and identified by MALDI-MS. Quantification was carried out with specific software

Table 1. The pI values, molecular weights and matches of the individual proteins is given

		pI	MW	Matches
O43175	D-3-phosphoglycerate dehydrogenase	6.69	57,369	9
O43488	Aflatoxin b1 aldehyde reductase	6.70	37,024	6
O95336	6-phosphogluconolactase	5.98	27,814	9
P00558	Phosphoglycerate kinase	8.12	44,967	8
P04075	Fructose-bisphosphate aldolase (muscle)	8.07	39,720	8
P09329	Ribose-phosphatepyrophosphate kinase I	6.95	35,194	5
P09972	Fructose-bis-phosphate aldolase (brain)	6.85	39,699	9
P12268	Inosine-5′-monophosphate dehydrogenase	6.98	56,225	6
P15121	Aldolasereductase	6.97	36,098	6
P18669	Phosphoglycerate mutase	7.21	28,768	9
P51570	Galactokinase	6.43	42,701	5

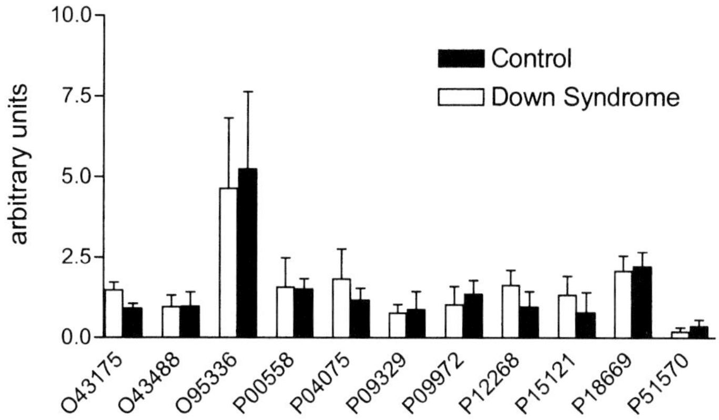

Fig. 2. The results of quantification (arbitrary units) of spots from two dimensional electrophoresis showed that no significant differences were found between controls and DS

Phosphoglycerate mutase, phosphoglycerate kinase 1; fructose-biphosphate aldolase A, fructose bisphosphate aldolase C; ribose-phosphate pyrophosphokinase 1; D-phosphoglycerate dehydrogenase, 6-phosphogluconolactonase; aflatoxin B1 aldehyde reductase 1, aldose reductase; inosine-5′-monophosphate dehydrogenase 2; galactokinase were comparable in fetal control and DS cortex (Fig. 2, Table 2).

Discussion

In previous reports we described aberrant glucose handling enzymes in adult DS patients brains (Labudova et al., 1999, 1999a; Lubec et al., 1999). Our current findings that key enzymes of glucose metabolism (phosphoglycerate mutase, phosphoglycerate kinase 1) were unchanged in fetal DS at the early second trimenon made us propose that aberrant glucose metabolism found in adult DS may be of secondary nature possibly assigned to "neurodegeneration". Also another pathway as the so-called pentose-phosphate shuttle, represented by 6-phosphogluconolactonase (Collard et al., 1999) was unaffected in fetal DS brain. Glycolysis was unchanged as reflected by fructose biphosphate aldolases A and C (Izzo et al., 1987; Buono et al., 1988).

Aberrant carbonyl handling was observed in adult DS and could be interpreted as an antioxidant reponse to increased oxidative stress but also as reflecting derangement of sugar carbonyls (Lubec et al., 1999b). Based upon our observations in fetal DS along with altered antioxidant parameters in fetal DS (unpublished results) we would suggest unaltered carbohydrate-related enzymes aflatoxin B1 aldehyde reductase 1 (aldoketoreductase 7; Ireland et al., 1998) and aldose reductase (AR; E.C. 1.1.1.21) as reflecting the patent carbohydrate metabolism in DS.

Table 2. Tabular presentation showing means and standard deviation of quantifications

		Down		Control	
		N	Mean ± SD	N	Mean ± SD
O43175	D-3-phosphoglycerate dehydrogenase	3	1.48 ± 0.24	4	0.91 ± 0.15
O43488	Aflatoxin b1 aldehyde reductase	5	0.96 ± 0.36	6	0.98 ± 0.44
O95336	6-phosphogluconolactase	5	4.64 ± 2.16	6	5.24 ± 2.40
P00558	Phosphoglycerate kinase	5	1.57 ± 0.92	6	1.51 ± 0.32
P04075	Fructosebisphosphate aldolase (muscle)	5	1.82 ± 0.94	6	1.17 ± 0.37
P09329	Ribose-phosphatepyrophosphate kinase I	5	0.77 ± 0.27	6	0.88 ± 0.56
P09972	Fructose-bisphosphate aldolase (brain)	5	1.03 ± 0.57	6	1.35 ± 0.43
P12268	Inosine-5′-monophosphate dehydrogenase	4	1.63 ± 0.47	5	0.96 ± 0.48
P15121	Aldolasereductase	5	1.33 ± 0.58	6	0.70 ± 0.62
P18669	Phosphoglycerate mutase	8	2.07 ± 0.48	6	2.21 ± 0.45
P51570	Galactokinase	4	0.19 ± 0.13	5	0.36 ± 0.20

Ribose-phosphate pyrophosphokinase 1 (syn: phosphoribosyl pyrophosphate synthetase 1) is utilized by both the novo and the salvage pathways by which endogenously formed or exogenously added pyrimidine, purine or pyridine bases are converted to the corresponding ribonucleoside monophosphates. A disease has been reported to be linked to ribose-phosphate pyrophosphokinase 1 (Roessler et al., 1991) characterized by excessive purine production, gout and uric acid urolithiasis. And indeed, also in DS hyperuricaemia has been reported by several groups (Epstein, 1992). It may therefore well be that protein levels were unchanged but a deficit or mutation leading to deranged activity may exist in fetal human brain of DS.

Increased red blood cell adenine nucleotides have been observed by several groups pointing to aberrant nucleotide metabolism in DS (Epstein, 1992). Determination of brain inosine-5′-monophosphate dehydrogenase 2 (IMP dehydrogenase 2; IMPD2) was therefore mandatory. It is the rate limiting enzyme in the de novo synthesis of guanine nucleotides and therefore is involved in the regulation of cell growth and may also have a role in the development of malignancy (Collart et al., 1988). Again, no changes at the protein levels were observed and it remains to be shown whether activity would be explaining nucleotide abnormalities in DS.

Finally, brain galactokinase, a member of the GHMP kinase family and GALK subfamily, the major enzyme for galactose metabolism, was not different between DS and controls.

The main finding of our study on carbohydrate handling enzymes is the absence of any deficient enzyme protein in fetal DS of the early second trimester, an observation that proposes that impaired carbohydrate metabolism may be a consequence of neurodegeneration inevitably developing in the adult brain of patients with DS. It remains open whether differences at the activity level can be found in fetal cortex of DS.

Acknowledgement

We are highly indebted to the Red Bull Company, Salzburg, for generous financial support.

References

Anneren KG, Korenberg JR, Epstein CJ (1987) Phosphofructokinase activity in fibroblasts aneuploid for chromosome 21. Hum Genet 76: 63–65

Berndt P, Hobohm U, Langen H (1999) Reliable automatic protein identification from matrix-assisted laser desorption mass spectrometric peptide fingerprints. Electrophoresis 20: 3521–3526

Bradford M (1976) A rapid and sensitive method for the quantitation of microgram quantities of protein utilizing the principle of protein-dye binding. Anal Biochem 72: 248–254

Buono P, Paolella G, Mancini FP, Izzo P, Salvatore F (1988) The complete nucleotide sequence of the gene coding for the human aldolase C. Nucl Acids Res 16: 4733–4737

Burger PC, Vogel FS (1973) The development of pathologic changes of Alzheimer's disease and senile dementia in patients with Down's syndrome. Am J Pathol 73: 457–476

Collard F, Collet JF, Gerin I, Veiga da Cunha M, Van Schaffingen E (1999) Identification of the cDNA encoding human 6-phosphogluconolactonase, the enzyme catalysing the second step of the pentose phosphate pathway. FEBS Lett 459: 223–226

Collart FR, Hubermann E (1988) Cloning and sequencing analysis of the human and chinese hamster inosine-5′-monophosphate dehydrogenase cDNAs. J Biol Chem 263: 15769–15772

Dani A, Pietrini P, Furey M, McIntosh CL, Horwitz B, Freo U, Alexander GE, Schapiro MB (1996) Brain cognition and metabolism of Down syndrome adults in association with development of dementia. Neuroreport 7: 2933–2936

Epstein CJ (1992) Down Syndrome (Trisomy 21). In: Scriver CR, Beaudet AL, Sly WS, Valle D (eds) The metabolic and molecular basis of inherited disease. McGraw Hill, New York, pp 749–794

Fountoulakis M, Langen H (1997) Identification of proteins by matrix-assisted laser desorption ionization mass spectrometry following in-gel digestion in low-salt, non-volatile buffer and simplified peptide recovery. Anal Biochem 250: 153–156

Haxby JV, Duara R, Grady CL, Cutler NR, Rapoport SI (1985) Relations between neuropsychological and cerebral metabolic asymmetries in early Alzheimer's disease. J Cereb Blood Flow Metab 5: 193–200

Hsia DY-Y, Justice P, Smith GF, Dowben RF (1971) Down's syndrome. A critical review of the biochemical and immunological data. Am J Dis Child 121: 153–171

Ireland LS, Harrison DJ, Neal GE, Hayes JD (1998) Molecular cloning, expression and catalytic activity of a human AKR7 member of the aldo-keto-reductase superfamily; evidence that the major 2-carboxybenzaldehyde reductase from human liver is a homologue of rat aflatoxin B1-aldehyde reductase. Biochem J 332: 21–34

Izzo P, Costanzo P, Lupo A, Rippa E, Paolella G, Salvatore F (1987) A new human species of aldolase A mRNA from fibroblasts. Eur J Biochem 164: 9–13

Labudova O, Kitzmueller E, Rink H, Cairns N, Lubec G (1999a) Increased phosphoglycerate kinase in the brains of patients with Down Syndrome but not with Alzheimer's disease. Clin Sci 96: 279–285

Labudova O, Cairns N, Kitzmueller E, Lubec G (1999) Impaired brain glucose metabolism in patients with Down Syndrome. J Neural Transm [Suppl] 57: 247–256

Langen H, Roeder D, Juranville J-F, Fountoulakis M (1997) Effect of the protein application mode and the acrylamide concentration on the resolution of protein spots separated by two-dimensional gel electrophoresis. Electrophoresis 18: 2085–2090

Lubec G, Labudova O, Cairns N, Fountoulakis M (1999) Increased glyceraldehyde-3-phosphate dehydrogenase levels in brain of patients with Down Syndrome. Neurosci Lett 260: 141–145

Lubec G, Labudova O, Cairns N, Berndt P, Langen H, Fountoulakis M (1999) Reduced aldehyde dehydrogenase in brain of patients with Down Syndrome. J Neural Transm [Suppl] 57: 21–40

Pietrini P, Dani A, Furey ML, Alexander GE, Freo U, Grady CL, Mentis MJ, Mangot D, Simon EW, Horwitz B, Haxby JV, Schapiro MB (1997) Low glucose metabolism during brain stimulation in older Down's Syndrome subjects at risk for Alzheimer's disease prior to dementia. Am J Psychiatry 154: 1063–1069

Roessler BJ, Palella TD, Heidler S, Becker MA (1991) Identification of distinct PRPS1 mutations in two patients with X-linked phosphoribosylpyrophosphate synthase superactivity. Clin Res 39: 267A–276A

Schapiro MB, Grady CL, Haxby JV (1992) Nature of mental retardation and dementia in Down's syndrome: study with PET, CT, and neuropsychology. Neurobiol Aging 13: 723–734

Van Keuren M, Drabkin H, Hart I, Patterson D, Vora S (1985) Regional mapping of liver type 6 phosphofructokinase to 21q22.3. Am J Hum Genet 37: A20–26

Vora S, Franke U (1981) Assignment of the human gene for liver-type 6-phosphofructokinase isoenzyme (PFK-L) to chromosome 21 by using somatic cell hybrids and monoclonal anti-L-antibody. Proc Natl Acad Sci USA 78: 3738–3742

Wisniewski KE, Wisniewski HM, Wen GY (1985) Occurrence of neuropathological changes and demential of Alzheimer's disease in Down's syndrome. Ann Neurol 17: 278–282

Authors' address: Prof. Dr. G. Lubec, Department of Pediatrics, University of Vienna, Währinger Gürtel 18, A-1090 Vienna, Austria, e-mail: gert.lubec@akh-wien.ac.at

Changes in nicotinic acetylcholine receptor subunits expression in brain of patients with Down syndrome and Alzheimer's disease

E. Engidawork[1], **T. Gulesserian**[1], **N. Balic**[2], **N. Cairns**[3], and **G. Lubec**[1]

[1] Department of Pediatrics, University of Vienna, and
[2] Clinical Institute for Medical and Chemical Laboratory Diagnostics, Vienna, Austria
[3] Department of Neuropathology, Institute of Psychiatry, King's College, London, United Kingdom

Summary. Cholinergic deficit associated with loss of nicotinic acetylcholine receptors (nAChRs) has been described in Alzheimer's disease (AD) by receptor binding assays, positron emission tomography and immunoblotting. However, little is known about the alteration of these receptors in a related disease, Down syndrome (DS) which might be of importance for therapeutic strategies. The protein levels of neuronal nAChR alpha and beta subunits in human postmortem brain samples (frontal cortex and cerebellum) of control, adult DS, and AD were investigated by making use of western blot analysis.

Two major bands at 26 and 45 kDa for $\alpha 3$, one at 50 kDa for $\alpha 4$ and $\beta 2$, and one at 45 kDa for $\alpha 7$ were detected by the respective antibodies. Specific alteration in individual subunits was also apparent in DS and AD. In frontal cortex, the 45 kDa $\alpha 3$ subunit was significantly increased in DS (121%) ($P < 0.05$) and AD (93%) ($P < 0.05$), whereas the 26 kDa, an isoform/truncated form of $\alpha 3$, displayed a reversed pattern. It was significantly decreased in DS (75%) ($P < 0.001$) and AD (52.6%) ($P < 0.05$). $\alpha 4$ was comparable in all groups by contrast, $\alpha 7$ was significantly decreased in AD (64%) ($P < 0.05$). In DS, however, although the levels tended to be lower (17.3%) the reduction was not significant. $\beta 2$ was unchanged in AD but showed a significant increase in DS frontal cortex (98.1%) ($P < 0.01$). In cerebellum, no significant alteration was observed in any of the subunits except $\beta 2$. It exhibited a significant increase (161%) ($P < 0.01$) in DS.

Derangement in expression of nAChRs is apparent in DS, as in AD that may have some relevance to DS neuropathology. Furthermore, the increase in $\beta 2$ expression indicate that these subunits may have more than a structural role. Hence, therapeutic strategies tailored towards these end might be of some benefit for cognitive enhancement in these disorders.

Introduction

Cholinergic neurons are involved in a variety of neuropsychic functions including learning, memory, arousal, sleep, and movement (Aigner and

Mishkin, 1986; Bartus et al., 1982) and abnormalities of these neuronal populations are responsible for the clinical manifestations of human neurological disorders (Brownell et al., 1970; Whithouse et al., 1982).

Acetylcholine is the neurotransmitter released from cholinergic neurons and its actions are subserved by two types of receptors; the muscarinic and nicotinic acetylcholine receptors (nAChRs). nAChRs are integral membrane glycoprotein macromolecules that function as ligand-gated ion channels in signal transmission. They belong to the superfamily of homologous neurotransmitter-gated channels that include glycine, $GABA_A$, and 5-HT_3 receptors. All members of the family are thought to have a common pentameric structure in which highly homologous subunits surround a central aqueous pore whose opening and closing is regulated by binding of the neurotransmitter (Unwin, 1989).

Nicotinic receptors at the muscle end plate are pentameric (Changeux, 1990) and the same is likely to be true for neuronal nAChRs (Cooper et al., 1991). Presently eleven genes that encode neuronal nAChRs subunits have been identified in rat and chickens (Heinemann et al., 1990; Patrick et al., 1993; Sargent, 1993). Eight are designated as α-type (α2–α9) and three as β-type (β2–β4 also called non α1–α3). However, only six α (α2–α7) and three β-type (β2–β4) subunits have been identified and cloned from human brain (McGehee and Role, 1995; Gotti et al., 1997). Two distinct classes of neuronal nAChRs can be distinguished by ligand binding studies (Sargent, 1993; McGehee and Role, 1995): the α-bungarotoxin sensitive, which are thought to function as homo-oligomers and possibly oligomers composed of α7, α8, or α9 subunits; α-bungarotoxin insensitive, which bind agonists such as nicotine, cytisine and epibatidine, includes α4 and α3 containing receptors (Wang et al., 1996; Stauderman et al., 1998). α4β2 are the predominant species and assemble in a two to three ratio to produce functional receptors (Cooper et al., 1991). α7 displays a high relative permeability to calcium, account for most of the high affinity for α-bungarotoxin and desensitize rapidly (Zhang et al., 1994). The more complex type is composed of α3 together with α5, β4, and sometimes β2 and is believed to be primarily expressed in autonomic neurons (Conroy and Berg, 1995); it may also be associated with noradrenergic input to the hippocampus and dopaminergic innervation of the striatum (Wonancot, 1997).

nAChRs are widely distributed in human brain (Gotti et al., 1997). In situ hybridization (Loborn et al., 1995) revealed expression of different nAChRs isoform mRNA in many neurons of the brain with regional difference in pattern of expression (Hellstrom-Lindahl et al., 1999). A number of studies have also indicated that neuronal nAChRs are involved in mechanisms for cognitive processes in the brain (Levin, 1992; Levin and Simon, 1998).

Down syndrome (DS) is the most frequent genetic cause of mental retardation, with an incidence of 1 in 700 pregnancies. According to the current concept, it is caused by over-expression of genes located within a particular segment of chromosome 21 and is associated with developmental abnormalities of the central nervous system that result in mental retardation and age dependent Alzheimer type neurodegeneration (de la Monte, 1999). It is of clinical interest to note that virtually all individuals with DS who survive beyond 35 years of age develop neuropathologic changes that closely resemble Alzheimer's disease (AD) (Mann, 1988; Lai and Williams, 1989). A finding that underscores the ties between the two disorders. Although reports indicate that there is an established reduction of nAChRs in postmortem AD brain tissue as measured by ligand binding, positron emission tomography and western blotting (Nordberg et al., 1990; Hellström-Lindahl et al., 1999; Guan et al., 2000), little is known about the change in these receptors in DS brain. Thus, we initiated the present study to address these issues which help to understand the underlying molecular abnormalities of the central nervous system in DS.

Material and methods

Human brain postmortem tissue

Postmortem brain samples (frontal cortex, superior frontal gyrus, and cerebellum) of patients with DS (n = 8; 2 females and 6 males; 55.9 ± 8.0 years old); AD (n = 10; 4 females and 6 males; 60.8 ± 8.4 years old); and controls (n = 10; 3 females and 7 males; 62.6 ± 8.0 years old) were obtained from the Medical Research Council's London Brain Bank for Neurodegenerative Diseases, Department of Neuropathology, Institute of Psychiatry, London, UK. The major cause of death in DS and AD was bronchopneumonia and heart disease in controls. The control brains were from individuals with no history of neurological or psychiatric illness whereas in all DS brains there was evidence of abundant beta A plaques and neurofibrillary tangles. The AD patients fulfilled the National Institute of Neurological and Communicative Disorders and Stroke and Alzheimer's Disease and Related Disorders Association (NINCDS/ADRDA) criteria for probable AD (Tierney et al., 1988). The histological diagnosis of AD was established and consistent with the consortium to establish a registry for Alzheimer disease (CERAD) criteria (Mirra et al., 1991) for a definite diagnosis of AD. Postmortem interval of brain dissection was 32.4 ± 20.7 for DS; 31.8 ± 28.3 for AD; and 37.3 ± 18.4 hours for controls. All samples were stored at −70°C and the freezing chain was never interrupted.

Sample preparation

Frozen brain samples were thawed on ice and homogenized in 4 volumes of TBS buffer (100 mM Tris, 150 mM NaCl, pH 7.5) containing protease inhibitors, (1 μg/ml each of antipain, pepstatin A, and leupeptin and 10 mM of phenylmethylsulfonyl fluoride) and 0.05% Tween 20. Samples were homogenized on ice for 30 seconds (6–8 strokes) with a Potter-Elvehjem homogenizer. The brain homogenate then was centrifuged for 10 min at 10,000 g and 4°C. To the supernatant, equal volume of sample buffer (125.5 mM Tris-HCl base, 70 mM sodium dodecyl sulfate, 0.001% Bromophenol blue, 20% glycerol, and 2% 2-mercaptoethanol, pH 6.8) was added and the mixture was heated at 95°C for 15 min. The concentration of protein was determined by the Bicinchoninic acid assay using bovine serum albumin as a standard (Pierce, USA).

Immunoblotting

Protein extracts containing either 50 μg (nAChR α_4, α_7), or 100 μg (nAChR α_3, β_2,) were subjected to sodium dodecyl sulfate-polyacrylamide gel electrophoresis (200 V for 3 hrs), and transferred to polyvinylidene fluoride membranes (Immobilon™-P, MilliPore, Bedford, USA) (192 V for 1 hr). The membranes were then washed two times for 5 min in Tris buffer (100 mM Tris-HCl, 150 mM NaCl, pH 7.5) and later blocked with blotto (1.5% non-fat dry milk in TBST (100 mM Tris-HCl, 150 mM NaCl, pH 7.5, 1 mM $MgCl_2$, 0.1% Tween-20)) overnight at 4°C. This was followed by incubation with primary monoclonal mouse anti-nAChR α_7 (M-220, SIGMA, St. Louis, USA) (1:500); monoclonal rat anti-nAChR α_4 antibody (MMS-613R, BAbCO, Richmond, USA) (1:1,000); polyclonal goat anti-nAChR α_3 antibody (sc-1771, Santa Cruz Biotechnology Inc. Santa Cruz, USA) (1:100) and polyclonal goat anti-nAChR β_2 antibody (sc-1449, Santa Cruz Biotechnology Inc. Santa Cruz, USA) (1:100) in blotto at room temperature for 3 hours on a rocking shaker. The membranes were then washed three times for 15 minutes with blotto prior to incubation with respective secondary horseradish peroxidase conjugated goat anti-mouse IgG (H + L) (1031-05, Southern Biotechnology Associates, Inc., Brimingham, USA); rabit anti-rat (A5795, SIGMA, St. Louis, USA) or rabbit anti-goat IgG (H + L) (AP106P, CHEMICON International, Inc., Temecula, USA) in blotto for one hour at room temperature on a rocking shaker. The membranes were washed again three times for 15 minutes followed by incubation with NEN chemiluminescence reagent for 1 minute and subsequent exposure to films (Kodak Blue XB-1, Rochester, USA). Developed films were scanned and densities of the immunoreactive bands were calculated by RFLP Scan 2.1 software program (Scanalytics, Billerica, USA).

Data analysis

Values are expressed as mean ± standard deviation of optical density. Inter-group differences were analyzed by non-parametric Mann-Whitney U-test and the significance was set at $P < 0.05$ for a two-tailed test. The relationship between age or postmortem interval and density of the proteins as well as the relation amongst the density of individual proteins was examined with linear regression analysis.

Results

Western blot analysis of nAChR subunits in human postmortem brain revealed the presence of nAChR immunoreactive (nAChR-IR) protein in frontal and cerebellar cortices. Two bands with apparent molecular size of 45 and 26 kDa were detected with anti-α3 antibody (Fig. 1A). The 45 and 26 kDa proteins may represent different isoforms of α3 subunit or the later may be a truncated form of the former. Anti α4 antibody produced one major band with size of 50 kDa that represents α4 subunit along with two minor bands at 35 and 30 kDa (Fig. 1B). α7 receptor migrated as a diffuse band ranging between 45–60 kDa with a major band at 45 kD (Fig. 1C). Likewise, bands that range in size between 15 and 75 kDa, with β2 migrating as 50 kDa protein, were obtained with anti-β2 antibody (Fig. 2).

Densitometric analysis of the two α3 nAChR-IR bands with varying molecular size revealed an intriguing result. In frontal cortex, the 45 kDa protein was significantly increased by ~121% and ~93% in DS ($P < 0.05$) and AD ($P < 0.05$), respectively. By contrast, the 26 kDa protein displayed a highly significant ($P < 0.001$) and significant reduction ($P < 0.05$) by ~75% and ~52.6% in DS and AD, respectively (Table 1). However, neither the total IR in both frontal cortex and cerebellum (Table 2) nor the individual proteins IR

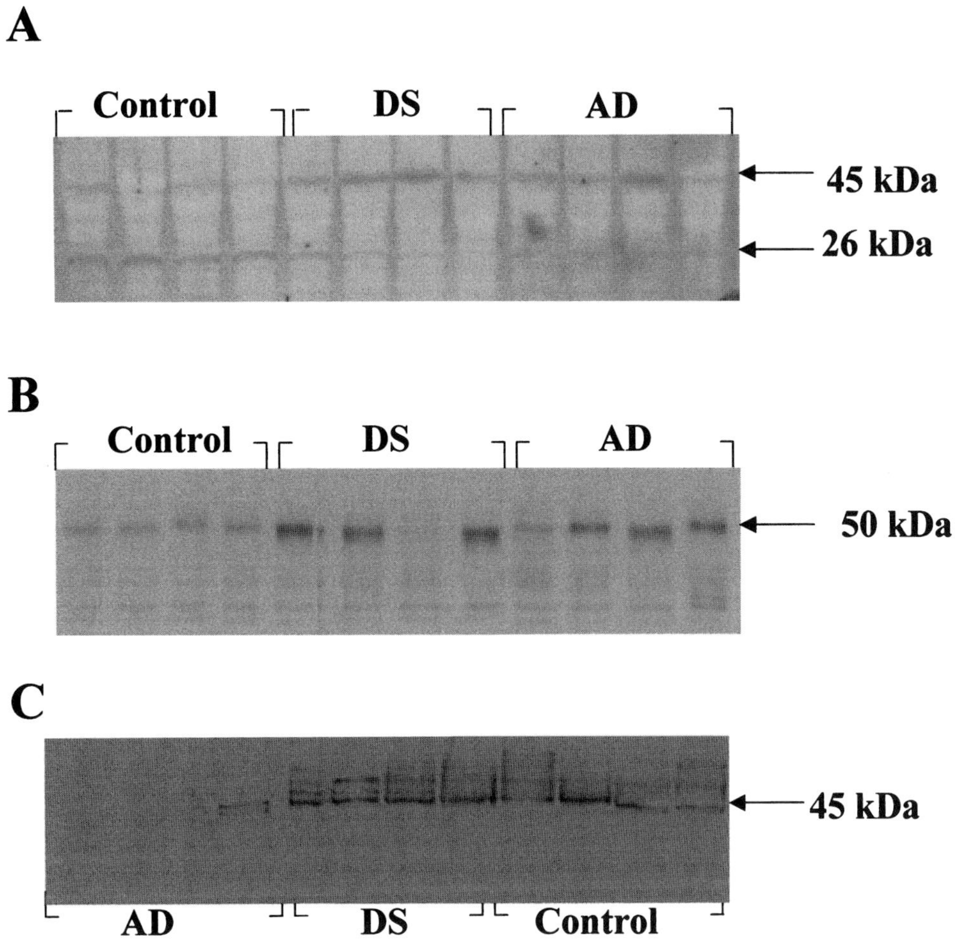

Fig. 1. Immunodetection of α nAChR subunits in frontal cortex. **A** α3, **B** α4 and **C** α7. Brain homogenate containing 100 and 50 μg protein/lane from the frontal cortex of 10 controls, 8 DS, and 10 AD brains were used for the immunodetection of α3, α4 and α7 nAChR subunits, respectively. The immunoblots of 4 controls, 4 DS and 4 AD cases were shown in the upper panels

in cerebellum (Table 1) of DS and AD showed significant difference. No significant change was observed in α4 nAChR-IR as well as total α4 nAChR-IR either in frontal cortex or cerebellum of the test subjects (Tables 1, 2). α7 nAChR-IR showed an insignificant reduction by ~17.3% in DS frontal cortex whereas decreased significantly by ~64% ($P < 0.05$) in AD compared to controls (Table 1). A similar pattern of reduction was noted in the same cortex of DS by ~21% and AD by ~65% ($P < 0.05$) when the total α7 nAChR-IR was quantified (Table 2). No significant difference, however, had been observed in the major as well as total α7 nAChR-IR in both DS and AD cerebellum (Tables 1, 2). β2 nAChR-IR was unchanged in AD patients but significantly increased by ~98.1% ($P < 0.01$) and ~161% ($P < 0.01$) in DS frontal cortex and cerebellum, respectively (Table 1). Relatively, a similar

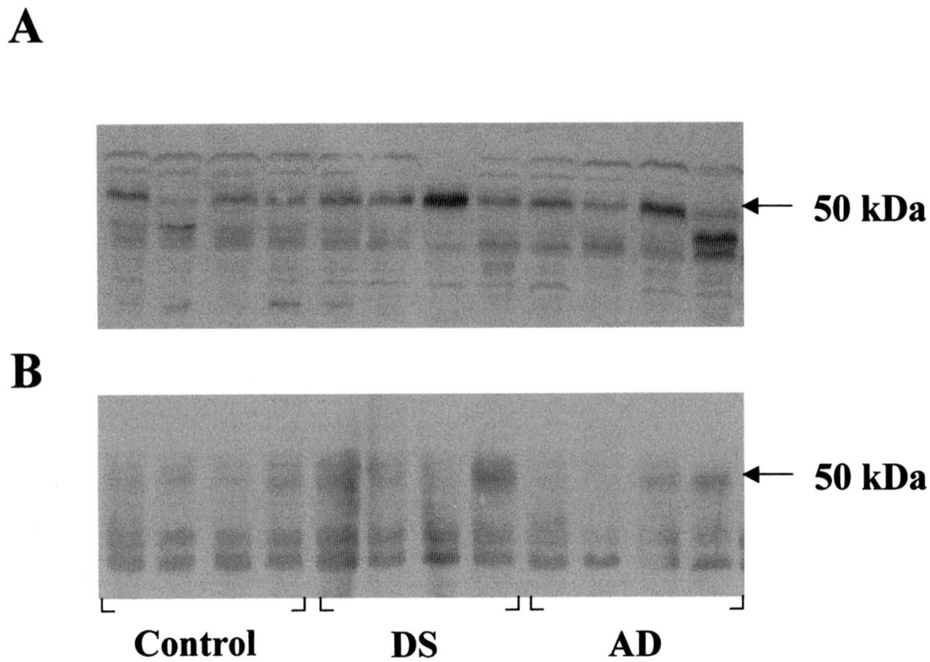

Fig. 2. Immunodetection of β2 nAChR subunits in frontal cortex (**A**) and cerebellum (**B**). Brain homogenate containing 100 μg protein/lane from the frontal cortex and cerebellum of 10 controls, 8 DS, and 10 AD brains were used for the immunodetection of β2 nAChR subunits. The immunoblots of 4 controls, 4 DS and 4 AD cases were shown in the upper panels

Table 1. Levels of individual nAChR subunits in two brain regions. Following western blot, immunoreactivity of 10 controls, 6–8 DS, and 8–10 AD cases were quantified by scanning densitometry. Values are expressed as mean ± standard deviation of arbitrary optical density and inter-group differences were analyzed by non-parametric Mann-Whitney U-test and the significance was set at $P < 0.05$ for a two-tailed test. *$P < 0.05$, **$P < 0.01$, ***$P < 0.001$ compared to controls

	Control	DS	AD
Frontal			
α3-nAChRs (45 kDa)	0.29 ± 0.11 (n = 10)	0.64 ± 0.31* (n = 8)	0.56 ± 0.33* (n = 8)
α3-nAChRs (26 kDa)	0.76 ± 0.47 (n = 10)	0.19 ± 0.09*** (n = 8)	0.36 ± 0.26* (n = 8)
α4-nAChRs	1.10 ± 0.55 (n = 9)	1.72 ± 1.40 (n = 7)	1.68 ± 0.62 (n = 8)
α7-nAChRs	1.39 ± 0.70 (n = 8)	1.15 ± 0.98 (n = 8)	0.50 ± 0.39* (n = 6)
β2-nAChRs	0.52 ± 0.24 (n = 10)	1.03 ± 0.44** (n = 8)	0.90 ± 0.49 (n = 8)
Cerebellum			
α3-nAChRs (45 kDa)	1.98 ± 0.88 (n = 9)	3.50 ± 1.76 (n = 6)	2.36 ± 1.56 (n = 9)
α3-nAChRs (26 kDa)	1.54 ± 1.06 (n = 9)	1.34 ± 0.68 (n = 6)	0.85 ± 0.51 (n = 9)
α4-nAChRs	0.97 ± 0.29 (n = 5)	2.09 ± 1.21 (n = 6)	1.88 ± 1.61 (n = 10)
α7-nAChRs	6.24 ± 2.09 (n = 9)	7.82 ± 0.88 (n = 6)	5.05 ± 2.88 (n = 9)
β2-nAChRs	0.21 ± 0.15 (n = 9)	0.55 ± 0.19** (n = 6)	0.13 ± 0.09 (n = 9)

Table 2. Levels of total nAChR-immunoreactive subunit proteins in two brain regions. Following western blot, immunoreactivity of 10 controls, 6–8 DS, and 8–10 AD cases were quantified by scanning densitometry. Values are expressed as mean ± standard deviation of arbitrary optical density and inter-group differences were analyzed by non-parametric Mann-Whitney U-test and the significance was set at $P < 0.05$ for a two-tailed test. *$P < 0.05$ compared to controls

	Control	DS	AD
Frontal			
α3-nAChRs	1.05 ± 0.53 (n = 10)	0.91 ± 0.33 (n = 8)	0.92 ± 0.50 (n = 8)
α4-nAChRs	1.94 ± 0.88 (n = 9)	2.24 ± 1.48 (n = 7)	2.31 ± 0.66 (n = 8)
α7-nAChRs	3.33 ± 1.47 (n = 8)	2.63 ± 2.23 (n = 8)	1.18 ± 1.08* (n = 6)
β2-nAChRs	1.30 ± 0.38 (n = 10)	2.27 ± 1.04* (n = 8)	1.74 ± 0.72 (n = 8)
Cerebellum			
α3-nAChRs	3.52 ± 1.55 (n = 9)	4.84 ± 1.62 (n = 6)	3.21 ± 1.82 (n = 9)
α4-nAChRs	1.70 ± 0.93 (n = 5)	2.38 ± 1.29 (n = 6)	1.81 ± 0.73 (n = 10)
α7-nAChRs	20.18 ± 8.19 (n = 9)	22.03 ± 4.21 (n = 6)	11.52 ± 4.83 (n = 9)
β2-nAChRs	1.31 ± 0.73 (n = 9)	1.90 ± 0.75* (n = 6)	1.02 ± 0.52 (n = 9)

pattern of increase in the total β2 nAChR-IR was obtained in DS frontal cortex ($P < 0.01$) and cerebellum ($P < 0.05$), but levels were unchanged in the corresponding regions of AD brains (Table 2). Linear regression analysis did not reveal any significant correlation between age or postmortem interval and levels of α3, α4, α7, and β2 nAChRs in both brain regions of the study populations. So did the correlation between α and β subunits.

Discussion

The present study was targeted at assessing levels of nAChRs in the normal human frontal cortex and cerebellum and the possible changes of nAChR expression in DS and AD. This is the first report to describe levels of nAChRs in DS although many reports have been published in AD. The nAChR blot produced bands with molecular size that span from 15 to 70 kDa. However, the molecular masses of nAChR subunits of different species reported from various laboratories range from 49–75 kDa (Sargent, 1993; Gotti et al., 1997). Bands which happened to be outside this range therefore may represent either splice variants or products obtained as a result of proteolysis. Indeed, northern blot analysis of α3 mRNA in PC12 cells revealed the presence of two bands designated as large and small size mRNAs (Nakayama et al., 2000) which is in line with our finding at protein level. Hence, the 26 kDa protein probably represent a protein obtained by alternative splicing, as there is further evidence for alternative splicing of the CHRNA3 gene (Mihovilovic et al., 1993).

Specific alterations in the different nAChR subtypes is apparent both in DS and AD. In previous studies, Guan et al. (2000) reported decreased α3 and α4, and unchanged β2 subunit levels in AD temporal cortex and hippocampus, and a decrease in α7 subunit in hippocampus but not in temporal cortex. In another study, the same group found decreased α7 subunit levels in frontal cortex of schizophrenic brain (Guan et al., 1999). A similar decrease in α4 subunits in AD temporal cortex was also reported by Burghaus et al. (2000), however, contrary to Guan et al. they found a decrease in α7 subunit in AD temporal cortex. In frontal cortex of AD patients, a decrease of α nAChRs was reported by Sugaya et al. (1990) and Schröder et al. (1991). In contrast, no significant change as compared to controls was reported by Flynn and Mash (1986). Our finding of decreased α3 isoform and α7, unchanged α4 and β2, and increased α3 subunit levels in AD frontal cortex is concordant with some and not with other findings mentioned above. The discrepancy may be due to selection of the areas used, compartments from which the sample is prepared or molecular diversity of nAChRs (Weiland et al., 2000).

The level of nAChRs is regulated by the expression of mRNAs, assembly of proteins, and protein turnover in the membrane. This level was reported to be affected by factors such as nerve growth factor, cAMP, nicotinic agonists, and temperature (Cooper et al., 1999; Nakayama et al., 2000). In rat phaeochromocytoma (PC12) cells, nAChR mRNA is regulated by nerve growth factor (NGF) and cAMP (Nakayama et al., 2000). cAMP down-regulates the expression of α3 nAChR mRNA via protein kinase A (PKA), an enzyme that phosphorylates ^{133}Ser of cAMP response element binding protein (CREB) which in turn, activates the expression of several genes (Sassone-Cosi, 1995; Nakayama et al., 2000). NGF can either cause up-regulation or down-regulation depending on the cell type (Nakayama et al., 2000). Both up-regulation and down-regulation are mediated via the Ras-MAPK cascade and in the later, activation of the Ras-MAPK pathway is followed by phosphorylation of CREB through activation of RSK2 (Xing et al., 1996). Thus, it appears that the cross-talk between the NGF and cAMP signaling pathways determine the level of nAChRs. Basal production of cAMP is reported to be decreased in DS and AD brains that could lead to impaired phosphorylation of CREB (Yamamoto-Sasaki et al., 1999; Lumbreras et al., 2000). Growth factor induced signaling cascade may also be impaired in diseases where oxidative stress may contribute to neuronal dysfunction (Zhang and Jope, 1999), such as DS and AD. Moreover, oxidative stress was shown to negatively modulate growth factor induced phosphorylation of CREB in PC12 cells (Zhang and Jope, 1991). Taken together, it seems likely that the increased α3 level of nAChRs may be associated with decreased CREB activation in DS and AD. This seems plausible, as one considers the fact that activation of CREB (Abel and Kandel, 1998) and the nicotinic systems (Albuquerque et al., 1997; Levin and Simon, 1998) are an important molecular components in the process of memory and learning. In contrast to the 45 kDa α3 protein, the 26 kDa isoform/truncated protein exhibited significant decrease in DS and AD frontal cortex. The significance of this reduction remains to be seen.

The reduction of α7 nAChRs may be caused by rapid trurnover of the protein due to loss of an appropriate binding partner for co-assembly (Blumenthal et al., 1997). Indeed, such receptor down-regulation seems to account for the observed reduction, as no change in NSE, a neuronal marker, levels were detected in AD frontal cortex (Gulesserian et al., 2000). This loss may decrease functional cholinergic synapses and cortical activity, and play a role in the cognitive impairment associated with AD. The unchanged levels of α4 subunit in DS and AD, even when the total (Table 2) as well as each individual bands are quantified (Table 1, data for minor bands not shown), suggest α4 may have less important role than α7 in cognitive functioning. The reduced level of α7 found in frontal cortex of schizophrenic patients (Guan et al., 1999) that manifest problems in learning efficiency and accuracy (Adler et al., 1998) reinforce the significant role played by α7 in learning and memory behaviors. DS is a disorder in which patients show cognitive deficits (Brugge et al., 1994). The failure to see any significant difference in DS frontal cortex α7 levels can be explained by the fact that cognitive deficit is not an invariable clinical feature of this syndrome (Brugge et al., 1994).

The finding in the present study that both the total as well as individual α nAChRs in cerebellum were comparable between the three groups (Tables 1, 2) may reflect that the decrease seems not to affect all brain areas. In AD nicotinic receptors do not decrease in subcortical areas like thalamus and striatum (Aubert et al., 1992) and cerebellar pathology is not a target which would be in agreement with our observation. No significant change in β2 subunits had also been observed in both regions of AD patients confirming the lack of sensitivity of these nicotinic receptors to AD pathology (Guan et al., 2000). However, there was a significant increase in cortical and cerebellar β2 levels in DS. Aged β2 knockout mice were reported to show a picture resembling neurodegenerative changes (Zoli et al., 1999). Based on this observation it is tempting to speculate that increased expression of β2 subunit may be one neuroprotective response elicited by the DS brain to limit neurodegeneration.

In conclusion, the current results although limited in scope because of the small number of subjects included, offer evidence for the existence of specific alteration of nAChRs in DS as observed in AD. The alteration of β2 levels in DS imply that this subunit may have more than a structural role. We are now following the changes in other subtypes of nAChRs in DS.

Acknowledgements

We are highly indebted to the Red Bull Company, Salzburg, Austria for generous financial support. EE is a recipient of scholarship from the Austrian Academic Exchange Service (ÖAD).

References

Abel T, Kandel E (1998) Positive and negative regulatory mechanisms that mediates long-term memory storage. Brain Res 26: 360–378

Adler LE, Olincy A, Waldo M, Harris JG, Griffith J, Stevens K et al (1998) Schizophrenia, sensory gating, and nicotinic receptors. Schizophr Bull 24: 189–202

Aigner TG, Mishkin M (1986) The effects of physostigmine and scopolamine on recognition memory in monkeys. Behav Neural Biol 45: 81–87

Albuquerque EX, Alkondon M, Periera EFR, Castro NG, Schrattenholz A, Barbosa CTF et al (1997) Properties of neuronal nicotinic acetylcholine receptors: pharmacological characterization and modulation of synatic function. J Pharmacol Exp Ther 280: 1117–1136

Aubert I, Araujo DM, Ce'cyre D, Robitaille Y, Gauthier S, Qurion R (1992) Comparative alterations of nicotinic and muscarinic binding sites in Alzheimer's disease and Parkinson's disease. J Neurochem 58: 529–541

Bartus RT, Dean RL III, Lippa AS (1982) The cholinergic hypothesis of geriatric memory dysfunction. Science 217: 408–417

Blumenthal EM, Conro WG, Romano SG, Kassner PD, Berg DK (1997) Detection of functional nicotinic receptors blocked by alpha-bungarotoxin on PC12 cells and dependence of their expression on posttranslational events. J Neurosci 17: 6094–6104

Brownell B, Oppenheimer DR, Hughes JT (1970) The central nervous system in motor neuron disease. J Neurol Neurosurg Psychiatry 33: 338–357

Brugge KL, Nichols SL, Salmon DP, Hill LR, Delis DC, Aaron L, Travner DA (1994) Cognitive impairement in adults with Down syndrome: similarities to early cognitive changes in Alzheimer's disease. Neurology 44: 232–238

Burghaus L, Schütz U, de Vos RAI, Steur ENHJ, Wevers A, Lindstrom J, Schröder H (2000) Quantitative assessement of nicotinic acetylcholine receptor proteins in the cereberal cortex of Alzheimer's patients. Mol Brain Res 76: 385–388

Changeux JP (1990) The Tips lecture: The nicotinic acetylcholine receptors: an allosteric protein prototype of ligand-gated ion channels. Trends Pharmacol Sci 11: 585–492

Conroy WG, Berg DK (1995) Neurons can maintain multiple classes of nicotinic acetylcholine receptors distinguished by different subunit compositions. J Biol Chem 270: 4424–4431

Cooper E, Couturier S, Ballivet M (1991) Pentameric structure and subunit stochiometry of a neuronal nicotinic acetylcholine receptor. Nature 350: 235–238

Cooper ST, Harkness PC, Baker ER, Millar NS (1999) Upregulation of cell surface $\alpha 4\beta 2$ neuronal nicotinic acetylchline receptors by lower temperature and expression of chimeric subunits. J Biol Chem 274: 27145–27152

de la Monte SM (1999) Molecular abnormalities of the brain in Down Syndrome: relevance to Alzheimer's neurodegeneration. J Neural Transm [Suppl] 57: 1–19

Flynn DD, Mash DC (1986) Characterization of L-[^{3}H]nicotine binding in human cerebral cortex: comparison between Alzheimer's disease and the normal. J Neurochem 47: 1948–1954

Gotti C, Fornasari D, Clementi F (1997) Human neuronal nicotinic receptors. Prog Neurobiol 53: 199–237

Guan Z, Zhang X, Blenow K, Nordberg A (1999) Decreased protein level of nicotinic receptor alpha 7 subunit in the frontal cortex from schizophrenic brain. Neuroreport 10: 1779–1782

Guan Z, Zhang X, Ravid R, Nordberg A (2000) Decreased levels of nicotinic receptor subunits in the hippocampus and temporal cortex of patients with Alzheimer's disease. J Neurochem 74: 237–243

Gulesserian T, Engidawork E, Cairns N, Lubec G (2000) Increased protein levels of serotonin transporter in frontal cortex of patients with Down syndrome. Neurosci Lett 296: 53–57

Heinemann S, Boulter J, Deneris E, Connolly J, Duvoisin R, Papke R, Patrick J (1990) The brain nicotinic acetylcholine receptor gene family. Prog Brain Res 86: 195–203

Hellström-Lindahl E, Mousavi M, Zhang X, Ravid R, Nordberg A (1999) Regional distribution of nicotinic receptors subunit mRNAs in human brain: comparison between Alzheimer and normal brain. Mol Brain Res 66: 94–103

Lai F, Williams RS (1989) A prospective study of Alzheimer's disease in Down's syndrome. Arch Neurol 46: 849–853

Levin ED (1992) Nicotinic systems and cognitive functions. Psychopharmacol 108: 417–431
Levin ED, Simon BB (1998) Nicotinic acetylcholine receptor involvement in cognitive function in animals. Pscychopharmacol 138: 217–230
Loborn CH, Wevers A, Dämgen K, Jeske A, Rontal D, Birtsch CH et al (1995) Cellular distribution in the rat telencephalon of mRNA encoding for the α3 and α4 subunits of nicotinic acetylcholine receptors. Mol Brain Res 30: 70–76
Lumbreras MA, Baamonde C, Lubec G, Cairns N, Crespo D, Martinez-Cue C et al (2000) (Abstract) Brain adenylyl cyclase and phosphorylase c signaling pathways in Alzheimer's disease and Down syndrome. Soc Neurosci 26: 32.13
Mann DM (1988) The pathological association between Down syndrome and Alzheimer's disease. Mech Aging Dev 43: 99–136
McGehee DS, Role LW (1995) Physiological diversity of nicotinic acetylcholine receptors expressed by vertebrate neurons. Annu Rev Physiol 57: 521–546
Mihovilovic M, Hulette C, Mittelstaedt J, Austin C, Roses AD (1993) Nicotinic neuronal acetylcholine receptor alpha-3 subunit transcription in normal and myasthenic thymus. Ann NY Acad Aci 681: 83–96
Mirra SS, Heyman A, McKeel D, Sumi S, Crain BJ (1991) The consortium to establish a registry for Alzheimer disease (CERAD). II. Standardization of neuropathological assessment of Alzheimer's disease. Neurology 41: 479–486
Nakayama H, Veno S, Ikeuchi T, Hatanaka H (2000) Regulation of alpha-3 nicotinic acetylcholine receptor subunit mRNA levels by NGF and cAMP in PC12 cells. J Neurochem 74: 1346–1354
Nordberg A, Hartvig P, Lilja A, Vitanen M, Amberla K, Lundqvist H, Anderson Y et al (1990) Decrease uptake and binding of ^{11}C-nicotine in brain of Alzheimer's patients as visualized by positron emission tomography. J Neural Transm [PD-Sect] 2: 215–224
Patrick J, Sequela P, Vernino S, Amador M, Luetje C, Dani JA (1993) Functional diversity of neuronal nicotinic acetylcholine receptors. Prog Brain Res 98: 113–120
Sargent PB (1993) The diversity of neuronal nicotinic acetylcholine receptors. Annu Rev Neurosci 16: 403–443
Sassone-Corsi P (1995) Transcription factors responsive to cAMP. Ann Rev Cell Dev Biol 11: 357–377
Schröder H, Giacobini E, Struble RG, Zilles K, Maelicke A (1991) Nicotinic cholinoceptive neurons of the frontal cortex are reduced in Alzheimer disease. Neurobiol Aging 12: 259–262
Stauderman KA, Mahaffy LS, Akong M, Velicelebi G, Chavez-Noriega LE, Crona JH et al (1998) Characterization of human recombinant neuronal nicotinic acetylcholine subunit combinations α2β4, α3β4, and α4β4 stably expressed in HEK293 cells. J Pharmacol Exp Ther 284: 777–789
Sugaya K, Giacobini E, Chiappinelli V (1990) Nicotinic acetylcholine receptor subtypes in human frontal cortex: changes in Alzheimer disease. J Neurosci Res 27: 349–359
Tierney MC, Fisher RH, Lewis AJ, Torzito ML, Snow WG, Reid DW et al (1988) The NINCDA/ADRDA work group criteria for the clinic diagnosis of probable Alzheimer's disease. Neurology 38: 359–364
Unwin N (1989) The structures of ion channels in membranes of excitable cells. Neuron 3: 665–676
Wang F, Gerzanich V, Wells GB, Anand R, Pang X, Keyser K, Lindstorm J (1996) Assembly of human neuronal nicotinic acetylcholine receptor α5 subunits with α3, β2, and β4 subunits. J Biol Chem 271: 17656–17665
Weiland S, Bertrand D, Leonard S (2000) Neuronal nicotinic acetylcholine receptors: from the gene to the disease. Behav Brain Res 113: 43–56
Whithouse PJ, Price DL, Struble RG, Clark AW, Coyle JT, Delong MR (1982) Alzheimer's disease and senile dementia: loss of neuron in the basal forebrain. Science 215: 1237–1239

Wonnacot S (1997) Presynaptic nicotinic acetylcholine receptors. Trends Neurosci 20: 92–98

Xing J, Ginty DD, Greenberg ME (1996) Coupling of the Ras-MAPK pathway to gene activation by RSK2, a growth factor regulated CREB kinase. Science 273: 959–963

Yamamoto-Sasaki M, Ozawa H, Saizo T, Rosler M, Riederer P (1999) Impaired phosphorylation of cAMP response element binding protein in the hippocampus of dementia of the Alzheimer type. Brain Res 824: 300–303

Zhang L, Jope RS (1999) Oxidative stress differentially modulates phosphorylation of ERK, p38 and CREB induced by NGF or EGF in PC12 cells. Neurobiol Aging 20: 271–278

Zhang ZW, Vijayaraghavan S, Berg DK (1994) Neuronal acetylcholine receptors that bind α-bungarotoxin with high affinity function as ligand-gated ion channels. Neuron 12: 167–177

Zoli M, Picciotto MR, Ferrari R, Cocchi D, Changeux JP (1999) Increased neurodegeneration during aging in mice lacking high-affinity nicotine receptors. EMBO J 18: 1235–1244

Authors' address: Prof. Dr. G. Lubec, Department of Pediatrics, University of Vienna, Währinger Gürtel 18-20, A-1090 Vienna, Austria, e-mail: Gert.Lubec@akh-wien.ac.at

Protein levels of human peroxiredoxin subtypes in brains of patients with Alzheimer's disease and Down Syndrome

S. H. Kim[1], **M. Fountoulakis**[2], **N. Cairns**[3], and **G. Lubec**[1]

[1] Department of Pediatrics, University of Vienna, Austria
[2] F. Hoffmann — La Roche, Basel, Switzerland
[3] Institute of Psychiatry, Brain Bank, King's College, United Kingdom

Summary. Human peroxiredoxin (Prx) play important roles in eliminating hydrogen peroxide generated during cellular mechanisms using electrons from thioredoxin (Trx). Oxidative stress induced by reactive oxygen species (ROS) such as hydrogen peroxide has been implicated in the pathogenesis of several neurodegenerative diseases. We applied the proteomic approach to study protein levels of three subtypes of human Prx in brain regions from patients with Alzheimer's disease (AD) and Down Syndrome (DS). Protein levels of Prx-I and Prx-II were significantly increased in AD and DS. Protein levels of Prx-III, a mitochondrial protein, however, were significantly decreased. We conclude that increased protein levels of Prx-I and Prx-II could provide protection against neuronal cell death induced by hydrogen peroxide. Decreased protein levels of Prx-III could be caused by mitochondrial damage shown in AD and DS. Showing upregulated Prx protein levels provides evidence for the involvement of ROS in the pathogenesis of AD and DS.

Introduction

Oxidative stress by reactive oxygen species (ROS) such as superoxide, hydrogen peroxide and hydroxyl radical can lead to neuronal cell death (apoptosis), a hallmark of neurodegenerative diseases such as Alzheimer's disease (AD) (Simonian and Coyle, 1996; Pappolla et al., 1992). There is some evidence for the neuronal cell death induced by oxidative stress in Down syndrome (DS) with AD-like pathology (Busciglio and Yankner, 1995; Jovanovic et al., 1998). However, cells evolved antioxidant defense systems to protect themselves against the toxicity of ROS and to convert ROS into less noxious compounds (Halliwell, 1999). Enzymatic defense against activated oxygen species involves a cooperative action of several enzymes. Superoxide dismutase (SOD) catalyses the dismutation of superoxide radical to hydrogen peroxide and catalase (CAT) and glutathione peroxidase (GPx) can scavenge hydrogen peroxide (Hayes and McLellan, 1999). Hydrogen peroxide produced during cellular mechanism can induce neuronal cell death via apoptosis (Whittemore et al., 1994, 1995).

Peroxiredoxin (Prx) proteins also serve an antioxidant function associated with removal of cellular hydrogen peroxide and are found ubiquitously in all living organisms from yeast to human (Chae et al., 1994a,b). Until now, six subfamilies of mammalian Prx proteins have been identified on the basis of their amino acid sequences and immunological properties (Chae et al., 1999; Kang et al., 1998a; Okado-Matsumoto et al., 2000; Zhou et al., 2000). Since the peroxidase activities of most mammalian Prx proteins (Prx-I, -II, -III, -IV and -V) were shown to rely on thioredoxin (Trx) as a source of reducing equivalents for the reduction of hydrogen peroxide, these proteins were also named Trx peroxidases (TPx) (Kang et al., 1998b). However Prx-VI (or 1-Cys Prx) was shown to use glutathione as electron donors (Kang et al., 1998a). Recently several studies provided evidence that Prx proteins could be critically involved in apoptosis. Prx-I (also named TPx II) prevented hydrogen peroxide accumulation in cells and Prx-II (also named TPx I, or TSA, thio-specific antioxidant, or PRP, protector protein) could promote neuronal cell survival (Zang et al., 1997; Ichimiya et al., 1997). Additionally, Prx-II was shown to prevent the opening of mitochondrial permeability transition (MPT) pore (Kowaltowski et al., 1998). Prx-V prevented ROS production stimulated by expression of p53 (Zhou et al., 2000).

These considerations prompted us to investigate the protein levels of three subtypes of human Prx (Prx-I, -II, and -III) in brain regions of patients with AD and DS. We applied the proteomic approach using two-dimensional (2-D) gel electrophoresis with subsequent matrix-assisted laser desorption ionization mass spectroscopy (MALDI-MS) identification and specific software for quantification of Prx proteins.

Methods

Brain samples preparation

Postmortem brain samples were obtained from the MRC London Brain Bank for Neurodegenerative Diseases, Institute of Psychiatry. All DS patients were karyotyped and possessed trisomy 21. A formal cognitive assessment of dementia in DS was not performed. In all DS brains there were abundant and extensive beta-amyloid deposits, neurofibrillary tangles and neuritic plaques. AD patients were selected prospectively and examined clinically and fulfilled the National Institute of Neurological Disorders and Stroke and Alzheimer's disease and Related Disorders Association criteria for probable AD (Tierney et al., 1988). The neuropathological diagnosis of "definite AD" was confirmed using the CERAD criteria (Mirra et al., 1991). Normal control brains were obtained from individuals with no history of neurological or psychiatric illness. The major cause of death was bronchopneumonia in AD and DS and heart disease in controls. Brain samples were dissected, coronal slices snap frozen and stored at 270°C until used. Seven brain regions (cerebellum, frontal, temporal, occipital, parietal cortex, thalamus and caudate nucleus) of patients with AD (n = 13, 58.54 ± 7.57 years old), DS (n = 9, 55.67 ± 7.48 years old) and controls (n = 18, 50.00 ± 16.94 years old) were used. The postmortem interval of brain dissection in AD, DS and controls was 30.31 = 26.17, 31.44 ± 19.56

and 38.78 ± 18.84 hours, respectively. The postmortem interval had a limited influence upon protein expression (data not shown) and the freezer storage time showed no effect. Brain tissue was suspended in 0.5 ml of sample buffer consisting of 40 mM Tris, 5 M urea (Merck, Darmstadt, Germany), 2 M thiourea (Sigma, St. Louis, MO, USA), 4% CHAPS (Sigma), 10 mM 1,4-dithioerythritol (Merck), 1 mM EDTA (Merck), 1 mM PMSF (Sigma) and 1 μg/ml of each pepstatin A, chymostatin, leupeptin and antipain. The suspension was sonicated for approximately 30 seconds and centrifuged at 10,000 × g for 10 minutes and the supernatant was centrifuged further at 150,000 × g for 45 minutes to sediment undissolved material. The protein concentration of the supernatant was determined by the Coomassie blue method (Bradford, 1976).

Two-dimensional (2-D) gel electrophoresis

The 2-D gel electrophoresis was performed essentially as reported (Langen et al., 1997). Samples of approximately 1.5 mg were applied on immobilized pH 3–10 nonlinear gradient strips (IPG, immobilized pH-gradient strips, Pharmacia Biotechnology, Uppsala, Sweden), at both, the basic and acidic ends of the strips. The proteins were focused at 300 V for 1 hour, after which the voltage was gradually increased to 3,500 V within 6 hours. Focusing was continued at 3,500 V for 12 hours and at 5,000 V for 48 hours. The second-dimensional separation was performed on 9–16% linear gradient polyacrylamide gels. After protein fixation with 40% methanol containing 5% phosphoric acid for 12 hours, the gels were stained with colloidal Coomassie blue (Novex, San Diego, CA, USA) for 48 hours. The molecular mass was determined by running standard protein markers at the right side of selected gels. The size markers (Gibco, Basel, Switzerland) covered the range 10–200 kDa. Isoelectric point (pI) values were used as given by the supplier of the IPG strips. The gels were destained with water and scanned in a Personal Densitometer (Molecular Dynamics, Sunnyvale, CA, USA). The images were processed using PhotoShop (Adobe) and PowerPoint (Microsoft) software.

Matrix-assisted laser desorption ionization mass spectroscopy (MALDI-MS)

MALDI-MS analysis was performed as described with minor modifications (Fountoulakis and Langen, 1997). Briefly, spots were excised, destained with 50% acetonitrile in 0.1 M ammonium bicarbonate and dried in a speedvac evaporator. The dried gel pieces were reswollen with 3 μl of 3 mM Tris-HCl, pH 8.8, containing 50 ng trypsin (Promega, Madison, WI, USA). After 15 minutes, 3 μl of water were added and left at room temperature for 12 hours. Two microliters of 30% acetonitrile, containing 0.1% trifluoroacetic acid were added, the content was vortexed, centrifuged for 3 minutes and sonicated for 5 minutes. One microliter was applied onto the dried matrix spot. The matrix consisted of 15 mg nitrocellulose (Bio-Rad) and 20 mg α-cyano 4 hydroxycinnamic acid (Sigma) dissolved in 1 ml acetone: isopropanol (1 : 1, v/v). A 0.5 μl portion of matrix solution was applied on the sample target. Specimen were analyzed in a time-of-flight Voyager Elite mass spectrometer (PerSeptive Biosystems, Cambridge, MA, USA) equipped with a reflectron. An accelerating voltage of 20 kV was used. Calibration was against total spots and background. The peptide masses were matched with the theoretical peptide masses of all proteins from all species of the SWISS-PROT database. For protein search, monoisotopic masses were used and a mass tolerance of 0.0075% was allowed.

Quantification of Prx proteins

In partial 2D-gel images including each subtype of Prx proteins and the neighbouring proteins, the percentage of the volume of the spot representing each subtype of Prx proteins was quantified as compared to the total proteins present using the ImageMaster 2D Elite software (Amersham Pharmacia Biotechnology). The previously constructed 2-D map of human brain proteins was referred to identify Prx subtypes (Langen et al., 1999).

Statistics

Results were expressed as means ± standard deviation (SD). Between-group differences were investigated by non-parametric Mann-Whitney U test. The significance was set at $P < 0.05$. All statistical analyses were performed by GraphPad Instat2 software, version 2.05.

Results

Identification of human brain proteins

Identification of human brain proteins based upon molecular weight, pI value and peptide matches performed by 2-D gel electrophoresis and MALDI-MS analysis was done in previous work (Langen et al., 1999).

Increased protein levels of Prx-I and Prx-II

Prx-I was significantly increased in AD (temporal cortex, occipital cortex and thalamus) and DS (thalamus) (Table 1). Prx-II was significantly increased in brain regions from patients with AD (thalamus) and DS (cerebellum, temporal cortex, thalamus and caudate nucleus) (Table 2). Although there is no statistical significance in other brain regions ($P > 0.05$), both Prx-I and Prx-II showed increased protein levels as compared to controls. The 2-D gel protein

Table 1. Prx-I protein levels in brain regions of patients with AD, DS and controls

	Control	AD	DS
Cerebellum	3.94 ± 2.25 (n = 9)	5.69 ± 3.21 (n = 9)	6.23 ± 2.69 (n = 6)
Frontal cortex	5.04 ± 1.83 (n = 9)	7.94 ± 3.60 (n = 6)	7.02 ± 3.65 (n = 4)
Temporal cortex	3.93 ± 1.80 (n = 14)	6.08 ± 2.15* (n = 10)	6.89 ± 3.94 (n = 8)
Occipital cortex	4.76 ± 1.97 (n = 10)	7.34 ± 1.88** (n = 10)	6.10 ± 1.64 (n = 8)
Parietal cortex	5.19 ± 1.67 (n = 6)	5.41 ± 3.64 (n = 3)	6.56 ± 1.16 (n = 5)
Thalamus	4.83 ± 2.25 (n = 10)	8.65 ± 4.46* (n = 8)	8.63 ± 2.20** (n = 7)
Caudate Nucleus	6.84 ± 2.66 (n = 5)	7.67 ± 2.37 (n = 7)	8.54 ± 6.23 (n = 7)

The proteins from seven brain regions of patients with DS and AD, and the control group were separated on 2-D gels and visualized following stain with colloidal Coomassie blue. In partial gel images, including Prx-I and the neighboring proteins, the intensities of the spots representing Prx-I were quantified, compared to the total proteins present. The quantification was performed using the ImageMaster 2D software. In each donor, the percentage of the volume of the spot representing Prx-I was determined. The statistical analyses were performed by Graphpad Instat2 software. Significant differences are indicated as follow: **$P < 0.01$; *$P < 0.05$

Table 2. Prx-II protein levels in brain regions of patients with AD, DS and controls

	Control	AD	DS
Cerebellum	19.33 ± 7.98 (n = 10)	23.20 ± 6.12 (n = 10)	29.31 ± 8.46* (n = 6)
Frontal cortex	20.43 ± 3.83 (n = 9)	17.76 ± 3.01 (n = 7)	20.82 ± 2.84 (n = 7)
Temporal cortex	14.53 ± 3.94 (n = 14)	17.09 ± 4.08 (n = 11)	20.72 ± 2.68** (n = 8)
Occipital cortex	17.57 ± 6.14 (n = 13)	17.98 ± 4.51 (n = 11)	19.21 ± 5.28 (n = 7)
Parietal cortex	20.26 ± 2.46 (n = 7)	19.62 ± 8.24 (n = 6)	20.00 ± 5.24 (n = 7)
Thalamus	11.11 ± 3.46 (n = 9)	18.75 ± 6.06* (n = 7)	18.58 ± 5.49** (n = 7)
Caudate Nucleus	11.81 ± 3.83 (n = 9)	12.08 ± 2.33 (n = 8)	18.67 ± 4.58 (n = 4)

The gels were prepared and the intensities of the spots representing Prx-II were quantified and analysed as described under legend to TABLE 1. The statistical analyses were performed by Graphpad Instat2 software. Significant differences are indicated as follow: ** $P < 0.01$; * $P < 0.05$

spots corresponding to Prx-I and Prx-II with the significant difference in AD or DS were presented in Fig. 1 or Fig. 2.

Decreased protein levels of Prx-III

As shown in Table 3, the protein levels of Prx-III located in mitochondria were significantly decreased in AD (occipital cortex and thalamus) and DS (frontal cortex). The protein levels of Prx-III were also decreased in other regions but without statistical significance.

Discussion

Oxidative stress induced by reactive oxygen species (ROS) such as superoxide, hydrogen peroxide and the highly reactive hydroxyl radical can lead to cell damage via apoptosis (or programmed cell death) that has been implicated in several neurodegenerative diseases such as AD. Superoxide and hydrogen peroxide are themselves poorly reactive and unable to directly damage DNA, lipids and proteins. However, excessive generation of them invariably accompanies molecular damage (Gutteridge, 1994). Several studies provided evidence for a role of hydrogen peroxide as a mediator of apoptosis in various cells (Pierce et al., 1991; Kanno et al., 1999; Chen et al., 2000; McBride et al., 1999). Hydrogen peroxide can also induce neuronal cell death in brain. The ability of hydrogen peroxide to cause apoptotic cell death was observed in cultured rat forebrain neurons (Hoyt et al., 1997). Exposure to low concentration of hydrogen peroxide induced cell death via apoptosis in cultured rat cortical neurons (Whittemore et al., 1994, 1995). Transcriptional regulation and intracellular signaling pathway could modulate hydrogen peroxide-induced cell death. Hydrogen peroxide-induced onset and execution of apoptotic cell death in mature rat brain oligodendrocytes in culture was accompanied by the induction and nuclear translocation of transcription

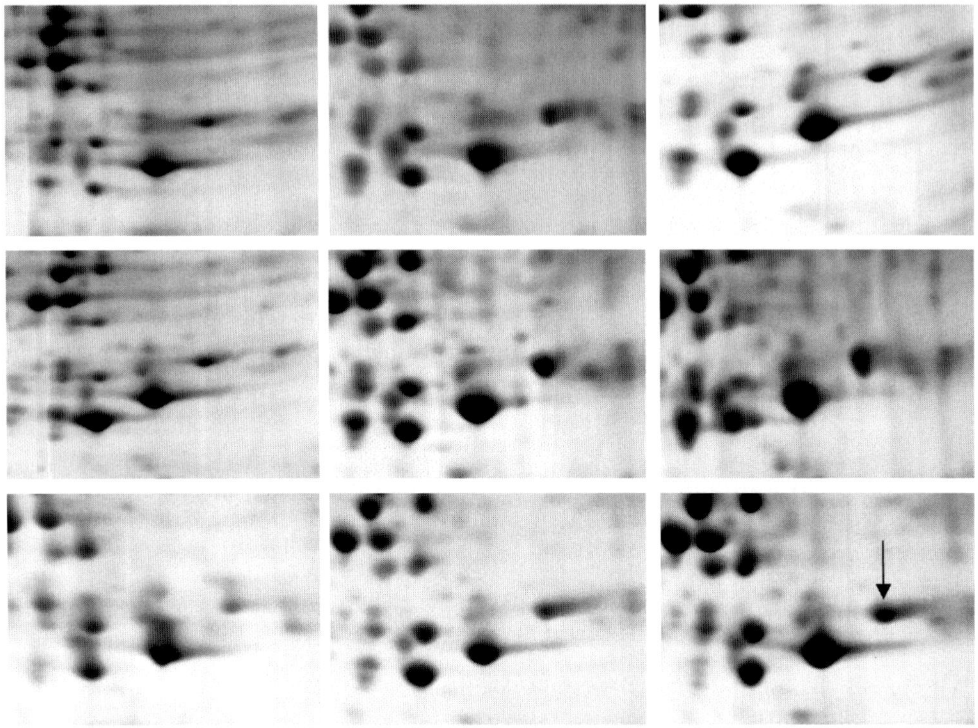

Fig. 1. Two-dimensional (2-D) gel images of Prx-I in brain regions with significant changes. The human brain proteins were extracted and separated on an immobilized pH 3–10 nonlinear gradient strip, followed by a 9–16% linear gradient polyacrylamide gel, as stated under experimental procedures. The gel was stained with Coomassie blue and the spots were analyzed by MALDI-MS. The partial 2-D gel images are shown and an arrow indicates the protein spot corresponding to Prx-I

Table 3. Prx-III protein levels in brain regions of patients with AD, DS and controls

	Control	AD	DS
Cerebellum	2.70 ± 1.01 (n = 13)	2.17 ± 1.06 (n = 11)	2.06 ± 0.76 (n = 6)
Frontal cortex	2.86 ± 1.14 (n = 9)	2.30 ± 0.56 (n = 7)	1.57 ± 0.77* (n = 7)
Temporal cortex	2.48 ± 1.00 (n = 14)	2.09 ± 0.92 (n = 11)	2.37 ± 0.73 (n = 9)
Occipital cortex	2.81 ± 1.05 (n = 13)	2.05 ± 0.69* (n = 11)	2.24 ± 0.52 (n = 8)
Parietal cortex	2.91 ± 0.47 (n = 7)	2.91 ± 1.31 (n = 7)	2.25 ± 1.31 (n = 8)
Thalamus	2.57 ± 0.91 (n = 10)	1.32 ± 0.38** (n = 8)	2.39 ± 0.90 (n = 7)
Caudate Nucleus	2.50 ± 0.79 (n = 8)	2.38 ± 0.89 (n = 8)	2.01 ± 0.81 (n = 8)

The gels were prepared and the intensities of the spots representing Prx-III were quantified and analysed as described under legend to TABLE 1. The statistical analyses were performed by Graphpad Instat2 software. Significant differences are indicated as follow: ** $P < 0.01$; * $P < 0.05$

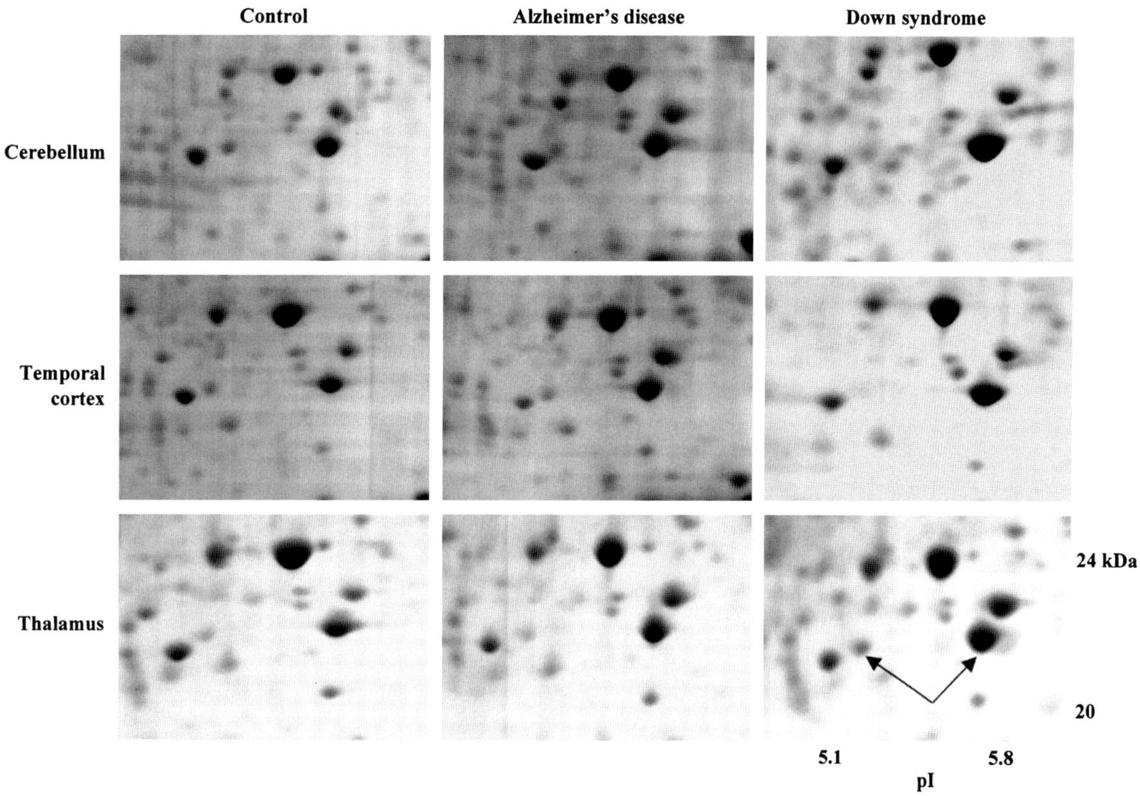

Fig. 2. Two-dimensional (2-D) gel images of Prx-II in brain regions with significant changes. The gels were prepared and the protein spots were analyzed as described under legend to Fig. 1. The partial 2-D gel images are shown and an arrow indicates the protein spot corresponding to Prx-II

factors AP-1 and NF-κB, both of which have been discussed as regulators of cell death and survival (Vollgraf et al., 1999). Oxidative stress by hydrogen peroxide in oligodendrocytes could lead to the onset of apoptosis, involving the transcriptional activation of the immediate-early genes, c-fos and c-jun (Richter-Landsberg and Vollgraf, 1998).

Several antioxidant enzymes including cytosolic copper-zinc SOD (Cu/ZnSOD or SOD1), mitochondrial manganese SOD (MnSOD or SOD2), GPx and CAT can control intracellular levels of ROS (Adler et al., 1999). A series of studies reported that antioxidant enzymes could be critically involved in neurodegenerative disease. A significant increase in erythrocyte SOD1 and CAT activities of patients with dementia of the Alzheimer type (DAT) was observed (Perrin et al., 1990). Mean erythrocyte SOD and GPx, but not CAT activities were significantly higher in the DS group than in controls (Brugge et al., 1999; Gerli et al., 1990). SOD1 and GPx activities of red blood cells were significantly elevated in DS patients without AD (Percy et al., 1990). The degeneration of DS neurons was prevented by treatment with free radical scavengers or CAT. Furthermore, DS neurons exhibited a three- to fourfold increase in intracellular ROS and elevated levels of lipid peroxidation preced-

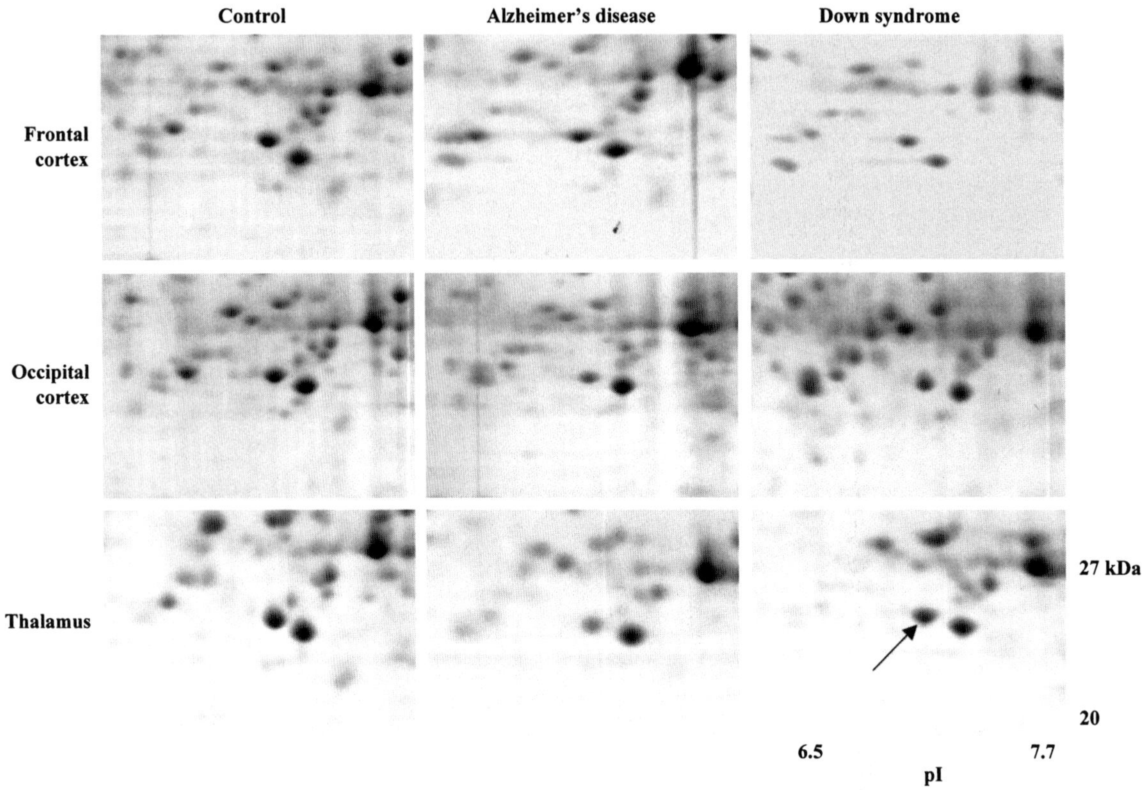

Fig. 3. Two-dimensional (2-D) gel images of Prx-III in brain regions with significant changes. The gels were prepared and the protein spots were analyzed as described under legend to Fig. 1. The partial 2-D gel images are shown and an arrow indicates the protein spot corresponding to Prx-III

ing neuronal death (Busciglio and Yankner, 1995). Although antioxidant enzymes can detoxify free radicals, these enzymes may not be able to cope with increasing oxidative stress in brains.

Mammalian Prx proteins play an important role in the antioxidant defense systems to eliminate hydrogen peroxide generated during cellular metabolism. Three major subtypes of Prx proteins have been identified in humans. The genes encoding Prx-I and Prx-II have been demonstrated to be identical to the genes encoding natural killer cell enhancing factor A and B (NKEF-A and -B). NKEF are cytosolic proteins found in human red blood cells and thought to be a major antioxidant, protecting red blood cells from oxidant injury (Shau et al., 1993; Shau and Kim, 1994). Several studies generated evidence for the antioxidant ability of NKEF proteins. Recombinant proteins encoded by the two human NKEF genes showed antioxidant function in the protection of protein and DNA from oxidant damage (Sauri et al., 1995). Human endothelial cell lines transfected with the coding region of NKEF-B cDNA were shown to be resistant to hydrogen peroxide (Shar et al., 1997). In addition, NKEF-B transfected cells displayed significantly lower levels of ROS and lipid peroxidation (Sarafian et al., 1997). NKEF-A is identical to proliferation-associated protein (PAG) and NKEF-B is identical to thiospecific antioxidant protein (TSA) (Kang et al., 1998b). Prx-III is the only protein located in mitochondria. Prx-III is also referred to as antioxidant protein 1 (AOP-1) or MER5 protein that was found in murine erythroleukemia cells (Tsuji et al., 1995; Yamamoto et al., 1989). MER5 is 91% similar to SP-22 protein that is a mitochondrial antioxidant protein in bovine adrenal cortex (Watabe et al., 1994, 1997). The induction of mitochondrial SP-22 mRNA and protein were enhanced by various oxidative stresses, including mitochondrial respiratory chain inhibitors, in vascular endothelial cells (Araki et al., 1999). SP-22 showed to have the ability to scavenge ROS, thus protecting radical-sensitive proteins from oxidation (Watabe et al., 1995).

Although mammalian Prx proteins are highly abundant in cells, different subtypes of Prx proteins have distinct tissue distribution profiles and are found in different subcellular locations. Differential expression of Prx-I and Prx-II was studied in cultured cell types from embryonic rat brain (Sarafian et al., 1998). Regional and cell-type specific expression of Prx-I and Prx-II was also examined in the cerebral cortex, cerebellum, basal ganglia, substantia nigra, and spinal cord of human brain (Sarafian et al., 1999). Prx-I was expressed primarily in astrocytes, while Prx-II was expressed exclusively in neurons. Restricted expression of Prx-I and Prx-II may contribute to the selective vulnerability of brain cell types to specific kinds of oxidative stress.

Several recent studies showed the anti-apoptotic activities of Prx proteins. Prx-I, could inhibit release of cytochrome c from mitochondria to cytosol and lipid peroxidation, and prevent hydrogen peroxide accumulation in cells, proposed to play a role as an endogenous regulator of apoptosis (Zang et al., 1997). The expression of mouse Prx-II in PC12 cells prolonged their survival in the absence of nerve growth factor (NGF) and serum, indicating that Prx-II could promote neuronal cell survival (Ichimiya et al., 1997). Additionally, Prx-II was shown to prevent mitochondrial permeability transition (MPT) that could be caused by ROS produced by the mitochondrial respiratory chain (Kowaltowski et al., 1998). Mouse Prx-V also was shown to prevent ROS production stimulated by expression of p53 (Zhou et al., 2000).

Our data could be interpreted in the context with that given above: Increased Prx-I in both disorders could reflect oxidative stress representing the antioxidant response, but also glial proliferation. We have shown in a previous publication that glial fibrillary acidic protein (GFAP), a marker for astrocytes, was remarkably increased in brain of a comparable cohort of AD and DS patients (Greber et al., 1999). As Prx-I is primarily expressed in astrocytes, glial proliferation may therefore account for increased Prx-I protein levels. Alternatively, increased Prx-I can be seen as an activated anti-apoptotic principle in AD and DS brains, where apoptosis is a hallmark of neuropathology. Increased Prx-II, however, exclusively expressed in neurons, clearly indicates significant upregulation of Prx-II protein despite of significant neuronal loss in brain regions studied (Cairns, 1999; de la Monte, 1999; Nagy, 1999; Sawa, 1999). Upregulation of Prx-II may not only reflect the antioxidant response to oxidative stress, but also mobilization of a direct neuroprotective principle (Ichimiya et al., 1997). Decreased protein levels of Prx-III could be either due to mitochondrial damage or simply to loss of brain cells. We observed mitochondrial impairment by the observation of decreased complex I and V proteins of the respiratory chain in a comparable cohort in comparable brain regions (Kim et al., submitted for publication). Concluding, we propose that Prx-protein system was activated in AD and DS brain most probably by increased hydrogen peroxide due to overexpression of SOD1, which has been well-documented (Fullerton et al., 1998). We furthermore propose the biological meaning of increased Prx-I and Prx-II as antioxidant response and anti-apoptotic principle, whereas the decreased Prx-III is interpreted as reflecting mitochondrial damage, which in turn is widely accepted in AD and DS. On the other hand, as Prx proteins are known to detoxify hydrogen peroxide, we may draw the conclusion that there is indeed excess production of hydrogen peroxide in both neurodegenerative disorders, either by overexpressed SOD1 in DS or by mitochondrial damage in AD.

Acknowledgments

The authors at the University of Vienna are highly indebted to the Red Bull Company, Salzburg, Austria, for generous financial support.

References

Adler V, Yin Z, Tew KD, Ronai Z (1999) Role of redox potential and reactive oxygen species in stress signaling. Oncogene 18: 222: 6104–6111

Araki M, Nanri H, Ejima K, Murasato Y, Fujiwara T, Nakashima Y, Ikeda M (1999) Antioxidant function of the mitochondrial protein SP-22 in the cardiovascular system. J Biol Chem 274: 2271–2278

Aylward EH, Habbak R, Warren AC, Pulsifer MB, Barta PE, Jerram M, Pearlson GD (1997) Cerebellar volume in adults with Down syndrome. Arch Neurol 54: 209–212

Bradford MM (1976) A rapid and sensitive method for the quantitation of microgram quantities of protein utilizing the principle of protein-dye binding. Anal Biochem 72: 248–254

Brugge K, Nichols S, Saitoh T, Trauner D (1999) Correlations of glutathione peroxidase activity with memory impairment in adults with Down syndromes. Biol Psychiatry 46: 1682–1689

Busciglio J, Yankner BA (1995) Apoptosis and increased generation of reactive oxygen species in Down's syndrome neurons in vitro. Nature 378: 776–779

Cairns NJ (1999) Neuropathology. J Neural Transm [Suppl] 57: 61–74

Chae HZ, Chung SJ, Rhee SG (1994a) Thioredoxin-dependent peroxide reductase from yeast. J Biol Chem 269: 27670–27678

Chae HZ, Robison K, Poole LB, Church G, Storz G, Rhee SG (1994b) Cloning and sequencing of thio-specific antioxidant from mammalian brain: alkyl hydroperoxide reductase and thio-specific antioxidant define a large family of antioxidant enzymes. Proc Natl Acad Sci USA 91: 7017–7021

Chae HZ, Kang SW, Rhee SG (1999) Isoforms of mammalian peroxiredoxin that reduce peroxides in presence of thioredoxin. Methods Enzymol 300: 219–226

Chen QM, Tu VC, Wu Y, Bahl JJ (2000) Hydrogen peroxide dose dependent induction of cell death or hypertrophy in cardiomyocytes. Arch Biochem Biophys 373: 242–248

de la Monte SM (1999) Molecular abnormalities of the brain in Down syndrome: relevance to Alzheimer's neurodegeneration. J Neural Transm [Suppl] 57: 1–19
Fountoulakis M, Langen H (1997) Identification of proteins by matrix-assisted laser desorption ionization-mass spectrometry following in-gel digestion in low-salt, nonvolatile buffer and simplified peptide recovery. Anal Biochem 250: 153–156
Fullerton HJ, Ditelberg JS, Chem SF, Sarco DP, Chan PH, Epstein CJ, Ferriero DM (1998) Copper/zinc superoxide dismutase transgenic brain accumulates hydrogen peroxide after perinatal hypoxia ischemia. Ann Neurol 44: 357–364
Gerli G, Zenoni L, Locatelli GF, Mongiat R, Piattoni F, Orsini GB, Montagnani A, Gueli MR, Gualandri V (1990) Erythrocyte antioxidant system in Down syndrome. Am J Med Genet [Suppl] 7: 272–273
Greber S, Lubec G, Cairns N, Fountoulakis M (1999) Decreased levels of synaptosomal associated protein 25 in the brain of patients with Down syndrome and Alzheimer's disease. Electrophoresis 20: 928–934
Gutteridge JM (1994) Biological origin of free radicals and mechanisms of antioxidant protection. Chem Biol Interact 91: 133–140
Halliwell B (1999) Antioxidant defence mechanisms: from the beginning to the end (of the beginning). Free Rad Res 31: 261–272
Hayes JD, McLellan LI (1999) Glutathione and glutathione-dependent enzymes represent a co-ordinately regulated defence against oxidative stress. Free Rad Res 31: 273–300
Hoyt KR, Gallagher AJ, Hastings TG, Reynolds IJ (1997) Characterization of hydrogen peroxide toxicity in cultured rat forebrain neurons. Neurochem Res 22: 333–340
Ichimiya S, Davis JG, O'Rourke DM, Katsumata M, Greene MI (1997) Murine thioredoxin peroxidase delays neuronal apoptosis and is expressed in areas of the brain most susceptible to hypoxic and ischemic injury. DNA Cell Biol 16: 311–321
Jovanovic SV, Clements D, MacLeod K (1998) Biomarkers of oxidative stress are significantly elevated in Down syndrome. Free Radic Biol Med 25: 1044–1048
Kang SW, Baines I, Rhee SG (1998a) Characterization of a mammalian peroxiredoxin that contains one conserved cysteine. J Biol Chem 273: 6303–6311
Kang SW, Chae HZ, Seo MS, Kim K, Baines IC, Rhee SG (1998b) Mammalian peroxiredoxin isoforms can reduce hydrogen peroxide generated in response to growth factors and tumor necrosis factor-alpha. J Biol Chem 273: 6297–6302
Kanno S, Ishikawa M, Takayanagi M, Takayanagi Y, Sasaki M (1999) Exposure to hydrogen peroxide induces cell death via apoptosis in primary cultured mouse hepatocytes. Biol Pharm Bull 22: 1296–1300
Kim SH, Vlkolinsky R, Cairns N, Fountoulakis M, Lubec G (2001) The reduction of NADH: ubiquinone oxidoreductase 24- and 75-kDa subunits in brains of patients with Down syndrome and Alzheimer's disease. Life Sci 68: 2741–2750
Kim SH, Vlkolinsky R, Cairns N, Lubec G (2000) Decreased protein levels of complex III core protein 1 and complex V beta chain. Cell Mol Life Sci 57: 1810–1816
Kowaltowski AJ, Netto LES, Vercesi AE (1998) The thiol specific antioxidant enzyme prevents mitochondrial permeability transition. J Biol Chem 273: 12766–12769
Langen H, Roder D, Juranville JF, Fountoulakis M (1997) Effect of protein application mode and acrylamide concentration on the resolution of protein spots separated by two-dimensional gel electrophoresis. Electrophoresis 18: 2085–2090
Langen H, Berndt P, Roder D, Cairns N, Lubec G, Fountoulakis M (1999) Two-dimensional map of human brain proteins. Electrophoresis 20: 907–916
McBride AG, Borutaite V, Brown GC (1999) Superoxide dismutase and hydrogen peroxide cause rapid nitric oxide breakdown, peroxynitrite production and subsequent cell death. Biochim Biophys Acta 1454: 275–288
Mirra SS, Heyman A, McKeel D, Sumi SM, Crain BJ, Brownlee LM, Vogel FS, Hughes JP, van Belle G, Berg L (1991) The Consortium to Establish a Registry for Alzheimer's Disease (CERAD), part II. Standardization of the neuropathologic assessment of Alzheimer's disease. Neurology 41: 479–486

Nagy Zs (1999) Mechanisms of neuronal death in Down's syndrome. J Neural Transm [Suppl] 57: 233–245

Okado-Matsumoto A, Matsumoto A, Fujii J, Taniguchi N (2000) Peroxiredoxin IV is a secretable protein with heparin-binding properties under reduced conditions. J Biochem 127: 493–501

Pappolla MA, Omar RA, Kim KS, Robakis NK (1992) Immunohistochemical evidence of oxidative [corrected] stress in Alzheimer's disease. Am J Pathol 140: 621–628

Percy ME, Dalton AJ, Markovic VD, McLachlan DR, Hummel JT, Rusk AC, Andrews DF (1990) Red cell superoxide dismutase, glutathione peroxidase and catalase in Down syndrome patients with and without manifestations of Alzheimer's disease. Am J Med Genet 35: 459–467

Perrin R, Briancon S, Jeandel C, Artur Y, Minn A, Penin F, Siest G (1990) Blood activity of Cu/Zn superoxide dismutase, glutathione peroxidase and catalase in Alzheimer's disease: a case-control study. Gerontology 36: 306–313

Pierce GB, Parchment RE, Lewellyn AL (1991) Hydrogen peroxide as a mediator of programmed cell death in the blastocyst. Differentiation 46: 181–186

Richter-Landsberg C, Vollgraf U (1998) Mode of cell injury and death after hydrogen peroxide exposure in cultured oligodendroglia cells. Exp Cell Res 244: 218–229

Sarafian TA, Rajper N, Grigorian B, Kim A, Shau H (1997) Cellular antioxidant properties of human natural killer enhancing factor B. Free Radic Res 26: 281–289

Sarafian TA, Huang C, Kim A, de Velli J, Shau H (1998) Expression of the antioxidant gene NKEF in the central nervous system. Mol Chem Neuropathol 34: 39–51

Sarafian TA, Verity MA, Vinters HV, Shih CC, Shi L, Ji XD, Dong L, Shau H (1999) Differential expression of peroxiredoxin subtypes in human brain cell types. J Neurosci Res 56: 206–212

Sauri H, Butterfield L, Kim A, Shau H (1995) Antioxidant function of recombinant human natural killer enhancing factor. Biochem Biophys Res Commun 208: 964–969

Sawa A (1999) Neuronal cell death in Down's syndrome. J Neural Transm [Suppl] 57: 87–97

Shau H, Kim A (1994) Identification of natural killer enhancing factor as a major antioxidant in human red blood cells. Biochem Biophys Res Commun 199: 83–88

Shau H, Gupta RK, Golub SH (1993) Identification of a natural killer enhancing factor (NKEF) from human erythroid cells. Cell Immunol 147: 1–11

Shau H, Kim AT, Hedrick CC, Lusis AJ, Tompkins C, Finney R, Leung DW, Paglia DE (1997) Endogenous natural killer enhancing factor-B increases cellular resistance to oxidative stresses. Free Radic Biol Med 22: 497–507

Simonian NA, Coyle JT (1996) Oxidative stress in neurodegenerative diseases. Annu Rev Pharmacol Toxicol 36: 83–106

Tierney MC, Fisher RH, Lewis AJ, Zorzitto ML, Snow WG, Reid DW, Nieuwstraten P (1988) The NINCDS-ADRDA Work Group criteria for the clinical diagnosis of probable Alzheimer's disease: a clinicopathologic study of 57 cases. Neurology 38: 359–364

Tsuji K, Copeland NG, Jenkins NA, Obinata N (1995) Mammalian antioxidant protein complements alkylhydroperoxide reductase (ahpC) mutation in Escherichia coli. Biochem J 307: 377–381

Vollgraf U, Wegner M, Richter-Landsberg C (1999) Activation of AP-1 and nuclear factor-kappaB transcription factors is involved in hydrogen peroxide-induced apoptotic cell death of oligodendrocytes. J Neurochem 73: 2501–2509

Watabe S, Kohno H, Kouyama H, Hiroi T, Yago N, Nakazawa T (1994) Purification and characterization of a substrate protein for mitochondrial ATP-dependent protease in bovine adrenal cortex. J Biochem (Tokyo) 115: 648–654

Watabe S, Hasegawa H, Takimoto K, Yamamoto Y, Takahashi SY (1995) Possible function of SP-22, a substrate of mitochondrial ATP dependent protease, as a radical scavenger. Biochem Biophys Res Commun 213: 1010–1016

Watabe S, Hiroi T, Yamamoto Y, Fujioka K, Hasegawa H, Yago N, Takahashi SY (1997) SP-22 is a thioredoxin-dependent peroxide reductase in mitochondria. Eur J Biochem 249: 52–60

Whittemore ER, Loo DT, Cotman CW (1994) Exposure to hydrogen peroxide induces cell death via apoptosis in cultured rat cortical neurons. Neuroreport 5: 1485–1488

Whittemore ER, Loo DT, Watt JA, Cotman CW (1995) A detailed analysis of hydrogen peroxide-induced cell death in primary neuronal culture. Neuroscience 67: 921–932

Yamamoto Y, Matsui Y, Natori S, Obinata M (1989) Cloning of a housekeeping-type gene (MER5) preferentially expressed in murine erythroleukemia cells. Gene 80: 337–343

Zang P, Liu B, Kang SW, Seo MS, Rhee SG, Obeid LM (1997) Thioredoxin peroxidase is a novel inhibitor of apoptosis with a mechanism distinct from that of Bcl-2. J Biol Chem 272: 30615–30618

Zhou Y, Kok KH, Chun AC, Wong CM, Wu HW, Lin MC, Fung PC, Kung H, Jin DY (2000) Mouse peroxiredoxin V is a thioredoxin peroxidase that inhibits p53-induced apoptosis. Biochem Biophys Res Commun 268: 921–927

Authors' address: Prof. Dr. G. Lubec, Department of Pediatrics, University of Vienna, Währinger Gürtel 18, A-1090 Vienna, Austria, e-mail: gert.lubec@akh-wien.ac.at

Effects of a single transdermal nicotine dose on cognitive performance in adults with Down syndrome

G. Bernert[1], **M. Sustrova**[2], **E. Sovcikova**[2], **R. Seidl**[1], and **G. Lubec**[1]

[1] Department of Pediatrics, University of Vienna, Vienna, Austria
[2] The Institute of Preventive and Clinical Medicine, Bratislava, Slovakia

Summary. Subjects with Down syndrome exhibit various types of cognitive impairment. Neuropathological and neurochemical studies revealed similarities between Down syndrome and Alzheimer's disease, cholinergic deficits being the most consistent findings. To explore the potential for cognitive enhancement utilizing nicotinic stimulation, 8 patients with Down syndrome (aged 18.5–31 years) received placebo and a single dose of transdermal nicotine (5 mg patch) over 2 h in a single-blind, within-subjects repeated measures design. Auditory event-related potentials (ERPs) and neuropsychological tests, comprising digit symbol performance subtest from WAIS-R and the Frankfurt Attention Inventory (FAIR) were performed. Effects of nicotine administration in Down syndrome individuals were a decrease of ERP-P3 latency in 7 of 8 subjects (electrode position Cz: 386.9 ± 24.0 ms vs. 363.1 ± 26.9.2 ms, placebo vs. nicotine, respectively; P = 0.058) and an increase of ERP-P3 amplitude in 6 of 8 subjects (electrode position Cz: 17.4 ± 5.5 vs. 18.0 ± 4.5 μV, placebo vs. nicotine respectively; P = 0.725). Neuropsychological tests exhibited improvements in digit symbol performance subtest in 4 of 8 subjects and 7 of 8 subjects in the Frankfurt Attention Inventory. These results suggest that stimulating central nicotinic receptors might have an acute cognitive benefit in young adult Down syndrome subjects.

Introduction

Down syndrome (DS) is a genetic disease with developmental brain abnormalities (Wisniewski and Kida, 1994) resulting in early mental retardation and precocious, age-dependent Alzheimer-type neurodegeneration (Wisniewski et al., 1985; Teller et al., 1996). A progressive decline in developmental quotient (DQ) or intelligence quotient (IQ) starts within the first year, followed by a plateau in mental age after the first decade of life (Melyn and White, 1973). Presenile dementia which is pathologically identical with Alzheimer disease occurs from 40 years of age in one third of DS patients (Wisniewski et al., 1985), but also nondemented older patients appear to manifest a selective pattern of neuropsychological reductions relative to younger adults with Down syndrome. They exhibit selective impairments on tests of recent memory (auditory verbal memory) and visuospatial functions with preservation of language (Haxby, 1989).

Neurochemical studies in DS revealed cholinergic deficits similar to those in Alzheimer's disease (AD) (Yates et al., 1980; Kish et al., 1989). Attempts to treat AD are based upon well-documented cholinergic deficit coupled with supporting evidence linking memory and the cholinergic system (Lawrence and Sahakian, 1998). These therapeutic attempts consist of 1) precursor loading to increase production of acetylcholine (ACh) at the presynaptic level, 2) receptor agonist strategies aimed at postsynaptic level and 3) attempts to increase the duration of action of ACh by preventing its breakdown by acetylcholine-esterase (AChE). At present time treatment options are limited primarily to the use of acetylcholinesterase inhibitors. However, evidence is accumulating that direct stimulation of nicotinic acetylcholine receptors (nAChR) may represent another therapeutic target (Newhouse et al., 1997). The nAChR is a protein complex that contains binding sites for ACh (the neurotransmitter) as well as binding sites for other types of drugs, and activation or inhibition of these sites can regulate sodium and calcium flux through these channels (Brioni et al., 1997). In addition to direct stimulation of nicotinic receptors (nAChR), nicotine may provide cascading effects via stimulation of release of a variety of transmitters involved in cognitive function, including ACh, dopamine, norepinephrine, serotonin, GABA, glutamate (McGehee et al., 1995; Brioni et al., 1997). Thus, cholinergic channel modulators (like nicotine) can produce their behavioral and cognitive effects both directly via activation if nAChRs and indirectly via transmitter systems other than ACh (Brioni et al., 1997), by enhancing synaptic transmission (Gray et al., 1996). In AD, nicotinic cholinergic receptors are substantially diminished in some parts of the brain (cortex and parahippocampal gyrus), as measured at postmortem examination (Perry et al., 1995) and in-vivo by PET studies (Nordberg et al., 1995), where loss of nicotinic receptor binding has been linked to cognitive performance. A small number of nicotine studies have been performed in AD (Jones et al., 1992; Wilson et al., 1995; Snaedal et al., 1996), including trials with transdermal delivery of nicotine using a patch preparation (Wilson et al., 1995; Snaedal et al., 1996; Parks et al., 1996; White and Levin, 1999). Nicotine administration resulted in a highly significant improvement in perception, accuracy and speed of responding in tasks of sustained attention, consistent with a role for the cholinergic system in attention. An improvement in verbal memory functions was noticed in one study (Parks et al., 1996), but must studies failed to demonstrate memory improvement (Jones et al., 1992; Wilson et al., 1995; Snaedal et al., 1996; White and Levin, 1999). Studies with AChE-inhibitors, such as tacrine and donezepil, also found improved attention but no consistent effect on memory (Lawrence and Sahakian, 1998). Reports of clinical improvements in Lewy body dementia, Parkinson's disease, attention deficit disorder, schizophrenia (for review see Lawrence and Sahakian, 1998) and Tourette's syndrome (Dursun et al., 1994) following nicotine administration made research into therapeutic nicotinic augmentation an area of increasingly active investigation. Although virtually all DS patient beyond 35 years of age develop neuropathological lesions and "cholinergic" deficit similar to AD, these two disorders cannot be considered as clinically identical as only a subgroub of DS individuals shows a typical progressive dementia of AD type. Down syndrome seems to include a genetically determined program of AD-like pathology and a detailed understanding might be crucial for understanding of AD pathogenesis, but mental retardation based on early neurodevelopmental abnormalities is present before AD-like neuropathology appears. Consequently, therapeutic attempts should neither be withheld from DS individuals, nor can findings of studies dealing with sporadic AD patients directly be extrapolated to DS.

To our knowledge, there is only one published study of cholinergic therapy in DS (Kishnani et al., 1999), reporting improvements in communication, expressive language, attention and mood stability following AChE-inhibition by Donezepil, but nothing is known about nicotine-agonistic effects.

We therefore decided to perform a study on the acute effects of nicotine administration via skin patches in young adult DS subjects with a neurophysiological and neuropsychological test battery.

Methods

Subjects

Eight patients (2 female and 6 male) with DS (karyotyped) aged between 18.5 and 31 years (mean age 22.7 ± 5.2 years) were studied. All subjects were recruited from family and non had ever lived in an institution. Medical histories were obtained and subjects were undergoing physical examination and laboratory tests to rule out unstable medical or active psychiatric illnesses. All subjects were nonsmokers and were caffeine and alcohol free for at least 24 hours before the study days. Informed and written consent either from the subjects or the appropriate responsible persons and authorities was obtained before the procedure. The study protocol was approved by the local Ethics Committee and were in accordance with the Helsinki Declaration of 1975, as revised 1983.

Experimental design

ERP recordings and neuropsychological tests were assessed in a single-blind randomised, placebo-controlled, repeated measures design. Assignment of subjects to the drug conditions (placebo/nicotine) was randomised in order to minimize practice effects.

Procedure

The two test sessions were seperated by at least 24 hours for each subject. The order of the medication (placebo/nicotine) was randomized and balanced across subjects. On a separate day, prior to testing, subjects were allowed to perform a ERP practice run, which was essentially in DS subjects. On study days, 5 mg transdermal nicotine patch or a placebo patch was attached. After 2 hours the patch was removed and the tests (ERPs and psychological tests) were performed in a fixed order for all sessions. Undesired effects of nicotine such as agitation, nausea/vomiting, dizziness, headache, tachycardia, hypertension, diarrhea, local itching were monitored and patch was removed immediately at minimal sign of their appearance.

Event-related potential (ERP) recording

Subjects were seated in a reclining chair in a quiet room and were instructed to close their eyes in order to avoid eye blink artifacts. Auditory ERPs were recorded following a simple active oddball procedure using stimuli presented binaurally at a rate of 0.5/sec and at an intensity of 70 dB SPL. Frequent stimuli (80%) were 1,000 Hz tones, infrequent (20%) stimuli were 2,000 Hz tones. Tone duration was 100 ms (10 ms rise, 80 ms plateau, 10 ms fall). Stimuli were presented in a randomized sequence of frequent and infrequent stimuli until recording contained 30 artifact-free responses to infrequent stimuli. The subject was instructed to raise the index finger on each occurrence of the target stimuli while ignoring the lower tones (background). All subjects had to perform a training until they mastered the task reliably and required frequent repetitions of the instructions ("High Tone = Move Finger"). The active electrodes were placed at Fz (frontal), Cz (central), Pz (parietal) according to the 10–20 International System, referenced to linked mastoids (A1–A2), with a forehead ground. In addition one electrode was placed at the outer canthus of the left eye and the second on the forehead above the eye to monitor horizontal and vertical eye movements. The impedances of the electrodes were maintained below 5 kOhm. The filter bandpass was 0.1–50 Hz. The analysis time was 900 ms with 100 ms prestimulus time. Automatic artifact rejection was used for amplitudes exceeding $\pm 100\,\mu V$ (EEG and EOG). In every run, stimuli were presented until the recording system (Nihon-Kohden, MEB 4200G) had collected 20 infrequent trials. For comparison the P3 wave peak latencies and the baseline-to-peak amplitude of P3 in response to target-stimuli were evaluated. Mean P3 latencies/amplitudes were calculated across the midline electrodes (Fz, Cz, Pz).

Neuropsychological procedures

Neuropsychological procedures consisted of two tests of attention/reaction time: the digit symbol subtest from the performance part of HAWIE-R (Tewes, 1991) (german version of WAIS-R), and the Frankfurt Attention Inventory (FAIR) (Moosbrugger and Oehlschlägel, 1996). The digit-symbol subtest (HAWIE-R) measures psychomotor coordination and speed of attentional resources. The paper-pencil FAIR-test measures directed attention as ability to visually discriminate patterns with high similarity in a precise and rapid manner, the diagnostically relevant values L ("Leistung" — performance: amount of test-items) and K ("Kontinuität" — continuity: amount of continuously sustained attention) provide a good retest-reliabilty (Crohnbachs alpha between 0.85 and 0.91). In addition the amount of errors can be calculated. Effects of treatment on P3 amplitude/latency and performance measures were evaluated with a non-parametric, repeated-measures Wilcoxon matched-pair signed-rank. All analyses were run with the SPSS-PC 10.0 statistical software package. The level of significance was set at 0.05.

Results

Table 1 summarises performance of each individual in the different tests (ERP, HAWIE-R, FAIR).

Table 1. Individual performance of 8 subjects with Down syndrome after administration of placebo or nicotine patch

Age/sex years	P3 Latency ms		P3 amplitude μV		HAWIE-R raw value		FAIR-L total performance		FAIR-K continuity	
	Placebo	Nicotine	Placebo	Nicotine	Placebo	Nicotine	Placebo	Nicotine	Placebo	Nicotine
21/M	380	370 (+)	20	17 (−)	28	34 (+)	75	122 (+)	53	83 (+)
31/M	395	375 (+)	13.5	15 (+)	10	18 (+)	20	24 (+)	7	11 (+)
31/M	380	330 (+)	25	20 (−)	18	21 (+)	9	39 (+)	1.6	18 (+)
22/F	440	400 (+)	15	17 (+)	18	18 (=)	1	47 (+)	0.02	26.3 (+)
20/M	390	330 (+)	12	16 (+)	12	11 (−)	11	21 (+)	2.9	6.3 (+)
19/M	360	390 (−)	22	27 (+)	19	23 (+)	51	180 (+)	34	132 (+)
19/F	370	340 (+)	22	20 (−)	28	19 (−)	115	40 (−)	73	16 (−)
20/M	370	380 (+)	10	12 (+)	33	29 (−)	27	57 (+)	9	27 (+)

P3 auditory event related potentials; *HAWIE-R* digit symbol subtest; *FAIR* Frankfurt Attention Inventory; (+) improved performance after nicotine vs. placebo; (−): impaired performance after nicotine vs. placebo; (=) equal performance after nicotine and placebo

P3 latency

The impact of nicotine was a decrease of P3 latency in 7/8 subjects (e.g. mean ± SD values at Cz: placebo vs. nicotine, 386.9 ± 24.0 vs. 363.1 ± 26.9 ms [P = 0.058].

P3 amplitude

Following nicotine, P3 amplitudes increased in 6/8 subjects. In 2/8 patients the amplitude seems to decrease, but determination of the true amplitude by the baseline to P3-peak method was disturbed by artifacts.

Neuropsychological procedures

Performance in the digit-symbol subtest (HAWIE-R) increased in 4/8, decreased in 3/8 and was equal in 1/8 subjects following nicotine compared to placebo (mean value 21.3 ± 6.6 vs. 21.3 ± 8.8, placebo vs. nicotine). FAIR performance increased in 7/8 subjects after nicotine (mean ± SD, L: 38.6 ± 39.4 vs. 66.3 ± 55.8; K: 22.4 ± 27.6 vs. 39.9 ± 44.3; placebo vs. nicotine). With the exception of a light nausea after the test session in a 19 years old women, no side effects occurred.

Discussion

The most obvious effect in response to acute adminstration of nicotine was an improvement of P3 latency and performance in the Frankfurt Attention

Inventory. The small number of participants means that a possible beneficial effect in certain tests can be assumed only for these 8 patients with no generalisability. However, when looking at Table 1, each subject exhibited an improved performance in at least one of the tests. Performance in digit-symbol subtest of widely used intelligence test HAWIE-R (german version of WAIS-R) did not show a consistent trend.

The common auditory oddball task seems to be exceptionally suitable for testing attentional cognitive abilities in DS subjects for several reasons. Offering some training, this easy task is well performed by almost all DS individuals over a wide range of cognitive abilities. The source of stimulation (auditory stimuli via earphones) is accepted with some pleasure and they can easily be motivated to concentrate on the task. The positive event-related potential (ERP) which occur 300 ms after the stimulus presentation (P3/P300 component) reflects brain activity related to higher brain function, such as attention and memory and has been widely used to investigate cognitive mechanisms of information processing and information processing deficits in a variety of pathological groups (Polich, 1996). In dementia, P3 amplitude reductions and latency prolongations have consistently been reported (Polich, 1996; Frodl-Bauch et al., 1999). In DS, passive and active oddball investigations showed longer P3 latencies than in normal subjects (St. Clair and Blackwood, 1985; Lincoln et al., 1986; Vieregge et al., 1992; Diaz and Zurron, 1995). Besides significant differences at younger ages probably reflecting inherent cerebral dysfunction, a relationship between age and P3 latency was determined as two linear regression lines, the interception of which occurred some 20 years earlier in DS than in controls (37 vs. 53 years), indicating the cognitive decline due to premature senility (St. Clair and Blackwood, 1985). Shorter latency of P3 in response to nicotine might reflect shorter time to evaluate different stimuli, including processes necessary for making a task-relevant decision, or more generally, can be considered as a consequence of the improved speed with which attentional resources can be allocated when immediate memory is updated, whereby P3 amplitude indexes the updating of the working memory. With respect to the neurochemical substrates and neuroanatomical generators of P3, there appears to be a direct triggering by glutamatergic neurotransmission in the temporoparietal junction, the parietal cortical areas and in the medial temporal lobe structures, areas which are affected in Alzheimer type neurodegeneration. An important modulator of the P3 activity is the cholinergic neurotransmission, which increases P3 amplitude and decreases P3 latency (Frodl-Bauch et al., 1999). All nicotine studies conducted in AD found improved attention following nicotine administration, but no consistent effect on memory. With respect to attention this report confirms acute attentional improvement in DS subjects. Again in AD studies, the issue was raised, mildly demented patients being most likely to benefit from such treatments. Virtually all DS subjects develop AD-like neuropathological changes from age of 35 years on (Wisniewski et al., 1985), and cholinergic deficits in brains of adult DS patients seem to be the result of a decline from normal levels at birth rather than a congenital brain cholinergic deficiency (Kish et al., 1989). This offers the possibility of early therapeutic prevention of the development of brain cholinergic deficiency and possibly associated cognitive decline in DS (Kish et al., 1989). Stimulation of nicotinic acetylcholine receptors might offer a good opportunity, as nicotine administration has been found to cause an increase in the number of nAChR, rather than causing a downregulation (Marks et al., 1985; Rowell and Maureen, 1997), although experimetally it appears that high nicotine concentration can produce a longer-lasting decrease in nAChR function (Rowell and Duggan, 1998); nicotine increases mRNA encoding vesicular acetylcholine transporter (Prendergast and Buccafusco, 1998); NGF levels are selectively longlasting up-regulated by nicotinic-receptor stimulation (French et al., 1999); nicotine protected rat cortical neurons against beta amyloid cytotoxicity (Kihara et al., 1998).

One study of cholinergic therapy in Down's syndrome is published so far (Kishnani et al., 1999). Improvements in communication, expressive language, attention and mood stability were noted following chronic acetylcholinesterase inhibition by donezepil. The improvement in attentional parameters in DS (present study and Kishnani et al., 1999) are encouraging. However, several limitations of our study suggest caution in interpretation of the results. Only 8 patients were studied which limits power of the findings and each patient only received a single nicotine dose. It is unclear whether the results would be obtained after long-term exposure or whether tolerance would develop to these effects. Furthermore, nicotine concentrations were not measured. Cognitive effects investigated can largely be attributed to wellknown attentional nicotinic effects, therefore investigation of other cognitive and behavioral domains are warranted. We are not convinced that higher nicotine doses are needed, because of known steep dose-response curve and rather narrow therapeutic index (Newhouse et al., 1997). Taking experimental findings into accout (Rowell and Maureen, 1997; Rowell and Duggan, 1998), low-dose or intermittend administration may be worth trying. This problem might be overcome in the future by use of nAChR subtype selective agonists or allosteric modulators of nAChR, exhibiting less gastrointestinal and autonomic toxicity.

Acknowledgements

The authors like to thank Pharmacia-Upjohn, Uppsala, Sweden for support of 5 mg Nicorette© transdermal patches. We are highly indebted to the Red Bull Company, Salzburg, for generous financial support.

References

Brioni JD, Decker MW, Sullivan JP, Arneric SP (1997) The pharmacology of (−)-nicotine and novel cholinergic channel modulators. Adv Pharmacol 39: 191–220

Diaz F, Zurron M (1995) Auditory evoked potentials in Down's syndrome. Electroencephalogr Clin Neurophysiol 96: 526–537

Dursun SM, Revely MA, Bird R, Stirton F (1994) Longasting improvement of Tourette's syndrome with transdermal nicotine. Lancet 344: 1577

French SJ, Humby T, Horner CH, Sofroniew MV, Rattray M (1999) Hippocampal neurotrophin and trk receptor mRNA levels are altered by local administration of nicotine, carbachol and pilocarpine. Mol Brain Res 67: 124–136

Frodl-Bauch T, Bottlender R, Hegerl U (1999) Neurochemical substrates and neuroanatomical generators of the event-related P300. Neuropsychobiol 40: 86–94

Gray R, Rajan AS, Radcliffe KA, Yakehiro M, Dani JA (1996) Hippocampal synaptic transmission enhanced by low concentration of nicotine. Nature 383: 713–716

Haxby JV (1989) Neuropsychological evaluation of adults with Down's syndrome: patterns of selective impairment in nondemented old adults. J Ment Defic Res 33: 193–200

Jones GMM, Sahakian BJ, Levy R, Warburton DM, Gray JA (1992) Effects of acute subcutaneous nicotine on attention, information processing and short-term memory in Alzheimer's disease. Psychopharmacology 108: 485–494

Kihara T, Shimohama S, Urushitani M, Sawada H, Kimura J, Kume T, Maeda T, Akaike A (1998) Stimulation of alpha 4 beta 2 nicotinic acetylcholine receptors inhibits beta-amyloid toxicity. Brain Res 792: 331–334

Kish S, Karlinsky H, Becker L, Gilbert J, Rebbetoy M, Chang LJ, DiStefano L, Hornykiewicz O (1989) Down's syndrome individuals begin life with normal levels of brain cholinergic markers. J Neurochem 52: 1183–1187

Kishnani PS, Sullivan JA, Walter BK, Spiridigliozzi GA, Doraiswamy PM, Krishnan KRR (1999) Cholinergic therapy for Down's syndrome. Lancet 353: 1064–1065

Knott VJ, Harr A, Ilivitsky V (1999) Separate and combined effecst of scopolamine and mecamylamine on human event-related brain potentials. Hum Psychopharmacol Clin Exp 14: 307–317

Lawrence AD, Sahakian BJ (1998) The cognitive psychopharmacology of Alzheimer's disease: focus on cholinergic systems. Neurochem Res 23: 787–794

Lincoln AJ, Courchesne E, Kilman BA, Galambos R (1986) Auditory ERPs and information processing in Down's syndrome children. In: McCallum WC, Zappoli Z, Denoth F (eds) Cerebral psychophysiology: studies in event-related potentials. Electroencephalogr Clin Neurophysiol [Suppl] 38: 492–495

Marks MJ, Stitzel JA, Collins AC (1985) Time course study of the effects of chronic nicotine infusion on drug response and brain receptors. J Pharmacol Exp Ther 235: 619–628

McGehee DS, Heath MJS, Gelber S, Devay P, Role LW (1995) Nicotine enhancement of fast excitatory synaptic transmission in CNS by presynaptic receptors. Science 269: 1692–1696

Melyn MA, White DT (1973) Mental and developmental milestones of noninstitutionalized Down's syndrome children. Pediatrics 52: 542–548

Moosbrugger H, Oehlschlägel J (1996) FAIR Frankfurter Aufmerksamkeits-Inventar. Hans Huber, Bern

Newhouse PA, Potter A, Levin ED (1997) Nicotinic systems and Alzheimer's disease: implications for therapeutics. Drugs Aging 11: 206–228

Nordberg A, Lundquist H, Hartvig P, Liljy A, Langström B (1995) Kinetic analysis of regional (S)(−)^{11}C-Nicotinic binding in normal and Alzheimer brains-in vivo assessment using positron emission tomography. Alz Dis Assoc Disord 9: 21–27

Parks RW, Becker RE, Rippey RF, Gilbert DG, Matthews JR, Kabatay E, Young CS, Vohs C, Danz V, Keim P, Collins GT, Zigler SS, Urycki PG (1996) Increased regional cerebral glucose metabolism and semantic memory performance in Alzheimer's disease: a pilot double blind transdermal nicotine positron emission tomography study. Neuropsychol Rev 6: 61–79

Perry EK, Morris CM, Court JA, Cheng A, Fairbairn AF, McKeith IG, Irving D, Brown A, Perry RH (1995) Alteration in nicotine binding sites in Parkinson's disease, Lewy body dementia and Alzheimer's disease: possible index of early neuropathology. Neuroscience 64: 385–395

Polich J (1996) Meta-analysis of P300 normative aging studies. Psychophysiology 33: 334–353

Prendergast MA, Buccafusco JJ (1998) (−)-Nicotine increases mRNA encoding G3PDH and the vesicular acetylcholine transporter in vivo. Neuroreport 9: 1261–1265

Rowell PP, Maureen L (1997) Dose-response relationship for nicotine-induced up-regulation of rat brain nicotine receptors. J Neurochem 68: 1982–1989

Rowell PP, Duggan DS (1998) Long-lasting inactivation of nicotine receptor function in vitro by treatment with concentrations of nicotine. Neuropharmacology 37: 103–111

Snaedal J, Johannesson T, Jonsson JE, Gylfadottir G (1996) The effects of nicotine in dermal plaster on cognitive functions in patients with Alzheimer's disease. Dementia 7: 47–52

St. Clair DF, Blackwood D (1985) Premature senility in Down's syndrome. Lancet Jul 6: 34

Teller JK, Russo C, DeBusk LM, Angelini G, Zaccheo D, Dagna-Bricarelli F, Scartezzini P, Bertolini S, Mann DM, Tabaton M, Gambetti P (1996) Presence of soluble amyloid B-peptide precedes amyloid plaque formation in Down's syndrome. Nat Med 2: 93–95

Tewes U (1991) Hamburg-Wechsler Intelligenztest für Erwachsene / Revision 1991. Hans Huber, Bern

Vieregge P, Verleger R, Schulze-Rava H, Kömpf D (1992) Late cognitive event-related in adult Down's syndrome. Biol Psychiatry 32: 1118–1134

White HK, Levin ED (1999) Four-week nicotine skin patch treatment effects on cognitive performance in Alzheimer's disease. Psychopharmacol 143: 158–165

Wilson AL, Langley LK, Monley J, Bauer T, Rottunda S, McFalls E, Kovera C, McCarten JR (1995) Nicotine patches in Alzheimer's disease: pilot study on learning, memory, and safety. Pharmacol Biochem Behav 51: 509–514

Wisniewski KE, Kida E (1994) Abnormal neurogenesis and synaptogenesis in Down syndrome brain. Dev Brain Dysfunct 7: 289–301

Wisniewski KE, Wisniewski HM, Wen GY (1985) Occurrence of neuropathological changes and dementia of Alzheimer's disease in Down's syndrome. Ann Neurol 17: 278–282

Yates CM, Simpson J, Maloney AFJ, Gordon A, Reid AH (1980) Alzheimer-like cholinergic deficiency in Down's syndrome. Lancet 2(8201): 979

Authors' address: Prof. G. Lubec, MD, Department of Pediatrics, University of Vienna, Währinger Gürtel 18–20, A-1090 Vienna, Austria, e-mail: gert.lubec@akh-wien.ac.at

The brain in Down syndrome

R. Seidl[1], **N. Cairns**[2], and **G. Lubec**[1]

[1] Department of Pediatrics, University of Vienna, Austria
[2] Department of Neuropathology, Institute of Psychiatry, King's College London, London, United Kingdom

Summary. Down syndrome (trisomy 21) is a genetic disease with developmental brain abnormalities resulting in early mental retardation and precocious, age dependent Alzheimer-type neurodegeneration. We tried to discuss the role of neurodevelopmental abnormalities in connection with aberrant expression of genes on chromosome 21 including amyloid precursor protein (APP), CuZn superoxide dismutase (SOD1) and glial-derived S100 beta protein for neurodegeneration in DS. In this model, alterations in developmental pathways due to aberrant gene expression can impair cellular homeostasis and predispose to neurodegeneration of certain brain regions and types of nerve cells, involving cholinergic, serotonergic and catecholaminergic transmission, by shifting balance toward a pro-apoptotic state.

Down syndrome is a genetic disease resulting from trisomy of chromosome 21. Besides characteristic dysmorphic features and an increased risk for congenital heart disease, specific leukaemias and immunological deficiencies, the most constant condition and of overriding clinical importance is brain involvement. Developmental brain abnormalities (Wisniewski and Kida, 1994) resulting in early mental retardation combine with precocious, age dependent Alzheimer-type neurodegeneration (Wisniewski, 1985)

With the rapid advances in molecular genetics, the DNA sequence of the chromosome 21 has been defined recently (Hattori et al., 2000). Ultimately, all the phenotypes should be a result of the presence of three copies of genes on chromosome 21. The gene catalogue now allows different approaches to study the molecular pathophysiology, a hypothesis-driven selection of different sets of candidate genes or a systematic investigation of the role of every resident gene without pre-existing hypotheses (Hattori et al., 2000). Considering highly complex and organized, developmentally regulated and age-dependent patterns of gene expression, molecular bases of DS phenotypes could be an interplay of several different genes including genes on other chromosomes, escaping explanations such as "a single responsible gene", "critical DS region" and also "gene dosage effect/imbalance".

In this review we tried to dissect the role of neurodevelopmental abnormalities in connection with already known candidate genes including amyloid precursor protein (APP), CuZn superoxide dismutase (SOD1) and glial-derived S100β protein for neurodegeneration in DS. As in the highly-related field of Alzheimer's disease (Smith and Perry, 1998), research now focuses not only on genetics and physiology of lesions (plaques and tangles), but also on more fundamental cellular processes, such as synaptic dysgenesis and oxidative imbalance. Virtually all DS individuals manifest progressive AD-like neuropathology beyond 35 years of age and clinical dementia occurs in one third beyond 40 years of age (Wisniewski et al., 1985), increasing to 75–90% of DS individuals after 60 years of age (Haxby and Schapiro, 1992). Nevertheless, irrespective of whether dementia occurs, a progressive loss of intellectual function and brain metabolism appears in many older individuals with DS: diminished long-term memory and impaired visuospatial construction in neuropsychological testing (Haxby, 1989); age dependent changes in P300 component of "event related potentials" after age of about 37 years (St. Clair and Blackwood, 1985); decreased glucose metabolism (positron emission tomography) in temporal and parietal cortical areas during "stress test" by audiovisual stimulation (Pietrini et al., 2000). This raises the question and also points to the difficulties in diagnosis of dementia in mentally retarded individuals, further complicated by occurrence of non-cognitive, behavioral symptoms mimicking functional decline in adults with DS (Geldmacher et al., 1997). In fact, some patients with DS who are thought to be demented are suffering from major depression (Warren et al., 1989).

Molecular biological alterations of Down syndrome seems to include or culminate in the accumulation of AD-like neuropathology, based on early neurodevelopmental abnormalities (Wisniewski and Kida, 1994). Abnormalities of neuronal differentiation, migration, synaptogenesis and maturation are detectable as early as from the 20th and 22nd week of intrauterine life (Epstein, 1995). Neuronal density is reduced in various areas of the cortex, including cortical interneurons and the basal nucleus, the dendritic tree is greater than normal in early infancy, but the excessive early outgrowth is followed by atrophy (Becker et al., 1986, 1991): the number of dendritic spines remains unchanged until 40 weeks of gestation; compared to controls, higher numbers of dendritic branches and intersections in infantile DS (<6 months of age) decrease steadily from the infantile to the juvenile age and dendritic length follows the same pattern. So, instead of a continuously expanding normal dendritic tree, DS neurons show a relatively expanded tree at 4 months of age, which then apparently stops growing and becomes atrophic. These perturbations of synaptogenesis may lead to an enhanced cell loss executed by apoptosis, clinically evident as global developmental retardation.

Pathological hallmarks common to AD and DS include β-amyloid plaques, dystrophic neurites associated with plaques and neurofibrillary tangles within nerve cell bodies. These markers could play a role in the disease, but there is no compelling evidence that they, by themselves, cause dementia (Mann et al., 1990). Whereas synapse density alone or in combination with plaque and tangle counts correlates well with psychometric indices (Terry et al., 1991). There are subtle differences in the distribution of lesions between elderly DS and AD. In DS, plaques and tangles are not evenly distributed in an affected area and there is a greater proportion of diffuse plaques not visualised by traditional staining methods. In AD, a higher proportion of more compact and ring-with-core deposits is observed. This finding points to the fact that high density of diffuse amyloid deposits may be important in DS but also limits a correlation of neurochemical and clinical findings with plaque and tangle count, whereas the clustering tendency of amyloid is important in sporadic AD (Armstrong, 1996; Cairns, 1999).

Changes that correlate best with the decline in cognitive function in AD include loss of pyramidal neurons and indicators of pyramidal connectivity such as synapse loss or neuritic alterations suggesting that compromised neuronal function precedes and is more widespread than lesion formation and neuronal death (Smith and Perry, 1998). Estimates of neuronal cell loss by conventional morphometry (Mann, 1988), or MRI (Aylward et al., 1999) found a similar pattern and magnitude in demented older DS subjects and AD compared with controls.

Whatever causes the death of the neurons or initiates neurodegeneration, there is accumulating evidence that apoptosis or programmed cell death is involved (Busciglio and Yankner, 1995; Sawa et al., 1997; Nagy and Esiri, 1997; Dela Monte et al., 1998; Stadelmann et al., 1999; Seidl et al., 1999, Anderson et al., 2000). Presence of TUNEL labeling, predominantly distributed in pyramidal cortical cell layers, morphologically apoptotic nuclei and high molecular weight DNA fragmentation are unequivocally suggestive of apoptotic cell death machinery in DS (Anderson et al., 2000).

Three classical pathways of apoptotic signaling have emerged (Dragovich et al., 1998). The first involves recruitment and activation of procaspases (e.g. casp-8) by adapters such as FADD or TRADD, that are bound to activated cell surface death receptors (e.g. FAS and TNFR1). The second is initiated by the withdrawal of growth factors and is regulated by members of the Bcl-2 family of proteins, resulting in cytochrome c release from mitochondria and activation of Apaf-1 (Yoshida et al., 1998; Kroemer, 1997). Apaf-1 in response to cytochrome c binding can form a complex with caspase-9 that initiates the subsequent caspase cascade (caspase-3). The third pathway is initiated by DNA damage, regulated by proteins such as p53. However p53 in turn can activate apoptosis by transcription-dependent (bax, p21/WAF1) and -independent mechanisms. Downstream caspases (like caspase-3) cleave and inactivate proteins crucial for the maintenance of cellular cytoskeleton, DNA repair, signal transduction and cell cycle control. A caspase-3-dependent deoxyribonuclease (CAD) links cytosolic caspase activation to apoptotic nuclear damage.

Several apoptosis-related proteins are altered in DS, including members of the bcl-2 gene family (Sawa et al., 1997; Nagy and Esiri, 1997), p53 protein, Fas (CD95/APO-1) receptor (Dela Monte et al., 1998; Seidl et al., 1999) and caspase-3 (Stadelmann et al., 1999). Single neurons with apoptotic morphology in DS (and AD) brain showed cytoplasmatic immunoreactivity for activated caspase-3, whereby apoptotic neurons were identified at a frequency of 1 in 1,100 to 5,000 neurons (Stadelmann et al., 1999). Apoptotic phenomena in DS are not merely associated with AD-pathology. In vitro cultured cortical neurons from fetal DS brain developed apoptosis after exhibiting an increase in intracellular reactive oxygen species (ROS) (Busciglio and Yankner, 1995). Bax expression was greater in 3 young DS subjects (mean age 8.6 years) without any AD pathology, than in 6 DS subjects with AD (Nagy and Esiri, 1997). JunD-mRNA has been found to be downregulated in fetal DS brain and JunD protein levels were remarkably reduced in frontal, temporal lobe and cerebellum in adult DS brains (Labudova et al., 1998). Induction of the transcription factor c-Jun, which is then further activated by the c-Jun N-terminal kinases/stress-activated protein kinase-1 (JNK/SAPK1), is an early response in neuronal injury (Herdegen, 1996). The fate of neurons is determined by a transcriptional network comprising c-Jun, its partners (activating transcription factor-2 and Jun D) and c-Jun N-terminal kinases (JNKs), where the action of c-Jun as a death factor might depend on downregulation of its partners ATF-2 and Jun D. Increased levels of Fas (CD95/Apo1) and p53 have independently been found in DS (Dela Monte et al., 1998; Seidl et al., 1999). Upregulation of Fas-ligand, which may be a key event in triggering neuronal death (Raoul et al., 2000), can be mediated via p53 (Owen-Schaub et al., 1995) and via activation of JNK. JNK phosphorylates c-Jun (see above), which can in turn activate transcription of Fas-L by binding to activating protein (AP)-1 sites in the promotor (see Raoul et al., 2000). Alternatively, Fas dependent apoptosis is triggered in response to reduced activity of Akt (protein kinase B) kinase (e.g. in the absence of trophic factors), where non-phosphorylated transcription factor forkhead-like protein (released from sequestration by 14-3-3 proteins) translocates to the nucleus and drives transcription of Fas-L gene (Raoul et al., 2000). In addition Akt/PKB mediates a broad range of cellular processes that are essential for multiple stages of neural development, including proliferation, cell cycle regulation, cellular metabolism, and viability through differential actions on multiple pro- and anti-apoptotic pathways. However, p53 mediated (transcriptional) activation of Fas links p53, a sensor of (DNA)-damage and essential for cell cycle arrest to Fas, an effector of apoptosis (Owen-Schaub et al., 1995).[33] Further uncovering the crosstalk between the components of different apoptotic signaling pathways should help in the future to untangle the role of p53, bcl2 protein family, JNK and Fas in neurodegeneration of DS brain.

The central question however is, what causes the death of neurons or initiates neurodegeneration? Is there a role for "candidate genes" APP, SOD1 and S100β?

The amyloid-β protein (Aβ), the principal component of the senile plaques, is produced from the amyloid-precursor protein (APP) by two proteolytic events. The smaller C-terminal part of APP is localized intracellularly (cytoplasm), whereas the larger N-terminal part is lumenal or extracellularly, containing glycosylation sites. A β-secretase (BACE: β-site APP cleaving enzyme) cleaves APP at the N terminus of Aβ (β site) in the lumen of the endoplasmatic reticulum, Golgi network or endosome, yielding a membrane-associated APP fragment of 99 aa (C99, carboxyl terminal of APP) and the soluble ectodomain of APP (sAPPβ). C99 (CTFβ) is then variably cleaved at a second site within the transmembrane domain of C99 (γ site) by an unknown γ-secretase to release Aβ, a peptide of 39–42 aa. In a non-amyloidogenic pathway, APP can alternatively be cleaved at the α-secretase cleavage site within the Aβ-domain of APP, to produce a C-terminal fragment of 83 aa (C83) and liberate the large soluble N-terminal domain sAPPα. C83 (CTFα) is also a substrate for γ-secretase, which liberates the small p3 peptide (less amyloidogenic and neurotoxic in tissue culture) (Howlett et al., 2000).

BACE2 (or Asp1/DRAP), a homolog gene of BACE1 (Asp2) on chromosome 21 at 21q22.3 (Saunders et al., 1999), cleaves APP at the β-site and also within the amyloid-β region of APP (internal Aβ-sites) (Farzan et al., 2000; Lin et al., 2000; Acquati et al., 2000). It is tempting to speculate that APP-overexpression (substrate) together with increased β-secretase activity (BACE2) results in (or even underlies) increased production of amyloidogenic Aβ fragments (Acquati et al., 2000). In adult brain however, BACE2 is expressed at a very low level compared to BACE1 (Bennet et al., 2000). While it is likely that BACE1 makes a significant contribution to the pool of Aβ, BACE2 in turn seems to limit the production of pathogenic Aβ in cells that express both BACE1 and BACE2. Possibly, overexpression of BACE2 shifts its own activitiy from limiting to producing amyloidogenic Aβ fragments, similar as in certain APP-mutations, which substantially increase the production of C99 by BACE2, even to levels greater than those produced by BACE1 (Farzan et al., 2000). However, while increased APP gene dosage (Podlisny et al., 1987; Celi et al., 1994) leads to overexpression of APP (Rumble et al., 1989) and presumably presence of soluble Aβ42 from 21 gestational weeks on (Teller et al., 1996), levels of BACE2 mRNA and protein in DS have not been reported so far. Until now it is known that soluble Aβ42 protein is present in the brains of DS subjects from 21 weeks of gestational weeks on, but undetectable in age-matched controls, and that Aβ42 deposition in plaques, which starts earliest in late childhood, precedes Aβ40 in the course of the deposition. Aβ42-labeled diffuse plaques were first detected in a 12-year-old DS subject (Lemere et al., 1996), tau-positive dystrophic neurites in compact Aβ42 plaques were first observed in one 15-year-old DS brain, whereas neurofibrillary tangles were first observed in a 29-year-old case (Stoltzner et al., 2000). The lack of Aβ42 deposits in DS until late childhood, even though levels of aggregable Aβ42 are high throughout life, suggests that age-related changes of sequestering or degradation of Aβ play a pivotal role in determining whether Aβ remains soluble or is deposited. On the other hand, dystrophy of neurites and/or synaptic abnormalities due to effects on membrane properties caused by abnormal APP axonal transport forced by APP overexpression could come first, inducing plaque generation in the absence of Aβ42 deposition (for review in terms of AD see Vickers et al., 2000). Recently, APP plaques (immunoreactivity to N-terminal APP) were found more abundant in DS cortex, while Aβ plaques (immunoreactivity to Aβ) predominated in early-onset familial and sporadic AD (Egensperger et al., 1999). APP immunostaining was observed in large neurons and in dystrophic neurites of senile plaques. In control brains, APP plaque immunoreactivity was absent. On the other hand, extracellular APP deposits, lacking conformation of amyloid fibrils, have also been found unrelated to degenerating neurites (Giaconne et al., 1989). So the exact time course and which factor is essential for the deposition of Aβ in the brain is still not yet established.

As the expression of APP increases at the time of neuronal differentiation, APP and its isoforms should play roles in developmental or synaptic plasticity (for review see Roßner et al., 1998). In DS, abnormalities of differentiation, synaptogenesis and maturation emerge in fetal and infant DS brain (Wisniwski and Kida, 1994; Epstein, 1995). Directly related to mental retardation are alterations of cortical lamination, reduced dendritic ramifications, and diminished synaptic formation. In early infancy however, the dendritic tree is greater than normal with increased number of spines. At 4 months of age, large neurons within cortical layer III and V exhibit an increased total length of the dendritic tree and a relatively expanded dendritic tree, with increased number of spines. This early proliferation and excessive outgrowth of the dendritic tree is followed by a progressive atrophy from the age of 11 months on (Becker et al., 1991), suggesting an incomplete postnatal synaptogenesis. Preferential localization of APP at nerve terminals in growing neurites is consistent with a role in early development. Cell-surface-expressed cellular APP has a neurite-promoting function, co-localizes with adhesion plaque components and participates in synaptic vesicle recycling, functions which are distinct from a trophic one of the secreted soluble APP derivates (Roßner et al., 1998; Neve et al., 2000). In DS, N- and C-terminal APP immunoreactivity (full length APP) was observed in both neurons and glial cells from the fetal period to infancy (Arai et al., 1997). The numerical density of APP-immunoreactive neurons is higher than that of controls in fetal DS brain (17–35 weeks of gestation) about 5 times increased in young DS (8 months to 9 years of age), going up to 12 times that of controls in adult DS (22–68 years) (Griffin et al., 1998). Suggesting that a percentage of APP may function as a cell surface receptor linked to Go signaling cascade, transducing signals from the extracellular matrix to the interior of the cell, APP overexpression in DS may disrupt the physiological function of APP (Neve et al., 2000). Isoform-specific changes in the expression pattern of cortical APP during postnatal development (animal studies) suggest distinct functional roles for the various APP isoforms in brain maturation and synaptogenesis. High levels of cortical APP695 mRNA (an isoform lacking the Kunitz type protease domain KPI) coincide with profiles of synaptophysin, MAP2 (microtubule associated protein 2), myelin basic protein and high rate of synapse formation, whereas high induction of APP751 mRNA is associated with the end of cellular growth and synaptogenesis. This counterbalance between protease activity (by secreted forms of APP751 and APP770) completing cellular growth and synaptogenesis on the one hand and protease inhibitor activity (by APP695) establishing mature connectional pattern of neurons on the other hand, might be important during early postnatal development (Roßner et al., 1998). Data on the isoform-specific pattern of APP (over-)expression in early DS would be of crucial interest. In brain of adult DS, mRNA levels of APP-isoforms (770, 751, 695) were equally increased, and the normally found proportionality of APP mRNA to that of tau mRNA was maintained for

APP695 versus three-repeat tau, but totally disrupted between APP751 and four-repeat tau (Omaya et al., 1994). In further animal models, overexpression of human APP isoforms led to increased numbers of synaptophysin immunoreactive presynaptic terminals and increased expression of GAP-43. Presence of human APP751 in transgenic mice reduced α-secretase activity and increased β-secretase activity, while APP695 increased α-secretase activity. Furthermore, generation of preamyloid plaques in transgenic mice expressing human APP751 (but not mice expressing APP695) resembled morphological characteristics seen in brains of young adults with DS (Higgins et al., 1993). Homozygous APPV717F transgenic mice overexpress a human beta-amyloid precursor protein, encoding a familial Alzheimer's disease mutation. Neuronal APP overexpression was found in cell somas of young mice, whereas older mice showed APP overexpression in dystrophic neurites. These age–related changes were accompanied by progressive increases in the expression of neuritotrophic cytokine S100β, evident months before the appearance of Aβ deposits (Sheng et al., 2000).

In conclusion, there is some reason for a hypothesis that APP overexpression and dysregulation of its isoform-specific pattern contributes to abnormalities of early postnatal development, altered synaptic plasticity througout life (e.g. neuritic dystrophy), and also enhances vulnerability and predisposes to neurodegeneration, before deposition of Aβ42 occurs later in life. In this model, alterations in developmental pathways due to aberrant gene expression can impair cellular homeostasis and predispose to neurodegeneration by shifting balance toward a pro-apoptotic state (Mehler et al., 2000).

Recently, APP as a signaling molecule (G protein coupled receptor), interaction with cell cycle protein APP-BP1 (binding protein 1) and a direct regulation and/or interaction between APP and programmed cell death have been stressed (Neve et al., 2000). Overexpression of wild-type APP causes apoptotic death of neurons, although to a lesser degree than does expresssion of FAD mutants of APP (for review see Neve et al., 2000). APP is cleaved by caspases in the C-terminus to generate a second peptide, called C31, which in turn is a potent inducer of apoptosis (Lu et al., 2000). APP is processed by caspases-6 and -8 in the intracellular portion of the protein, in a site distinct from those processed by secretases (Pellegrini et al., 1999). Overexpression of APP695 in rat hippocampus induced caspase-3 activation, and other inflammatory responses including microglial activation (Masumura et al., 2000).

S100 beta (S100β), encoded on the subtelomeric position of chromosome 21, is an astrocyte-derived calcium binding protein and has been shown to promote growth and neurite extension in cultured glial and neuronal cells. Active S100β is markedly elevated in Alzheimer brain, largely attributed to plaque-associated asctrocytes, and has therefore been implicated in dystrophic neurite formation, plaque evolution as well as in the evolution of neurofibrillary tangles (see Sheng et al., 1997). According to a gene dosage effect, S100β mRNA and protein is increased in DS brain at all ages from fetus to adulthood (Marks et al., 1996; Griffin et al., 1998). About twice as many S100β-immunoreactive astrocytes were found in DS at all ages, S100β exhibiting a normal age-related increase. In contrast to that the numerical density of β-APP-immunoreactive neurons started at an increased level in fetal brains (17–35 weeks of gestation), was 5-times that of controls in young DS (8 months to 9 years) and 12 times that of controls in adult DS (22–68 years) (Griffin et al., 1998). It remains open whether ~1.5 times overexpression of S100β may contribute to pathogenesis of dendritic abnormalities in DS. In S100β transgenic mice (however expressing a much higher human gene copy number), the neurons in the hippocampus develop dendritic branches at an accelerated rate (Whitaker-Azmitia et al., 1997), resembling the relatively expanded dendritic tree in early-infantile DS. Nevertheless, three subjects with mirror duplications of chromosome 21 who were trisomic for most of chromosome 21, but monosomic for distal 21q22.3, including S100β, exhibited all stigmata of DS, including mental retardation (Pangalos et al., 1992).

Superoxide dismutase I (SOD1; CuZn-SOD), the key enzyme in the metabolism of oxygen free radicals is encoded on 21q22.1. Individuals with DS have three copies of the SOD1 gene and correspondingly, increased levels of SOD1 protein and activity (Castagne et al., 1999; Gulesserian et al., 2001). SOD1 catalyzes the dismutation reaction of superoxide anions to hydrogen peroxide and if increased SOD1 is not counterbalanced by (adaptively) increased activity of hydrogen peroxide handling enzymes, glutathion peroxidase and catalase, increased hydrogen peroxide tissue levels could be expected and where shown (DeHaan et al., 1997). In vitro cultured cortical neurons from fetal DS brain underwent apoptotic cell death after generating increased intracellular reactive oxygen species (ROS), which could be prevented by the application of free radical scavengers (Busciglio and Yankner, 1995). Although there is accumulating indirect evidence that oxidative stress plays a role in degenerative pathologies in DS, there is no direct proof that increased SOD1 activity leads to e.g. increased lipid peroxidation. On the other hand, not only oxidative stress (ROS) but also the cellular redox status as a whole modulates mechanisms regulating neuronal fate (Castagne et al., 1999). Links between APP and ROS support the involvement of oxidative stress as an active factor in Aβ-mediated neuropathology (for review see Miranda et al., 2000). High APP levels do not appear to be associated with amyloid deposition unless they exist in conjunction with excess free radical generation, disturbing neuronal homeostasis (Friedlich et al., 1994). Possible mechanisms are that free radicals may potentiate neurotoxicity of Aβ, increase Aβ formation and/or impair the ability to remove soluble Aβ.

A single case reported by Prasher et al. (1998) might illustrate the importance of APP and/or SOD1: a 78-year-old woman with partial trisomy 21 exhibited no evidence of Alzheimer's disease on neuropsychological, magnetic resonance imaging and neuropathological assessment. The gene sequence of APP and SOD was present in only two copies, whereas loci including S100β and presumably BACE2 were present in three copies.

Although neuritic plaques and neurofibrillary tangles are the hallmark lesions, they may not be sufficient to cause cognitive impairment. The key pathological change is the gradual degeneration of nerve cells and the related loss of specific synaptic connections. Crucially, this degenerative process is selective for certain brain regions and types of nerve cells, involving cholinergic, serotonergic and catecholaminergic transmission.

Cholinergic deficits are observed in both AD and DS (Yates et al., 1980) and have led to a generation of cholinergic analogues and cholinesterase inhibitors for the treatment of AD and DS (Kishnani, 1999). Neurochemical studies of other subcortical neurotransmitter systems revealed sometimes quite different results in DS and AD (Yates et al., 1983, 1986; Mann and Yates, 1986; Godridge et al., 1987; Seidl et al., 1999). In summary, deficits of serotonin (5-HT), noradrenaline and dopamine have been reported in various brain regions in DS. In the case of amino acid neurotransmitters, reflecting cortical neuronal groups, attention has been focused mainly on AD, only limited information in DS is controversial. One study reported a deficit of glutamate (aspartate) in the hippocampus, with a tendency to lower levels in other subcortical regions (amygdala, caudate nucleus), and GABA tended to reduced levels also in the hippocampus (Reynolds et al., 1988). Other studies found significantly reduced aspartate and glutamate levels in the parahippocampal gyrus (Risser et al., 1997) or no significant reduction in any of the brain regions (Seidl et al., 2001).

Of all neurotransmitter networks in the mammalian brain, the serotonin (5-HT) system is the phylogenetically oldest and most expansive (Jacobs and Fornal, 1995). Early expression of 5-HT in developing midbrain raphe neurons and their projecting terminals prior to synaptogenesis indicates that it is an important regulator of morphogenetic activities during early embryonic development, including cell proliferation, migration and differentiation (Lauder, 1993). In the adult brain, removel of serotonin results in loss of synapses, reduction in the expression of synaptophysin (marker for synapses) and MAP-2 (marker for dendrites) and also decreased levels of S100β, indicative of a return to an immature, undifferentiated brain stage. Besides the role as neurotransmitter, serotonin functions also as a trophic factor, regulating maturation of target cells (for review see Azmitia, 1999). In addition, there is experimental evidence, that concurrent loss of cholinergic and monoaminergic (5-HT > NA) activity may contribute to the disturbances in cortical processing and cognition/behavior and therefore combined cholinergic-monoaminergic stimulation may provide an effective treatment option (Dringenberg, 2000).

Following reports of serotonin deficiency in blood, cerebrospinal fluid and platelets, as well as decreased 5-HT platelet uptake in DS (see Epstein, 1995), significant reductions of 5-HT have been reported in most telencephalic brain regions (Yates et al., 1986; Godridge et al., 1987; Seidl et al., 1999; Risser et al., 1997). Neuronal loss in 5-HT projection side, midbrain dorsal raphe nuclei has also been identified (Mann et al., 1985). Reversal of hypotonia by administration of 5-hydroxytryptophan (Bazelon et al., 1967), reduction of self-injurious behavior by dietary increase in serotonin (Gedye, 1990, 1991), but also no beneficial effect on intelligence development in response to 5-hydroxytryptophan (Weise, 1974) or increased risk for infantile spasms (Coleman, 1971) have been reported. Recently, treatment with selective serotonin-reuptake inhibitor (SSRI) medication in six DS patients, aged 23 to 63 years, showed improvement in behaviors and on objective measures, such as workplace productivity. These DS patients presented with functional decline in adult life, but non-cognitive symptoms were prominent and included aggression, social withdrawal, compulsive behaviors and cognitive dysfunction at varying degrees. Authors concluded, that treatment trials with SSRIs may, therefore, be warranted in such cases (Geldmacher et al., 1997). In an earlier study, low dose antidepressant treatment combined with a serotonin-enhancing diet in a mentally handicapped adult with Down syndrome showing signs of Alzheimer-type dementia decreased aggressive behavior (Gedye, 1991). However, clinical, biochemical and/or neuropathological data supporting a serotonergic dysfunction in infant and childhood DS brain have not been reported.

In conclusion, there is increasing evidence that the APP gene dosage effect in DS cannot be ruled out as a specific risk factor for neurodegeneration based on neurodevelopmental abnormalities. Additional factors like SOD1 associated ROS-generation and/or altered cellular redox status may compromise the ability of regional neuronal subpopulations for neural network formation and/or cause enhanced vulnerability to a host of various stressors. In this model, alterations in developmental pathways due to aberrant gene expression can impair cellular homeostasis and predispose to neurodegeneration by shifting balance toward a pro-apoptotic state (Mehler and Gokhan, 2000). Research in Down syndrome has had and will have a major impact on our understanding of brain function during preclinical stages of Alzheimer-type dementia and other neurodegenerative diseases and is a direct challenge to therapeutic nihilism for this unique group of patients (Schapiro et al., 1992).

Acknowledgement

The support of our Down syndrome project by the Red Bull Company, Salzburg, is highly appreciated.

References

Acquati F, Accarino M, Nucci C, Fumagelli P, Jovine L, Ottolenghi S, Taramelli R (2000) The gene encoding DRAP (BACE2), a glycosylated transmembrane protein of the aspartic family, maps to Down critical region. FEBS Lett 468: 59–64

Anderson AJ, Stoltzner S, Lai F, Su J, Nixon RA (2000) Morphological and biochemical assessment of DNA damage and apoptosis in Down syndrome and Alzheimer disease, and effect of postmortem tissue archival on TUNEL. Neurobiol Aging 21: 511–524

Arai Y, Suzuki A, Mizuguchi, Takashima S (1997) Developmental and aging changes in the expression of amyloid precursor protein in Down syndrome brains. Brain Dev 19: 290–294

Armstrong RA (1996) Correlation between the morphology of diffuse and primitive beta-amyloid (A-beta) deposits and the frequency of associated cells in Down's syndrome. Neuropathol Appl Neurobiol 22: 527–530

Aylward EH, Li Q, Honeycutt NA (1999) MRI volumes of the hippocampus and amygdala in adults with Down's syndrome with and without dementia. Am J Psychiatry 156: 564–568

Azmitia EC (1999) Serotonin neurons, neuroplasticity, and homeostasis of neural tissue. Neuropsychopharm 21: 33S–45S

Bazelon M, Paine RS, Coeiw VA, Hunt P, Houck JC, Mahanand D (1967) Reversal of hypotonia in infants with Down's syndrome by administration of 5-hydroxytryptophan. Lancet 1: 1130–1133

Becker LE, Armstrong DL, Chan F (1986) Dentritic atrophy in children with Down's Syndrome. Ann Neurol 20: 520–526

Becker LE, Mito T, Takashima S, Onodera K (1991) Growth and development of the brain in Down syndrome. Prog Clin Biol Res 373: 133–152

Bennet BD, Babu-Khan S, Loeloff R, Louis JC, Curran Eileen, Citrol M, Vassar R (2000) Expression analysis of BACE2 in brain and peripheral tissues. J Biol Chem 275: 20647–20651

Busciglio J, Yankner BA (1995) Apoptosis and increased generation of reactive oxygen species in Down Syndrome. Nature 378: 776–779

Cairns NJ (1999) Neuropathology. J Neural Transm [Suppl] 57: 61–74

Castagne V, Gautschi M, Lefevre K, Posada A, Clarke PGH (1999) Relationships between neuronal death and the cellular redox status. Focus on the developing nervous system. Progr Neurobiol 59: 397–423

Celi FS, Cohen MM, Antonarakis SE, Wertheimer E, Roth J, Shuldiner AR (1994) Determination of gene dosage by a quantitative adaptation of the polymerase chain reaction (gd-PCR): rapid detection of deletions and duplications of gene sequences. Genomics 21: 304–310

Coleman M (1971) Infantile spasms associated with 5-hydroxytryptophan administered in patients with Down's syndrome. Neurol 21: 911

De Haan JB, Wolvetang EJ, Christiano F, Ianello R, Bladier C, Kelner MJ, Kola I (1997) Reactive oxygen species and their contribution to pathology in Down syndrome. Adv Pharmacol 38: 379–402

DelaMonte SM, Sohn YK, Ganju N, Wands JR (1998) p53- and CD95-associated apoptosis in neurodegenerative diseases. Lab Invest 78: 401–411

Dragovich T, Rudin CM, Thompson CB (1998) Signal transduction pathways that regulate cell survival and cell death. Oncogene 17: 3207–3213

Dringenberg HC (2000) Alzheimer's disease: more than a cholinergic disorder — evidence that cholinergic-monoaminergic interactions contribute to EEG slowing and dementia. Behav Brain Res 115: 235–249

Egensperger R, Weggen S, Ida N, Multhaup G, Schnabel R, Beyreuther K, Bayer TA (1999) Reverse relationship between β-amyloid precursor protein and β-amyloid peptide plaques in Down's syndrome versus sporadic/familial Alzheimer's disease. Acta Neuropathol 97: 113–118

Epstein CJ (1995) Down Syndrome (Trisomy 21). In: Scriver SR, Beaudet AL, Sly WS, Valle D (eds) The metabolic and molecular bases of inherited disease. McGraw-Hill, New York, pp 749–794

Farzan M, Schnitzler CE, Vasilieva N, Leung N, Choe H (2000) BACE2, a β-secretase homolog, cleaves at the β site and within the amyloid-β region of the amyloid-β precursor protein. PNAS 97: 9712–9717

Friedlich AL, Butcher LL (1994) Involvement of free oxygen radicals in β-amyloidosis: an hypothesis. Neurobiol Aging 15: 443–455

Gedye A (1990) Dietary increase in serotonin reduces self-injurious behaviour in a Down's syndrome adult. J Ment Defic Res 34: 195

Gedye A (1991) Serotonergic treatment for aggression in a Down's syndrome adult showing signs of Alzheimer's disease. J Ment Defic 35: 247–258

Geldmacher DS, Lerner AJ, Voci JM, Noelker EA, Somple LC, Whitehouse PJ (1997) Treatment of functional decline in adults with Down syndrome using selective serotonin-reuptake inhibitor drugs. J Geriatr Psychiat Neurol 10: 99–104

Giaccone G, Tagliavini F, Linoli G, Bouras C, Frigerio L, Frangione B, Bugiano O (1989) Down patients: extracellular preamyloid deposits precede neuritic degeneration and senile plaques. Neurosci Lett 97: 232–238

Godridge H, Reynolds GP, Czudek C, Calcutt NA, Benton M (1987) Alzheimer-like neurotransmitter deficits in adult Down's syndrome brain tissue. J Neurol Neurosurg Psychiatry 50: 775–778

Griffin WST, Sheng JG, McKenzie LE, Royston MC, Gentleman SM, Brumback RA, Cork LC, Del Bigio MR, Roberts GW, Mrak RE (1998) Life-long overexpression pf S100β in Down's syndrome: implications for Alzheimer pathogenesis. Neurobiol Aging 19: 401–405

Gulleserian T, Seidl R, Hardmeier R, Cairns N, Lubec G (2001) Superoxide dismutase SOD1, encoded on chromosome 21, but not SOD2 is overexpressed in brains of patients with Down syndrome. J Invest Med 49: 41–46

Haxby JV (1989) Neuropsychological evaluation of adults with Down's syndrome: patterns of selective impairment in nondemented old adults. J Ment Defic Res 33: 193–197

Haxby JV, Schapiro MB (1992) Longitudinal study of neuropsychological functions in older adults with Down syndrome. In: Epstein C, Nadel L (eds) Down syndrome and Alzheimer disease. Wiley Liss, New York, pp 35–50

Herdegen T (1996) Jun, Fos, and CREB/ATF transcription factors in the brain: control of gene expression under normal and pathophysiological conditions. Neuroscientist 2: 153–161

Higgins LS, Catalano R, Quon D, Cordell B (1993) Transgenic mice expressing human βAPP751, but not mice expressing βAPP695, display early Alzheimer's disease-like histopathology. Ann NY Acad Sci 695: 224–227

Howlett DR, Simmons DL, Dingwall C, Christie G (2000) In search of an enzyme: the β-secretase of Alzheimer's disease is an aspartic proteinase. Trends Neurosci 23: 565–570

Jacobs B, Fornal C (1995) Serotonin and behavior, a general hypothesis. In: Bloom F, Kupfer D (eds) Psychopharmacology: the fourth generation of progess. Raven Press, New York, pp 461–469

Kishnani PS, Sullivan JA, Walter BK, Spiridigliozzi GA, Doraiswamy PM, Krishnan KRR (1999) Cholinergic therapy for Down's syndrome. Lancet 353: 1064–1065

Kroemer G (1997) The proto-oncogene Bcl-2 and its role in regulating apoptosis. Nature Med 3: 614–620

Labudova O, Krapfenbauer K, Moenkemann H, Rink H, Kitzmüller E, Cairns N, Lubec G (1998) Decreased transcription factor JunD in brains of patients with Down syndrome. Neurosci Lett 252: 159–162

Lauder JM (1993) Neurotransmitters as growth regulatory signals: role of receptors and second messengers. Trends Neurosci 16: 233–240

Lemere CA, Blustzjan JK, Yamaguchi H, Wisniewski T, Saido TC, Selkoe DJ (1996) Sequence of deposition of heterogenous amyloid-β-peptides and ApoE in Down syndrome: implications for initial events in amyloid plaque formation. Neurobiol Dis 3: 16–32

Lin X, Koelsch G, Wu S, Downs D, Dashti A, Tang J (2000) Human aspartic protease memapsin 2 cleaves the β-secretase site of β-amyloid precursor protein. PNAS 97: 1456–1460

Lu DC, Rabizadeh S, Chandra S, Shayya RF, Ellerby LM, Ye X, Salvesen GS, Koo EH, Bredesen DE (2000) A second cytotoxic proteolytic peptide derived from amyloid$^{\beta}$ — peptide precursor. Nature Med 6: 397–404

Mann DA (1988) The pathological association between Down syndrome and Alzheimer disease. Mech Ageing Dev 43: 99–136

Mann DM, Yates PO, Marcyniuk B, Rawndra CR (1985) Pathological evidence for neurotransmitter deficits in Down's syndrome of middle age. J Ment Defic Res 29: 125–135

Mann DMA, Yates PO (1986) Neurotransmitter deficits in Alzheimer's disease and in other dementing disorders. Hum Neurobiol 5: 147–158
Mann DMA, Royston MC, Ravindra CR (1990) Some morphological observations on the brains of patients with Down's syndrome: their relationship to age and dementia. J Neurol Sci 99: 153
Marks A, O'Hanlon D, Lei M, Percy ME, Becker LE (1996) Accumulation of S100β mRNA and protein in cerebellum during infancy in Down syndrome and control subjects. Mol Brain Res 36: 343–348
Masumura M, Hata R, Nishimura I, Uetsuki T, Sawada T, Yoshikawa K (2000) Caspase-3 activation and inflammatory responses in rat hippocampus inoculated with a recombinant adenovirus expressing the Alzheimer amyloid precursor protein. Mol Brain Res 80: 219–227
Mehler MF, Gokhan S (2000) Mechanisms underlying neural cell death in neurodegenerative diseases: alterations of a developmentally-mediated cellular rheostat. Trends Neurosci 23: 599–605
Miranda S, Opazo C, Larrondo LF, Munoz FJ, Ruiz F, Leighton F, Inestrosa NC (2000) The role of oxidative stress in the toxicity induced by amyloid β-peptide in Alzheimer's disease. Progr Neurobiol 62: 633–648
Nagy Z, Esiri MM (1997) Apoptosis-related protein expression in the hippocampus in Alzheimer's disease. Neurobiol Aging 18: 565–571
Neve RL, McPhie DL, Chen Y (2000) Alzheimer's disease: a dysfunction of amyloid precursor protein. Brain Res 886: 54–66
Owen-Schaub LB, Zhang W, Cusack JC (1995) Wild-type human p53 and a temperature-sensitive mutant induce Fas/APO-1 expression. Mol Cell Biol 15: 3032–3040
Oyama F, Cairns NJ, Shimada H, Oyama R, Titani K, Ihara Y (1994) Down' syndrome: upregulation of β-amyloid protein precursor and τ mRNAs and their defective coordination. J Neurochem 62: 1062–1066
Pangalos C, Theophile D, Sinet PM, Marks A, Stamboulieh-Abazis D, Chettouch Z, Prieur M, Verellen C, Rethore M-O, Lejeune J, Delaber J-M (1992) No significant effect of monosomy for distal 21q22.3 on the Down syndrome phenotype in 'mirror' duplications of chromosome 21. Am J Hum Genet 51: 1240–1250
Pellegrini L, Passer BJ, Tabaton M, Ganjei JK, D'Adamio L (1999) Alternative, non-secretase processing of Alzheimer's β-amyloid precursor protein during apoptosis by caspase-6 and -8. J Biol Chem 274: 21011–21016
Pietrini P, Alexander GE, Furey ML, Hampel H, Guazelli M (2000) The neurometabolic landscape of cognitive decline: in vivo studies with positron emission tomography in Alzheimer's disease. Int J Psychophysiol 37: 87–98
Podlisny MB, Lee G, Selkoe DJ (1987) Gene dosage of the amyloid beta precursor protein in Alzheimer's disease. Science 238(4827): 669–671
Prasher VP, Farrer MJ, Kessling AM, Fisher EMC, West RJ, Barber PC, Butler (1998) Molecular mapping of Alzheimer-type dementia in Down's syndrome. Ann Neurol 43: 380–383
Raoul C, Pettmann B, Henderson CE (2000) Active killing of neurons during development and following stress: a rol for p75^{NTR} and Fas? Corr Opin Neurobiol 10: 111–117
Reynolds GP, Warner CEJ (1988) Amino acid neurotransmitter deficits in adult Down's syndrome brain tissue. Neurosci Lett 94: 224–227
Risser D, Lubec G, Cairns N, Herrera-Marschitz M (1997) Excitatory amino acids and monoamines in parahippocampal gyrus and frontal cortical pole of adults with Down syndrome. Life Sci 60: 1231–1237
Roßner S, Ueberham U, Schliebs R, Schliebs R, Perez-Polo JR, Bigl V (1998) The regulation of amyloid precursor protein metabolism by cholinergic mechanisms and neurotrophin receptor signaling. Progr Neurobiol 56: 541–569
Rumble B, Retallack R, Hilbich C, Simms G, Multhaup G, Martins R, Hockey A, Montgomery P, Beyreuther K, Masters CL (1989) Amyloid A4 protein and its

precursor in Down's syndrome and Alzheimer's disease. N Engl J Med 320: 1446–1452

Saunders AJ, Kim T-W, Tanzi RE, Fan W, Bennet BD, Babu-Khan S, Luo Y, Louis J-C, McCaleb M, Citron M, Vassar R, Richards WG (1999) BACE maps to chromosome 11 and a BACE homolog, BACE2, reside in the obligate Down syndrome region of chromosome 21. Science 286: 1255–1256

Sawa A, Oyama F, Cairns NJ, Amano N, Matsushita M (1997) Aberrant expression of bcl-2 gene family in Down's syndrome brains. Mol Brain Res 48: 53–59

Schapiro MB, Haxby JV, Grady CL (1992) Nature of mental retardation and dementia in Down syndrome: study with PET, CT, and neuropsychology. Neurobiol Aging 13: 723–734

Seidl R, Fang-Kircher S, Bidmon B, Cairns N, Lubec G (1999) Apoptosis-associated proteins p53 and APO-1/Fas (CD95) in brains of adult patients with Down syndrome. Neurosci Lett 260: 9–12

Seidl R, Kaehler ST, Prast H, Singewald N, Cairns N, Gratzer M, Lubec G (1999) Serotonin (5-HT) in brains of adult patients with Down syndrome. J Neural Transm [Suppl] 57: 221–232

Seidl R, Cairns N, Singewald N, Kaehler ST, Lubec G (2001) Cortical GABA deficit in Alzheimer's disease but not in adult Down syndrome. Naunyn Schmiedebergs Arch Pharmacol 363: 139–145

Sheng JG, Mrak RE, Griffin WST (1997) Glial-neuronal interactions in Alzheimer's disease: progressive association of IL-1α^+ microglia and S100β^+ astrocytes with neurofibrillary tangle stage. J Neuropathol Exp Neurol 56: 285–290

Sheng JG, Mrak RE, Bales KR (2000) Overexpression of the neuritotrophic cytokine S100β precedes the appearance of neuritic beta-amyloid plaques in APPV717F mice. J Neurochem 74: 295–301

Smith MA, Perry G (1998) What are the facts and artifacts of the pathogenesis and etiology of Alzheimer disease. J Chem Neuroanat 16: 35–41

St. Clair DF, Blackwood D (1985) Premature senility in Down's syndrome. Lancet Jul 6: 34

Stadelmann C, Deckwerth TL, Srinivasan A, Bancher C, Brück W, Jellinger K, Lassmann H (1999) Activation of caspase-3 in single neurons and autophagic granules of granulovacuolar degeneration in Alzheimer's disease. Evidence for apoptotic cell death. Am J Pathol 155: 1459–1466

Stoltzner S, Grenfell TJ, Mori C, Wisniewski KE, Wisniewski TM, Selkoe DJ, Lemere CA (2000) Temporal accrual of complement proteins in amyloid plaques in Down's syndrome with Alzheimer's disease. Am J Pathol 156: 489–499

The chromosome 21 mapping and sequencing consortium (2000) The DNA sequence of human chromosome 21. Nature 405: 311–319

Teller JK, Russo C, DeBusk LM, Angelini G, Zaccheo D, Dagna-Bricarelli F, Scartezzini P, Bertolini S, Mann DM, Tabaton M, Gambetti P (1996) Presence of soluble amyloid β-peptide precedes amyloid plaque formation in Down's syndrome. Nat Med 2: 31–32

Terry RD, Masliah E, Salmon DP, Butters N, DeTeresa R, Hill R, Hansen LA, Katzman R (1991) Physical basis of cognitive alterations in Alzheimer's disease: synapse loss is the major correlate of cognitive impairment. Ann Neurol 30: 572–580

Vickers JC, Dickson TC, Adlard PA Saunders HL, King CE, McCormack G (2000) The cause of neuronal degeneration in Alzheimer's disease. Progr Neurobiol 60: 139–165

Warren AC, Holroyd S, Folstein P (1989) Major depression in Down's syndrome. Br J Psychiatry 155: 202–207

Weise P, Koch R, Shaw KNF, Rosenfeld MJ (1974) The use of 5-HTP in the treatment of Down's syndrome. Pediatr 54: 165–167

Whitaker-Azmitia PM, Wingate M, Borella A, Gerlai R, Roder J, Azmitia EC (1997) Transgenic mice overexpressing the neurotrophic factor S-100 beta show neuronal cytoskeletal and behavioral signs of altered aging processes: implications for Alzheimer's disease and Down's syndrome. Brain Res 776: 51–60

Wisniewski KE, Kida E (1994) Abnormal neurogenesis and synaptogenesis in Down syndrome brain. Dev Brain Dysfunct 7: 289–301

Wisniewski KE, Wisniewski HM, Wen GY (1985) Occurrence of neuropathological changes and dementia of Alzheimer's disease in Down syndrome. Ann Neurol 17: 278–282

Yates CM, Simpson J, Maloney AFJ, Gordon A, Reid AH (1980) Alzheimer-like cholinergic deficiency in Down syndrome. Lancet 2(8201): 979

Yates CM, Simpson J, Gordon A, Maloney AFJ, Allison Y, Ritchie IM, Urquhart A (1983) Catecholamines and cholinergic enzymes in pre-senile and senile Alzheimer-type dementia and Down's syndrome. Brain Res 280: 119–126

Yates CM, Simpson J, Gordon A (1986) Regional brain 5-hydroxytryptamine levels are reduced in senile Down's syndrome as in Alzheimer's disease. Neurosci Lett 65: 189–192

Yoshida H, Kong Y-Y, Yoshida R (1998) Apaf1 is required for mitochondrial pathways of apoptosis and brain development. Cell 94: 739–750

Authors' address: Prof. Dr. G. Lubec, Department of Pediatrics, University of Vienna, Währinger Gürtel 18-20, A-1090 Vienna, Austria, e-mail: gert.lubec@akh-wien.ac.at

Decreased levels of ARPP-19 and PKA in brains of Down syndrome and Alzheimer's disease

S. H. Kim[1], **A. C. Nairn**[2], **N. Cairns**[3], and **G. Lubec**[1]

[1] Department of Pediatrics, University of Vienna, Austria
[2] Laboratory of Molecular and Cellular Neuroscience, Rockefeller University, New York, USA
[3] Institute of Psychiatry, Brain Bank, King's College, United Kingdom

Summary. ARPP-19 (cAMP-regulated phosphoprotein of Mr = 19,000) is a substrate for cAMP-dependent protein kinase (PKA). ARPP-19 is found in all brain regions but the function of ARPP-19 is not fully elucidated yet. We detected a downregulated sequence with 100% homology with ARPP-19 in temporal cortex of patients with Down syndrome (DS) as compared to controls, but not in Alzheimer's disease (AD) using differential display-polymerase chain reaction (DD-PCR). We subsequently determined protein levels of ARPP-19 in temporal cortex and cerebellum by immunoblotting and observed significant reduction of ARPP-19 in DS (temporal cortex) and AD (cerebellum). We also observed decreased activities of PKA in DS (temporal cortex and cerebellum) and AD (temporal cortex). These findings suggest that decreased ARPP-19 along with decreased activities of PKA is involved in pathomechanisms of both neurodegenerative disorders. Furthermore, these findings provide first evidence for an impaired mechanism of cAMP-related signal transduction and phosphorylation in both dementing disorders.

Introduction

ARPP-19 (cAMP-regulated phosphoprotein of Mr = 19,000) is a substrate for cAMP-dependent protein kinase (PKA) (Horiuch et al., 1990). The phosphorylation level of ARPP-19 is increased by activators of adenylyl cyclase and cAMP (Girault et al., 1988) and expression of mRNA encoding ARPP-16/19 was observed in caudate nucleus, putamen, nucleus accumbens and neocortex. Additionally, high levels of ARPP-16/19 were found in the cerebellar cortex, where it was confined to deeper layers (Brene et al., 1994). The expression of ARPP-19 proteins has been shown to differ markedly with ARPP-16 and the levels of ARPP-19 were highest in the embryo and decreased during development (Girault et al., 1990; Horiuch et al., 1990). The regional distribution of phosphoproteins has been described in rat brain using immunohistochemical techniques (Girault et al., 1990). In contrast to the specific distribution of ARPP-16, which appears to be restricted to striatal and cortical neurons, ARPP-19 is found in all brain regions and in peripheral organs of the adult rat (Girault et al., 1988).

Increased intracellular levels of cAMP result in the activation of PKA, which appears to be the major mediator of cAMP responses in mammalian cells and phosphorylates a variety of proteins in nervous tissue as well as ARPP-19 (Nestler and Greengard, 1983; Browning et al., 1985). Mammalian brain is an unusually rich source of protein phosphorylation systems. These include diverse phosphoproteins as substrate, protein kinases, phosphoprotein phosphatases and modulators (Nairn et al., 1985). A number of phosphoproteins have been shown to act as intracellular messengers in the mammalian central nervous system and the phosphorylation and dephosphorylation of phosphoproteins constitute important pathways for the regulation of neuronal function (Greengard, 1987; Walaas and Greengard, 1991). Regulation of phosphorylation of specific substrate proteins by PKA appears to be a general mechanism by which neurotransmitters produce physiological effects in specific target neurons. However the function of ARPP-19, a substrate for PKA, is not yet known in neurodegenerative diseases. In the present study, using differential display-polymerase chain reaction (DD-PCR), we found a downregulated sequence with 100% homology to ARPP-19 mRNA in temporal cortex of patients with DS. This finding along with several data proposing the involvement of cAMP signaling in the pathology of neurodegenerative diseases made us examine the protein levels of ARPP-19 and the activities of PKA in brain of DS and AD compared to controls.

Materials and methods

Human brain samples

Postmortem human brain samples (temporal cortex and cerebellum) were obtained from the MRC London Brain Bank for Neurodegenerative Diseases, Institute of Psychiatry. All the DS patients were karyotyped and possessed trisomy 21. A formal cognitive assessment of dementia in DS was not performed. In all DS brains there were abundant and extensive beta-amyloid deposits, neurofibrillary tangles and neuritic plaques. The AD patients fulfilled the National Institute of Neurological Disorders and Stroke and Alzheimer's disease and Related Disorders Association criteria for probable AD (Tierney et al., 1998). The neuropathological diagnosis of "definite AD" was confirmed using the CERAD criteria (Mirra et al., 1991). Normal brains obtained from individuals with no history of neurological or psychiatric illness were used as controls. The major cause of death was bronchopneumonia in DS and AD and heart disease in controls. The fresh brain was dissected, coronal slices snap frozen and stored at −70°C until use.

Isolation of mRNA

Brain tissues (temporal cortex) were ground using a mortar and pestle under liquid nitrogen and lysed in RNAzol B (Molecular Research Center, Newark, USA). The homogenates added 0.1 volume of chloroform were shaked vigorously for 15 seconds, stored on ice for 5 minutes and centrifuged at 12,000 g for 15 minutes at 4°C. Total RNA remained in the upper aqueous phase was precipitated by mixing with an equal volume of isopropanol. The mixtures were incubated for 15 minutes at 4°C and centrifuged at 12,000 g for 15 minutes at 4°C. Total RNA pellets were washed with 75% ethanol, dried and dissolved in RNase-free water. To prevent the contamination of chromosome DNA, mRNA was isolated using Quick Prep Micro mRNA purification kit (Pharmacia Biotech, Uppsala, Sweden) containing oligo(dT)-cellulose according to the manufacturer's instructions. Concentrations of total RNA and mRNA were calculated spectrophotometrically with absorbance at 260 nm.

Differential display-polymerase chain reaction (DD-PCR)

DD-PCR was performed using the RNAimage kit (GenHunter Corp., Nashville, USA). Briefly, the isolated mRNA (100 ng) was reverse transcribed in reverse transcriptase buffer (25 mM Tris-Cl, pH 8.3, 37.6 mM KCl, 1.5 mg $MgCl_2$ and 5 mM DTT) containing 5 unit/μl of MMLV-reverse transcriptase, 20 μM dNTP mix and 0.2 μM of each one-base-anchored oligo(dT) primer (-G, -A or -C). The reverse transcription mixture was used for PCR in dilution of 1:10. Subsequent PCR (20 μl) was performed in PCR buffer (10 mM Tris-Cl, pH 8.4, 50 mM KCl, 1.5 mM $MgCl_2$ and 0.001% gelatin) containing 2 μM dNTP, 0.2 μM of one-base-anchored oligo(dT) primer, 0.2 μM of arbitrary primer, 0.2 μl of α-[^{33}P]dATP (2000 Ci/mmole) and 0.05 unit/μl of AmpliTaq DNA Polymerase (Perkin-Elmer). The thermocycler (GeneAmp PCR System 9700, Perkin-Elmer) was programmed as follows: 40 cycles at 94°C for 30 seconds, 40°C for 2 minutes, and 72°C for 30 seconds, and terminated with a final extension at 72°C for 5 minutes. ^{33}P-labeled PCR products were separated on 6% denaturing polyacrylamide gel for 3.5 hours at 60 watts constant power. The blotted gel on a piece of 3 M paper was dried under vacuum at 80°C for 1 hour. The autoradiogram oriented with the dried gel was exposed and developed. The bands of interest were cut from the dried gel, eluted by boiling in water and reamplified by PCR with the same set of primers in the same condition. The reamplified cDNA fragments were cloned in PCR-TRAP vector using PCR-TRAP cloning system (GenHunter) according to the manufacturer's instructions. DNA sequencing was performed at MWG-BIOTECH (Ebersberg, Germany) and the sequence was aligned in all EMBL libraries.

Immunoblot

Brain tissues were suspended in homogenate buffer consisting of 10 mM Tris-HCl (pH 7.5), 150 mM NaCl, 0.05% (v/v) Tween 20, 4% CHAPS, 1 mM EDTA and a mixture of protease inhibitors, 1 mM PMSF and 1 μg/ml of each pepstatin A, chymostatin, leupeptin and antipain for immunoblotting. The suspension was homogenized in a Potter-Elvehjem homogenizer at 4°C and centrifuged at 10,000 × g for 10 minutes. The BCA protein assay kit (Pierce, USA) determined the concentration of protein in the supernatant. Samples (10 μg) were mixed with the same volume of sample buffer (125.5 mM Tris, 70 mM SDS, 0.001% bromophenol blue, 20% glycerol, 2% 2-mercaptoethanol, pH 6.8), incubated at 95°C for 15 minutes and loaded onto a 12.5% ExcelGel SDS homogeneous gel (Amersham Pharmacia Biotech, Sweden). The electrophoresis was performed with Multiphor II Electrophoresis System (Amersham Pharmacia Biotech, Sweden). Proteins separated on the gel were transferred onto PVDF membrane (Millipore, Bedford, USA) and the membrane was blocked in blocking buffer (10 mM Tris-HCl, pH 7.5, 150 mM NaCl, 0.1% Tween 20, 0.1% $MgCl_2$ and 1% non-fat dry milk). The membrane was incubated for 2 hours at room temperature with rabbit antisera against bovine ARPP-19. It was kindly supplied by P. Greengard (Laboratory of Molecular and Cellular Neuroscience, Rockefeller University, NY). After 3 times washing for 15 minutes with blocking buffer, membranes were probed with a 1:1,000 dilution of secondary antibody, anti-rabbit IgG(H + L), coupled to horseradish peroxidase (Southern Biotechnology Associates, Inc., Alabama, USA) for 1 hour. The membrane was washed 3 times for 15 minutes and developed with the Western blot chemiluminescence reagents (NEN™ Life Science Products, Inc., USA). The densities of signaling bands were measured by RFLPscan version 2.1 software program (Scanalytics, USA).

PKA activity assay

For the determination of PKA activities, aliquots of brain samples were homogenized by ultrasound in an ice bath in nine volumes of 5 mM Tris acetate buffer (pH 7.4) containing 10 mM mercaptoethanol. The 10,000 × g supernatant of the homogenate was used for enzyme activity measurement. We used the SpinZyme Non-radioactive PKA assay kit (Pierce) according to the manufacturer's protocols. Briefly, samples were incubated with reaction mixture containing the peptide substrate labeled with a fluorescent dye. The reaction mixture was applied to a separation unit containing the affinity membrane. The phosphorylated product specifically bound onto the membrane while unreacted substrate passed through the support when washed with binding buffer using centrifugation. After transferring the membrane containing bucket to a new receptacle, the bound phosphorylated substrate was eluted from the membrane with elution buffer. Quantitation of both the non-phosphorylated and the phosphorylated substrate was accomplished by measuring absorbance at 570 nm and fluorescence readings with excitation at 573 nm and emission at 589 nm.

Statistics

Between group differences were calculated by non-parametric Mann-Whitney U test using GraphPad Instat2 program and the level of significance was considered at $P < 0.05$. All results are presented as mean ± standard deviation (SD).

Results

The decreased mRNA level of ARPP-19 in DS

To detect deteriorated genes in human brain with neurodegenerative disorders we studied the mRNA expression pattern of temporal region from patients with DS and AD as compared to controls using DD-PCR. DD-PCR was performed using a total of 24 primer combinations. As shown in Fig. 1, the arrowhead-indicated cDNA bands that were absent in DS but expressed in AD were excised from the gel, reamplified, subcloned and subsequently sequenced. Alignment of the nucleic acid sequence with the sequence obtained from EMBL was shown that 157-base cDNA fragment showed 100% homology with ARPP-19 mRNA sequence (Fig. 2). In parietal cortex of DS and AD, we observed the low expression level of cDNA fragment with high homology to ARPP-19 mRNA using DD-PCR with same primers and same condition (data not shown).

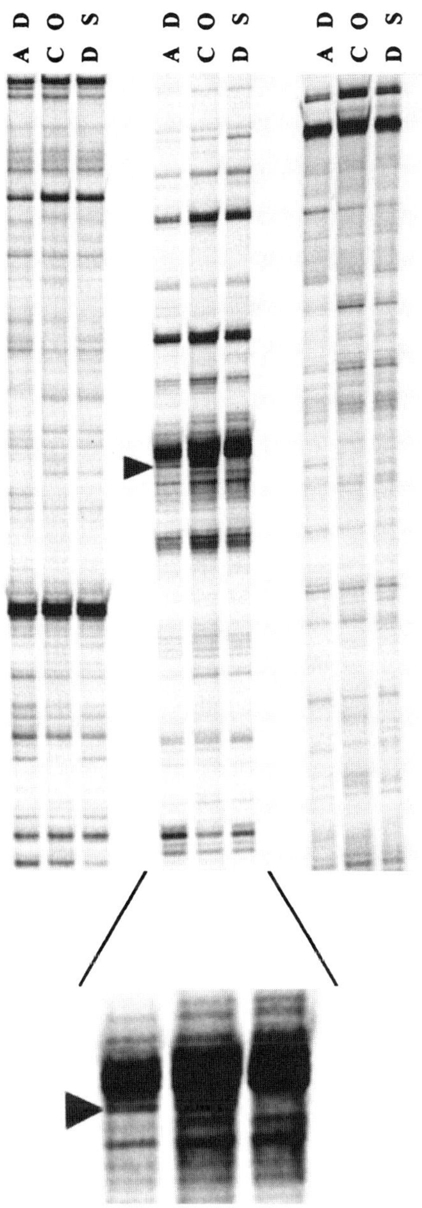

Fig. 1. Differential display to present mRNAs expressed in AD and DS compared to controls (CO). The arrowhead indicates a differentially expressed 157-base cDNA fragment that was amplified using one-base anchored oligo(dT) primer (5′-AAGCTTTTTT TTTTTA-3′) and arbitrary primer (5′-AAGCTTCGACTGT-3′)

The decreased protein levels of ARPP-19 in DS and AD

We investigated the protein levels of ARPP-19 in temporal cortex and cerebellum. The protein levels of ARPP-19 were significantly reduced in DS (temporal cortex) and AD (cerebellum) as shown in Table 1. Age and postmortem interval of brain samples used in immunoblotting varied, but no

```
                    1          11         21         31
157-base cDNA       actgaaaggg ccaagagtaa atgccttgtt ttgttttttt
                    :::::::::: :::::::::: :::::::::: ::::::::::
ARPP-19 mRNA        actgaaaggg ccaagagtaa atgccttgtt ttgttttttt
                    4996       5006       5016       5026

                    41         51         61         71
157-base cDNA       gttttgtttt gttttgcttt ttgttaaaac atgtctatag
                    :::::::::: :::::::::: :::::::::: ::::::::::
ARPP-19 mRNA        gttttgtttt gttttgcttt ttgttaaaac atgtctatag
                    5036       5046       5056       5066

                    81         91         101        111
157-base cDNA       agttggcagt taatgctgaa tttgtcaaat acccccttcca
                    :::::::::: :::::::::: :::::::::: ::::::::::
ARPP-19 mRNA        agttggcagt taatgctgaa tttgtcaaat acccccttcca
                    5076       5086       5096       5106

                    121        131        141        151
157-base cDNA       aaattatact tgtatttaaa aaataaatgg atctacc
                    :::::::::: :::::::::: :::::::::: :::::::
ARPP-19 mRNA        aaattatact tgtatttaaa aaataaatgg atctacc
                    5116       5126       5136       5146
```

Fig. 2. Sequence alignment of 157-base cDNA fragment found by DD-PCR with the sequence obtained from EMBL. The sequence of 157-base cDNA fragment has 100% homology with ARPP-19 mRNA sequence

Table 1. ARPP-19 protein levels in brain regions of patients with DS, AD and controls

	Control	DS	AD
Temporal cortex			
ARPP-19	5.62 ± 2.28	1.36 ± 1.14*	5.88 ± 4.15
N	6	3	4
Age (years)	39.17 ± 16.11	57.33 ± 11.50	60.25 ± 6.95
Postmortem interval (hours)	33.67 ± 12.26	44.33 ± 23.97	18.25 ± 17.17
Cerebellum			
ARPP-19	3.12 ± 3.04	2.22 ± 2.78	0.83 ± 0.59*
N	6	4	5
Age (years)	62.00 ± 9.84	58.50 ± 10.47	60.80 ± 6.83
Postmortem interval (hours)	34.17 ± 12.70	30.00 ± 29.70	17.20 ± 8.38

*$P < 0.05$

Table 2. PKA activities in brain regions of patients with DS, AD and controls

	Control	DS	AD
Temporal cortex	0.55 ± 0.13	0.26 ± 0.09***	0.36 ± 0.17*
Cerebellum	0.35 ± 0.05	0.20 ± 0.08**	0.38 ± 0.09
N	10	9	9
Age (years)	45.10 ± 19.05	55.67 ± 7.48	59.00 ± 8.52
Postmortem interval (hours)	29.70 ± 11.95	31.44 ± 19.56	27.11 ± 19.70

*** $P < 0.001$; ** $P < 0.01$; * $P < 0.05$

correlation was seen between either age or postmortem interval and ARPP-19 protein levels (data not shown). Protein data obtained from immunoblotting confirmed our differential display findings of deteriorated ARPP-19 mRNA levels in temporal cortex of DS.

The decreased activities of PKA in DS and AD

Since ARPP-19 is a substrate for PKA, we also determined activities of PKA. The activities of PKA were decreased in temporal cortex of DS and AD as well as in cerebellum of DS (Table 2). Age and postmortem interval of brain samples had no influence upon PKA activities (data not shown).

Discussion

We observed significantly decreased protein levels of ARPP-19 in temporal cortex of DS confirming DD-PCR results showing deteriorated mRNA level of ARPP-19. The significantly decreased protein levels of ARPP-19 were also observed in cerebellum of AD.

ARPP-19, found in all brain regions, is one of cAMP-regulated phosphoproteins such as ARPP-16, ARPP-21, and DARPP-32 (dopamine and cAMP-regulated phosphoprotein). Intracellular cAMP plays key roles as second messenger signals in modulating myriad physiological processes. Since the synthesis of cAMP is regulated by adenylyl cyclase that acts through G protein-coupled cell-surface receptors with other extracellular stimuli, the level of cAMP is determined by the activities of adenylyl cyclase. Vasoactive intestinal peptide (VIP) and forskolin as activators of adenylate cyclase increased the level of phosphorylation of ARPP-19 (Girault et al., 1988). Several studies reported the involvement of cAMP signaling pathway in the pathology of AD and in an animal model of DS. Basal production of cAMP was shown to be impaired and the response of adenylyl cyclase to the stimulation of beta-adrenoceptors with isoprenaline and of the catalytic subunit with forskolin were both severely depressed in the hippocampus of segmental trisomy (Ts65Dn) mouse, a model for mental retardation in DS (Dierssen et al., 1996, 1997). Reduced cAMP signal transduction was reported in postmortem hippocampus of demented old individuals revealing the significant reduction in basal as well as stimulated adenylyl cyclase activity (Ohm et al., 1989). The significantly reduced activity of adenylyl cyclase was observed in frontal and/or temporal cortex of AD and confirmed in hippocampus and cerebellum of AD (Ross et al., 1993; Schnecko et al., 1994). Reduced levels of brain-specific adenylyl cyclase subtypes were also reported in patients with dementia of the Alzheimer type (DAT). Immunoreactivities of adenylyl cyclase-I and -II were significantly decreased, but those of adenylyl cyclase-IV and -V/VI were unchanged in parietal cortex membranes in DAT (Yamamoto et al., 1996, 1997).

Decreased cAMP levels due to the reduced activity of adenylyl cyclase could subsequently result in decreased activity of PKA. PKA, a major intracellular receptor for cAMP, can mediate the effects due to changes in the intracellular concentration of this second messenger and phosphorylate a variety of proteins in nervous tissue (Horiuch et al., 1990). In this study, we observed decreased activities of PKA in temporal cortex of DS and AD as well as in cerebellum of DS. Synapsin I, a neuronal phosphoprotein enriched in presynaptic nerve terminals, is also regulated by PKA and indeed lower levels of synapsin I were observed in AD hippocampus (Perdahl et al., 1984). Reduced in vitro phosphorylation of synapsin I (site 1) was also reported in AD postmortem tissues (Parks et al., 1991). PKA also appears to modulate the function of nuclear factors that bind to DNA sequences present in the promoter regions of cAMP-induced genes. Most of these genes contain one or a few cAMP-responsive elements (CREs) and cAMP response element binding protein (CREB) is one of them. In DAT, the significantly decreased immunoreactivity of phosphorylated CREB was reported (Yamamoto-Sasaki et al., 1999). CREB-dependent transcription appears essential for long-term memory consolidation in animals (Frank and Greenberg, 1994). Hippocampal long-term potentiation (LTP) is thought to serve as an elementary mechanism for the establishment of certain forms of explicit memory in the mammalian brain. Inhibitors of PKA blocked longer lasting stage (L-LTP), which is one of LTP stages and requires protein synthesis. Furthermore, analogs of cAMP induced potentiation that blocked naturally induced L-LTP (Frey et al., 1993). Activation of presynaptic PKA was shown to be required for induction of cerebellar LTP (Linden and Ahn, 1999). The effect of Alzheimer amyloid beta peptides on the impairment of hippocampal LTP was reported and in Ts65Dn mouse the decreased LTP was also observed (Siarey et al., 1999). On the basis of these studies, the effect of protein phosphorylation on neuronal function could include biochemical events that underlie short-term and long-term memory in neurodegenerative diseases showing dementia.

In conclusion, we observed decreased activity of PKA and its substrate, ARPP-19, in brains of patients with DS and AD. These findings suggest that decreased expression of ARPP-19 could be related to decreased activities of PKA, although decreased PKA maybe due to reduced cAMP levels caused by dysfunction of adenylyl cyclase in neurodegenerative diseases. Alternatively, decreased ARPP-19 may well be due to decreased PKA secondarily, a suggestion that is less probable if seen in the context of our results. Moreover these findings may indicate the contribution of ARPP-19 and PKA to impaired phosphorylation and signaling, relevant for the pathophysiology of both disorders.

Acknowledgments

The authors at the University of Vienna are highly indebted to the Red Bull Company, Salzburg, Austria, for generous financial support.

References

Brene S, Lindefors N, Ehrlich M, Taubes T, Horiuchi A, Kopp J, Hall H, Sedvall G, Greengard P, Persson H (1994) Expression of mRNAs encoding ARPP-16/19, ARPP-21, and DARPP-32 in human brain tissue. J Neurosci 14: 985–998

Browning MD, Huganir R, Greengard P (1985) Protein phosphorylation and neuronal function. J Neurochem 45: 11–23

Dierssen M, Vallina IF, Baamonde C, Lumbreras MA, Martinez-Cue C, Calatayud SG, Florez J (1996) Impaired cyclic AMP production in the hippocampus of a Down syndrome murine model. Brain Res Dev Brain Res 95: 122–124

Dierssen M, Vallina IF, Baamonde C, Garcia-Calatayud S, Lumbreras MA, Florez J (1997) Alternations of central noradrenergic transmission in Ts65Dn mouse, a model for Down syndrome. Brain Res 749: 238–244

Frank DA, Greenberg ME (1994) CREB: a mediator of long-term memory from mollusks to mammals. Cell 79: 5–8

Frey U, Huang Y-Y, Kandel ER (1993) Effects of cAMP stimulate a late stage of LTP in hippocampal CA1 neurons. Science 260: 1661–1664

Girault J-A, Shalaby IA, Rosen NL, Greengard P (1988) Regulation by cAMP and vasoactive intestinal peptide of phosphorylation of specific proteins in striatal cells in culture. Proc Natl Acad Sci USA 85: 7790–7794

Girault J-A, Horiuchi A, Gustafson EL, Rosen NL, Greengard P (1990) Differential expression of ARPP-16 and ARPP-19, two highly related cAMP-regulated phosphoproteins, one of which is specially associated with dopamine-innervated brain region. J Neurosci 10: 1124–1133

Greengard P (1987) Neuronal phosphoproteins. Mediators of signal transduction. Mol Neurobiol 1: 81–119

Horiuch A, Williams KR, Kurihara T, Nairn AC, Greengard P (1990) Purification and cDNA cloning of ARPP-16, a cAMP-regulated phosphoprotein enriched in basal ganglia, and of a related phosphoprotein, ARPP-19. J Biol Chem 265: 9476–9484

Linden DJ, Ahn S (1999) Activation of presynaptic cAMP-dependent protein kinase is required for induction of cerebellar long-term potentiation. J Neurosci 19: 10221–10227

Mirra SS, Heyman A, McKeel D, Sumi SM, Crain BJ, Brownlee LM, Vogel FS, Hughes JP, van Belle G, Berg L (1991) The Consortium to Establish a Registry for Alzheimer's Disease (CERAD), part II. Standardization of the neuropathologic assessment of Alzheimer's disease. Neurology 41: 479–486

Nairn AC, Hemmings HC Jr, Greengard P (1985) Protein kinases in the brain. Ann Rev Biochem 54: 931–976

Nestler EJ, Greengard P (1983) Protein phosphorylation in the brain. Nature 305: 583–588

Ohm TG, Bohl J, Lemmer B (1989) Reduced cAMP-signal transduction in postmortem hippocampus of demented old people. Prog Clin Biol Res 317: 501–509

Parks KM, Sugar JE, Haroutunian V, Bierer L, Perl D, Wallace WC (1991) Reduced in vitro phosphorylation of synapsin I (site 1) in Alzheimer's disease postmortem tissues. Brain Res Mol Brain Res 9: 125–134

Perdahl E, Adolfsson R, Alafuzoff I, Albert KA, Nestler EJ, Greengard P, Winblad B (1984) Synapsin I (protein I) in different brain regions in senile dementia of Alzheimer type and in multi-infarct dementia. J Neural Transm 60: 133–141

Ross BM, McLaughin M, Roberts M, Milligan G, McCulloch J, Knowler JT (1993) Alternations in the activity of adenylate cyclase and high affinity GTPase in Alzheimer's disease. Brain Res 622: 35–42

Schnecko A, Witte K, Bohl J, Ohm T, Lemmer B (1994) Adenylyl cyclase activity in Alzheimer's disease brain: stimulatory and inhibitory signal transduction pathways are differently affected. Brain Res 644: 291–296

Siarey RJ, Carlson EJ, Epstein CJ, Balbo A, Rapoport SI, Galdzicki Z (1999) Increased synaptic depression in the Ts65Dn mouse, a model for mental retardation in Down syndrome. Neuropharmacology 38: 1917–1920

Tierney MC, Fisher RH, Lewis AJ, Zorzitto ML, Snow WG, Reid DW, Nieuwstraten P (1998) The NINCDS-ADRDA Work Group criteria for the clinical diagnosis of probable Alzheimer's disease: a clinicopathologic study of 57 cases. Neurology 38: 359–364

Walaas SI, Greengard P (1991) Protein phosphorylation and neuronal function. Pharmacol Rev 43: 299–349

Yamamoto M, Ozawa H, Saito T, Frolich L, Riederer P, Takahata N (1996) Reduced immunoreactivity of adenylyl cyclase in dementia of the Alzheimer type. Neuroreport 7: 2965–2970

Yamamoto M, Ozawa H, Saito T, Hatta S, Riederer P, Takahata N (1997) Ca2+/CaM-sensitive adenylyl cyclase activity is decreased in the Alzheimer's brain: possible relation to type I adenylyl cyclase. J Neural Transm 104: 721–732

Yamamoto-Sasaki M, Ozawa H, Saito T, Rosler M, Riederer P (1999) Impaired phosphorylation of cyclic AMP response element binding protein in the hippocampus of dementia of the Alzheimer type. Brain Res 824: 300–303

Authors' address: Prof. Dr. G. Lubec, Department of Pediatrics, University of Vienna, Währinger Gürtel 18, A-1090 Vienna, Austria, e-mail: gert.lubec@akh-wien.ac.at

Increased protein levels of heterogeneous nuclear ribonucleoprotein A2/B1 in fetal Down syndrome brains

S. H. Kim[1], **M. Dierssen**[2], **J. C. Ferreres**[3], **M. Fountoulakis**[4], and **G. Lubec**[1]

[1] Department of Pediatrics, University of Vienna, Austria
[2] Medical and Molecular Genetics Center-IRO, Hospital Duran i Reynals, Barcelona, and
[3] Department of Pathology UDIAT-CD, Corporacis Sanit'ria Parc Tauli, Barcelona, Spain
[4] F. Hoffmann — La Roche, Basel, Switzerland

Summary. Heterogeneous nuclear ribonucleoproteins (hnRNPs) are predominantly nuclear RNA-binding proteins that form complexes with RNA polymerase II transcripts. These proteins play pivotal roles in transcription, pre-mRNA processing in the nucleus, cytoplasmic mRNA translation and its turnover. In addition, hnRNPs have been shown to be essential for embryonic development of Drosophila. Here we studied the protein levels of hnRNPs (A2/B1, H and H′) in fetal brain with Down syndrome (DS; n = 5) compared to controls (n = 7). We used two-dimensional (2-D) gel electrophoresis, matrix-assisted laser desorption ionization mass spectroscopy (MALDI-MS) and specific software for quantification. hnRNP A2/B1 was significantly increased in fetal DS brain (13.52 ± 4.50) compared to controls (9.16 ± 1.35), but both hnRNP H and H′ were unchanged. Increased hnRNP A2/B1 in fetal DS brain may represent high activity of RNA processing such as RNA trafficking and telomere protection, and/or it could contribute to abnormal development of DS brains. Furthermore, comparable expression of hnRNP H and H′ suggest a specific upregulation of hnRNP A2/B.

Introduction

The primary transcripts of eukaryotic RNA polymerase II, termed heterogeneous nuclear RNAs (hnRNAs) or pre-mRNAs, undergo a complex and highly regulated series of events in the nucleus as they mature into functional, cytoplasmic mRNA (Darnell, 1982; Nevins, 1983).

Newly transcribed mRNAs (hnRNAs or pre-mRNA) are associated with heterogeneous nuclear ribonucleoproteins (hnRNPs) that were described as a major group of chromatin-associated RNA-binding proteins (Dreyfuss et al., 1993). The structure formed upon the association of these proteins with a pre-mRNA molecule is known as an hnRNP complex (Dreyfuss et al., 1993; Gorlach et al., 1993).

Studies of hnRNP proteins were performed with hnRNP complexes and proteins isolated from nuclear extracts using sedimentation methods. More recent approaches have identified proteins which are associated with nuclear poly(A)$^+$ RNA in vivo by using UV light to covalently cross-link the protein to the RNA of these complexes in living cells and subsequently examining the proteins which have become cross-linked to the RNA. The most sensitive technique for proteins bound to pre-mRNA was made possible by generating monoclonal antibodies against UV-cross-linked RNA-binding proteins. By immunoprecipitation from nucleoplasm under conditions which preserve the integrity of the pre-mRNA, these reagents can in a specific and sensitive manner isolate RNA-protein complexes containing an assortment of ~20 major nuclear proteins (Pinol-Roma et al., 1988). These proteins were denoted hnRNP A1 through hnRNP U. Until now, Approximately 30 abundant proteins have been identified by two-dimensional (2-D) gel electrophoresis of human hnRNP complexes, and the structures of many of these proteins are known (Dreyfuss et al., 1993; Krecic and Swanson, 1999). Cloning of genes and subsequent structural and functional analyses demonstrated that individual hnRNP proteins contain RNA-binding motifs that possess different RNA sequence binding preferences (Dreyfuss et al., 1993).

Many hnRNPs are abundant nuclear proteins that shuttle between the nucleus and cytoplasm and possible functions of hnRNPs in nucleocytoplasmic mRNA export and cytoplasmic mRNA trafficking pathways have been reviewed (Nakielny and Dreyfuss, 1997; Carson et al., 1998). In addition, these proteins have important functions in pre-mRNA processing. The functional relatedness of hnRNP proteins and splicing factors is underscored by the observation that alternative splicing can be regulated by the interplay between an abundant hnRNP A1 and another RNP motif protein, ASF/SF2 (Mayeda and Krainer, 1992).

Studies in non-human vertebrate cells have shown that there is a high degree of conservation, with respect both to the assortment of hnRNP proteins as well as their individual sequences. An important role in embryonic development has recently been demonstrated for Drosophila hnRNP A/B protein, hrp40, which has been shown to be essential for ventral dorsal-ventral axis development of the Drosophila oocyte (Kelley, 1993; Matunis et al., 1993).

Unambiguous identification on hnRNPs in fetal brain with Down syndrome (DS) using 2-D gel electrophoresis and matrix-assisted laser desorption ionization mass spectroscopy (MALDI-MS) made us evaluate the protein levels of three hnRNPs (A2/B1, H and H′) in controls and fetal DS brains for understanding the transcriptional splicing controls in brain development of DS.

Methods

Brain samples preparation

Brain tissues of controls and fetus with DS were obtained from Medical and Molecular Genetics Center-IRO, Hospital Duran i Reynals, Barcelona, Spain. Cerebral cortex of controls (n = 7, male/female = 6/1, 18.79 ± 2.23 weeks) and fetuses with DS (n = 5, male/female = 4/1, 19.80 ± 2.28 weeks) were used for the studies. The brain samples were stored at −70°C until use. Brain tissue was suspended in 0.5 ml of sample buffer consisting of 40 mM Tris, 5 M urea (Merck, Darmstadt, Germany), 2 M thiourea (Sigma, St. Louis, MO, USA), 4% CHAPS (Sigma), 10 mM 1,4-dithioerythritol (Merck), 1 mM EDTA (Merck), 1 mM PMSF (Sigma) and 1 μg/ml of each pepstatin A, chymostatin, leupeptin and antipain. The suspension was sonicated for approximately 30 seconds and centrifuged at 10,000 × g for 10 minutes and the supernatant was centrifuged further at 150,000 × g for 45 minutes to sediment undissolved material. The protein concentration of the supernatant was determined by the Coomassie blue method (Bradford, 1976).

Two-dimensional (2-D) gel electrophoresis

The 2-D gel electrophoresis was performed essentially as reported (Langen et al., 1997). Samples of 2 mg were applied on immobilized pH 3–10 nonlinear gradient strips (IPG, immobilized pH-gradient strips, Pharmacia Biotechnology, Uppsala, Sweden) in sample cups at their basic and acidic ends. Focusing stated at 200 V and the voltage was gradually increased to 5,000 V at 3 V/min and kept constant for a further 24 hours (approximately 180,000 kVh totally). The second-dimensional separation was performed on 9–16% SDS gradient polyacrylamide gels. The gels (180 × 200 × 1.5 mm) were run at 40 mA per gel. After protein fixation with 40% methanol containing 5% phosphoric acid for 12 hours, the gels were stained with colloidal Coomassie blue (Novex, San Diego, CA, USA) for 48 hours. Molecular masses were determined by running standard protein markers (Gibco, Basel, Switzerland), covering the range 10–200 kDa. Isoelectric point (pI) values were used as given by the supplier of the IPG strips. Excess of dye was washed out from the gels with H_2O and the gels were scanned in an Agfa DUOSCAN densitometer (resolution 200). Electronic images of the gels were recorded using PhotoShop (Adobe) and PowerPoint (Microsoft) software.

Matrix-assisted laser desorption ionization mass spectroscopy (MALDI-MS)

MALDI-MS analysis was performed as described elsewhere (Fountoulakis and Langen, 1997) with some modifications. The spots were excised with a spot picker and placed into 96-well microtiter plates. Each spot was destained with 100 μl of 30% acetonitrile in 50 mM ammonium bicarbonate and dried in a speedvac evaporator. Each dried gel pieces was rehydrated with 4 μl of 3 mM Tris-HCl, pH 9.0, containing 50 ng trypsin (Promega, Madison, WI, USA). After 16 hours at room temperature, 7 μl of H_2O were added to each gel piece and the samples were shaken in for 10 minutes. Four μl of 50% acetonitrile, containing 0.3% trifluoroacetic acid, the standard peptides des-Arg-bradykinin (Sigma, 904.4681 Da) and adrenocorticotropic hormone fragment 18–39 (Sigma, 2,465.1989 Da), in H_2O were added to each gel piece and shaken for 10 minutes. The application of the samples was performed with a SymBiot I sample processor (PE Biosystems, Framingham, MA, USA). 1.5 μl of the peptide mixture was simultaneously applied with 1 μl of matrix, consisting of a saturated solution of α-cyano-4-hydroxycinnamic acid (Sigma) in 50% acetonitrile, containing 0.1% trifluoroacetic acid Samples were analyzed in a time-of-flight mass spectrometer (Reflex 3, Bruker Analytics, Bremen, Germany). An accelerating voltage of 20 kV was used. Peptide matching and protein searches were performed automatically. The peptide masses were compared to the theoretical peptide masses of all available proteins from all species. Monoisotopic masses were used and a mass tolerance of 0.0025% was allowed. The algorithm used for determining the probability of a false positive match with a given MS-spectrum is described elsewhere (Berndt et al., 1999).

Quantification of proteins

In partial 2D-gel images including each protein corresponding hnRNPs and the neighbouring proteins, the percentage of the volume of the spot representing each protein was quantified as compared to the total proteins present using the ImageMaster 2D Elite software (Amersham Pharmacia Biotechnology).

Statistics

Results were expressed as means ± standard deviation (SD). Between-group differences were investigated by non-parametric Mann-Whitney U test. The level of significance was set at $P < 0.05$. All statistical analyses were performed by GraphPad Instat2 software, version 2.05.

Results

Identification of hnRNPs in fetal brain

Fetal brain proteins separated by 2-D gel electrophoresis were identified by MALDI-MS analysis based upon molecular weight, pI value and peptide matches. In Fig. 1, the protein spots corresponding to hnRNPs (A2/B1, H and H′) were presented and the data generated were shown in Table 1.

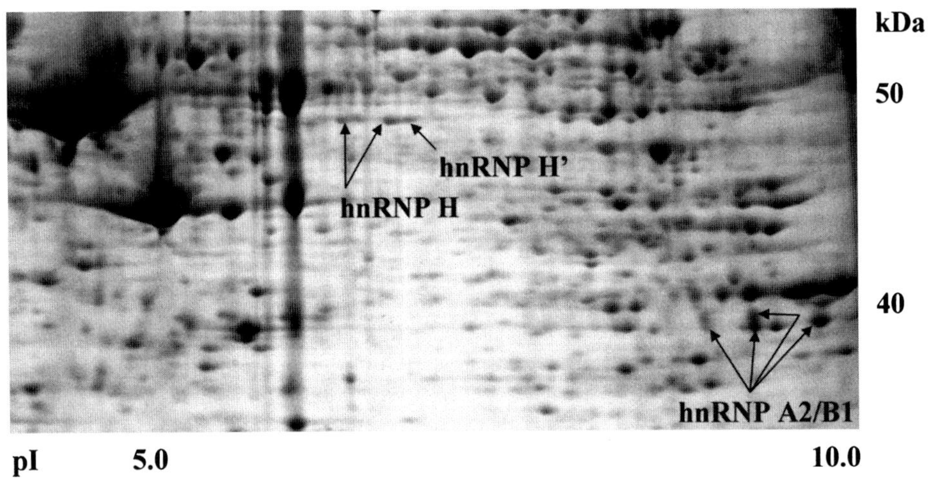

Fig. 1. The partial 2-D gel images of fetal brain. Fetal brain proteins were extracted and separated on an immobilized pH 3–10 nonlinear gradient strip, followed by a 9–16% linear gradient polyacrylamide gel, as stated under experimental procedures. The gel was stained with Coomassie blue and the spots were analyzed by MALDI-MS. Arrows indicate hnRNPs

Table 1. Identification of hnRNPs in fetal brain

Components		pI	MW	Matches	Probability
hnRNP	A2/B1	9.78	37,463	11	1.04E-16
	H	6.26	49,483	9	3.83E-13
	H′	6.26	49,517	5	7.43E-06

Proteins separated by 2-D electrophoresis was identified by MALDI-MS, following digestion with trypsin. The peptide masses were matched with the theoretical peptide masses of all proteins from all species of the SWISS-PROT database

Protein levels of hnRNPs

hnRNP A2/B1 (4 spots), hnRNP H (2 spots) and hnRNP H′ (1 spot) were quantified and the sum value of volume percentage of each spot was used in statistics. The protein levels of hnRNP A2/B1 were significantly increased in fetal DS brain compared to controls, but the protein levels of both hnRNP H and H′ were unchanged (Table 2). The 2-D gels including hnRNP A2/B1 were presented in Fig. 2.

Discussion

The structures, RNA-binding properties, abundance and intracellular trafficking of hnRNPs suggest that they have important functions in mRNA metabolism, and these proteins have been demonstrated to play a role in embryonic development. However, until now, a role of individual hnRNPs has not been elucidated yet in fetal control brains or DS. Here we report that the protein levels of hnRNP A2/B1 were significantly increased in cerebral cortex of fetal DS brain with unchanged protein levels of both hnRNP H and H′.

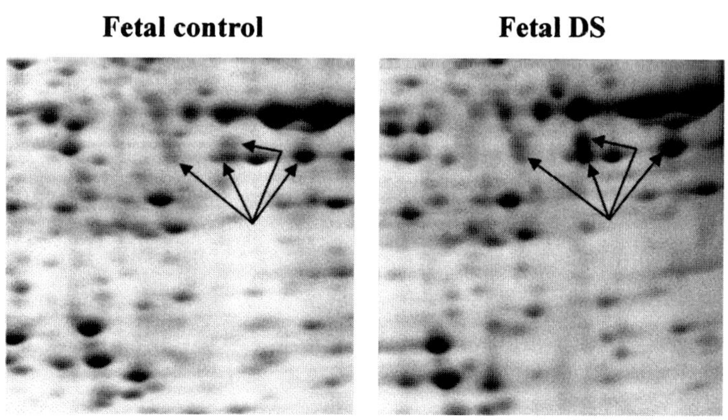

Fig. 2. The partial 2-D gel images of hnRNP A2/B1 with significant difference. Arrows indicate hnRNP A2/B1

Table 2. The protein levels of hnRNPs in fetal brain

Components		Control	DS
hnRNP	A2/B1	9.16 ± 1.35	13.52 ± 4.50*
	H	2.41 ± 0.74	2.77 ± 1.18
	H′	1.24 ± 0.55	1.30 ± 0.62

Proteins from cerebral cortex from fetal DS and controls were separated on 2-D gels and visualized following stain with colloidal Coomassie blue. In partial gel images, including each protein and the neighboring proteins, the intensities of the spots representing each protein were quantified with ImageMaster 2D software, compared to the total proteins present. The statistical analyses were performed by Graphpad Instat2 software. Significant differences are indicated as follow: $^{*}P < 0.05$

Alternative splicing of A2/B1 gene produces abundant A2, less abundant B1. The distribution of A2 and B1 are not constant throughout the rat forebrain and this diversity could reflect the post-transcriptional regulation of cell-specific gene expression of neuronal cells (Mizukami et al., 2000). hnRNP A2 is enriched in brain and immunostaining of oligodendrocytes indicated that the protein is found in both the nucleus and cytoplasm, where it appears in granules in the distal processes and myelin compartment, consistent with a role in RNA trafficking (Carson et al., 1998). In addition, hnRNP A2 was shown to selectively bind the cytoplasmic transport sequence of myelin basic protein mRNA hnRNP A2, especially the 21 nucleotide RNA trafficking signal that is necessary and sufficient for transport of RNA along microtubules in oligodendrocytes (Hoek et al., 1998; Know et al., 1999). In mutation analysis, hnRNP A2 response element was shown to be involved in cytoplasmic RNA trafficking (Munro et al., 1999). hnRNP A2/B1 has been also identified to bind specifically to single stranded vertebrate telomeric repeat TTAGGGn (McKay and Cooke, 1992). Furthermore, in functional assays, B1 was binding to telomeric repeats in a tandem fashion and protect them from a nuclease and promote telomerase activity (Kamma et al., 2001).

The hnRNP H family has been reported to play a role in RNA processing such as splicing and polyadenylation. hnRNP H could activate a c-src alternative exon in neuronal cells as a component of a splicing enhancer complex (Chou et al., 1999). DSEF-1, a member of the hnRNP H family, is an auxiliary factor that assists in the assembly of the general 3′-end processing factors onto the core elements of the polyadenylation signal (Bagga et al., 1998).

In a study comparing total cellular mRNA, rRNA and DNA levels in fibroblasts and lymphocytes derived from normal and DS individuals, there was no significant difference (McQillan and Choo, 1992). However, differences in ribosomal protein genes expression between normal and segmental trisomy mouse, Ts65Dn, were reported using the serial analysis of gene expression technique and implied the involvement in development and maintenance of DS phenotypes (Chrast et al., 2000).

Considering upregulation of hnRNP A2/B1 in dysregulated cell populations (Montuenga et al., 1998) and abnormal ribosomal biogenesis in DS, increased expression of hnRNP A2/B1 protein in fetal DS brain might be a requirement to sustain the higher activity of pre-RNA processing mechanisms (such as RNA trafficking) and telomere protection from a nuclease during brain development, and/or it could play a role for abnormal development of DS brains. Furthermore, unchanged expression of hnRNP H and H′ points to the specificity of hnRNP A2/B1 overexpression in fetal DS brains.

References

Bagga PS, Arhin GK, Wilusz J (1998) DSFE-1 is a member of the hnRNP H family of RNA-binding proteins and stimulates pre-mRNA cleavage and polyadenylation in vitro. Nucl Acids Res 26: 5343–5350

Bendt P, Hobohm U, Langen H (1999) Reliable automatic protein identification from matrix-associated laser desorption/ionization mass spectrometric peptide fingerprints. Electrophoresis 20: 3521–3526

Bradford MM (1976) A rapid and sensitive method for the quantitation of microgram quantities of protein utilizing the principle of protein-dye binding. Anal Biochem 72: 248–254

Carson JH, Kwon S, Barbarese E (1998) RNA trafficking in myelinating cells. Curr Opin Neurobiol 8: 607–612

Chou MY, Rooke N, Turck CW, Black DL (1999) hnRNP H is a component of a splicing enhancer complex that activates a c-src alternative exon in neuronal cells. Mol Cell Biol 19: 69–77

Chrast R, Scott HS, Papasavvas MP, Rossier C, Antonarakis ES, Barras C, Davisson MT, Schmidt C, Estivill X, Dierssen M, Pritchard M, Antonarakis SE (2000) The mouse brain transcriptome by SAGE: differences in gene expression between P30 brains of the partial trisomy 16 mouse model of down syndrome (Ts65Dn) and normals. Genome Res 10: 2006–2021

Darnell JE (1982) Variety in the level of gene control in eukaryotic cells. Nature 297: 365–371

Dreyfuss G, Matunis MJ, Pinol-Roma S, Burd CG (1993) hnRNP proteins and the biogenesis of mRNA. Annu Rev Biochem 62: 289–321

Fountoulakis M, Langen H (1997) Identification of protein by matrix-assisted laser desorption ionization-mass spectroscopy following in gel digestion in low-salt non-volatile buffer and simplified peptide recovery. Anal Biochem 250: 153–156

Gorlach M, Burd CG, Portman DS, Dreyfuss G (1993) The hnRNP proteins. Mol Biol Rep 18: 72–78

Hoek KS, Kidd GJ, Carson JH, Smith R (1998) hnRNP A2 selectively binds the cytoplasmic transport sequence of myelin basic protein mRNA. Biochemistry 37: 7021–7029

Kamma H, Fujimoto M, Fujiwara M, Matsui M, Horiguchi H, Hamasaki M, Satoh H (2001) Interaction of hnRNP A2/B1 isoforms with telomeric ssDNA and the in vitro function. Biochem Biophys Res Commun 280: 625–630

Kelley RL (1993) Initial organization of the Drosophila dosoventral axis depends on an RNA-binding protein encoded by the squid gene. Genes Dev 7: 948–960

Krecic AM, Swanson MS (1999) hnRNP complexes: composition, structure, and function. Curr Opin Cell Biol 11: 363–371

Kwon S, Barbarese E, Carson JH (1999) The cis-acting RNA trafficking signal from myelin basic protein mRNA and its cognate trans-acting ligand hnRNP A2 enhance cap-dependent translation. J Cell Biol 147: 247–256

Langen H, Roeder D, Juranville J-F, Fountoulakis M (1997) Effect of the protein application mode and the acrylamide concentration on the resolution of protein spots separated by two-dimensional gel electrophoresis. Electrophoresis 18: 2085–2090

Matunis EL, Kelley R, Dreyfuss G (1993) Essential role for a heterogeneous nuclear ribonucleoprotein (hnRNP) in oogenesis: hrp40 is absent from the germ line in the dorsoventral mutant squid. Proc Natl Acad Sci USA 91: 2781–2784

Mayeda A, Krainer AR (1992) Regulation of alternative pre-mRNA splicing by hnRNP A1 and splicing factor SF2. Cell 68: 365–375

McKay SJ, Cooke H (1992) hnRNP A2/B1 binds specifically to single stranded vertebrate telomeric repeat TTAGGGn. Nucl Acids Res 20: 6461–6464

McQuillan CI, Choo KH (1992) Comparison of total cellular DNA, mRNA, and rRNA levels between normals and Down syndrome patients. J Inherit Metab Dis 15: 112–120

Mizukami K, Kamma H, Ishikawa M, Dreyfuss G (2000) Immunohistochemical study of the hnRNP A2 and B1 in the rat forebrain. Neuroreport 11: 3099–3102

Montuenga LM, Zhou J, Avis I, Vos M, Martinez A, Cuttitta F, Treston AM, Sunday M, Mulshine JL (1998) Expression of heterogeneous nuclear ribonucleoprotein A2/B1 changes with critical stages of mammalian lung development. Am J Respir Cell Mol Biol 19: 554–562

Munro TP, Magee RJ, Kidd GJ, Carson JH, Barbarese E, Smith LM, Smith R (1999) Mutational analysis of a heterogeneous nuclear ribonucleoprotein A2 response element for RNA trafficking. J Biol Chem 274: 34389–34395

Nakielny S, Dreyfuss G (1997) Nuclear export of proteins and RNAs. Curr Opin Cell Biol 9: 420–429

Nevins JR (1983) The pathway of eukaryotic mRNA formation. Ann Rev Biochem 52: 441–466

Pinol-Roma S, Choi YD, Matunis MJ, Dreyfuss G (1988) Immunopurification of heterogeneous nuclear ribonucleoprotein particles reveals an assortment of RNA-binding protein. Genes Dev 2: 215–227

Authors' address: Prof. Dr. G. Lubec, Department of Pediatrics, University of Vienna, Währinger Gürtel 18, A-1090 Vienna, Austria, e-mail: gert.lubec@akh-wien.ac.at

Decreased protein levels of stathmin in adult brains with Down syndrome and Alzheimer's disease

M. S. Cheon[1], **M. Fountoulakis**[2], **N. J. Cairns**[3], **M. Dierssen**[4], **K. Herkner**[1], and **G. Lubec**[1]

[1] Department of Pediatrics, University of Vienna, Austria
[2] F. Hoffmann — La Roche, Basel, Switzerland
[3] Institute of Psychiatry, Brain Bank, King's College, United Kingdom
[4] Medical and Molecular Genetics Center-IRO, Hospital Duran i Reynals, Barcelona, Spain

Summary. Stathmin, distributed in neurons with high abundance, acts as an intracellular relay, integrating various transduction pathways triggered by extracellular signals and it is involved in physiological regulation of microtubule destabilization. Stathmin has been also shown to be a critical molecule in pathology of neurodegeneration such as Alzheimer's disease (AD), particularly, in neurofibrillary tangle (NFT) formation. Here we evaluated protein levels of stathmin in adult brain from patients with AD and Down syndrome (DS) showing AD-like pathology by applying proteomic technologies with two-dimensional (2-D) gel electrophoresis, matrix-assisted laser desorption ionization mass spectroscopy (MALDI-MS) identification and specific software for quantification of proteins. Significantly decreased protein levels of stathmin were observed in frontal (2.12 ± 1.17, $n = 6$) and temporal (3.05 ± 2.81, $n = 10$) cortices of AD compared to controls (frontal cortex: 4.41 ± 1.70, $n = 8$; temporal cortex: 5.26 ± 2.26, $n = 13$). Stathmin was also significantly decreased in frontal (2.47 ± 1.11, $n = 7$) and temporal (2.02 ± 1.18, $n = 9$) cortices of DS. We also investigated stathmin levels in fetal brain. Stathmin was not significantly changed between fetal DS brain and controls. We suggest that the decreased protein level of stathmin in brains is associated with tangle formation and microtubule instability in DS as well as AD, but stathmin is not involved in the abnormal development of fetal DS brain.

Introduction

Stathmin is an ubiquitous, phylogenetically conserved phosphoprotein (Maucuer et al., 1993) and unevenly distributed in the brain with the highest expression in neurons (Peschanski et al., 1993). It is a major substrate for cyclin-dependent kinase (cdk) kinase and protein kinase A (PKA), and acts as an intracellular relay, integrating various transduction pathways triggered by extracellular signals (Sobel, 1991).

Stathmin could be a critical molecule in pathologic cases where aberrant signal transduction plays a major role, as is suggested for neurofibrillary tangle (NFT) formation. The stathmin protein concentration was reduced in Alzheimer's disease (AD) neocortex and the alteration of the stathmin protein concentration was negatively correlated with tangle numbers but not with plaque numbers (Jin et al., 1996). On the other hand, in contradicting reports increased immunoreactivity of stathmin was demonstrated in AD hippocampal neurons (Saitoh et al., 1993). Stathmin has recently shown to be also involved in physiological regulation of microtubule dynamics that is important for many cellular processes including intracellular transport, maintenance of cell shape and cell polarity (Belmont and Mitchison, 1996).

These promoted us to investigate the protein levels of stathmin in brains from patients with AD and Down syndrome (DS) with AD-like pathology. Two-dimensional (2-D) gel electrophoresis with subsequent matrix-assisted laser desorption ionization mass spectroscopy (MALDI-MS) identification and specific software for quantification of proteins was used. We also investigated fetal DS brain to find out whether the alteration of stathmin begins already in prenatal life.

Methods

Brain samples preparation

Postmortem brain samples were obtained from the MRC London Brain Bank for Neurodegenerative Diseases, Institute of Psychiatry. All DS patients were karyotyped and possessed trisomy 21. A formal cognitive assessment of dementia in DS was not performed. In all DS brains there were abundant and extensive beta-amyloid deposits, neurofibrillary tangles and neuritic plaques. AD patients were selected prospectively and examined clinically and fulfilled the National Institute of Neurological Disorders and Stroke and Alzheimer's disease and Related Disorders Association criteria for probable AD (Tierney et al., 1988). The neuropathological diagnosis of "definite AD" was confirmed using the CERAD criteria (Mirra et al., 1991). Normal control brains were obtained from individuals with no history of neurological or psychiatric illness. The major cause of death was bronchopneumonia in AD and DS and heart disease in controls. Brain samples were dissected, coronal slices snap frozen and stored at −70°C until used. Seven brain regions (cerebellum, frontal, temporal, occipital, parietal cortex, thalamus and caudate nucleus) of patients with AD (n = 13, 58.54 ± 7.57 years old), DS (n = 9, 55.67 ± 7.48 years old) and controls (n = 18, 50.00 ± 16.94 years old) were used. The postmortem interval of brain dissection in AD, DS and controls was 30.31 ± 26.17, 31.44 ± 19.56 and 38.78 ± 18.84 hours, respectively.

Brain tissues of controls and fetus with DS were obtained from Dr. Mara Dierssen (Medical and Molecular Genetics Center-IRO, Hospital Duran i Reynals, Barcelona, Spain). Cerebral cortex of controls (n = 7, male/female = 6/1, 18.79 ± 2.23 weeks) and fetuses with DS (n = 8, male/female = 6/2, 19.81 ± 2.00 weeks) was used.

Brain tissue was suspended in 0.5 ml of sample buffer consisting of 40 mM Tris, 5 M urea (Merck, Darmstadt, Germany), 2 M thiourea (Sigma, St. Louis, MO, USA), 4% CHAPS (Sigma), 10 mM 1,4-dithioerythritol (Merck), 1 mM EDTA (Merck), 1 mM PMSF (Sigma) and 1 μg/ml of each pepstatin A, chymostatin, leupeptin and antipain. The suspension was sonicated for approximately 30 seconds and centrifuged at 10,000 × g for 10 minutes and the supernatant was centrifuged further at 150,000 × g for 45 minutes to sediment undissolved material. The protein concentration of the supernatant was determined by the Coomassie blue method (Bradford, 1976).

Two-dimensional (2-D) gel electrophoresis

The 2-D gel electrophoresis was performed essentially as reported (Langen et al., 1997). Adult samples of approximately 1.5 mg were applied on immobilized pH 3–10 nonlinear gradient strips (IPG, immobilized pH-gradient strips, Pharmacia Biotechnology, Uppsala, Sweden), at both, the basic and acidic ends of the strips. The proteins were focused at 300 V for 1 hour, after which the voltage was gradually increased to 3,500 V within 6 hours. Focusing was continued at 3,500 V for 12 hours and at 5,000 V for 48 hours. The second-dimensional separation was performed on 9–16% linear gradient polyacrylamide gels. After protein fixation with 40% methanol containing 5% phosphoric acid for 12 hours, the gels were stained with colloidal Coomassie blue for 48 hours. The molecular mass was determined by running standard protein markers at the right side of selected gels. The size markers (Gibco, Basel, Switzerland) covered the range 10–200 kDa. Isoelectric point (pI) values were used as given by the supplier of the IPG strips. The gels were destained with H_2O and scanned in a Personal Densitometer (Molecular Dynamics, Sunnyvale, CA, USA). The images were processed using PhotoShop (Adobe) and PowerPoint (Microsoft) software.

For fetal brains, samples of 2 mg were applied, and focusing stated at 200 V and the voltage was gradually increased to 5,000 V at 3 V/min and kept constant for a further 24 hours (approximately 180,000 kVh totally). The second-dimensional separation was performed on 9–16% SDS gradient polyacrylamide gels, and the gels (180 × 200 × 1.5 mm) were run at 40 mA/gel. The gels were scanned in an Agfa DUOSCAN densitometer (resolution 200).

Matrix-assisted laser desorption ionization mass spectroscopy (MALDI-MS)

MALDI-MS analysis was performed as described with minor modifications (Fountoulakis and Langen, 1997). Briefly, spots from gels for adult brain were excised, destained with 50% acetonitrile in 0.1 M ammonium bicarbonate and dried in a speedvac evaporator. The dried gel pieces were reswollen with 3 µl of 3 mM Tris-HCL, pH 8.8, containing 50 ng trypsin (Promega, Madison, WI, USA). After 15 minutes, 3 µl of H_2O were added and left at room temperature for 12 hours. Two microliters of 30% acetonitrile, containing 0.1% trifluoroacetic acid were added, the content was vortexed, centrifuged for 3 minutes and sonicated for 5 minutes. One microliter was applied onto the dried matrix spot. The matrix consisted of 15 mg nitrocellulose (Bio-Rad) and 20 mg α-cyano 4 hydroxycinnamic acid (Sigma) dissolved in 1 ml acetone: isopropanol (1:1, v/v). A 0.5 µl portion of matrix solution was applied on the sample target. Specimen were analyzed in a time-of-flight Voyager Elite mass spectrometer (PerSeptive Biosystems, Cambridge, MA, USA) equipped with a reflectron. An accelerating voltage of 20 kV was used. Calibration was against total spots and background. The peptide masses were matched with the theoretical peptide masses of all proteins from all species of the SWISS-PROT database. For protein search, monoisotopic masses were used and a mass tolerance of 0.0075% was allowed.

For fetal brains, the spots were excised with a spot picker and placed into 96-well microtiter plates. Each spot was destained with 100 µl of 30% acetonitrile in 50 mM ammonium bicarbonate. Each dried gel pieces was rehydrated with 4 µl of 3 mM Tris-HCL, pH 9.0, containing 50 ng trypsin. After 16 hours at room temperature, 7 µl of H_2O

were added to each gel piece and the samples were shaken in for 10 minutes. Four μl of 50% acetonitrile, containing 0.3% trifluoroacetic acid, the standard peptides des-Arg-bradykinin (Sigma, 904.4681 Da) and adrenocorticotropic hormone fragment 18–39 (Sigma, 2,465.1989 Da), in H_2O were added to each gel piece and shaken for 10 minutes. The application of the samples was performed with a SymBiot I sample processor (PE Biosystems, Framingham, MA, USA). 1.5 μl of the peptide mixture was simultaneously applied with 1 μl of matrix, consisting of a saturated solution of α-cyano-4-hydroxycinnamic acid (Sigma) in 50% acetonitrile, containing 0.1% trifluoroacetic acid Samples were analyzed in a time-of-flight mass spectrometer (Reflex 3, Bruker Analytics, Bremen, Germany). Peptide matching and protein searches were performed automatically. Monoisotopic masses were used and a mass tolerance of 0.0025% was allowed. The algorithm used for determining the probability of a false positive match with a given MS-spectrum is described elsewhere (Berndt et al., 1999).

Quantification of proteins

In partial 2D-gel images including each protein corresponding to stathmin and the neighbouring proteins, the percentage of the volume of the spot representing each protein was quantified as compared to the total proteins present using the ImageMaster 2D Elite software (Amersham Pharmacia Biotechnology).

Statistics

Results were expressed as means ± standard deviation (SD). Between-group differences were investigated by non-parametric Mann-Whitney U test. The significance was set at $P < 0.05$. All statistical analyses were performed by GraphPad Instat2 software, version 2.05.

Results

Identification of human brain proteins

As shown in Table 1, stathmin was identified in human adult (Langen et al., 1999) and fetal brain based upon molecular weight, pI value and peptide matches by 2-D gel electrophoresis and MALDI-MS.

Table 1. Identification of stathmin

	pI	MW	Matches
Adult brain	5.90	17,291	6
Fetal brain	5.88	17,160	4

Proteins separated by 2-D gel electrophoresis were identified by MALDI-MS, following digestion with trypsin. The peptide masses were matched with the theoretical peptide masses of all proteins from all species of the SWISS-PROT database

Table 2. The protein levels of stathmin in adult brain with DS and AD

	Control (n)	AD (n)	DS (n)
Cerebellum	1.86 ± 1.64 (10)	1.57 ± 0.91 (10)	1.11 ± 0.95 (5)
Frontal cortex	4.41 ± 1.70 (8)	2.12 ± 1.17* (6)	2.47 ± 1.11* (7)
Temporal cortex	5.26 ± 2.26 (13)	3.05 ± 2.81* (10)	2.02 ± 1.18*** (9)
Occipital cortex	3.27 ± 1.71 (13)	2.41 ± 0.85 (11)	2.16 ± 0.93 (8)
Parietal cortex	4.57 ± 1.68 (6)	3.04 ± 0.69 (5)	3.62 ± 2.46 (7)
Thalamus	1.84 ± 2.16 (8)	2.68 ± 2.04 (6)	0.93 ± 0.78 (5)
Caudate Nucleus	1.70 ± 1.23 (4)	2.05 ± 1.21 (7)	1.20 ± 0.81 (8)

Proteins from seven brain regions of patients with DS and AD, and the control group were separated on 2-D gels and visualized following stain with colloidal Coomassie blue. In partial gel images, including stathmin and the neighboring proteins, the intensities of the spots representing stathmin were quantified, compared to the total proteins present. The quantification was performed using the ImageMaster 2D software. In each donor, the percentage of the volume of the spot representing stathmin was determined. Results were represented as means ± SD (quantified sample number). The statistical analyses were performed by Graphpad Instat2 software. The level of significance was: $^{***}P < 0.001$; $^{*}P < 0.05$

Protein levels of stathmin in brains with DS and AD

Significantly decreased protein levels of stathmin were observed in frontal and temporal cortices of DS and AD (Table 2 and Fig. 1). Although there was no statistical significance, stathmin was also decreased in other regions (occipital cortices, parietal cortices, and cerebellum) of both disorders. In fetal DS brain, however, stathmin was not significantly changed (Table 3).

Discussion

In this study, protein levels of stathmin in frontal and temporal cortices of adult brain with DS and AD were significantly decreased as compared to controls. Stathmin plays a key role in signal transduction pathways involved in cell proliferation and differentiation, and is a major substrate for cdk kinase and PKA. In our previous study, protein levels and activities of cdk were remarkably and significantly decreased in DS brains (Bernert et al., 1996). In

Table 3. The protein levels of stathmin in fetal brain with DS

	Control	DS
Cerebral cortex	10.44 ± 1.64	11.54 ± 1.04

The gels were prepared and the intensities of the spots representing each protein were quantified and analysed as described under legend to Table 2

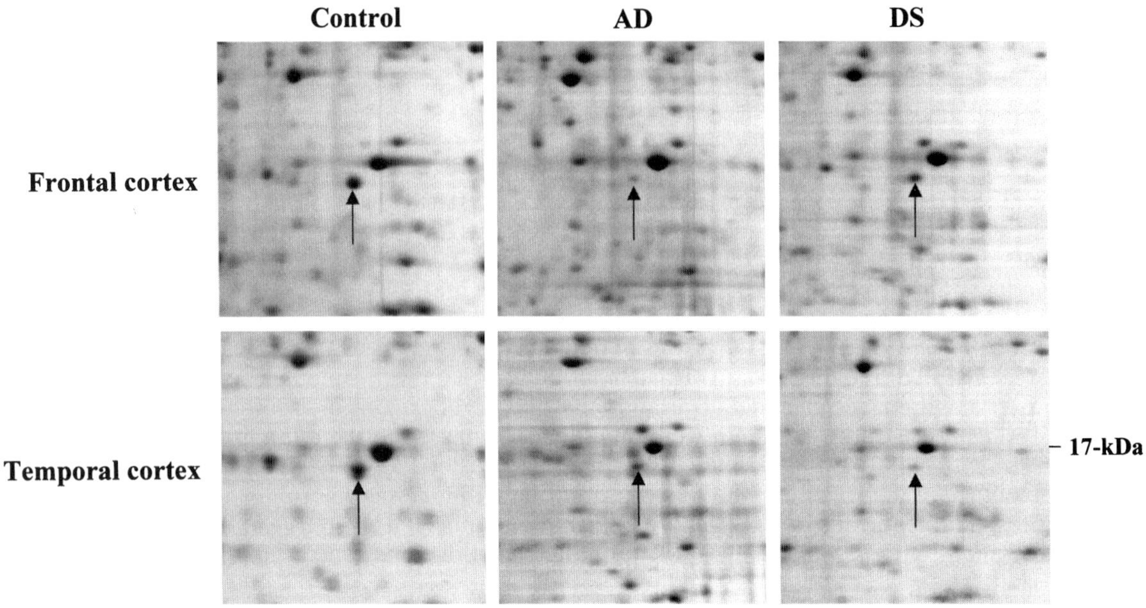

Fig. 1. Two-dimensional (2-D) gel images of stathmin in adult brain regions with significant changes. The human adult brain proteins were extracted and separated on an immobilized pH 3–10 nonlinear gradient strip, followed by a 9–16% linear gradient polyacrylamide gel, as stated under experimental procedures. The gel was stained with Coomassie blue and the spots were analyzed by MALDI-MS. The partial 2-D gel images are shown and an arrow indicates the protein spot corresponding to stathmin with significant changes

addition, significantly reduced activities of PKA were observed in temporal cortex and cerebellum of DS and temporal cortex of AD (unpublished data). Considering significant decrease of cdk, PKA, and stathmin, a major substrate for these kinases, we propose deterioration of the relay function, integrating diverse intracellular signaling pathways involved in regulation of cell proliferation and differentiation in both diseases. These findings could also be in agreement with data providing evidence for cell cycle disturbances in AD and the possibility of this disturbance for the development of pathological changes and the promotion of apoptosis in neurodegenerative diseases (Nagy et al., 1998).

Interestingly, stathmin has been shown to be a critical molecule in pathology of neurodegenerative disease such as AD, particularly, in NFT formation. Stathmin is widely distributed in the cytoplasm of cortical neurons in normal brain (Amat et al., 1991). In AD brain, however, stathmin was found in the cytoplasm of some of the tangle-bearing neurons with high density, although total protein levels of stathmin were reduced. (Jin et al., 1996). In addition, stathmin protein concentration (or stathmin-positive neuronal numbers) in AD had a negative correlation with the number of NFT, but not with plaques (Jin et al., 1996). Moreover, the protein levels of SCG10, a member of neuronal growth-associated proteins (nGAPs) including stathmin, had a positive correlation with the number of NFT, but not with plaques (Okazaki et al., 1995). However, mRNA levels of GAP-43, another nGAP, were diminished as a function of NFT numbers (Coleman et al., 1992). In agreement with the present showing the decreased protein level of stathmin in brains of AD and DS, it is likely that the aberrant expression of several nGAPs is associated with tangle formation in both AD and DS showing AD-like pathology.

Recently, it has been found that stathmin physically interacts with tubulin and causes microtubule destabilization (Walczak, 2000). Microtubules are prominent cytoskeletal elements within the neuron, and essential for differentiation, growth, and maintenance of axons and dendrites. In addition, the size, shape, and compartmentalization of neurons must be generated and supported by the cytoskeletal systems of intracellular transport such as the microtubule-based transport system. Therefore, we suggest that decreased stathmin in neurodegenerative diseases could lead to the disruption and disorganization (or imbalance) of microtubules resulting in improper function related to transport along microtubules.

Stathmin was significantly changed in adult DS and AD brain, however, protein levels of stathmin in fetal DS brain were comparable to controls. We therefore conclude that deranged stathmin may not be involved in the abnormal development of fetal DS brain and may be a phenomenon linked to or reflect neurodegeneration and impaired signaling.

References

Amat JA, Fields KL, Schubart UK (1991) Distribution of phosphoprotein p19 in rat brain during ontogeny: stage-specific expression in neurons and glia. Dev Brain Res 60: 205–218

Belmont LD, Mitchison TJ (1996) Identification of a protein that interacts with tubulin dimers and increases the catastrophe rate of microtubules. Cell 84: 623–631

Bendt P, Hobohm U, Langen H (1999) Reliable automatic protein identification from matrix-associated laser desorption/ionization mass spectrometric peptide fingerprints. Electrophoresis 20: 3521–3526

Bernert G, Nemethova M, Herrera-Marschitz M, Cairns N, Lubec G (1996) Decreased cyclin dependent kinase in brain of patients with Down syndrome. Neurosci Lett 216: 68–70

Bradford MM (1976) A rapid and sensitive method for the quantitation of microgram quantities of protein utilizing the principle of protein-dye binding. Anal Biochem 72: 248–254

Coleman PD, Kazee AM, Lapham L, Eskin T, Rogers K (1992) Reduced GAP-43 message levels are associated with increased neurofibrillary tangle density in the frontal association cortex (area 9) in Alzheimer's disease. Neurobiol Aging 13: 631–639

Fountoulakis M, Langen H (1997) Identification of protein by matrix-assisted laser desorption ionization-mass spectroscopy following in-gel digestion in low-salt nonvolatile buffer and simplified peptide recovery. Anal Biochem 250: 153–156

Jin LW, Masliah E, Iimoto D, Deteresa R, Mallory M, Sundsmo M, Mori N, Sobel A, Saitoh T (1996) Neurofibrillary tangle-associated alteration of stathmin in Alzheimer's disease. Neurobiol Aging 17: 331–341

Langen H, Roeder D, Juranville JF, Fountoulakis M (1997) Effect of the protein application mode and the acrylamide concentration on the resolution of protein spots separated by two-dimensional gel electrophoresis. Electrophoresis 18: 2085–2090

Langen H, Berndt P, Roder D, Cairns N, Lubec G, Fountoulakis M (1999) Two-dimensional map of human brain proteins. Electrophoresis 20: 907–916

Maucuer A, Moreau J, Mechali M, Sobel A (1993) The stathmin gene family: phylogenetic conservation and development regulation in Xenopus. J Biol Chem 268: 16420–16429

Mirra SS, Heyman A, McKeel D, Sumi S, Crain BJ (1991) The consortium to establish a registry for Alzheimer's disease. Neurology 41: 479–486

Nagy Z, Esiri MM, Smith AD (1998) The cell division cycle and the pathophysiology of Alzheimer's disease. Neuroscience 87: 731–739

Okazaki T, Wang H, Masliah E, Cao M, Johnson SA, Sundsmao M, Saitoh T, Mori N (1995) SCG10, a neuronal-specific growth-associated protein in Alzheimer's disease. Neurobiol Aging 16: 883–894

Peschanski M, Doye V, Hirsch E, Marty L, Dusart I, Manceau V, Sobel A (1993) Stathmin: cellular localization of a major phosphoprotein in the adult rat CNS, with a note on human forebrain. J Comp Neurol 337: 655–668

Saitoh T, Horsburgh K, Masliah E (1993) Hyperactivation of signal transduction system in Alzheimer's disease. Ann NY Acad Sci 695: 34–41

Sobel A (1991) Stathmin: a relay phosphoprotein for multiple signal transduction? Trends Biochem Sci 16: 301–305

Tierney MC, Fisher RH, Lewis AJ, Torzitto ML, Snow WG, Reid DW, Nieuwstraaten P, Van Rooijen LAA, Derks HJGM, Van Wijk R, Bischop A (1988) The NINCDA-ADRDA work group criteria for the clinical diagnosis of probable Alzheimer's disease. Neurology 38: 359–364

Walczak CE (2000) Microtubule dynamics and tubulin interacting proteins. Curr Opin Cell Biol 12: 52–56

Authors' address: Prof. Dr. G. Lubec, Department of Pediatrics, University of Vienna, Währinger Gürtel 18, A-1090 Vienna, Austria, e-mail: gert.lubec@akh-wien.ac.at

Molecular neuropathology of transgenic mouse models of Down syndrome

N. J. Cairns

Department of Neuropathology, Institute of Psychiatry, King's College London, De Crespigny Park, London, United Kingdom

Summary. Down syndrome (DS) is a complex, clinically heterogeneous disorder which shows both impairment of neurodevelopement and the neurodegenerative changes of Alzheimer's disease (AD). The phenotype of DS is caused by triplication of chromosome 21 and transgenic mouse models have been developed, and are being created, that carry single genes and chromosomal segments to excess. For example, transgenic mice containing additional copies of the amyloid precursor protein (APP) gene, have been useful in producing the Aβ deposition characteristic of AD and DS, but not the cytoskeletal changes that are the hallmarks of these human disorders. Such models are useful in replicating aspects of pathogenesis and allow for the testing of therapeutic agents to restore impaired function. Segmental trisomic mouse models, which survive to adulthood and possess three copies of multiple genes responsible for the DS phenotype, such as Ts1Cje and Ts65Dn, have been used to explore aspects of neurodevelopment and neurodegeneration. These animal models show some but not all the pathological, biochemical, and transcriptional changes seen in DS. They also have the advantage of allowing for the testing of therapeutic agents to restore impaired function. Analysis of the transcriptome and proteome of fetal and adult DS indicates that there is a complex relationship between gene dosage, gene and protein expression, and that data from animal models will need to be compared and evaluated in the light of data obtained from DS tissue.

Introduction

The clinical phenotype of Down syndrome (DS), the most common form of mental retardation, is the result of trisomy 21. In addition to abnormal development of the central and peripheral nervous systems, there are multiple abnormalities in other organ systems. DS is also remarkable because in those individuals who live to adulthood there are also the stigmata of Alzheimer's disease, the most common neurodegenerative cause of dementia (Lacey-Casem and Oster-Granite, 1994; Cairns, 1999; Bayer et al., 2001). Thus, any animal model that claims to mimic the human disease needs to account for both neurodevelopmental abnormalities and neurodegenerative disease. Data obtained from gene expression and proteomic studies using human brain tissue is a benchmark against which data from in vitro and animal model systems may be compared. Although it is unlikely that an extra copy of single genes on chromosome 21 contributes to all of the changes observed in the DS brain, trisomic mouse models have been useful in elucidating the contribution of single genes and segments of chromosome 21 to dysregulation and neurodegeneration in DS. Transgenic models of multiple genes in excess acting synergistically are more likely to generate the DS phenotype but generating such models is a challenge.

The Down syndrome critical region

The examination of chromosome 21 in subjects with DS reveals that all or parts of the chromosome are triplicated. The Down Syndrome Critical Region (DSCR) has been mapped to 21q22.3 and, if triplicated, results in the abnormalities of DS (Rahman et al., 1989). Although most cases show an extra copy of the distal segment of 21q, no narrow region of chromosome 21 appears to be entirely causative in the development of the DS phenotype (Shapiro, 1999). Heterogeneity in clinical phenotype may reflect the size and site of triplicated segments, but much further work is required to show the precise contribution of each gene or segment to the DS phenotype.

The triplication of 21q including the superoxide dismutase (SOD1) gene led to the hypothesis that the extra copy of the gene in DS would lead to increased SOD1 activity and be involved in pathogenesis. However, it does not appear that an extra copy of this gene necessarily leads to increased cellular activity of the enzyme. Similarly, mRNA levels of the protooncogene ets-2 coded by a locus between D21S17 and DS21S55 are not over expressed in cardiac tissue from children with congenital heart disease and DS when compared to chromosomally balanced children with congenital heart disease (Greber-Platzer et al., 1999b). In contrast, the human single-minded (SIM) gene, located on chromosome 21q22, has been implicated in abnormal development and impairment of learning because the homologue in Drosophila has been show to be a master regulator of the early development of the fly central nervous system (Nambu et al., 1991; Ema et al., 1999). The expression pattern of this gene suggests that it may be involved in neurodevelopmental abnormality (Dahmane et al., 1995). Two novel genes, DSCR5 and DSCR6, have been isolated from the critical region and are candidates for the pathogenesis of DS, although the function of these genes remains to be elucidated (Rahman et al., 1989; Shibuya et al., 2000). Thus, although a number of genes on chromosome 21 are triplicated their precise role has yet to be determined and they are not necessarily markers of DS or essential for abnormal development or degeneration.

Mouse models of Down syndrome

Many of the morphological features observed in trisomy 16 (Ts16) mice are similar to those described in DS indicating that specific genes may affect the phenotype rather than there being a direct gene dosage effect. Unfortunately, Ts16 mice die in utero and any observed effects may be restricted to abnormalities of development rather than neurodegeneration that is seen in adult life. There are basic alterations in the central nervous system including adhesion, migration, proliferation, a reduction in cell numbers and extracellular space is increased (for a review see Lacey-Casem and Oster-Granite, 1994). The fine structural studies of the hippocampus and dorsal root ganglia of Ts16 embryos do not show significant morphological differences to normal control mice and thus their usefulness as a model of the fine structural changes in adult DS and AD is limited (Lane et al., 1996). This difficulty may be overcome by transplanting fetal Ts16 tissue into adult mouse brain and observing the progression of the graft (Richards et al., 1991; Stoll et al., 1993; Nelson et al., 1997; Lushchekina et al., 1999) or as immortalized neural cells (Frederiksen et al., 1996). Two other partial trisomy 16 mice that survive to adulthood have been used. Ts65Dn corresponds to region HSA21 which includes the amyloid precursor protein (APP) gene, glutamate receptor, ionotropic, kainite 1 (GRIK1), superoxide dismutase 1 (SOD1), phosphoribosylglycinamide formyltransferase (GART), simple-minded 2 (Sim1), dual-specificity tyrosine-(Y)-phosphorylation regulated kinase 1a (DRYK1), E26 avian leukemia oncogene (ETS2), and myxovirus (influenza virus) resistence-1 (MX1) genes (Sago et al., 2000). These transgenic mice show some macroscopic and microscopic evidence of selective neuronal loss, dendritic and synaptic anomalies, early developmental delay, reduced birth weight, muscular trembling, male sterility, and learning impairment (Escorihuela et al., 1995; Coussons-Read and Crnic, 1996; Kaufmann and Moser, 2000; Hyde et al., 2001b). A smaller region of chromosome 16, from the SOD1 gene MX1, is included in Ts1Cje. Both sexes are fertile and the morphological and behavioral characteristics are similar to those of Ts65Dn. There are, however, subtle differences in cytoarchitecture and behavior. Learning deficits are less severe than those of Ts65Dn mice. The loss of cholinergic neurons from the basal forebrain, a striking feature of Ts16 (Holtzman et al., 1992b) and Ts65Dn (Holtzman et al., 1996; Granholm et al., 2000), is not observed in Ts1Cje (Sago et al., 1998). These studies indicate that over expression of at least some of the genes in the region APP to SOD1 are necessary for the impairment of some learning tasks as well as selective loss of cholinergic neurons in Ts65Dn. In addition, gene knockout mice are being use to identify the function of genes on chromosome 21 and these, too, will be useful in helping to illuminate the relationship between genes on chromosome 21 with the pathology of DS (Kola and Hertzog, 1997).

Single gene transgenic mice

It is unlikely that any one single gene will cause a single DS phenotypic characteristic. However, animal models with additional copies of single genes may yield novel insights into pathogenesis. Although DS is a complex disease with several genes over expressed it is likely that multiple cellular processes will be affected. Two genes on chromosome 21, SOD1 and APP have been investigated as single gene transgenic mice. Superoxide dismutase 1, an enzyme that metabolizes oxygen free radicals, is over expressed in some transgenic mice but not all studies have confirmed these observations (Galdzicki et al., 1993; London et al., 1997; Huang et al., 2000a). A gene dosage effect has been observed for APP. Constitutively, APP is cleaved by α-secretase, but cleavage by β-secretases (BACE1 and BACE2) and γ-secretase produces a pathological peptide, Aβ, a 39–42 amino acid peptide that aggregates to form β-pleated sheets (Hussain et al., 2000). These aggregates form Aβ plaques in the brains of Alzheimer's disease patients and adult DS patients but two to three decades earlier in the latter. These Aβ deposits induce cytoskeletal changes in neurons by mechanisms that are not fully understood. The resulting neurofibrillary tangles and neuritic plaques contain paired helical filaments that are composed of the microtubule-associated protein tau which is hyperphosphorylated. These changes lead to neuronal dysfunction, loss of synapses, and eventually neuronal loss and are the substrate for the dementia seen in DS and AD (Higgins et al., 1994; Bayer et al., 2001). A number of transgenic mice expressing the amyloid precursor protein gene have been developed including the 751-amino acid form of the human APP which develops extracellular Aβ-immunoreactive deposits co-localizing with dystrophic neurites, reactive astrocytes, and microglial cells, but without the paired helical filaments of adult AD or DS (Higgins et al., 1994). These models are useful for investigating pathogenesis as well as facilitating testing the efficacy and side effects of anti-Aβ compounds.

Neurodevelopmental abnormalities

A great advantage of an animal model, while attempting to replicate some or all of the changes in the neuropathology, neurochemistry, neurophysiology, cognition and behavior of the human form of the disease, is that it offers the opportunity of therapeutic intervention. Typically, following the generation of different transgenic mice as models of DS, there have been studies that have characterized various phenotypic traits. It has become apparent that selective loss of neuronal populations, biochemical and physiological alterations as well as changes in learning and behavior reflect those abnormalities that make up the spectrum of neurodevelopmental changes seen in DS patients (Lacey-Casem and Oster-Granite, 1994).

Triplication of genes on chromosome 21 produces a complex pattern of expression in DS tissues and ongoing studies suggest that a gene dosage effect is naïve (Greber-Platzer et al., 1999a,b) and the comparison between the human pattern of gene expression with that of the transgenic mice is an urgent task. Analysis of the global gene expression profile of normal and Ts65Dn mice brain at postnatal day 30 (P30) has been undertaken and 14 ribosomal protein genes, of which nine were under expressed, were among 330 statistically significant differences between normal male and Ts65Dn male brains (Chrast et al., 2000b). One gene, the single-minded gene, located on chromosome 21 has been implicated in the abnormal neurodevelopment of DS. The expression pattern of murine mSim2 gene in Ts1Cje has been investigated and is over expressed in the zona limitans of the diencephalon (Vialard et al., 2000) and contributes to histopathological abnormalities and reduced exploratory behavior (Chrast et al., 2000a). In the Ts16 mouse, dysregulation of a developmentally regulated gene, APP, has been observed not only from tissue to tissue but also during development in the same tissue indicating that aneuploidy may result in part from disordered expression of developmentally regulated genes (Holtzman et al., 1992a). These results should be seen in the context of altered gene expression in fetal DS brain as revealed by the gene hunting technique of subtractive hybridization where trisomy 21 leads to the deterioration of gene expression and impairment of transport, carriers, channels, signaling and known metabolic and hormone imbalances and the dys-coordinated expression of transcription factors (Kitzmueller et al., 1999). Such studies will be invaluable for establishing the altered pattern of expression in aneuploid brain cells in comparison with normal cells.

The cellular biochemistry that underlies the neurodevelopmental consequences of chromosomal triplication in Ts16 mice has been investigated. Nuclear magnetic resonance (NMR) spectroscopy has been used to evaluate the cytosolic compounds and membrane phospholipids in Ts16 and euploid (control) murine brain at fetal day 15. Proton NMR levels of brain tissue extracts demonstrated decreased levels of choline and increased levels of myo-inositol (MI) in Ts16 brains compared with controls (Yao et al., 2000). These observations are consistent with the cholinergic deficits seen in the basal forebrain of transgenic models of DS (Holtzman et al., 1992b; Burke et al., 1994). Compared with normal brains, Ts16 brains also have higher levels of creatine, adenosine, tyrosine, and anionic phospholipid membrane components, such as phosphatidylinositol and phosphatidylethanolamine (Yao et al., 2000). These observations indicate that multiple biochemical pathways are affected in Ts16 brain development.

Murine trisomy 16 has been used to model neurodevelopmental changes seen in DS. Cell lines obtained from the cerebral cortex of normal and trisomy 16 fetal mice have been shown to express neuronal markers (Allen et al., 2000). Neuronal uptake of H3-choline was shown to be 50% lower in the Ts16 neuronal cell lines compared to controls. These results indicate that there is cholinergic dysfunction in cultured cells as well as primary cultures of TS16 brain tissue. Other structures affected in DS and AD are also altered in trisomy 16 mice models. For example, cultured hippocampal neurons from Ts16 mice have an abnormal action potential due to reduced inward sodium current (Galdzicki et al., 1993). These abnormal electrical properties of neurons may alter the hippocampus during development.

Inflammatory markers are found in association with abnormal and degenerating neurons in the DS brain and these have been associated with increased interferon sensitivity resulting in high levels of interferon inducible gene products including Fas and complement factor C3 that could result in a damaging inflammatory reaction in the brain. Consistent with this possibility, trisomy 16 mouse fetus has been shown to have increased whole brain interferon-γ and Fas receptor immunoreactivity and cultured neurons have increased levels of caspase 1 activity (Hallam et al., 2000). These findings suggest that an interferon-mediated inflammatory response in the trisomy brain may result in loss of neurons.

Oxidative stress has been proposed as a mechanism causing neuronal dysfunction and premature death in AD, DS, and trisomy 16 mice. Evidence to support this hypothesis comes from studies of primary cultures of cerebral cortex from fetal Ts16 mice. In these cortical cultures, the effects of increased production of reactive oxygen species, possibly due to increased SOD1, GluR5, or APP gene dosage, are exacerbated by an insufficient or missing antioxidant response (Scortegagna et al., 1998).

Abnormalities in neurodevelopment occur outside the central nervous system. Using protein gene product (PGP) 9.5 immunohistochemistry, it has been shown that the development and differentiation of the gastroenteric nervous system is delayed and the myenteric nervous plexus does not appear until embryonic day 18 in Ts16 mice in comparison to embryonic day 14 in normal littermates (Li et al., 2000). Thus, trisomy 16 mice embryos are markedly retarded in the gastroenteric nervous system development.

Neurodegenerative changes

Adult DS patients show the stigmata of Alzheimer's disease: abundant Aβ deposits, neurofibrillary tangles, neuritic plaques, neuronal loss and astrocytosis (Bayer et al., 2001). No animal model has perfectly replicated all aspects of AD pathology. Single gene transgenic mice have replicated some of the lesions seen in AD, notably mice with additional copies of the APP gene. Oxidative damage has been implicated in the ageing process of several neurodegenerative diseases. For example, mutations in the SOD1 gene, leading to loss of function, have been found in about 10% of cases of amyotrophic lateral sclerosis, a disease leading to selective loss of motor neurons. An extra copy of the SOD1 gene in DS might be considered to provide some additional protection against free radical damage during ageing. However, mice over expressing SOD1 have been rather less successful in producing an altered phenotype comparable to that seen in DS. Thus, it is likely that some genes that are triplicated in DS will have a more significant effect on neurodegenerative processes than other genes. Both developmental abnormalities and age-related neurodegenerative changes have been modeled in transplanted fetal Ts16 neurons from the basal forebrain (Holtzman et al., 1992b) and in the transgenic mouse (Ts65Dn) (Holtzman et al., 1996; Granholm et al., 2000). These models have been useful for demonstrating the selective atrophy of cholinergic neurons in the basal forebrain, an area of particular vulnerability in both AD and DS. Additional research will be required to demonstrate the relative contribution of single genes and the synergy or dysregulation of multiple genes that lead to neurodevelopmental abnormalities and neurodegeneration in DS.

Oxidative stress caused by increased mitochondrial generation of radical oxygen species has been implicated as a mechanism leading to cell death in neurodegenerative diseases. The Ts16 model has been used to show that there is an increased basal generation of superoxide in Ts16 neurons which may be caused by deficient complex I of the mitochondrial electron transport chain leading to impaired mitochondrial energy metabolism and neuronal death (Schuchmann and Heinemann, 2000b). Also, diminished glutathione levels in Ts16 neurons affect mitochondrial-mediated neuronal death and, therefore, elevating intracellular glutathione concentration may have a neuroprotective effect in DS and AD (Schuchmann and Heinemann, 2000a). Homozygous mice with a two- to five-fold elevation of CuZnSOD in various tissues showed a slight reduction of life span from that of nontransgenic littermate controls. These results suggest that the constitutive and ubiquitous over expression of CuZnSOD alone is insufficient to extend the life spans of transgenic mice (Huang et al., 2000a).

Hippocampal neurons from Ts16 mouse, AD and DS die at an accelerated rate and that this neuronal death may be caused by apoptosis. The nuclei of dying Ts16 neurons are pyknotic and undergo DNA fragmentation, as revealed by terminal transferase-mediated dUTP nick end-labeling (Bambrick and Kruegar, 2001). Secondly, accelerated Ts16 neuronal death is prevented by inhibition of the caspase family of proteases, which are thought to act at an obligatory step in the apoptotic pathway (Bambrick and Kruegar, 2001). These results indicate that over expression of genes on mouse chromosome 16 may lead to caspase-mediated apoptosis.

One of the macroscopic observations of the DS brain is reduced cerebellar volume and this atrophy is also seen in the Ts65Dn mouse (Hyde et al., 2001a). As the cerebellum is involved in motor learning it is likely that loss of volume of this structure contributes to impairment of motor learning. Memory impairment is associated with hippocampal pathology and medial temporal lobe structures are typically atrophic in both AD and adult DS brains. Although Ts65Dn mice do not show marked changes in the volume of the hippocampus in comparison to controls, there is a significant reduction in the number of neurons in subfields of the hippocampus including the dentate gyrus and higher numbers in the CA3 subfield (Insausti et al., 1998). These quantitative changes may contribute to the memory impairment observed in Ts65Dn mice.

Microscopically, a progressive loss of cholinergic neurons in the basal forebrain of Ts65Dn has been observed and coupled with deficits in spatial memory tasks in mice over 6 months of age (Granholm et al., 2000). In a study by Hyde et al. (2001) three types of motor task were used (peg running, accelerating rotorod, and rotating rod). The Ts65Dn mice learned the peg running task as well as controls, and learned the accelerating rotorod and rotating rod tasks as well as the controls indicating that this model of DS did not show the expected impairment of motor learning. Increased expression of voltage-activated calcium channels in cultured hippocampal neurons of Ts16 mice suggest that alterations in long-term depression and long-term potentiation may contribute to mental retardation in DS (Galdzicki et al., 1998). Long-term potentiation, a model of learning, has been investigated in the hippocampus of young (2 months) and older (9 months) Ts65Dn mice and found to be reduced in comparison to age-matched controls indicating that altered synaptic plasticity may also contribute to mental retardation in DS (Siarey et al., 1997). Other behavioral responses, including pain, have been studied in the Ts65Dn mouse. In a model of tonic pain, Ts65Dn mice showed depressed sensitivity to nociception and in the tail-flick test they showed longer latencies than controls, but no differences among groups were observed in morphine responses (Martinez-Cue et al., 1999). The contrasting motor outcomes of these studies may be attributed to methodological differences and different time points used to measure neuronal loss and behavioral and cognitive changes.

Using quantitative electron microscopy, 30% less asymmetric synapses were counted in the temporal cortex of Ts65Dn mice than in controls (Kurt et al., 2000). These results indicate that excitatory synapses are preferentially affected in Ts65Dn mice and that there is an attempt to compensate for the deficit of asymmetric synapses by increasing the contact zone area of existing synapses (Kurt et al., 2000).

Rescue of function by therapeutic intervention

Acetylcholine is one of the major neurotransmitter systems affected in Alzheimer's disease and DS. Both disorders have loss of cholinergic neurons from the nucleus basalis of Meynert and trisomy transgenic mice show similar pathology in the basal forebrain. There have been attempts to enhance the survival of these populations of neurons by infusing neurotrophic factors. Brain-derived neurotrophic factor (BDNF) has been shown to attenuate the loss of cholinergic septal/diagonal band neurons following excitotoxic lesions of their target neurons within the hippocampus of postnatal day 10 rat pups (Burke et al., 1994). Nerve growth factor (NGF) rescues atrophic cholinergic neurons and prevents the shrinking of remaining cholinergic neurons (Holtzman et al., 1993). In addition, both BDNF and NGF induce a sprouting of cholinergic processes within the hippocampus (Burke et al., 1994). Thus, intracerebroventricular injections of NGF, and to a lesser degree BDNF, prevent loss of developing basal forebrain cholinergic neurons following the removal of normal target cells. Ts65Dn mice have also been used to study the effects of lithium on myo-inositol, and other metabolites, which are elevated in the frontal cortex, hippocampus, and brain stem, but not cerebellum (Shetty et al., 2000). Following lithium treatment brain myo-inositol was reduced (Huang et al., 2000b).

A large cluster of genes encoding components of interferon-α, -β, and -γ are carried by human chromosome 21 and mouse chromosome 16 and DS cells display increased sensitivity to interferon action. Trisomy 16 neurons display decreased viability in culture when compared to normal euploid cells and are more sensitive to the anti-cellular effects of interferon. The increased apoptosis seen in Ts16 neurons may be almost completely prevented by anti-γ-interferon IgG (Hallam and Maroun, 1998). Also, in utero anti-interferon IgG treatment of mice pregnant with trisomy 16 fetuses results in an improvement in fetal growth and partial interferon receptor knockout trisomy (PIRKOT) mouse fetus also shows improved growth and increased numbers of viable cortical neurons (Maroun et al., 2000). These data indicate that endogenous interferon plays an important role in the premature death of trisomy neurons and that therapeutic intervention may reverse some of these changes.

Discussion

Chromosome 21 is the smallest human autosome yet important genetic and familial neurodegenerative disorders map to it including familial Alzheimer's disease (APP gene mutations), familial amyotrophic lateral sclerosis (SOD1 mutations), and Down syndrome (trisomy 21). A number of increasingly sophisticated transgenic mouse models have been, and are being, developed to study the effects of gene knockouts, single genes in excess, chromosome segments, and clusters of genes to determine their biological function and their role in abnormal development of the central nervous system and neurodegeneration (Lacey-Casem and Oster-Granite, 1994; Kola and Hertzog, 1997; Sago et al., 2000). Transgenic mice models have been useful in demonstrating gain- and loss-of-function caused by specific mutations and facilitate the use of therapeutic agents to modulate aspects of neurodevelopment or neurodegeneration.

The Ts65Dn model has been particularly useful in modeling specific aspects of neurodevelopment and neurodegeneration and pharmacological intervention may restore loss of function. For example, neurotrophic factors have been shown to reverse cholinergic deficits, pathology common to AD and DS, in the basal forebrain of trisomic mice. APP transgenic mice have also been particularly useful in the development of anti-Aβ compounds and strategies, including immunization (Morgan et al., 2000). As new genes are characterized on chromosome 21 efforts will continue to be made to create transgenics that carry single and multiple genes in excess. Gene expression data from the transgenic models will have to be compared with the complex pattern of expression that is seen in DS. It is a salutary lesson that increased expression, as seen in chromosomal disorders like DS, does not automatically lead to a gain of function but rather a complex pattern of expression and dysregulation.

References

Allen DD, Martin J, Arriagada C, Cardenas AM, Rapoport SI, Caviedes R, Caviedes P (2000) Impaired cholinergic function in cell lines derived from the cerebral cortex of normal and trisomy 16 mice. Eur J Neurosci 12: 3259–3264

Bambrick LL, Kruegar BK (2001) Neuronal apoptosis in mouse trisomy 16: mediation by caspases. J Neurochem 72: 1769–1772

Bayer TA, Wirths O, Majtenyi K, Hartmann T, Multhaup G, Beyreuther K, Czech C (2001) Key factors in Alzheimer's disease: β-amyloid precursor protein processing, metabolism and intraneuronal transport. Brain Pathol 11: 1–11

Burke MA, Mobley WC, Cho JY, Wiegand SJ, Lindsay RM, Mufson EJ, Kordower JH (1994) Loss of developing cholinergic basal forebrain neurons following excitotoxic lesions of the hippocampus: rescue by neurotrophins. Exp Neurol 130: 178–195

Cairns NJ (1999) Neuropathology. J Neural Transm [Suppl] 57: 61–74

Chrast R, Scott HS, Madani R, Huber L, Wolfer DP, Prinz M, Aguzzi A, Lipp HP, Antonarakis SE (2000a) Mice trisomic for bacterial artificial chromosome with the single-minded 2 gene (Sim2) show phenotypes similar to some of those present in the partial trisomy 16 mouse models of Down syndrome. Hum Mol Genet 9: 1853–1864

Chrast R, Scott HS, Papasavvas MP, Rossier C, Antonarakis ES, Barras C, Davisson MT, Schmidt C, Estivill X, Dierssen M, Pritchard M, Antonarakis SE (2000b) The mouse brain transcriptome by SAGE: differences in gene expression between P30 brains of the partial trisomy 16 mouse model of Down syndrome (Ts65Dn) and normals. Genome Res 10: 2006–2021

Coussons-Read ME, Crnic LS (1996) Behavioral assessment of the Ts65Dn mouse, a model for Down syndrome: altered behavior in the elevated plus maze and open field. Behav Genet 26: 7–13

Dahmane N, Charrong G, Lopes C, Yaspo ML, Maunoury C, Decorte L, Sinet PM, Bloch B, Delabar JM (1995) Down syndrome critical region contains a gene homologous to Drosophila sim expressed during rat and human central nervous system development. Proc Natl Acad Sci USA 92: 9191–9195

Ema M, Ikegami S, Hosoya T, Mimura J, Ohtani H, Nakao K, Katsuki M, Galdzicki Z (1999) Mild impairment of learning and memory in mice over expressing the mSim2 gene located on chromosome 16: an animal model of Down's syndrome. Hum Mol Genet 8: 1409–1415

Escorihuela RM, Fernandezteruel A, Vallina IF, Baamonde C, Lumbreras MA, Dierssen M, Tobena A, Florez J (1995) A behavioral assessment of Ts65Dn mice: a putative Down syndrome model. Neurosci Lett 199: 143–146

Frederiksen K, Thorpe A, Richards SJ, Warers J, Dunnett SB, Sandberg BE (1996) Immortalized neural cells from trisomy 16 mice as models for Alzheimer's disease. Ann NY Acad Sci 777: 415–420

Galdzicki Z, Coan E, Rapoport SI (1993) Cultured hippocampal neurons from trisomy 16 mouse, a model for Down syndrome, have an abnormal action-potential due to a reduced inward sodium current. Brain Res 604: 69–78

Galdzicki Z, Coan EJ, Rapoport SI, Stoll J (1998) Increased expression of voltage-activated calcium channels in cultured hippocampal neurons from mouse trisomy 16, a model for Down syndrome. Mol Brain Res 56: 200–206

Granholm ACE, Sanders LA, Crnic LS (2000) Loss of cholinergic phenotype in basal forebrain coincides with cognitive decline in a mouse model of Down's syndrome. Exp Neurol 161: 647–663

Greber-Platzer S, Schatzmann-Turhani D, Cairns N, Lubec G (1999a) Expression of the transcription factor ets-2 in brain of patients with Down syndrome: evidence against the overexpression-gene dosage hypothesis. J Neural Transm [Suppl] 57: 269–281

Greber-Platzer S, Schatzmann-Turhani D, Wollenek G, Lubec G (1999b) Evidence against the current hypothesis of "gene dosage effects" of trisomy 21: ets-2, encoded

on chromosome 21, is not overexpressed in hearts of patients with Down syndrome. Biochem Biophys Res Commun 254: 395–399
Hallam DM, Maroun LE (1998) Anti-gamma interferon can prevent the premature death of trisomy 16 mouse cortical neurons in culture. Neurosci Lett 252: 17–20
Hallam DM, Capps NL, Travelstead AL, Brewer GJ, Maroun LE (2000) Evidence for an interferon-related inflammatory reaction in the trisomy 16 mouse brain leading to caspase 1 mediated neuronal apoptosis. J Neuroimmunol 110: 66–75
Higgins LS, Holtzman DM, Rabin J, Mobley WC, Cordell B (1994) Transgenic mouse brain histopathology resembles early Alzheimer's disease. Ann Neurol 35: 598–607
Holtzman DM, Bayney RM, Li YW, Berger CN, Epstein CJ, Mobley WC (1992a) Dysregulation of gene expression in mouse trisomy 16: an animal model of Down syndrome. EMBO J 11: 619–627
Holtzman DM, Li YW, DeArmond SJ, McKinley MP, Gage FH, Epstein CJ, Mobley WC (1992b) Mouse model of neurodegeneration: atrophy of basal forebrain cholinergic neurons in trisomy 16 transplants. Proc Natl Acad Sci USA 89: 1383–1387
Holtzman DM, Li Y, Chen K, Gage FH, Epstein CJ, Mobley WC (1993) Nerve growth factor reverses neuronal atrophy in a Down syndrome model of age-related neurodegeneration. Neurology 43: 2668–2673
Holtzman DM, Santucci D, Kilbridge J, Chua-Couzens J, Fontana DJ, Daniels SE, Johnson RM, Chen K, Sun YL, Carlson E, Alleva E, Epstein CJ, Mobley WC (1996) Developmental abnormalities and age-related neurodegeneration in a mouse model of Down syndrome. Proc Natl Acad Sci USA 93: 13333–13338
Huang TT, Carlson EJ, Gillespie AM, Shi YP, Epstein CJ (2000a) Ubiquitous overexpression of CuZn superoxide dismutase does not extend life span in mice. J Gerontol A Biol Sci Med Sci 55: B5–B9
Huang W, Galdzicki Z, van Gelderen P, Balbo A, Chikhale EG, Schapiro MB, Rapoport SI (2000b) Brain myo-inositol level is elevated in Ts65Dn mouse and reduced after lithium treatment. Neuroreport 11: 445–448
Hussain I, Powell DJ, Howlett DR, Chapman GA, Gilmour L, Murdock PR, Tew DG, Meek TD, Chapman C, Schneider K, Ratcliffe SJ, Tattersall D, Testa TT, Southan C, Ryan DM, Simmons DL, Walsh FS, Dingwall FS, Christie G (2000) ASP1 (BACE2) cleaves the amyloid precursor protein at the beta-secretase site. Mol Cell Neurosci 16: 609–619
Hyde LA, Crnic LS, Pollock A, Bickford PC (2001a) Motor learning in Ts65Dn mice: a model for Down syndrome. Dev Psychobiol 38: 33–45
Hyde LA, Frisone DF, Crnic LS (2001b) Ts65Dn mice, a model for Down syndrome, have deficits in context discrimination learning suggesting impaired hippocampal function. Behav Brain Res 118: 53–60
Insausti AM, Megias M, Crespo D, Cruz-Orive LM, Dierssen M, Vallina IF, Insausti R, Florenz J (1998) Hippocampal volume and neuronal number in Ts65Dn mice: a murine model of Down syndrome. Neurosci Lett 253: 175–178
Kaufmann WE, Moser HW (2000) Dendritic anomalies in disorders associated with mental retardation. Cereb Cortex 10: 981–991
Kitzmueller E, Labudova O, Rink H, Cairns N, Lubec G (1999) Altered gene expression in fetal Down syndrome brain as revealed by the gene hunting technique of subtractive hybridization. J Neural Transm [Suppl] 57: 99–124
Kola I, Hertzog PJ (1997) Animal models in the study of the biological functions of genes on human chromosome 21 and their role in the pathophysiology of Down syndrome. Hum Mol Genet 6: 1713–1727
Kurt MA, Davies DC, Kidd M, Dierssen M, Florenz J (2000) Synaptic deficit in the temporal cortex of partial trisomy 16 (Ts65Dn) mice. Brain Res 858: 191–197
Lacey-Casem ML, Oster-Granite ML (1994) The neuropathology of the trisomy 16 mouse. Crit Rev Neurobiol 8: 293–322

Lane NJ, Balbo A, Rapoport SI (1996) A fine structural study of the hippocampus and dorsal root ganglion in mouse trisomy 16, a model of Down's syndrome. Cell Biol Int 20: 673–680

Li JC, Busch LC, Kuhnel W (2000) Immunohistochemical study on gastroenteric nervous system in trisomy 16 mice: an animal model of Down syndrome. World J Gastroenterol 6: 793–799

London J, Paris P, Nabarra B, Casanova M, Paly E, Paul V, Brice P, Gilet R, Toyama K, Therond P, Sinet PM (1997) Human SOD-1 (hSOD-1) transgenic mice: a murine model for the study of trisomy 21 and aging. Cytogenet Cell Genet 79: 49

Lushchekina EA, Schtal T, Shlibs R, Goldammar A, Beck M, Bigl V (1999) Can neurografts from mice with chromosome 16 trisomy serve as a model of Alzheimer disease? Bull Exp Biol Med 128: 1145–1149

Maroun LE, Heffernan TN, Hallam DM (2000) Partial IFN-alpha/beta and IFN-gamma receptor knockout trisomy 16 mouse fetuses show improved growth and cultured neuron viability. J Interferon Cytokine Res 20: 197–203

Morgan D, Diamond DM, Gottschall PE, Ugen KE, Dickey C, Hardy J, Duff K, Jantzen P, DiCarlo G, Wilcock D, Connor K, Hatcher J, Hope C, Gordon M, Arendash GW (2000) A beta peptide vaccination prevents memory loss in an animal model of Alzheimer's disease. Nature 408: 982–985

Martinez-Cuẹ C, Baamonde C, Lumbreras MA, Vallina IF, Dierssen M, Florenz J (1999) A murine model for Down syndrome shows reduced responsiveness to pain. Neuroreport 10: 1119–1122

Nambu JR, Lewis JO, Wharton KA, Cruz-Orive LM (1991) The Drosophila single-minded gene encodes a helix-loop-helix protein that acts as a master regulator of CNS midline development. Cell 67: 1157–1167

Nelson PG, Fitzgerald S, Rapoport SI, Neale EA, Galdzicki Z, Dunlap V, Bowers L, von Agoston D (1997) Cerebral cortical astroglia from the trisomy 16 mouse, a model for Down syndrome, produce neuronal cholinergic deficits in cell culture. Proc Natl Acad Sci USA 94: 12644–12648

Rahman Z, Blouin JL, Creaugoldberg N, Watkins PC, Mattei JF, Poissonnier M, Prieur M, Chettouh Z, Nicole A, Aurias A, Sinet PM, Delabar JM (1989) Critical role of the D21S55 region on chromosome 21in the pathogenesis of Down syndrome. Proc Natl Acad Sci USA 86: 5958–5962

Richards SJ, Waters JJ, Beyreuther K, Masters CL, Wischik CM, Sparkman DR, White CL, Abraham CR, Dunnett SB (1991) Transplants of mouse trisomy 16 hippocampus provide a model of Alzheimer's disease neuropathology. EMBO J 10: 297–303

Sago H, Carlson EJ, Smith DJ, Kilbridge J, Rubin EM, Mobley WC, Epstein CJ, Huang W (1998) Ts1Cje, a partial trisomy 16 mouse model for Down syndrome, exhibits learning and behavioral. Proc Natl Acad Sci USA 95: 6256–6261

Sago H, Carlson EJ, Smith DJ, Rubin EM, Crnic LS, Epstein CJ (2000) Genetic dissection of region associated with behavioral abnormalities in mouse models for Down syndrome. Pediatr Res 48: 606–613

Schuchmann S, Heinemann U (2000a) Diminished glutathione levels cause spontaneous and mitochondrial-mediated cell death in neurons from trisomy 16 mice: a model of Down's syndrome. J Neurochem 74: 1205–1214

Schuchmann S, Heinemann U (2000b) Increased mitochondrial superoxide generation in neurons from trisomy 16 mice: a model of Down's syndrome. Free Rad Biol Med 28: 235–250

Scortegagna M, Galdzicki Z, Rapoport SI, Hanbauer I (1998) In cortical cultures of trisomy 16 mouse brain the upregulated metallothionen-I/II fails to respond to H_2O_2 exposure or glutamate receptor stimulation. Brain Res 787: 292–298

Shapiro BL (1999) The Down syndrome critical region. J Neural Transm [Suppl] 57: 41–60

Shetty HU, Siarey RJ, Galdzicki Z, Stoll J, Rapoport SI (2000) Ts65Dn mouse, a Down syndrome model, exhibits elevated myo-inositol in selected brain regions and peripheral tissues. Neurochem Res 25: 431–435

Shibuya K, Kudoh J, Minoshima S, Kawasaki K, Asakawa S, Shimizu N (2000) Isolation of two novel genes, DSCR5 and DSCR6, from Down syndrome critical region on human chromosome 21q22.2. Biochem Biophys Res Commun 271: 693–698

Siarey RJ, Stoll J, Rapoport SI, Galdzicki Z (1997) Altered long-term potentiation in the young and old Ts65Dn mouse, a model for Down syndrome. Neuropharmacology 36: 1549–1554

Stoll J, Balbo A, Ault B, Rapoport SI, Fine A (1993) Long-term transplants of mouse trisomy 16 hippocampal neurons, a model for Down's syndrome, do not develop Alzheimer's disease neuropathology. Brain Res 610: 295–304

Vialard F, Toyama K, Vernoux S, Carlson EJ, Sinet PM, Rahmani Z (2000) Overexpression of mSim2 gene in the zona limitans of the diencephalon of segmental trisomy 16 Ts1Cje fetuses, a mouse model for trisomy 21: a novel whole-mount based RNA hybridization study. Dev Brain Res 121: 73–78

Yao FS, Caserta MT, Wyrwicz AM (2000) In vitro H-1 and P-31 NMR spectroscopic evidence of multiple aberrant biochemical pathways in murine trisomy 16 brain development. Int J Dev Neurosci 18: 833–841

Author's address: Dr. N. J. Cairns, Department of Neuropathology, Institute of Psychiatry, King's College London, De Crespigny Park, London, SE5 8AF, United Kingdom, e-mail: n.cairns@iop.kcl.ac.uk

Down syndrome patients start early prenatal life with normal cholinergic, monoaminergic and serotoninergic innervation

B. Lubec[1], **B. C. Yoo**[2], **M. Dierssen**[3], **N. Balic**[4], and **G. Lubec**[2]

[1] Department of Neonatology, and
[2] Department of Pediatrics, University of Vienna, Vienna, Austria
[3] Medical and Molecular Genetics Center-IRO, Hospital Duran i Reynals, Barcelona, Spain
[4] Institute of Medical and Chemical Laboratory Diagnostics, University of Vienna, Vienna, Austria

Summary. Information on the fetal brain in Down syndrome (DS) is limited. In particular, there is no systematic study available on cholinergic, monoaminergic or serotoninergic innervation in the early second trimester. It was therefore the aim of the study to investigate whether deficits of any of these systems known to occur in adults with DS, was present at this early phase.

For this purpose we determined markers for neuronal density (neuron-specific enolase, NSE), for cholinergic innervation (vesicular acetylcholine transporter, VAChT), for monoaminergic innervation (vesicular monoamine transporter 2, VMAT2; tyrosine hydroxylase, TH) and for the serotoninergic system (serotonine transporter, SERT) in brain of control and DS fetuses in the early second trimester using immunoblotting.

Values for all neurotransmission systems were measurable at this time point of human development and comparable in control and DS fetuses.

We conclude that during the second trimester DS patients do not differ in terms of immunoreactivity for all markers studied. This first study on that subject warrants further investigations for the determination of the time point when neurotransmitter deficits in DS brain are starting, a hallmark most important for pathogenesis and pharmacotherapy.

Abbreviations

DS Down syndrome, *NSE* neuron-specific enolase, *SERT* serotonine transporter, *VAChT* vesicular acetylcholine transporter, *VMAT2* vesicular monoamine transporter 2, *TH* tyrosine hydroxylase.

Introduction

Development of the central nervous system in the fetus with Down syndrome (DS) is seriously and regularly impaired, although no characteristic morphological pattern has been reported yet. Macroscopically, DS brains are small, rounded, foreshortened, exhibiting a steep rise of the occipital lobes, show extreme narrowing of the superior temporal gyri, incomplete opercularization with exposure of the insular cortex and reduced secondary sulcation (De la Monte, 1999). Microscopically, aberrant migration, synaptogenesis and differentiation can be observed as early as from the second trimester (Takashima et al., 1981).

Nothing is known to our knowledge about the nature of neuronal deficits in the fetus with DS and no fair evaluation of neuronal loss was published so far, however important the anecdotal information on neuronal loss is (Lane et al., 1996). There is also no study on the individual cell lineage in terms of cholinergic, monoaminergic and in particular serotoninergic although regular impairment of these neurotransmitter systems has been widely accepted (Schapiro et al., 1987; Sacks and Smith, 1989). However, one study provided evidence for the absence of cholinergic deficits showing comparable brain levels of acetylcholineesterase (AchE) and choline acetyltransferase (ChAT) at the activity level in DS brain of infants 3 months to 1 year of age (Kish et al., 1989), and there is another report on one individual with normal cortical cholinergic marker enzymes at the 38th gestational week (Brooksbank et al., 1989).

The rationale for our study not only was limited information in literature but also the fact that one could expect changes at an early time point in human development: aberrant migration as well as the existing impairment of the concerted actions of transcription factors (TFs) in adult DS brain may well have led to deranged differentiation into monoaminergic or cholinergic neurons (Goelet et al., 1986; Arenander and de Vellis, 1982).

The aim of our study was to show whether in DS brain a) neuronal loss took place at this early stage by quantifying neuron-specific enolase (NSE), a marker of neuronal density, b) whether monoaminergic deficits were present by performing quantification of the markers tyrosine hydroxylase (TH), the vesicular monoamine transporter 2 (VMAT2) and the serotonine transporter (SERT) and c) whether deranged cholinergic innervation was detectable using the cholinergic marker vesicular acetylcholine transporter (VAChT).

We provided evidence that cholinergic, monoaminergic and particularly serotoninergic deficits did not exist at the brain protein level and given time point of brain development in DS, which inevitably develops later in adult DS brain.

Materials and methods

Human fetal brain

Cerebral cortices of fetuses with DS (n = 8, 2 females and 6 males, 19.03 ± 2.33 weeks old) and controls (n = 6, 1 female and 5 males, 19.94 ± 1.97 weeks old) were obtained from Medical and Molecular Genetics Center-IRO, Hospital Duran i Reynals Avia. Castelldefels Km 2,7 08907 Hospitalet de Llobregat Barcelona, Spain. All samples were stored at −70°C and the freezing chain was never interrupted.

Immunobloting

Frozen brain tissues were thawed and suspended in homogenate buffer consisting of 10mM Tris-HCl (pH 7.5), 150mM NaCl, 0.05% (v/v) Tween 20, 1mM PMSF and a mixture of protease inhibitors (1 μg/ml of each pepstatin A, chymostatin, leupeptin and antipain). The suspension was homogenized in a Potter-Elvehjem homogenizer and centrifuged at 4,000 × g for 10 minutes at 4°C. After determination of the protein concentration of the supernatant using BSA protein assay kit (Pierce, USA), the supernatants were mixed with the same volume of SDS-sample buffer consisting of 125.5mM Tris-HCl (pH 6.8), 70mM SDS, 0.001% bromophenol blue, 20% glycerol, 2% 2-mercaptoethanol, incubated at 95°C for 15 minutes and subjected to SDS-PAGE. SDS-PAGE was performed with 12.5% homogeneous gel. To identify the molecular weight of the detected protein, size markers (Bio-Rad, USA) were used. Proteins separated on the gel were transferred onto PVDF membranes (Millipore, USA) and membranes were washed 2 times for 5 minutes with TBS buffer (10mM Tris-HCl pH 7.5 and 150mM NaCl). After blocking in 1.5% non-fat dry milk in TBS buffer containing 0.1% Tween 20 and 1mM $MgCl_2$, membranes were probed for 2 hours at room temperature with either anti-VAChT polyclonal (C20; Santa Cruz, USA) (1:100), anti-VMAT2 polyclonal (C20; Santa Cruz, USA) (1:100), anti-TH monoclonal (Novus, USA) (1:200), anti-SERT polyclonal (Oncogene, USA) (1:2,000) or anti-NSE monoclonal (Chemicon, USA) (1:3,000). After primary antibody incubation, membranes were washed 3 times for 15 minutes with blocking buffer and probed for 1 hour at room temperature with a 1:2,000 dilution of either horseradish peroxidase conjugated anti-rabbit or anti-mouse IgG antibodies (Southern Biotechnology Associates, Inc., USA). Membranes were washed 3 times for 15 minutes and developed with the western blot chemiluminescence reagent (NEN™ Life Science Product, Inc., USA). Developed films were scanned.

Statistical analyses

Optical densities of immunoreactive protein bands on the films were calculated by RFLP scan 2.1 software program (Scanalytics, USA). All results were expressed as means ± standard deviations. Between-group differences were calculated by non-parametric Mann-Whitney U-test and within-group-correlations were done using the Spearman rank-coefficient. The significance was set at the $P < 0.05$ level and all statistical analyses were performed by the InStat2 software.

Results

As shown in Fig. 1A, the immunoreactive bands of NSE, VAChT, VMAT2 and TH in fetal cerebral cortex appeared with a molecular mass of approximately 47, 40, 40 and 60kDa, respectively. Two immunoreactive SERT bands were detected with different molecular mass, approximately 35 and 30kDa (Fig. 1A). In both fetal DS and control groups, an additional 30kDa immunoreactive SERT band appeared as a band with very weak density compared to 35kDa immunoreactive band.

Comparable expression levels of all proteins studied were noted in fetal DS cerebral cortex (Fig. 1B). Although NSE, VAChT and SERT levels showed tendency to decease in fetal DS by 22% to 41% compared to controls, the difference did not reach significance level due to the variation among the

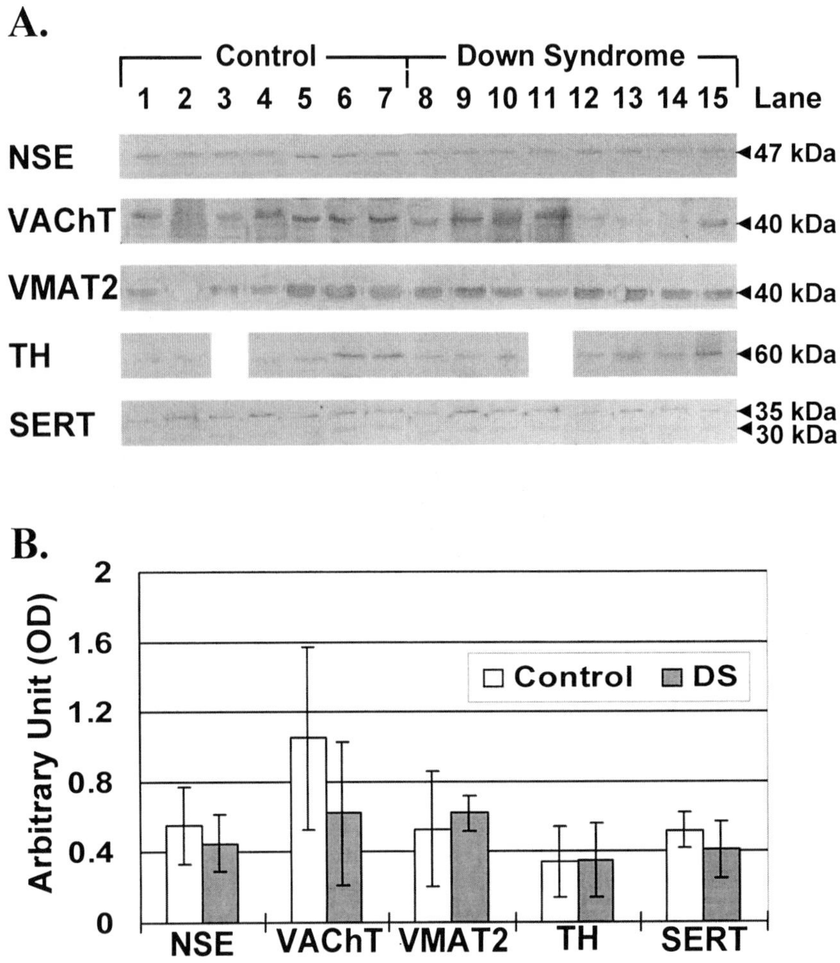

Fig. 1. Unaltered protein levels of NSE, VAChT, VMAT2, SERT and TH in cerebral cortex of the fetal DS. **A** Immunoblot analysis of fetal DS brain proteins. Proteins from fetal DS brains (Lane 1–7) and controls (Lane 8–15) were subjected to SDS-PAGE and immunoblotted with either anti-NSE monoclonal, anti-VAChT polyclonal, anti-VMAT2 polyclonal, anti-SERT polyclonal or anti-TH monoclonal. The major immunoreactivities of NSE, VAChT, VMAT2, TH and SERT appear as bands with a molecular mass of approximately 47, 40, 40, 60 and 35 kDa, respectively. Immunoreactivity of fetal SERT appears as an additional band with low molecular mass (approximately 30 kDa). Lane 4 is for 36 weeks old control and all data from this lane were not used for statistical analysis. **B** Protein levels of NSE, VAChT, VMAT2, SERT and TH in fetal DS brain. All protein levels were expressed as the mean ± standard deviation of immunoreactive band density. The statistical significance was set at the $P < 0.05$ level. All protein levels did not show any significant difference in fetal DS brain

results. Normalized VAChT, TH and SERT levels against NSE and VMAT2 also did not show any difference in fetal DS cerebral cortex (data not shown). Only the immunoreactive level of minor species SERT with the low molecular mass (30 kDa) and its normalized level against VMAT2 were decreased in fetal DS cerebral cortex (both $P < 0.05$, data not shown). However, total immunoreactive SERT (Fig. 1B) and its normalized levels against NSE and VMAT2 (data not shown) did not show any significant difference in fetal DS cerebral cortex.

Discussion

In contrast to the few anecdotal histological case reports (Lane et al., 1996), no neuronal loss could be detected in fetal DS brain using monoclonal antibody against NSE — protein as a marker for neuronal density. The result showed comparable NSE level in fetal DS brain indicating the absence of neuronal loss (Fig. 1). Although neuronal density was comparable, we had to show whether the relation of individual innervation systems was shifted.

Perry and co-workers (1986) described and summarized the development of the cholinergic system in the human system adding further evidence at the cholinergic receptor level to the observation that the nucleus of Meynert is considered the major source of cholinergic axons projecting to the neocortex and it has been demonstrated that this pathway is present at the very early stage of human development i.e. 8 weeks (Candy et al., 1985) with AChE and ChAT activity levels comparable to those of adults. Yew and Chan (1999) recently detected by immuno-histochemical methods AChE — as early as from 7th week of gestation. In our study we therefore confirmed the presence of cortical cholinergic innervation by the determination of VAChT, as we considered this marker more appropriate as AChE, which is expressed not only in brain tissue and furthermore, VAChT may not be susceptible to metabolic tissue factors as an enzyme marker. VAChT, which transports acetylcholine into synaptic vesicles, was used as it is known to be consistent with anatomical, pharmacological and histochemical information on the distribution of functional cholinergic neurons in the brain (Schafer et al., 1994). Using anti-VAChT monoclonal we detected no differences of VAChT level between fetal control and DS cortex (Fig. 1). The unequivocal findings of impaired cholinergic innervation found in adult DS as a hallmark of disease has to be seen as a secondary phenomenon most probably due to the slowly developing AD affecting virtually all DS patients from the fourth decade (Malamud, 1972).

Yew and Chan (1999) also provided evidence for the early appearance of the monoaminergic system revealing that monoamine oxidase protein is detectable in layer II, III and IV of the visual cortex from the 18th week of gestation and the development of MAO activity was found to be comparable to that in rodents (Catravas et al., 1977). For our studies we considered TH and the VMAT2 reliable markers widely used for the evaluation of monoaminergic innervation. TH is the rate limiting enzyme in the biosynthetic pathways of the catecholamines L-dopa, dopamine, norepinephrine and epinephrine and TH-immunoreactive cells are found as early as during the first trimester of gestation (Sundstrom et al., 1993).

VMAT2 is ontogenetically conserved and localized in monoaminergic cell groups of the CNS including brain stem, sensory relay nuclei, limbic structures, basal ganglia and hypothalamic nuclei, all brain regions known to be affected in DS (Hansson et al., 1998). VMAT2 is present along axons, soma and dendrites (Hoffman et al., 1998) and this transporter is a major determinant for regulation of monoamine levels.

We conclude that both marker proteins TH and VMAT2 are expressed at the protein level at the early second trimester in human brain and that protein levels were comparable between control and DS brains (Fig. 1). Again in this system, the fetus with DS starts the second trimester with normal monoaminergic systems. While TH may reflect monoaminergic neuronal density, the transporter VMAT2 would be expected to be also dysregulated if any monoaminergic impairment was present as described by several groups (Risser et al., 1997).

The neurotransmitter of the serotoninergic system is known to be present in growth cones even before synaptogenesis occurs (Ivgy-May et al., 1994). Studies on the ontogenesis of 5-HT neurones indicated that 5-HT positive cells were observed at week 10 of gestation (Shen et al., 1986). Moreover, serotonin (5-HT) itself plays a role in the development process (Lauder, 1990). Several authors have shown that 5-HT acts as a differentiation and growth factor (Lipton and Kater, 1989; Hamon et al., 1989; Mattson, 1998). Findings of decreased brain levels of 5-HT in adult DS therefore were challenging and would be able to explain a series of brain defects in that disorder (Seidl et al., 1999). Aberrant serotoninergic transporter expression reinforced the hypothesis that the serotoninergic transporter is involved in the pathogenesis of DS (Gulesserian et al., 2000). In the present study we used the serotoninergic marker SERT, a serotonin transporter, for the evaluation of serotoninergic density. Immunoreactive level of minor species SERT with the low molecular mass (30kDa) was decreased in fetal DS cerebral cortex ($P < 0.05$, data not shown). We suspect that this minor species is an accumulated or pathogenetically modified protein of SERT. Its biological meaning in development human brain is still not clear. However, total immunoreactive SERT (Fig. 1B) and its normalized levels against NSE and VMAT2 (data not shown) did not show any significant difference in fetal DS cerebral cortex.

The findings of comparable levels in controls and the fetus with DS (Fig. 1) urgently warrants studies of the neurotransmitter 5-HT and its metabolites per se in fetal brain before the attractive 5-HT hypothesis for DS pathogenesis is abandoned.

This report showing that the fetus in the second trimester presents without neuronal loss, without deterioration of cholinergic, monoaminergic or serotoninergic system, must be seen as a stimulus to further put the emphasis on studies elucidating the fetal brain in DS at different systemic levels and different time points.

Acknowledgments

We are highly indebted to the Red Bull company, Salzburg, Austria, for generous support of the study.

References

Arenander AT, de Vellis J (1982) Glial-released proteins in neural intercellular communication: molecular mapping, modulation, and influence on neuronal differentiation. Prog Clin Biol Res 79: 243–269

Brooksbank BW, Walker D, Balazs R, Jorgensen OS (1989) Neuronal maturation in the foetal brain in Down's syndrome. Early Hum Dev 18: 237–246

Candy JM, Perry EK, Perry RH, Bloxham CA, Thompson J, Johnson M, Oakley AE, Edwardson JA (1985) Evidence for the early prenatal development of cortical cholinergic afferents from the nucleus of Meynert in the human foetus. Neurosci Lett 61: 91–95

Catravas GN, Takenaga J, McHale CG (1977) Effect of chronic administration of morphine on monoamine oxidase activity in discrete regions of the brain of rats. Biochem Pharmacol 26: 211–214

De la Monte SM (1999) Molecular abnormalities of the brain in Down syndrome: relevance to Alzheimer's neurodegeneration. J Neural Transm [Suppl] 57: 1–19

Goelet P, Castellucci VF, Schacher S, Kandel ER (1986) The long and the short of long-term memory—a molecular framework. Nature 322: 419–422

Gulesserian T, Engidawork E, Cairns N, Lubec G (2000) Increased protein levels of serotonin transporter in frontal cortex of patients with Down syndrome. Neurosci Lett 296: 53–57

Hamon M, Bourgoin S, Chanez C, De Vitry F (1989) Do serotonin and other neurotransmitters exert a trophic influence on the immature brain? In: Evrard P, Minkowski A (eds) Nestle nutrition workshop series: developmental neurobiology. Raven Press, New York, pp 171–183

Hansson SR, Hoffman BJ, Mezey E (1998) Ontogeny of vesicular monoamine transporter mRNAs VMAT1 and VMAT2. I. The developing rat central nervous system. Brain Res Dev Brain Res 110: 135–158

Hoffman BJ, Hansson SR, Mezey E, Palkovits M (1998) Localization and dynamic regulation of biogenic amine transporters in the mammalian central nervous system. Front Neuroendocrinol 19: 187–231

Ivgy-May N, Tamir H, Gershon MD (1994) Synaptic properties of serotonergic growth cones in developing rat brain. J Neurosci 14: 1011–1029

Kish S, Karlinsky H, Becker L, Gilbert J, Rebbetoy M, Chang LJ, DiStefano L, Hornykiewicz O (1989) Down's syndrome individuals begin life with normal levels of brain cholinergic markers. J Neurochem 52: 1183–1187

Lane NJ, Balbo A, Rapoport SI (1996) A fine structural study of the hippocampus and dorsal root ganglion in mouse trisomy 16, a model of Down's syndrome. Cell Biol Int 20: 673–680

Lauder JM (1990) Ontogeny of the serotonergic system in the rat: serotonin as a developmental signal. Ann NY Acad Sci 600: 297–314

Lipton SA, Kater SB (1989) Neurotransmitter regulation of neuronal outgrowth, plasticity and survival. Trends Neurosci 12: 265–270

Malamud N (1972) Neuropathology of organic brain sydromes associated with aging. In: Gartz CM (ed) Aging and the brain. Nestec Ltd., Vevey; Raven Press, New York, p 63

Mattson MP (1998) Neurotransmitters in the regulation of neuronal cytoarchitecture. Brain Res 472: 179–212

Perry EK, Smith CJ, Atack JR, Candy JM, Johnson M, Perry RH (1986) Neocortical cholinergic enzyme and receptor activities in the human fetal brain. J Neurochem 47: 1262–1269

Risser D, Lubec G, Cairns N, Herrera-Marschitz M (1997) Excitatory amino acids and monoamines in parahippocampal gyrus and frontal cortical pole of adults with Down syndrome. Life Sci 60: 1231–1237

Sacks B, Smith S (1989) People with Down's syndrome can be distinguished on the basis of cholinergic dysfunction. J Neurol Neurosurg Psychiatry 52: 1294–1295

Schafer MK, Weihe E, Varoqui H, Eiden LE, Erickson JD (1994) Distribution of the vesicular acetylcholine transporter (VAChT) in the central and peripheral nervous systems of the rat. J Mol Neurosci 5: 1–26

Schapiro MB, Kay AD, May C, Ryker AK, Haxby JV, Kaufman S, Milstien S, Rapoport SI (1987) Cerebrospinal fluid monoamines in Down's syndrome adults at different ages. J Ment Defic Res 31: 259–269

Seidl R, Kaehler ST, Prast H, Singewald N, Cairns N, Gratzer M, Lubec G (1999) Serotonin (5-HT) in brains of adult patients with Down syndrome. J Neural Transm [Suppl] 57: 221–232

Shen WZ, Luo ZB, Zheng DR, Yew DT (1989) Immunohistochemical studies on the development of 5-HT (serotonin) neurons in the nuclei of the reticular formations of human fetuses. Pediatr Neurosci 15: 291–295

Sundstrom E, Kolare S, Souverbie F, Samuelsson EB, Pschera H, Lunell NO, Seiger A (1993) Neurochemical differentiation of human bulbospinal monoaminergic neurons during the first trimester. Brain Res Dev Brain Res 75: 1–12

Takashima S, Becker LE, Armstrong DL, Chan F (1981) Abnormal neuronal development in the visual cortex of the human fetus and infant with down's syndrome. A quantitative and qualitative Golgi study. Brain Res 225: 1–21

Yew DT, Chan WY (1999) Early appearance of acetylcholinergic, serotoninergic, and peptidergic neurons and fibers in the developing human central nervous system. Microsc Res Tech 45: 389–400

Authors' address: Prof. Dr. G. Lubec, Department of Pediatrics, University of Vienna, Währinger Gürtel 18, A-1090 Vienna, Austria, e-mail: gert.lubec@akh-wien.ac.at

Expression profiles of proteins in fetal brain with Down syndrome

M. S. Cheon[1], **M. Fountoulakis**[2], **M. Dierssen**[3], **J. C. Ferreres**[4], and **G. Lubec**[1]

[1] Department of Pediatrics, University of Vienna, Austria
[2] F. Hoffmann — La Roche, Basel, Switzerland
[3] Medical and Molecular Genetics Center-IRO, Hospital Duran i Reynals, Barcelona, and [4] Department of Pathology UDIAT-CD, Corporacis Sanit'ria Parc Tauli, Barcelona, Spain

Summary. Proteomics is a powerful tool for evaluating differential protein expression comparing hundreds of proteins simultaneously. In the current study we performed "gene hunting" at the protein level and identified and quantified 10 protein spots in control and Down syndrome (DS) fetal brains. Using two-dimensional (2-D) electrophoresis of fetal brain proteins with subsequent MALDI-identification and quantification with specific software, we identified a series of poorly known proteins, in part hypothetical and orphans or poorly documented proteins. Hypothetical protein DKFZp564D177.1-human (fragment), one of these proteins was identified in fetal brain and was significantly decreased in DS (0.61 ± 0.44, $n = 7$) compared to controls (3.43 ± 1.83, $n = 7$). Septin 6, previously shown to be associated with synaptic vesicles, was present in all of 7 controls, but only in 1 out of 6 DS brains. We suggest that decreased protein levels of hypothetical protein DKFZp564D177.1-human (fragment) and lower prevalence of septin 6 could be involved in the maldevelopment of fetal DS brains. The other 8 proteins (WD repeat protein 1, novel protein highly similar to septin 2 homolog, septin 5, septin 2, DJ37E16.5 (novel protein similar to nitrophenylphosphatases from various organism), hypothetical 30.2 kDa protein, neuronal protein NP25, and DC7 protein (vacuolar sorting protein 29)) were comparable between controls and DS but could be identified in fetal and DS cortex, thus proposing them as tentative brain proteins.

Introduction

Down syndrome (DS), or trisomy 21, is the most common genetic cause of mental retardation. Changes in the neuropathology, neurochemistry, neurophysiology, and neuropharmacology of DS patients' brains indicate that there is abnormal development and maintenance of CNS structure and function

(Wisniewski and Rabe, 1986; Becker et al., 1991; Cairns, 1999; Nagy, 1999; Epstein, 2001). Although a large series of genes and proteins have been found to be deranged in adult DS brain, information on the proteome of the fetal brain and in particular the proteome of fetal DS brain is limited. We therefore aimed to contribute to the fetal brain proteome performing "gene hunting" at the protein level, generating brain protein maps and investigating individual spots on two-dimensional (2-D) gels qualitatively and quantitatively using proteomics: Brain proteins are separated by 2-D gel electrophoresis, identified by matrix-assisted laser desorption ionization mass spectroscopy (MALDI-MS) based upon molecular weight, pI value and peptides are matched with SWISS-PROT database, and quantified with specific software for evaluating brain protein levels.

The field of proteomics in brain research is becoming increasingly important as genome sequences are being completed and annotated (Dutt and Lee, 2000). Moreover, working at the protein level gives more direct informations than working at the RNA level, in particular, as the protein is mainly the functioning part and there is a long and unpredictable way from RNA to protein expression. We therefore decided to systematically generate brain maps and search the proteome of fetal control and DS for differential protein expression and already at this time point our first adult brain protein maps (Langen et al., 1999) are updated to many hundreds (unpublished results).

In this study we selected 10 spots from 2-D gels representing proteins for characterization and quantification in control and DS fetal brain. We detected proteins, which were either hypothetical protein, orphans without assigned function and not even described in brain or poorly documented proteins. The findings of two deranged proteins in fetal DS brain made us report this observation.

Methods

Brain sample preparation

Brain tissues of controls and fetus with DS were obtained from Dr. M. Dierssen (Medical and Molecular Genetics Center-IRO, Hospital Duran i Reynals, Barcelona, Spain) and Dr. J. C. Ferreres (Department of Pathology UDIAT-CD, Corporacis Sanit'ria Parc Tauli, Barcelona, Spain). Cerebral cortex of controls (n = 7, male/female = 6/1, 18.79 ± 2.23 weeks of gestation) and fetuses with DS (n = 8, male/female = 6/2, 19.81 ± 2.00 weeks of gestation) were used.

Brain tissue was suspended in 0.5 ml of sample buffer consisting of 40 mM Tris, 5 M urea (Merck, Darmstadt, Germany), 2 M thiourea (Sigma, St. Louis, MO, USA), 4% CHAPS (Sigma), 10 mM 1,4-dithioerythritol (Merck), 1 mM EDTA (Merck), 1 mM PMSF (Sigma) and 1 μg/ml of each pepstatin A, chymostatin, leupeptin and antipain. The suspension was sonicated for approximately 30 seconds and centrifuged at 10,000 × g for 10 minutes and the supernatant was centrifuged further at 150,000 × g for 45 minutes to sediment undissolved material. The protein concentration of the supernatant was determined by the Coomassie blue method (Bradford, 1976).

Two-dimensional (2-D) gel electrophoresis

The 2-D gel electrophoresis was performed essentially as reported (Langen et al., 1997). Samples of 2 mg were applied on immobilized pH 3–10 nonlinear gradient strips (IPG, immobilized pH-gradient strips, Pharmacia Biotechnology, Uppsala, Sweden) in sample cups at their basic and acidic ends. Focusing stated at 200 V and the voltage was gradually increased to 5,000 V at 3 V/min and kept constant for a further 24 hours (approximately 180,000 kVh totally). The second-dimensional separation was performed on 9–16% SDS gradient polyacrylamide gels. The gels (180 × 200 × 1.5 mm) were run at 40 mA per gel. After protein fixation with 40% methanol containing 5% phosphoric acid for 12 hours, the gels were stained with colloidal Coomassie blue (Novex, San Diego, CA, USA) for 48 hours. Molecular masses were determined by running standard protein markers (Gibco, Basel, Switzerland), covering the range 10–200 kDa. Isoelectric point (pI) values were used as given by the supplier of the IPG strips. Excess of dye was washed out from the gels with H_2O and the gels were scanned in an Agfa DUOSCAN densitometer (resolution 200). Electronic images of the gels were recorded using PhotoShop (Adobe) and PowerPoint (Microsoft) software.

Matrix-assisted laser desorption ionization mass spectroscopy (MALDI-MS)

MALDI-MS analysis was performed as described elsewhere (Fountoulakis and Langen, 1997) with some modifications. The spots were excised with a spot picker and placed into 96-well microtiter plates. Each spot was destained with 100 µl of 30% acetontrile in 50 mM ammonium bicarbonate and dried in a speedvac evaporator. each dried gel pieces was rehydrated with 4 µl of 3 mM Tris-HCl, pH 9.0, containing 50 ng trypsin (Promega, Madison, WI, USA). After 16 hours at room temperature, 7 µl of H_2O were added to each gel piece and the samples were shaken in for 10 minutes. Four µl of 50% acetonitrile, containing 0.3% trifluoroacetic acid, the standard peptides des-Arg-bradykinin (Sigma, 904.4681 Da) and adrenocorticotropic hormone fragment 18–39 (Sigma, 2,465.1989 Da), in H_2O were added to each gel piece and shaken for 10 minutes. The application of the samples was performed with a SymBiot I sample processor (PE Biosystems, Framingham, MA, USA). 1.5 µl of the peptide mixture was simultaneously applied with 1 µl of matrix, consisting of a saturated solution of α-cyano-4-hydroxycinnamic acid (Sigma) in 50% acetonitrile, containing 0.1% trifluoroacetic acid Samples were analyzed in a time-of-flight mass spectrometer (Reflex 3, Bruker Analytics, Bremen, Germany). An accelerating voltage of 20 kV was used. Peptide matching and protein searches were performed automatically. The peptide masses were compared to the theoretical peptide masses of all available proteins from all species. Monoisotopic masses were used and a mass tolerance of 0.0025% was allowed. The algorithm used for determining the probability of a false positive match with a given MS-spectrum is described elsewhere (Berndt et al., 1999).

Quantification of proteins

In partial 2-Dgel images including each protein and the neighbouring proteins, the percentage of the volume of the spot representing each protein was quantified as compared to the total proteins present using the ImageMaster 2D Elite software (Amersham Pharmacia Biotechnology).

Statistics

Results were expressed as means ± standard deviation (SD). Between-group differences were investigated by non-parametric Mann-Whitney U test. The Fisher's exact test was used to evaluate the prevalence of proteins in controls versus DS. The level of significance was set at $P < 0.05$. All statistical analyses were performed by GraphPad Instat2 software, version 2.05.

Results

Protein identification in fetal brain

Fetal brain proteins separated by 2-D gel electrophoresis were identified by MALDI-MS analysis based upon molecular weight, pI value and peptide matches. The protein spots evaluated in this study were presented in Fig. 1 and the data generated were shown in Table 1.

Profiles of protein expression levels in fetal DS brain compared to controls

Protein expression levels in fetal DS brains were presented compared to controls in Table 2. In the case that proteins were not well-separated or located ambiguously, proteins were not quantified for correctness. In case of septin 6, since it was not regularly expressed in fetal DS brains, we studied the prevalent rate of this protein. The statistically significant lower prevalence of

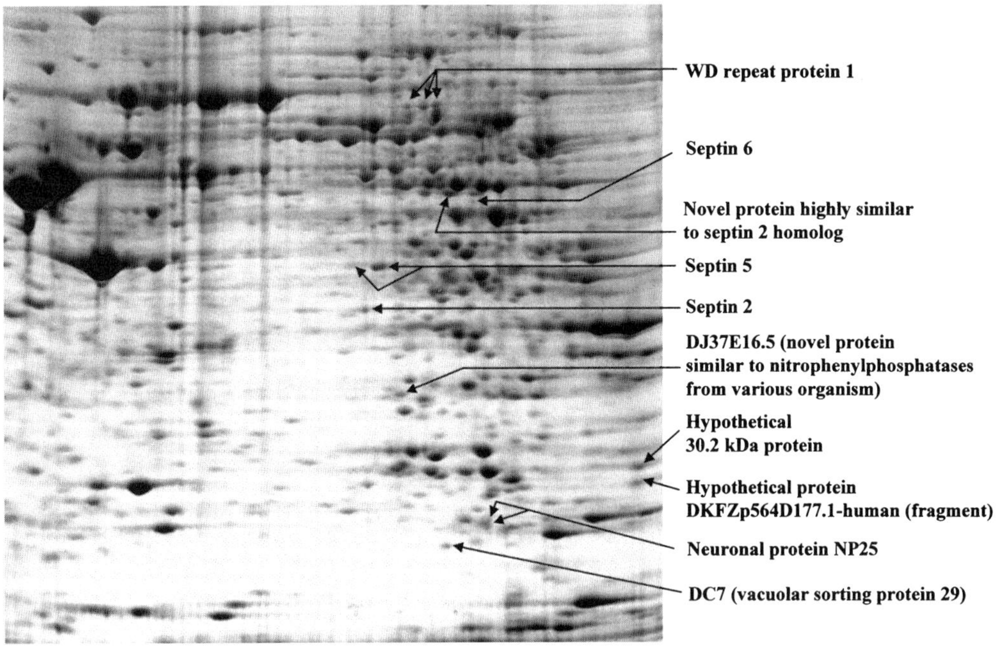

Fig. 1. The partial 2-D gel images of fetal brain. Fetal brain proteins were extracted and separated on an immobilized pH 3–10 nonlinear gradient strip, followed by a 9–16% linear gradient polyacrylamide gel, as stated under experimental procedures. The gel was stained with Coomassie blue and the spots were analyzed by MALDI-MS. Arrows indicate proteins used in this study

Table 1. Identification of proteins in fetal brain

Protein (SWISS-PROT accession number)	pI	MW	Matches	Probability
WD repeat protein 1 (O75083)	6.64	66,836	10	5.58E-14
Septin 6 (Q14141)	6.80	48,910	6	2.00E-06
Novel protein highly similar to septin 2 homolog (Q9NVA2)	6.80	49,652	5	2.92E-05
Septin 5 (Q99719)	6.65	43,205	6	8.85E-09
Septin 2 (Q15019)	6.59	41,689	6	7.88E-08
DJ37E16.5 (novel protein similar to nitrophenylphosphatases from various organism) (Q9UGY2)	6.51	32,077	8	2.64E-11
Hypothetical 30.2-kDa protein (Q9UFN0)	10.28	30,396	5	1.10E-04
Hypothetical protein DKFZp564D177.1 — human (fragment) (T17302)*	10.28	30,396	6	1.00E-04
Neuronal protein NP25 (Q9R1Q8)	7.37	22,627	6	1.15E-09
DC7 protein (vacuolar sorting protein 29) (Q9UBQ0)	6.79	20,663	6	2.34E-09

Proteins separated by 2-D gel electrophoresis were identified by MALDI-MS, following digestion with trypsin. The peptide masses were matched with the theoretical peptide masses of all proteins from all species of the SWISS-PROT database. *from GenBank

septin 6 was demonstrated in fetal DS brains (Table 3). The protein levels of hypothetical protein DKFZp564D177.1-human (fragment) were significantly decreased in fetal DS brains (Fig. 2), but the other 8 proteins studied were comparable between groups.

Table 2. The protein levels in fetal brain with DS

Protein	Control (n)	DS (n)
WD repeat protein 1	3.03 ± 1.66 (7)	3.54 ± 1.32 (6)
Novel protein highly similar to septin 2 homolog	2.08 ± 0.69 (7)	2.39 ± 0.70 (5)
Septin 5	10.74 ± 2.47 (7)	11.56 ± 3.35 (5)
Sepin 2	1.07 ± 0.63 (7)	1.54 ± 0.48 (4)
DJ37E16.5 (novel protein similar to Nitrophenylphosphatases from various organism)	1.81 ± 0.35 (7)	1.64 ± 0.38 (5)
Hypothetical 30.2-kDa protein	1.49 ± 1.18 (7)	0.79 ± 1.22 (8)
Hypothetical protein DKFZp564D177.1 — human (fragment)	3.43 ± 1.83 (7)	0.61 ± 0.44* (7)
Neuronal protein NP25	1.62 ± 0.54 (7)	1.39 ± 0.76 (6)
DC7 protein (vacuolar sorting protein 29)	1.27 ± 0.64 (7)	1.16 ± 0.39 (8)

Proteins from cerebral cortex from fetal DS and controls were separated on 2-D gels and visualized following stain with colloidal Coomassie blue. In partial gel images, including each protein and the neighboring proteins, the intensities of the spots representing each protein were quantified with ImageMaster 2D software, compared to the total proteins present. The statistical analyses were performed by Graphpad Instat2 software. Significant differences are indicated as an asterisks indicating $p < 0.01$

Table 3. The prevalence of proteins in fetal brain with DS

Protein	Control	DS
Septin 6	7/7	1/6*

The Fisher's exact test was used to evaluate the prevalence of proteins in controls versus DS. Results were indicated as automatically detected gel number with basic sensitivity/studied gel number. The significance was set at $P < 0.05$. The significant difference is indicated as an asterisk

Discussion

We identified a series of poorly described or –documented proteins in our fetal brain maps. While we may propose all these proteins to be tentative brain proteins (Table 4, Fig. 1), two out of them may be of significant importance for impaired development of the brain in DS.

One major finding was the significant decrease of hypothetical protein DKFZp564D177.1-human (fragment) in cerebral cortex of the DS fetus. The amino acids sequence of hypothetical protein DKFZp564D177.1-human (fragment) was just submitted to GenBank (accession number: T17302) and neither function nor structure have been reported so far and we suggest that this abundant brain protein may well play a role in human brain development and may be involved in the maldevelopment of the DS brain.

The second prominent result was the observation of deficient human septin 6 (Q14141) with concomitant unaltered other septins or septin–like proteins; novel protein highly similar to septin 2 homolog, septin 5 (cell division control related protein 1 (cdcrel-1), peanut-like protein 1) and septin 2 (nedd5 protein homolog). Septins, a family of evolutionarily conserved cytoskeletal GTPase enzymes, are required for the cytokinesis stage of cell division, cell polarity, and exocytosis in diverse organism (Longtine et al., 1996). Septins show distinct patterns of cytoskeletal and membrane associa-

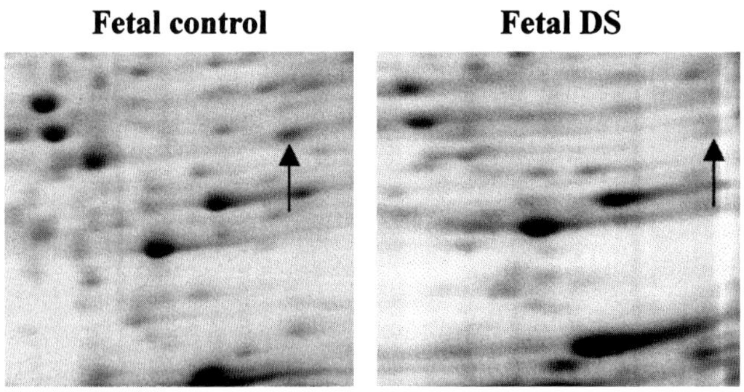

Fig. 2. The partial 2-D gel images of hypothetical protein DKFZp564D177.1-human (fragment) with significant difference. Arrows indicate hypothetical protein DKFZp564D177.1-human (fragment)

Table 4. Proposed function of proteins unchanged in fetal brain

Protein	Proposed function	References
WD repeat protein 1	Interactions with actin	Adler et al. (1999)
Novel protein highly similar to septin 2 homolog	Cytokinesis or exocytosis	(direct submission to EMBL database)
Septin 5	Cytokinesis	Yagi et al. (1998)
Septin 2	Cytokinesis	Nagase et al. (1995)
DJ37E16.5 (novel protein similar to nitrophenylphosphatases from various organism)	?	(direct submission to EMBL database)
Hypothetical 30.2-kDa protein	?	(direct submission to EMBL database)
Neuronal protein NP25	Belongs to the calponin family	Ren et al. (1994)
DC7 protein (vacuolar sorting protein 29)	Retrograde transport of protein from Endosomes to the trans-Golgi network	Edgar and Polak (2000)

tion (Xie et al., 1999) and comparative immunohistochemical studies revealed that levels of expression of each septin showed overlapping but distinct distribution in mouse brain (Kinoshita et al., 2000). The differential expression and subcellular localization of septins may indicate that individual members of the septin family in brain may serve several different roles in the brain.

The septin 6, deficient in fetal DS, was shown to be associated with synaptic vesicles in various brain regions particularly in hippocampal areas CA1, CA3 and dentate gyrus, structures mainly affected in DS neuropathology (Kinoshita et al., 2000). Based upon our results that synaptic structures are seriously deteriorated in brains of patients with DS (Yoo et al., 2001) along with the observation that also other septins co-purified with synaptic vesicles or inhibited exocytosis by interacting with syntaxin (Caltagarone et al., 1998; Beites et al., 1999), we draw the conclusion that the septin 6 may represent, reflect or lead to deteriorated synaptic functions as e.g. synaptic vesicle transport in fetal DS cortex. Moreover, as some mammalian homologs of septin proteins are found in dendrites, the finding of deranged septin 6 may well be in agreement with a role for septins in the formation or maintenance of spines (Walikonis et al., 2000), structures which are seriously impaired from the 19th gestational week in DS brain (unpublished results) as reflected by deficient drebrin at that time of development.

Taken together, we identified a series of proteins in human fetal brain assigning them to tentative brain proteins. Both deficient proteins, DKFZp564D177.1-human (fragment) and septin 6, may play a pathogenetic role in the abnormal development of fetal DS brain and, more specifically, septin 6 may be leading to or reflecting deteriorated synaptic structures including dendritic spines, a hallmark of DS brain.

References

Adler HJ, Winnicki RS, Gong TW, Lomax MI (1999) A gene upregulation in the acoustically damaged chick basilar papilla encodes a novel WD40 repeat protein. Genomics 56: 59–69

Becker L, Mito T, Takashima S, Onodera K (1991) Growth and development of the brain in Down syndrome. Prog Clin Biol Res 373: 133–152

Beites CL, Xie H, Bowser R, Trimble WS (1999) The septin CDCrel-1 binds syntaxin and inhibits exocytosis. Nat Neurosci 2: 434–439

Bendt P, Hobohm U, Langen H (1999) Reliable automatic protein identification from matrix-associated laser desorption/ionization mass spectrometric peptide fingerprints. Electrophoresis 20: 3521–3526

Bradford MM (1976) A rapid and sensitive method for the quantitation of microgram quantities of protein utilizing the principle of protein-dye binding. Anal Biochem 72: 248–254

Cairns NJ (1999) Neuropathology. J Neural Transm [Suppl] 57: 61–74

Caltagarone J, Rhodes J, Honer WG, Bowser R (1998) Localization of a novel septin protein, hCDCrel-1, in neurons of human brain. Neuroreport 9: 2907–2912

Dutt MJ, Lee KH (2000) Proteomic analysis. Curr Opin Biotechnol 11: 176–179

Edgar AJ, Polak JM (2000) Human homologues of yeast vacuolar protein sorting 29 and 35. Biochem Biophys Res Commun 277: 622–630

Epstein CJ (2001) Down syndrome (trisomy 21). In: Scriver CR et al (eds) The metabolic and molecular bases of inherited disease. McGraw-Hill, New York, pp 1223–1256

Fountoulakis M, Langen H (1997) Identification of protein by matrix-assisted laser desorption ionization-mass spectroscopy following in-gel digestion in low-salt nonvolatile buffer and simplified peptide recovery. Anal Biochem 250: 153–156

Kinoshita A, Noda M, Kinoshita M (2000) Differential localization of septins in the mouse brain. J Comp Neurol 428: 223–239

Langen H, Roeder D, Juranville JF, Fountoulakis M (1997) Effect of the protein application mode and the acrylamide concentration on the resolution of protein spots separated by two-dimensional gel electrophoresis. Electrophoresis 18: 2085–2090

Langen H, Berndt P, Roder D, Cairns N, Lubec G, Fountoulakis M (1999) Two-dimensional map of human brain proteins. Electrophoresis 20: 907–916

Longtine MS, DeMarin DJ, Valencik ML, Al-Awar OS, Fares H, de Virgilio C, Pringle JR (1996) The septins: roles in cytokinesis and other processes. Curr Opin Cell Biol 8: 106–119

Nagase T, Seki N, Tanaka A, Ishikawa K, Nomura N (1995) Prediction of the coding sequences of unidentified human genes. IV. The coding sequences of 40 new genes (KIAA0121-KIAA0160) deduced by analysis of cDNA clones from human cell line KG-1. DNA Res 2: 167–174

Nagy Zs (1999) Mechanisms of neuronal death in Down's syndrome. J Neural Transm [Suppl] 57: 233–245

Ren WZ, Ng GY, Wang RX, Wu PH, O'Dowd BF, Osmond DH, George SR, Liew CC (1994) The identification of NP25: a novel protein that is differentially expressed by neuronal subpopulations. Brain Res Mol Brain Res 22: 173–185

Walikonis RS, Jensen ON, Mann M, Provance DW, Mercer JA, Kennedy MB (2000) Identification of proteins in the postsynaptic density fraction by mass spectrometry. J Neurosci 20: 4069–4080

Wisniewski HM, Rabe A (1986) Discrepancy between Alzheimer-type neuropathology and dementia in persons with Down's syndrome. Ann NY Acad Sci 477: 247–260

Xie H, Surka M, Howard J, Trimble WS (1999) Characterization of the mammalian septin H5: distinct patterns of cytoskeletal and membrane association from other septin proteins. Cell Motil Cytoskeleton 43: 52–62

Yagi M, Zieger B, Roth GJ, Ware J (1998) Structure and expression of the human septin gene HCDCREL-1. Gene 212: 229–236

Yoo BC, Cairns N, Fountoulakis M, Lubec G (2001) Synaptosomal proteins, beta-soluble N-ethylmaleimide-sensitive factor attachment protein (beta-SNAP), gamma-SNAP and synaptotagmin I in brain of patients with Down syndrome and Alzheimer's disease. Dement Geriatr Cogn Disord 12: 219–225

Authors' address: Prof. Dr. G. Lubec, Department of Pediatrics, University of Vienna, Währinger Gürtel 18, A-1090 Vienna, Austria, e-mail: gert.lubec@akh-wien.ac.at

Expression patterns of chaperone proteins in cerebral cortex of the fetus with Down Syndrome: dysregulation of T-complex protein 1

B. C. Yoo[1], **M. Fountoulakis**[2], **M. Dierssen**[3], and **G. Lubec**[1]

[1] Department of Pediatrics, University of Vienna, Vienna, Austria
[2] Pharmaceutical Research, Genomics Technologies, F. Hoffmann-La Roche, Ltd., Basel, Switzerland
[3] Medical and Molecular Genetics Center-IRO, Hospital Duran i Reynals, Barcelona, Spain

Summary. Chaperone proteins reduce stress-induced denaturation and aggregation of intracellular proteins and exert protective actions by interfering with the stress-induced apoptotic pathway in the neuronal cells. Although many studies have focused on the critical role of chaperones in protein folding and their relevance in protein conformational diseases such as Alzheimer's disease (AD) and prion disease, limited information is available on their expression pattern in AD-related disease, Down Syndrome (DS).

In a previous study, we reported differentially expressed molecular chaperone proteins in the brains of adult DS patients. This report demonstrated the relevance of the aberrant expression patterns to biochemical and neuropathological abnormalities in adult DS brain.

To investigate whether this aberrant expression patterns already exist in second trimester DS brain, we performed two-dimensional polyacrylamide gel electrophoresis (2-DE) and matrix-associated laser desorption ionization mass spectroscopy (MALDI-MS) using fetal DS brain. We unambiguously identified and quantified twelve heat shock proteins (HSPs), three glucose regulated proteins (GRPs) and eight T-complex 1 (TCP-1) subunits. Unlike in adult brain with DS, these studies provide evidence for the normal response of most molecular chaperone proteins at the given time point of DS brain development. However, chaperonin protein TCP-1 alpha and beta subunits showed a significant decrease ($P < 0.05$) in second trimester brain with DS.

Comparable response of chaperone proteins in fetal DS brain suggests that not heat-shock induced proteins but rather chaperonin protein TCP-1 may contribute significantly to neuropathogenesis in the early second trimester DS brain.

This study extends observations of distinct expression patterns of molecular chaperones in DS brain and we discuss the relevance of aberrant TCP-1 expression patterns to developmental DS brain at the early stage.

Abbreviations

AD Alzheimer's disease, *DS* Down Syndrome, *GRP* glucose regulated protein, *HSP* heat shock protein, *MALDI-MS* matrix-associated laser desorption ionization mass spectroscopy, *TCP*-1 T-complex protein 1, *2-DE* two-dimensional gel electrophoresis.

Introduction

Molecular chaperones are proteins whose role is to mediate the folding of certain other polypeptides and, in some instances, their assembly into oligomeric structures, but which are not components of these final structures (Ellis, 2000). The prominent chaperones include heat shock protein (HSP) 70 family, HSP 40 family, HSP 90 family, the small HSPs (sHSPs) and the chaperonins (Agashe and Hartl, 2000).

Chaperone proteins reduce stress-induced denaturation and aggregation of intracellular proteins and exert protective actions by interfering with the stress-induced apoptotic pathway in the neuronal cells (Hartl, 1996), therefore, most molecular chaperone studies in neurodegenerative diseases such as Alzheimer's disease (AD) have focused on their role in amyloid fibril formation.

The expression of sHSP alpha crystallin B has been proposed to reflect the defensive response to diminish amyloid fibril formation and subsequent toxicity (Shinohara et al., 1993; Renkawek et al., 1994). Like sHSPs, HSP 40 and HSP 70 have been reported to inhibit self-assembly of poly-glutamine proteins into amyloid like fibrils (Muchowski et al., 2000). Furthermore, recent studies revealed that glucose regulated protein (GRP) 78 also binds to amyloid precursor protein (APP) and decrease $A\beta_{1-40}$ and $A\beta_{1-42}$ secretion (Yang et al., 1998). In contrast to other chaperones, HSC 73 and HSP 90 appear to be involved in proteasome function and especially HSC 73 effectively interacts with cytoplasmic domain of APP in the presence of proteasome inhibitors (Kouchi et al., 1999). These findings suggest that chaperones may function in the conformational maintenance of proteasome and also raises the possibility of aberrant protein recognition by proteasome. Recently, interest has been directed to the modulatory role of the stress proteins on apoptosis. Although an increasing bulk of evidence shows that HSPs and GRPs can provide protective effect against cell death (Mehlen et al., 1996; Mosser et al., 1997; Reddy et al., 1999), in some situations various HSPs have been found to accelerate apoptotic processes (Samali et al., 1999; Xanthoudakis et al., 1999).

The relevance of molecular chaperones in AD neuropathology has been recognized (Hartl, 1996; Milner and Campbell, 1996), however, little was known about these chaperones in AD-related Down Syndrome (DS) brain.

DS is a genetic disease with AD-like neuropathology manifested later in life. As deposition of Aβ and neuronal loss mediated via apoptosis are features of both AD and DS, knowing the expression of molecular chaperones under pathological conditions might be essential to understand how folding defects can lead to disease. Therefore, we previously studied expression patterns of molecular chaperone proteins in the brains with adult DS patients (Yoo et al., 2001a). Accordingly, HSP 70RY, heat shock cognate (HSC) 71 and GRP 75 showed a significant decrease in DS temporal cortex whereas HSP 70.1 and GRP 78 were significantly increased in cerebellum. Whilst T-complex 1 (TCP-1) epsilon subunit showed a significant decrease in parietal cortex, a similar extent of increase as that observed in cerebellum was obtained in parietal levels of GRP 78. Alpha-crystallin B, HSP 60 and GRP 94 did not show any detectable changes in expression patterns. This report presented the first approach to quantify nine different chaperones simultaneously at the protein level in different brain regions, and provided evidence for aberrant chaperone expression patterns in adult brain with DS and its relevance in DS neuropathology.

To investigate whether this aberrant expression patterns already exist in fetal brain with DS, we performed two-dimensional polyacrylamide gel electrophoresis (2-DE) and matrix-assisted laser desorption ionization mass spectroscopy (MALDI-MS) using fetal DS brain. Unlike in adult DS brain, these studies provided evidence for the normal response of most molecular chaperone proteins at an early time point of DS brain development. This comparable response of chaperone proteins in fetal DS brain suggests that not heat-shock induced proteins but chaperonin protein TCP-1 may contribute significantly to neuropathogenesis in the early second trimester DS brain.

These studies extend observations of distinct expression patterns of molecular chaperones in DS brain and we discuss the relevance of aberrant TCP-1 expression patterns to neuropathological abnormalities of developmental DS brain at the early stage.

Materials and methods

Human fetal brain

Cerebral cortices of fetuses with DS ($n = 8$, 2 females and 6 males, 19.81 ± 2.00 weeks old) and controls ($n = 7$, 1 female and 6 males, 18.79 ± 2.23 weeks old) were obtained from Medical and Molecular Genetics Center-IRO, Hospital Duran i Reynals Avia. Castelldefels Km 2,7 08907 Hospitalet de Llobregat Barcelona, Spain. All samples were stored at −70 degree and the freezing chain was never interrupted.

Two-dimensional gel electrophoresis

Two-dimensional gel electrophoresis was performed essentially as reported (Langen et al., 1997). Samples of 2 mg protein were applied on immobilized pH 3–10 nonlinear gradient strips in sample cups at their basic and acidic ends. Focusing started at 200 V and the voltage was gradually increased to 5,000 V at 3 V/min and kept constant for a further

24 h (approximately 180,000 kVh totally). The second-dimensional separation was performed on 9–16% SDS gradient polyacrylamide gels. The gels (180 × 200 × 1.5 mm) were run at 40 mA per gel. After protein fixation for 12 h in 40% methanol, containing 5% phosphoric acid, the gels were stained with colloidal Coomassie blue (Novex, San Diego, CA, USA) for 48 h. Molecular masses were determined by running standard protein markers (Gibco, Basel, Switzerland), covering the range 10–200 kDa. pI values were used as given by the supplier of the IPG strips. Excess of dye was washed out from the gels with H_2O and the gels were scanned in an Agfa DUOSCAN densitometer (resolution 200). Electronic images of the gels were recorded using Photoshop (Adobe) and PowerPoint (Microsoft) software. Protein spots were outlined (first automatically and then manually) and quantified using the ImageMaster 2D Elite software (Amersham Pharmacia Biotechnology). The percentage of the volume of the spots representing a certain protein was determined in comparison with the total proteins present in the 2-D gel.

Matrix-assisted laser desorption ionization mass spectroscopy (MALDI-MS)

MALDI-MS analysis was performed as described elsewhere (Fountoulakis and Langen, 1997) with some modifications. The spots were excised with a spot picker and placed into 96-well microtiter plates. Each spot was destained with 100 μl of 30% acetonitrile in 50 mM ammonium bicarbonate and dried in a speedvac evaporator. Each dried gel piece was rehydrated with 4 μl of 3 mM Tris-HCl, pH 9.0, containing 50 ng trypsin (Promega, Madison, WI, USA). After 16 h at room temperature, 7 μl of H_2O were added to each gel piece and the samples were shaken in for 10 min. Four μl of 50% acetonitrile, containing 0.3% trifluoroacetic acid and the standard peptides des-Arg-bradykinin (Sigma, 904.4681 Da) and adrenocorticotropic hormone fragment 18–39 (Sigma, 2,465.1989 Da), in water were added to each gel piece and shaken for 10 min. The application of the samples was performed with a SymBiot I sample processor (PE Biosystems, Framingham, MA, USA). 1.5 μl of the peptide mixture was simultaneously applied with 1 μl of matrix, consisting of a saturated solution of α-cyano-4-hydroxycinnamic acid (Sigma) in 50% acetonitrile, containing 0.1% trifluoroacetic acid. Samples were analyzed in a time-of-flight mass spectrometer (Reflex 3, Bruker Analytics, Bremen, Germany). An accelerating voltage of 20 kV was used. Peptide matching and protein searches were performed automatically. The peptide masses were compared to the theoretical peptide masses of all available proteins from all species. Monoisotopic masses were used and a mass tolerance of 0.0025% was allowed. The algorithm used for determining the probability of a false positive match with a given MS-spectrum is described elsewhere (Berndt et al., 1999).

Statistical analyses

Percentage volume of total protein was obtained from quantifying protein spots on 2-DE map. Between-group differences were calculated by non-parametric Mann-Whitney U-test and within-group-correlations were done using the Spearman rank-coefficient. The significance was set at the $P < 0.05$ level.

Results

Identification and quantification of chaperone proteins on 2-DE map of fetal brain

Fetal brain proteins were applied on 2-D gel. The protein spots were visualized by staining with colloidal Coomassie blue. Figure 1 shows a representative gel of brain proteins separated by 2D from the cerebral cortex of control subject. The protein spots were analyzed by MALDI-MS, following in-gel digestion. The peptide masses were matched with the theoretical peptide masses of all proteins of the SWISS-PROT database. We used internal standards to correct the measured peptide mass reducing thus, the windows of mass tolerance and increasing the confidence of identification. Approximately 400 proteins were successfully identified (Data not shown) and among them twelve HSPs, three GRPs and eight TCP-1 subunits were identified (Table 1) and quantified using the Image Master 2D Elite software (Fig. 2). The expression levels of molecular chaperones were determined as a percentage volume of total proteins in the area of interest (AOI). The SWISS-PROT accession numbers, names of the quantified all chaperone proteins and four AOI used in the quantification are given in Fig. 1.

The majority of the proteins was represented by one spot. This one spot-one protein motif was violated in many cases. For example, TCP-1 beta subunit, HSP 60 and HSP 27 were represented by two spots with different pI value indicating the existence of different isoforms (Fig. 1). Among the chaperone proteins quantified in the present study, HSP 90-alpha and HSP 90-beta were represented as one smeared protein in pI range 4.77 to 4.80 and abundantly expressed HSP 70 family proteins, HSC 71, HSP 70.1, HSP 70RY, heat shock 70kDa protein b′ and heat shock 70kDa protein 2 formed unseparated protein complex in pI range 5.52 to 6.22 (Fig. 1).

Expression patterns of chaperone proteins in fetal brain with DS

Quantitative analysis demonstrated normal expression pattern of all HSPs and GRPs in fetal brain with DS (Fig. 2).

With regard to HSP 70 family including HSP 60, HSP 70.1, HSP 70RY, heat shock 70kDa protein 1-hom, heat shock 70kDa protein b′, heat shock 70kDa protein 2 and HSC 71, comparable levels to those of controls were obtained.

The same patterns of GRP proteins including GRP 75, GRP 78 and GRP 94 were noted. Neither HSP 90 family proteins HSP 90-alpha and -beta nor membrane-bound chaperone prohibitin showed any different expression level in fetal DS.

In fetal DS, although the difference failed to reach significance level owing to the variation among the results, among sHSPs alpha crystallin B showed an increase by 70%. However, HSP 27 showed no significant difference.

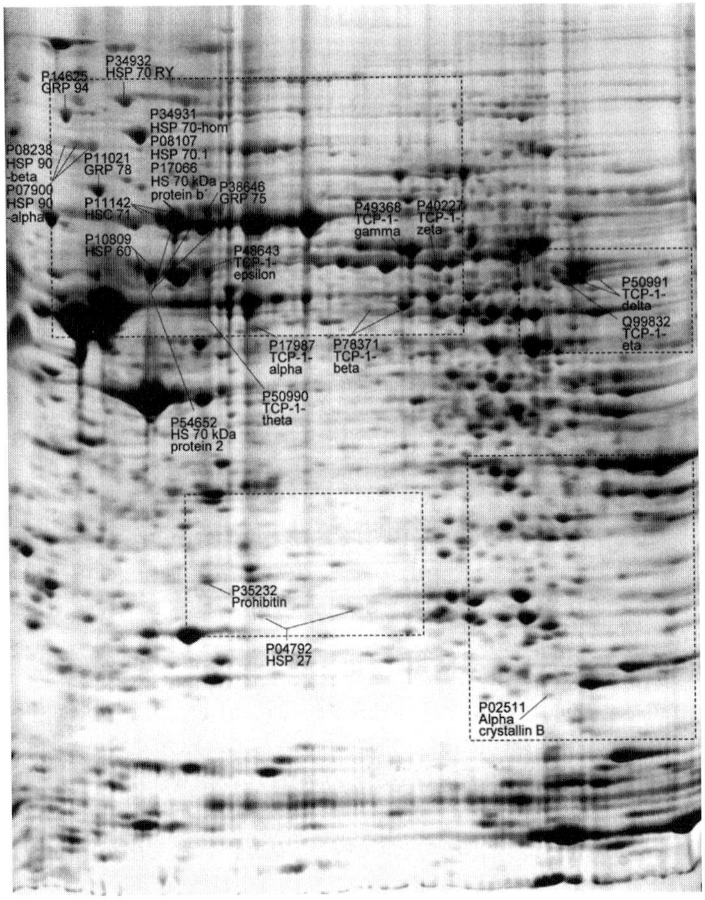

Fig. 1. 2-DE map of human fetal brain proteins obtained from cerebral cortex of a control. The 2-DE was performed in an immobilized pH 3–10 nonlinear gradient strip, followed by the second-dimensional separation on 9–16% linear gradient polyacrylamide gels, and separated proteins were detected by colloidal Coomassie blue staining. Approximately 400 spots were analyzed by MALDI-MS. Identified molecular chaperones are designated with their SWISS-PROT accession numbers and the names referring to Table 1. Four squares in figure indicate AOI used in the quantification of proteins

Aberrant expression of TCP-1 subunits

We successfully identified and quantified all eight TCP-1 subunits. Among them alpha and beta subunits were found to be decreased in fetal DS cerebral cortex (both $P < 0.05$) (Fig. 2). Also delta, epsilon, zeta, theta and eta subunit levels tended to decrease in fetal DS by 15% to 40% compared to controls but the difference did not reach significance level due to the variation among the results.

Correlation of molecular chaperone levels with age

Linear regression analysis revealed no differential expression of molecular chaperone proteins between 17 to 22 weeks of gestation age (data not shown).

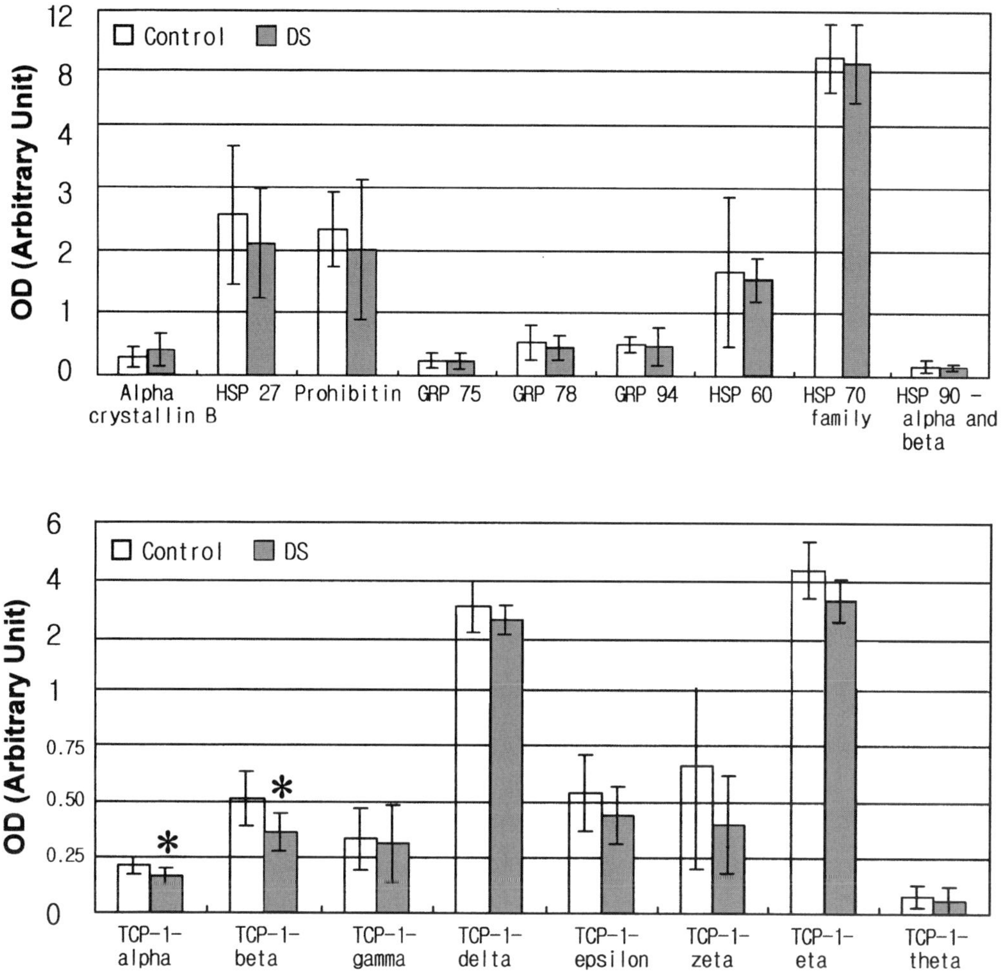

Fig. 2. Expression levels of molecular chaperone proteins in cerebral cortex of the fetus with DS. All molecular chaperones studied on the 2-DE gels from cerebral cortices of fetuses with DS and age-matched controls were quantified using the ImageMaster 2D Elite software and compared to each other employing non-parametric Mann-Whitney U-test in order to detect different expression levels of those proteins. Molecular chaperone expression level was determined as the percentage volume of total proteins present in AOI. Among the chaperone proteins quantified in the present study, HSP 90-alpha and HSP 90-beta were represented as one smeared protein in pI range 4.77 to 4.80 and abundantly expressed HSP 70 family proteins, HSC 71, HSP 70.1, HSP 70 RY, heat shock 70 kDa protein 1-hom, heat shock 70 kDa protein b′ and heat shock 70 kDa protein 2 formed unseparated protein complex in pI range 5.52 to 6.22, therefore, these proteins were considered as one HSP 70 family protein and quantified. Two different pI isoforms of HSP 60, HSP 27, TCP-1-beta and -delta were quantified individually and Fig. 2 shows only sum of levels of these isoforms. The columns and bars indicate the means and standard deviations of the expression levels of each molecular chaperone. In comparison with controls, * ($P < 0.05$) is considered significantly decreased expression

Table 1. MALDI-MS analysis. Proteins from the parietal region of human fetal brain were extracted and separated by 2-DE, as described under Materials and Methods. Approximately 400 proteins were identified by MALDI-MS, following trypsin digestion. The identified protein spots were characterized by their SWISS-PROT accession numbers. All identified and quantified molecular chaperones were indicated in Fig. 1. The theoretical Mr and pI values, as well as the matching peptides and the probability of assignment of a wrong identity are given

					MALDI-MS	
No.[a]	Abbr. name	Protein name	pI	MW (Da)	Matches	Probability
P02511	SW:CRAB_ HUMAN	Alpha crystallin B (Rosenthal fiber component)	7.83	20,146	5	8.77E-06
P04792	SW:HS27_ HUMAN	Heat shock 27 kDa protein (HSP 27)	8.14	22,427	7	1.84E-08
P07900	SW:HS9A_ HUMAN	Heat shock protein HSP 90-alpha	4.77	84,888	10	2.22E-12
P08107	SW:HS71_ HUMAN	Heat shock 70 kDa protein 1 (HSP 70.1)	5.41	70,294	13	4.55E-22
P08238	SW:HS9B_ HUMAN	Heat shock protein HSP 90-beta	4.80	83,453	9	5.17E-10
P10809	SW:P60_ HUMAN	Mitochondrial matrix protein P1 (HSP 60)	5.64	61,187	8	2.63E-06
P11021	SW:GR78_ HUMAN	78 kDa glucose regulated protein precursor (GRP 78) (bip)	4.87	72,185	14	2.78E-24
P11142	SW:HS7C_ HUMAN	Heat shock cognate 71 kDa protein (HSC 71)	5.26	71,082	11	8.86E-17
P14625	SW:ENPL_ HUMAN	Endoplasmin precursor (94 kDa glucose regulated protein) (GRP 94)	4.59	92,696	13	5.11E-21
P17066	SW:HS76_ HUMAN	Heat shock 70 kDa protein 6 (Heat shock 70 kDa protein b′)	5.80	71,209	6	4.99E-07

P17987	SW:TCPA_HUMAN	T-complex protein 1, alpha subunit (TCP-1-alpha) (cct-alpha)	6.00	60,818	10	2.50E-11
P34931	SW:HS7H_HUMAN	Heat shock 70 kDa protein 1-hom (HSP 70-hom)	5.87	70,754	6	3.30E-07
P34932	SW:HS74_HUMAN	Heat shock 70 kDa protein 4 (HSP 70 RY)	4.97	85,330	12	1.18E-15
P35232	SW:PHB_HUMAN	Prohibitin	5.55	29,842	8	1.39E-12
P38646	SW:GR75_HUMAN	Mitochondrial stress-70 protein (75 kDa glucose regulated protein) (GRP 75)	6.22	74,018	9	6.49E-12
P40227	SW:TCPZ_HUMAN	T-complex protein 1, zeta subunit (TCP-1-zeta) (cct-zeta)	6.66	58,443	9	1.74E-12
P48643	SW:TCPE_HUMAN	T-complex protein 1, epsilon subunit (TCP-1-epsilon) (cct-epsilon)	5.43	60,088	6	3.88E-07
P49368	SW:TCPG_HUMAN	T-complex protein 1, gamma subunit (TCP-1-gamma) (cct-gamma)	6.61	60,862	10	8.88E-11
P50990	SW:TCPQ_HUMAN	T-complex protein 1, theta subunit (TCP-1-theta) (cct-theta)	5.50	60,166	7	7.06E-09
P50991	SW:TCPD_HUMAN	T-complex protein 1, delta subunit (TCP-1-delta) (cct-delta)	7.57	58,315	8	2.48E-09
P54652	SW:HS72_HUMAN	Heat shock-related 70 kDa protein 2 (Heat shock 70 kDa protein 2)	5.52	70,262	5	2.72E-05
P78371	SW:TCPB_HUMAN	T-complex protein 1, beta subunit (TCP-1-beta) (fragment)	6.27	22,924	9	9.86E-13
Q99832	SW:TCPH_HUMAN	T-complex protein 1, eta subunit (TCP-1-eta) (cct-theta)	7.62	59,842	10	3.78E-15

[a] SWISS-PROT Number

Discussion

By the proteomics approach employing 2-DE and MALDI-MS analysis, we successfully identified chaperone proteins including twelve HSPs, three GRPs and all eight subunits of TCP-1 from approximately 400 proteins on the 2-D map of human fetal brain. Protein spots were measured as percentage volume of total proteins and this measurement allowed the observation of distinct expression patterns of chaperone proteins in fetal brain with DS indicating the relevance of these proteins for aberrant developmental pathways.

Comparable molecular chaperone expression in fetal DS brain

All HSP family proteins studied did not show any difference in fetal DS brain (Fig. 2). Based on the role of HSP 70 in protecting cells from apoptosis (Mosser et al., 1997, 2000), it has been suggested that loss of HSP 70 family proteins particularly, HSP 70RY and HSC 71 may play a greater role in neuronal death associated with adult DS (Yoo et al., 2001a). We, however, failed to observe aberrantly expressed proteins of HSP 70 family in fetal DS brain. This might implicate that finding in the adult may be simply related to AD neurodegeneration which appears from the 4th decade.

In previous studies (Labudova et al., 1999; Lubec et al., 1999), it has been confirmed that glucose metabolism in adult DS is impaired. Deranged glucose metabolism as well as induced cellular stress may account for the increased expression of GRPs observed in adult DS brain (Hori et al., 1996; Yoo et al., 2001a). Interestingly, we did not find any expressional change of GRP family in fetal DS (Fig. 2). But no information is available on the glucose handling in fetal brain with DS.

HSP 60 is proposed to participate in activation of caspase-3 by acting as a chaperone to promote maintenance of procaspase-3 in a protease sensitive state (Samali et al., 1999; Xanthoudakis et al., 1999). The protein was further shown to be translocated to the cytosol during apoptosis to amplify the caspase cascade (Samali et al., 1999). The lack of detectable change in expression of HSP 60 in both adult (Yoo et al., 2001a) and fetal DS brain (Fig. 2) may be associated with its passive role in apoptosis and recycling property (Xanthoudakis et al., 1999), and moreover, there is evidence for the absence of increased apoptosis in fetal DS (Engidawork et al., 2001).

2-DE analysis did not reveal any significant change of sHSPs expressions in fetal DS (Fig. 2). Although the difference failed to reach significance level owing to the variation among the results, among sHSPs alpha crystallin B showed an increase like in adult DS (Yoo et al., 2001a). Alpha crystallin B protects cell from death by modulating reactive oxygen species and glutathione levels (Mehlen et al., 1996). Increased expression of alpha crystallin B has been reported in temporal and frontal cortices of AD brains (Shinohara et al., 1993; Yoo et al., 2001b), and in reactive astrocytes, microglia and oligodendrocytes (Shinohara et al., 1993; Renkawek et al., 1994). The increased alpha crystallin B in both adult and fetal DS may reflect either glial effort to maintain viability of neurons or simply glial proliferation.

Aberrant expression of TCP-1 in fetal DS brain

We previously reported dysregulated TCP-1 epsilon subunit in adult DS brain (Yoo et al., 2001a) and we therefore investigated levels of all measurable individual TCP-1 subunits in fetal cerebral cortex with DS. Expression levels of TCP-1 alpha and beta subunits showed a decrease in fetal DS (both $P < 0.05$) (Fig. 2). Although other TCP-1 subunits did not show statistical difference, delta, epsilon, zeta, theta and eta subunit levels tended to decrease in fetal DS by 15% to 40% compared to controls (Fig. 2).

TCP-1 is a cytosolic chaperonin with a toroidal structure and it is assembled from eight distinct but related subunits in the range of 57–62 kDa (Kubota et al., 1991; Lewis et al., 1992; Rommelaere et al., 1993). Both genetic (Ursic et al., 1991; Chen et al., 1991) and biochemical evidence (Lewis et al., 1992; Yaffe et al., 1992; Gao et al., 1993; Melki and Cowan, 1993) implicate that TCP-1 is involved in the facilitated folding of actin and tubulin. It is suggested that each TCP-1 subunit has evolved a specific independent function while maintaining the common property of ATPase activity (Kubota et al., 1994). The specific functions of individual TCP-1 subunits may be associated with the provision of different polypeptide substrate binding sites, and thus individual subunits of TCP-1 may play different roles in substrate interaction (Liou and Willison, 1997). Roobol and Carden (1999) provided evidence that TCP-1 subunits are not only components of the hetero-oligomeric chaperonin particle but exist as significant populations of free subunits or smaller oligomers in cells. Furthermore, it has been suggested the important function of TCP-1 for growth and development (Iijima et al., 1998), and abundant existence of TCP-1 alpha subunit in neurites of mouse P19 neurons (Roobol et al., 1995; Roobol and Carden, 1999). Taken together, a decreasing tendency of most individual TCP-1 subunits especially alpha and beta subunits in fetal DS brain strongly supports an important role of this protein for development of the human brain at the early second trimester.

In interphase and mitotic cells TCP-1 is localized within the centrosome which serves as an initiation site for microtubule growth (Brown et al., 1996). Since TCP-1 is involved in microtubule growth, its mutation or abnormal expression may lead to mitotic and meiotic abnormalities of the cell cycle. If cell cycle regulatory mechanisms fail, the cell cycle is allowed to progress into the G_2 phase. Nagy (1999) suggested that the cell cycle in DS neurons can be arrested at the G_2/M transition and at this point may undergo apoptosis via the p53/Bax related pathway. Interestingly, through sensitive differential hybridization approach, Dittmar and co-workers showed that the expression of TCP-1 appeared to be specifically linked with the S to G_2/M phase transition of the cell cycle (Dittmar et al., 1997). Furthermore, recently it has been reported that the existence of a common mechanism of tumor suppression and apoptosis shared by p53, p21 (Waf1) and SIAH-1 involved regulation of

TCP-1 (Roperch et al., 1999). We therefore propose that decreased TCP-1 epsilon subunit expression in brains of fetus as well as adult (Yoo et al., 2001a) with DS may lead to arrest of cells at the G_2/M transition point and disruption of the cytoskeleton, and subsequently cells may undergo apoptosis.

In conclusion, among all molecular chaperones studied only TCP-1 did show expressional abnormalities in fetal DS brain. Our results suggest that most chaperone proteins are not likely to be involved in pathological changes of fetal DS brain at the early stage of development. However, aberrant TCP-1 expression may shed light on an important neuropathological role of this protein in the developing brain at the early second trimester.

Acknowledgment

The authors at the University of Vienna are highly indebted to the Red Bull Company, Salzburg, for financial support of the study.

References

Agashe VR, Hartl FU (2000) Roles of molecular chaperones in cytoplasmic protein folding. Semin Cell Dev Biol 11: 15–25

Berndt P, Hobohm U, Langen H (1999) Reliable automatic protein identification from matrix-assisted laser desorption/ionization mass spectrometric peptide fingerprints. Electrophoresis 20: 3521–3526

Brown CR, Doxsey SJ, Hong-Brown LQ, Martin RL, Welch WJ (1996) Molecular chaperones and the centrosome. A role for TCP-1 in microtubule nucleation. J Biol Chem 271: 824–832

Chen X, Sullivan DS, Huffaker TC (1994) Two yeast genes with similarity to TCP-1 are required for microtubule and actin function in vivo. Proc Natl Acad Sci USA 91: 9111–9115

Dittmar G, Schmidt G, Kopun M, Werner D (1997) Mapping of G2/M-phase prevalences of chaperon-encoding transcripts by means of a sensitive differential hybridization approach. Cell Biol Int 21: 383–391

Ellis RJ (2000) Moledular chaperones ten years. Introduction. Semin Cell Dev Biol 11: 1–5

Engidawork E, Balic N, Juranville J-F, Dierssen M, Lubec G (2001) Unaltered expression of Fas (CD95/APO-1), Caspase-3, Bcl-2 and Annexins in brains of fetal Down syndrome: evidence against increased apoptosis (this volume)

Fountoulakis M, Langen H (1997) Identification of proteins by matrix-assisted laser desorption ionization-mass spectrometry following in-gel digestion in low-salt, non-volatile buffer and simplified peptide recovery. Anal Biochem 250: 153–156

Gao Y, Thomas JO, Chow RL, Lee GH, Cowan NJ (1992) A cytoplasmic chaperonin that catalyzes beta-actin folding. Cell 69: 1043–1050

Gao Y, Vainberg IE, Chow RL, Cowan NJ (1993) Two cofactors and cytoplasmic chaperonin are required for the folding of alpha- and beta-tubulin. Mol Cell Biol 13: 2478–2485

Hartl FU (1996) Molecular chaperones in cellular protein folding. Nature 381: 571–579

Hori O, Matsumoto M, Kuwabara K, Maeda Y, Ueda H, Ohtsuki T, Kinoshita T, Ogawa S, Stern DM, Kamada T (1996) Exposure of astrocytes to hypoxia/reoxygenation enhances expression of glucose-regulated protein 78 facilitating astrocyte release of the neuroprotective cytokine interleukin 6. J Neurochem 66: 973–979

Iijima M, Shimizu H, Tanaka Y, Urushihara H (1998) A dictyostelium discoideum homologue to Tcp-1 is essential for growth and development. Gene 213(1–2): 101–106
Kouchi Z, Sorimachi H, Suzuki K, Ishiura S (1999) Proteasome inhibitors induce the association of Alzheimer's amyloid precursor protein with Hsc73. Biochem Biophys Res Commun 254: 804–810
Kubota H, Hynes G, Carne A, Ashworth A, Willison K (1994) Identification of six Tcp-1-related genes encoding divergent subunits of the TCP-1-containing chaperonin. Curr Biol 4: 89–99
Labudova O, Kitzmueller E, Rink H, Cairns N, Lubec G (1999) Increased phosphoglycerate kinase in the brains of patients with Down's syndrome but not with Alzheimer's disease. Clin Sci (Colch) 96: 279–285
Langen H, Roeder D, Juranville J-F, Fountoulakis M (1997) Effect of the protein application mode and the acrylamide concentration on the resolution of protein spots separated by two-dimensional gel electrophoresis. Electrophoresis 18: 2085–2090
Lewis VA, Hynes GM, Zheng D, Saibil H, Willison K (1992) T-complex polypeptide-1 is a subunit of a heteromeric particle in the eukaryotic cytosol. Nature 358: 249–252
Liou AK, Willison KR (1997) Elucidation of the subunit orientation in CCT (chaperonin containing TCP1) from the subunit composition of CCT micro-complexes. EMBO J 16: 4311–4316
Lubec G, Labudova O, Cairns N, Fountoulakis M (1999) Increased glyceraldehyde 3-phosphate dehydrogenase levels in the brain of patients with Down's syndrome. Neurosci Lett 260: 141–145
Mehlen P, Kretz-Remy C, Preville X, Arrigo AP (1996) Human hsp27, Drosophila hsp27 and human alphaB-crystallin expression-mediated increase in glutathione is essential for the protective activity of these proteins against TNFalpha-induced cell death. EMBO J 15: 2695–2706
Melki R, Vainberg IE, Chow RL, Cowan NJ (1993) Chaperonin-mediated folding of vertebrate actin-related protein and gamma-tubulin. J Cell Biol 122: 1301–1310
Milner CM, Campbell RD (1990) Structure and expression of the three MHC-linked HSP70 genes. Immunogenetics 32: 242–251
Mosser DD, Caron AW, Bourget L, Denis-Larose C, Massie B (1997) Role of the human heat shock protein hsp70 in protection against stress-induced apoptosis. Mol Cell Biol 17: 5317–5327
Mosser DD, Caron AW, Bourget L, Meriin AB, Sherman MY, Morimoto RI, Massie B (2000) The chaperone function of hsp70 is required for protection against stress-induced apoptosis. Mol Cell Biol 20: 7146–7159
Muchowski PJ, Schaffar G, Sittler A, Wanker EE, Hayer-Hartl MK, Hartl FU (2000) Hsp70 and hsp40 chaperones can inhibit self assembly of polyglutamine proteins into amyloid-like fibrils. Proc Natl Acad Sci USA 97: 7841–7846
Nagy Z (1999) Mechanisms of neuronal death in Down's syndrome. J Neural Transm [Suppl] 57: 233–245
Reddy RK, Lu J, Lee AS (1999) The endoplasmic reticulum chaperone glycoprotein GRP94 with Ca(2+)-binding and antiapoptotic properties is a novel proteolytic target of calpain during etoposide-induced apoptosis. J Biol Chem 274: 28476–28483
Renkawek K, Voorter CE, Bosman GJ, van Workum FP, de Jong WW (1994) Expression of alpha B-crystallin in Alzheimer's disease. Acta Neuropathol (Berl) 87: 155–160
Rommelaere H, Van Troys M, Gao Y, Melki R, Cowan NJ, Vandekerckhove J, Ampe C (1993) Eukaryotic cytosolic chaperonin contains t-complex polypeptide 1 and seven related subunits. Proc Natl Acad Sci USA 90: 11975–11979
Roobol A, Carden MJ (1999) Subunits of the eukaryotic cytosolic chaperonin CCT do not always behave as components of a uniform hetero-oligomeric particle. Eur J Cell Biol 78: 21–32
Roobol A, Holmes FE, Hayes NV, Baines AJ, Carden MJ (1995) Cytoplasmic chaperonin complexes enter neurites developing in vitro and differ in subunit composition within single cells. J Cell Sci (Pt 4): 1477–1488

Roperch JP, Lethrone F, Prieur S, Piouffre L, Israeli D, Tuynder M, Nemani M, Pasturaud P, Gendron MC, Dausset J, Oren M, Amson RB, Telerman A (1999) SIAH-1 promotes apoptosis and tumor suppression through a network involving the regulation of protein folding, unfolding, and trafficking: identification of common effectors with p53 and p21(Waf1). Proc Natl Acad Sci USA 96: 8070–8073

Samali A, Cai J, Zhivotovsky B, Jones DP, Orrenius S (1999) Presence of a pre-apoptotic complex of pro-caspase-3, Hsp60 and Hsp10 in the mitochondrial fraction of jurkat cells. EMBO J 18: 2040–2048

Shinohara H, Inaguma Y, Goto S, Inagaki T, Kato K (1993) Alpha B crystallin and HSP28 are enhanced in the cerebral cortex of patients with Alzheimer's disease. J Neurol Sci 119: 203–208

Ursic D, Culbertson MR (1991) The yeast homolog to mouse Tcp-1 affects microtubule-mediated processes. Mol Cell Biol 11: 2629–2640

Xanthoudakis S, Roy S, Rasper D, Hennessey T, Aubin Y, Cassady R, Tawa P, Ruel R, Rosen A, Nicholson DW (1999) Hsp60 accelerates the maturation of pro-caspase-3 by upstream activator proteases during apoptosis. EMBO J 18: 2049–2056

Yaffe MB, Farr GW, Miklos D, Horwich AL, Sternlicht ML, Sternlicht H (1992) TCP1 complex is a molecular chaperone in tubulin biogenesis. Nature 358: 245–248

Yang Y, Turner RS, Gaut JR (1998) The chaperone BiP/GRP78 binds to amyloid precursor protein and decreases Abeta40 and Abeta42 secretion. J Biol Chem 273: 25552–25555

Yoo BC, Vlkolinsky R, Engidawork E, Cairns N, Fountoulakis M, Lubec G (2001a) Differential expression of molecular chaperones in brain of patients with Down Syndrome. Electrophoresis 22: 1233–1241

Yoo BC, Kim SH, Cairns N, Fountoulakis M, Lubec G (2001b) Deranged expression of molecular chaperones in brains of patients with Alzheimer's disease. Biochem Biophys Res Commun 280(1): 249–258

Authors' address: Prof. Dr. G. Lubec, Department of Pediatrics, University of Vienna, Währinger Gürtel 18-20, A-1090 Vienna, Austria, e-mail: Gert.Lubec@akh-wien.ac.at

β-Amyloid precursor protein, ETS-2 and collagen alpha 1 (VI) chain precursor, encoded on chromosome 21, are not overexpressed in fetal Down syndrome: further evidence against gene dosage effect

E. Engidawork[1], **N. Balic**[2], **M. Fountoulakis**[3], **M. Dierssen**[4], **S. Greber-Platzer**[1], and **G. Lubec**[1]

[1] Department of Pediatrics, and
[2] Institute of Medical and Chemical Laboratory Diagnostics, AKH, University of Vienna, Vienna, Austria
[3] F. Hoffman-La Roche, Basel, Switzerland
[4] Medical and Molecular Genetics Center, IRO Hospital Duran i Reynals, Barcelona, Spain

Summary. Down syndrome (DS) is the most common human chromosomal abnormality caused by an extra copy of chromosome 21 and characterized clinically by somatic anomalies, mental retardation and precocious dementia. The phenotype of DS is thought to result from overexpression of a gene or genes located on the triplicated chromosome or chromosome region. Reports that challenge this notion, however, have been published. To add to this body of evidence, the expression of β-amyloid precursor protein (APP), ETS-2 and collagen α1 (VI) chain precursor, encoded on chromosome 21, was investigated in fetal brain by western blot and two-dimensional electrophoresis (2-DE). Western blot detected APP and ETS-2 that migrated at ~75 and 50 kDa, respectively. Subsequent densitometric analysis of APP and ETS-2 immunoreactivity did not produce any significant change between controls and DS. Since the metabolic fate of APP determines the propensity of amyloid β production, the expression of the secreted forms of APP (sAPP) had been examined. Neither the expression of sAPPα nor sAPPβ showed any detectable changes among the two groups. Collagen α1 (VI) chain precursor, a protein resolved as a single spot on 2D gel was identified by matrix associated laser desorption ionization mass spectroscopy. Quantitative analysis of this spot using the 2D Image Master software revealed a significant decrease in fetal DS ($P < 0.01$) compared to controls. Linear regression analysis did not show any correlation between protein levels and age. The current data suggest that overexpression per se can not fully explain the DS phenotype.

Introduction

Down syndrome (DS) is the most common birth defect with an incidence of 1 in 700 live births. Although it is characterized by several inconsistent anomalies, mental retardation remains an invariable hallmark of the disorder (Epstein, 1995). DS occurs when an extra copy of chromosome 21 is present in the genome either as a free chromosome or as part of a fusion chromosome resulting from translocation. More than 95% of all individuals with DS have trisomy 21 and the remainder are mosaics or have translocations (Epstein, 1995; de Haan et al., 1997).

Chromosome 21 is the smallest of the human autosomes constituting about 1–1.5% of the human genome and 225 genes are embedded within the 33.8 Mb of its genomic DNA (Hattori et al., 2000). The major part of chromosome 21 is the long arm (21q) which harbors almost all genes of known function and is essential for normal development and function (Epstein, 1995). Phenotype-genotype correlation of cases with partial trisomy 21 led to the proposal that there is a region of about 4 Mb between DNA markers D21S17 and ETS-2 that, if triplicated, is responsible for the clinical features of DS. This region was termed Down syndrome critical region (DSCR) (Antonarakis, 1998). This notion is criticized on the ground that different critical regions are identified by different investigators and no rationale exists to exclude other transcribing loci (Shapiro, 1999). Nevertheless, there is consensus that the distal half of the long arm of chromosome 21, which accounts for at least 85% of its transcribing capacity is of pathogenic importance in development of the DS phenotype.

Several proteins of medical importance are encoded on chromosome 21. β amyloid precursor protein (APP) is a transmembrane glycoprotein with multiple isoforms (APP695, APP751 and APP770) generated by differential splicing of gene that maps to the region 21q21.2 of chromosome 21 (Antonarakis, 1998; Kang et al., 1987; Yoshikai et al., 1990). Processing of APP through the amyloidogenic pathway generates amyloid beta (Aβ) (Mills and Reiner, 1999), the major component of senile plaques that characterizes Alzheimer's disease. The vast majority of DS patients develop abundant Aβ-plaques by the fourth decade of life (Wisniewski et al., 1985) and this has been linked to overexpression of the APP gene (Golaz et al., 1991). Human ETS-2 is a proto-oncogene and transcription factor that defines the distal limit of the DSCR and encoded by ETS-2 gene, which is a member of the ETS gene family. The ETS family of transcription factors are characterized by the presence of a highly conserved domain of 85 amino acids termed the ETS domain that mediates specific binding to DNA purine rich sequences with an invariant GGA/TA core element. ETS family members play a critical role in the transcriptional regulation of many genes and the ETS-DNA binding motif has been found in the regulatory control regions of numerous genes (Wasylyk, 1993). ETS-2 is ubiquitously expressed in human tissues including the brain (Bhat, 1987). In mice, its cortical expression is increased during organogenesis and suggested to play an important role in brain development (Maraoulakou, 1994). In addition, transgenic mice that overexpress ETS-2 are shown to display skeletal abnormalities reminiscent of those seen in DS (Sumarsono et al., 1996). Another protein of interest is collagen type VI. This protein is a heterotrimer of three genetically distinct α1, α2 and α3 chains (Jander et al., 1985). Whilst the α1 and α2 chains are encoded by genes located on chromosome 21 within the region thought to be critical for congenital heart defects, the gene encoding the α3 chains is located on chromosome 2 (Weil et al., 1988). Collagen type VI anchors large interstitial structures, such as nerves, blood vessels and collagen fibrils into surrounding connective tissue. It is also involved in cell migration and differentiation, and may play a role in bridging cells with the extracellular matrix (Knupp and Squire, 2001). Genetic mutation of collagen type VI is linked to Bethlem myopathy, an inherited disorder characterized by early childhood onset of generalized muscle weakness and wasting (Lamande et al., 1999).

According to the gene dosage hypothesis, the chromosomal imbalance brought about by an aneuploid state would have a gene dosage effect on the loci present on the chromosome or chromosome segment leading to a 1.5 fold increase in concentrations or activities of proteins in trisomic cells (Epstein, 1995). Previous reports from our group challenged this hypothesis (Greber-Platzer et al., 1998, 1999) and the present study is undertaken to gather some more evidence. To this end, the expression of APP, ETS-2 and collagen α1 (VI) chain precursor encoded on chromosome 21 is investigated in the cerebral cortex of control and DS fetal brains by using western blotting and two-dimensional electrophoresis.

Material and methods

Fetal brain tissue

Brain tissue (cerebral cortex) of aborted control fetuses with no obvious abnormalities (n = 7; 1 female and 6 males with mean gestational age of 18.8 ± 2.2 weeks) and DS (n – 9; 2 females and 7 males with mean gestational age of 19.6 ± 2.0 weeks) was used for the experiment. The brain samples were obtained from the Medical and Molecular Genetics Center-IRO, Hospital Duran i Reynals, Barcelona, Spain. All samples were stored at −70°C and the freezing chain was never interrupted.

Western Blot

Sample preparation

Frozen brain samples were thawed on ice and homogenized in 4 volumes of TBS buffer (100 mM Tris, 150 mM NaCl, pH 7.5) containing protease inhibitors, (1 µg/ml each of antipain, pepstatin A, and leupeptin and 10 mM of phenylmethylsulfonyl fluoride) and 0.05% Tween 20. Samples were homogenized on ice for 30 seconds (6–8 strokes) with a Potter-Elvehjem homogenizer. The brain homogenate then was centrifuged for 10 min at 10,000 × g and 4°C. To the supernatant, equal volume of sample buffer (125.5 mM Tris-HCl base, 70 mM sodium dodecyl sulfate, 0.001% Bromophenol blue, 20% glycerol, and 2% 2-mercaptoethanol, pH 6.8) was added and the mixture was heated at 95°C for 15 min. The concentration of protein was determined by the Bicinchoninic acid assay using bovine serum albumin as a standard (Pierce, USA).

Immunoblotting

Protein extracts containing 30 μg of ETS-2, APP/Aβ or 15 μg sAPP (secreted form of APP) were subjected to sodium dodecyl sulfate-polyacrylamide gel electrophoresis (200 V for 3 hrs), and transferred to polyvinylidene fluoride membranes (Immobilon™-P, MilliPore, Bedford, USA) (192 V for 1 hr). The membranes were then washed two times for 5 min in Tris buffer (100 mM Tris-HCl, 150 mM NaCl, pH 7.5) and later blocked with blotto (1.5% non-fat dry milk in TBST (100 mM Tris-HCl, 150 mM NaCl, pH 7.5, 1 mM $MgCl_2$, 0.1% Tween-20)) overnight at 4°C followed by incubation with polyclonal rabbit anti-ETS-2 antibody (Cat # sc-351, Santa Cruz Biotechnology, Santa Cruz, USA) (1:500), monoclonal mouse anti-APP/Aβ antibody (Cat # M066-3, Medical & Biological Laboratories Co.Ltd) (1:1,000) and anti-sAPP antibody (Cat # MAB5228, CHEMICHON International, Inc, Temecula, CA, USA) (1:500) at room temperature for 2 hours on a rocking shaker. The membranes were washed three times for 15 minutes in blocking solution. Later membranes were incubated with goat anti-rabbit IgG (H + L) (Cat # 4050-05), goat anti mouse IgG (H + L) (Cat # 1031-05) and goat anti-mouse IgG2a (Cat # 1080-05) (Southern Biotechnology Associates Inc Birmingham, USA), horseradish peroxidase conjugated secondary antibodies, in blotto (1:2,000) on a rocking shaker at room temperature for one hour. The membranes were washed again three times for 15 minutes followed by incubation with NEN chemiluminescence reagent for 1 minute and subsequent exposure to films (Kodak Blue XB-1, Rochester, USA). Developed films were scanned and densities of the immunoreactive bands were calculated by RFLP Scan 2.1 software program (Scanalytics, Billerica, USA). Protein size markers (Cat # 161-0372) were purchased from BIORAD.

Two dimensional gel electrophoresis (2-DE)

Brain tissue was suspended in 0.5 ml of sample buffer consisting of 40 mM Tris, 5 M urea (Merck, Darmstadt, Germany), 2 M thiourea (Sigma, St. Louis, MO, USA), 4% CHAPS (Sigma), 10 mM 1,4-dithioerythritol (Merck), 1 mM EDTA (Merck) and a mixture of protease inhibitors, 1 mM PMSF and 1 μg of each pepstatin A, chymostatin, leupeptin and antipain. The suspension was sonicated for approximately 30 sec and centrifuged at 150,000 g for 45 min. The protein content in the supernatant was determined by the Coomassie blue method (Bradford, 1976).

2-DE was performed essentially as reported (Langen et al., 1997). Samples of 2 mg protein were applied on immobilized pH 3–10 nonlinear gradient strips in sample cups at their basic and acidic ends. Focusing started at 200 V and the voltage was gradually increased to 5,000 V at 3 V/min and kept constant for a further 24 h (approximately 180,000 kVh totally). The second-dimensional separation was performed on 9–16% SDS gradient polyacrylamide gels. The gels (180 × 200 × 1.5 mm) were run at 40 mA per gel. After protein fixation for 12 h in 40% methanol, containing 5% phosphoric acid, the gels were stained with colloidal Coomassie blue (Novex, San Diego, CA, USA) for 48 h. Molecular masses were determined by running standard protein markers (Gibco, Basel, Switzerland), covering the range 10–200 kDa. pI values were used as given by the supplier of the IPG strips. Excess of dye was washed out from the gels with H_2O and the gels were scanned in an Agfa DUOSCAN densitometer (resolution 200). Electronic images of the gels were recorded using Photoshop (Adobe) and PowerPoint (Microsoft) software.

Matrix-associated laser desorption ionization mass spectroscopy (MALDI-MS)

MALDI-MS analysis was performed as described elsewhere (Fountoulakis and Langen, 1997) with some modifications. The spots were excised with a spot picker and placed into

96-well microtiter plates. Each spot was destained with 100μl of 30% acetonitrile in 50mM ammonium bicarbonate and dried in a speedvac evaporator. Each dried gel piece was rehydrated with 4μl of 3mM Tris-HCl, pH 9.0, containing 50ng trypsin (Promega, Madison, WI, USA). After 16h at room temperature, 7μl of H_2O was added to each gel piece and the samples were shaken in for 10min. Four μl of 50% acetonitrile, containing 0.3% trifluoroacetic acid and the standard peptides des-Arg-bradykinin (Sigma, 904.4681 Da) and adrenocorticotropic hormone fragment 18–39 (Sigma, 2,465.1989 Da), in water were added to each gel piece and shaken for 10min. The application of the samples was performed with a SymBiot I sample processor (PE Biosystems, Framingham, MA, USA). 1.5μl of the peptide mixture was simultaneously applied with 1μl of matrix, consisting of a saturated solution of α-cyano-4-hydroxycinnamic acid (Sigma) in 50% acetonitrile, containing 0.1% trifluoroacetic acid. Samples were analyzed in a time-of-flight mass spectrometer (Reflex 3, Bruker Analytics, Bremen, Germany). An accelerating voltage of 20kV was used. Peptide matching and protein searches were performed automatically. The peptide masses were compared to the theoretical peptide masses of all available proteins from all species. Monoisotopic masses were used and a mass tolerance of 0.0025% was allowed. The algorithm used for determining the probability of a false positive match with a given MS-spectrum is described elsewhere (Berndt et al., 1999).

Quantification of protein spots

Protein spots were outlined (first automatically and then manually) and quantified using the ImageMaster 2D Elite software (Amersham Pharmacia Biotechnology). The percentage of the volume of the spots representing a certain protein was determined in comparison with the total proteins present in the 2-D gel.

Statistical analysis

Values are expressed as mean ± standard deviation of optical density for western blotting and percentage of the spot volume in each particular gel after subtraction of the background values for 2-DE. Inter-group differences were analyzed using non-parametric Mann-Whitney U-test and within-group correlation was done using linear regression analysis. The level of significance was set at $P < 0.05$.

Results

Western blot

The immunoblots revealed a diffuse APP immunoreactive bands that range in molecular size from 45 to 80kDa, with the major band migrating at ~75kDa (Fig. 1A). The immunoreactive bands were detected by an antibody raised against amino acid residues 1–16 of Aβ that recognizes only APP and Aβ (which is too small to be detected in our setting). Among the bands detected, there appeared to be a triplet with molecular size of ~70–80kDa. The triplet is unlikely to represent the different isoforms of APP, since APP695 is supposed to be the only species present in fetal brain (Oyama et al., 1994). Hence, bands other than the major band may represent posttranslationally modified form, truncated form or fragments of APP formed by the alternative non-

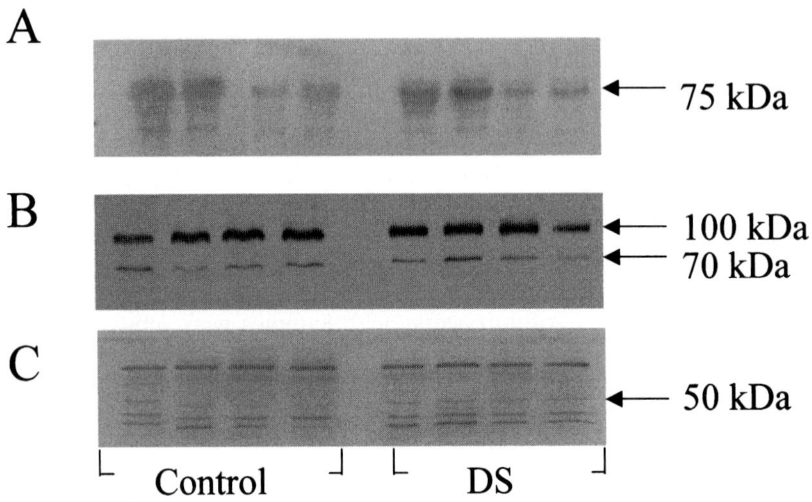

Fig. 1. Immunodetection of APP (**A**), sAPP (**B**) and ETS-2 (**C**) in cerebral cortex of control and DS fetuses. Brain homogenate containing 30 (APP and ETS-2) and 15 (sAPP) μg protein/lane from the cerebral cortices of 7 controls and 8 DS brains were used for the immunodetection of APP, sAPP and ETS-2. The immunoblots of 4 controls (lanes 1–4) and 4 DS (lanes 5–8) cases were shown in the upper panels

secretase pathway (Pellegrini et al., 1999). Quantitative analysis of APP-immunoreactivity did not produce any significant change between controls and DS, although the mean immunoreactivity of the 75 kDa protein as well as total APP-immunoreactivity tended to increase in DS by ~40% and 10.4%, respectively (Table 1). There was a great individual variability in levels of

Table 1. Levels of gene products encoded on chromosome 21 in cerebral cortex of fetal control and DS brain. Following western blot, APP, sAPP and ETS-2 immunoreactivity (IR) of 6–7 controls and 7–8 DS cases were quantified by scanning densitometry. Collagen α1 (VI) chain precursor was resolved and identified by 2-DE and MALDI-MS. This was followed by quantification using the 2D Image Master. Values are expressed as mean ± standard deviation of relative percent level for collagen α1 (VI) chain precursor and arbitrary optical density for the others. Inter-group differences were analyzed using non-parametric Mann-Whitney U-test and the significance was set at $P < 0.05$ for a two-tailed test. $^{**}P < 0.01$

Protein	Control	DS
Total APP-IR	1.62 ± 0.78 (n = 6)	1.79 ± 1.39 (n = 7)
APP	0.62 ± 0.21 (n = 6)	0.87 ± 0.51 (n = 7)
Total sAPP	4.88 ± 2.15 (n = 7)	5.09 ± 1.55 (n = 8)
SAPPα	4.09 ± 2.13 (n = 7)	4.32 ± 1.63 (n = 8)
SAPPβ	0.78 ± 0.62 (n = 7)	0.77 ± 0.49 (n = 8)
SAPPα/sAPPβ	8.19 ± 5.62 (n = 7)	7.99 ± 4.54 (n = 8)
Total ETS-2 IR	5.07 ± 1.60 (n = 7)	4.19 ± 1.22 (n = 8)
ETS-2	0.25 ± 0.21 (n = 7)	0.22 ± 0.11 (n = 8)
Collagen α1 (VI) chain precursor	5.66 ± 2.10 (n = 6)	2.47 ± 0.92 (n = 7)**

APP in DS that may account for the observed mean difference. The proportion of sAPPα and sAPPβ produced by the α-secretase and β-secretase, respectively indicates the propensity of Aβ production. Furthermore, since it is APP processing rather than expression which is being regulated (Howlett et al., 1995), to support our finding we determined the expression of sAPP as described in Material and Methods. Western blot produced two bands that migrated at ~100 kDa and ~70 kDa representing sAPPα and sAPPβ, respectively (Fig. 1B). Neither the levels of sAPPα nor sAPPβ did show any significant change between the two groups. So did the sum of sAPPα and sAPPβ as well as the ratio of the two secreted forms (Table 1).

The transcription factor ETS-2 ran on SDS (sodium dodecylsulphate) gel at ~50 kDa along with several bands that span from ~25 to 120 kDa (Fig. 1C). Bands of higher molecular mass than the full length ETS-2 may represent either posttranslationally modified ETS-2 species or protein complexes, whereas those of low molecular mass could be proteolytic products. No apparent difference was detected in the immunoreactivity of the 50 kDa protein as well as total ETS-2 between the study populations (Table 1).

2-DE

The protein spots were visualized by staining with colloidal Coomassie blue. The protein spots were analyzed by MALDI-MS, following in-gel digestion. The peptide masses were matched with the theoretical peptide masses of all proteins of the SWISS-PROT database. We used internal standards to correct the measured peptide mass thus reducing the windows of mass tolerance and increasing the confidence of identification. Approximately 389 proteins were resolved from fetal brain (data not shown, see elsewhere in this volume), among which collagen α1 (VI) chain precursor was represented by one spot (Fig. 2). The data obtained from MALDI analysis indicated that the

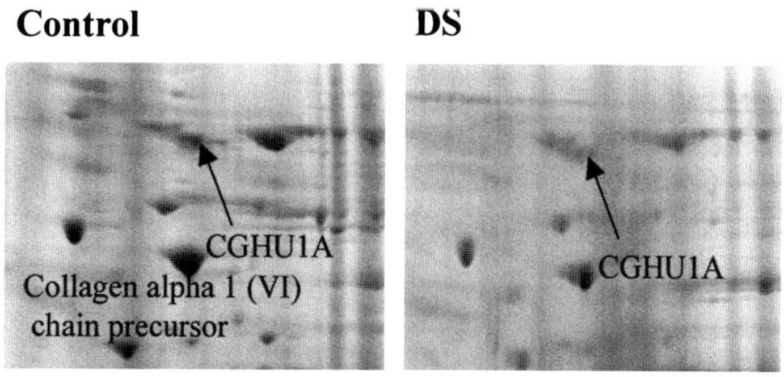

Fig. 2. Partial two-dimensional gel image of human fetal brain proteins. The gels were prepared and the protein was analyzed as described under Material and methods. The spot corresponding to collagen type VI alpha 1 chain precursor is designated with its SWISS-PROT number

spot has a molecular size of 109,712 Da, pI value of 5.15, matching peptide of 11 and the probability of wrongly assign the spot was 6.23E-14. The identified collagen α1 (VI) chain precursor was quantified using the Image Master 2D Elite software and its expression level was determined as a percentage volume of total proteins in the area of interest. The relative percent level of collagen α1 (VI) chain was significantly decreased by ~56% in DS ($P < 0.01$) as compared to controls (Table 1 and Fig. 2). Linear regression analysis did not show any significant correlation between expression of the investigated proteins and age (data not shown).

Discussion

Different hypotheses have been put forward to explain the DS phenotype. These include the "gene dosage effect" hypothesis, the "amplified developmental stability" hypothesis, and the "molecular misreading concept" (Epstein, 1995; Shapiro, 1983; Pritchard and Kola, 1999). The "gene dosage effect" hypothesis states that the DS phenotype is a direct result of overexpression of genes located on the triplicated chromosome or chromosome region. This paper attempts to challenge the gene dosage hypothesis by presenting evidence that shows otherwise.

Immunoreactivity to the various domains of APP was observed in both neurons and glial cells from fetal period to infancy reflecting the role of APP in development (Arai et al., 1997). This immunoreactivity disappears in childhood and reappears in adulthood only in DS patients. The early reappearance of APP-immunoreactivity in adults with DS is linked to a gene dosage effect (Arai et al., 1997), as APP gene maps to the long arm of chromosome 21. Confirming the gene dosage effect, APP mRNA was reported to be overexpressed in fetal (Neve et al., 1988) and adult (Oyama et al., 1994) DS brains. In addition, the numerical density of APP-immunoreactive neurons was found to be slightly higher in fetal DS brain than controls and the increase becomes much more significant with age (Griffin et al., 1998). In the present study, at protein level, however, we could not replicate the gene dosage effect observed at mRNA level, since no statistically significant difference was observed in brain APP expression between control and DS fetuses. The metabolic fate of APP is one of the key factors in the pathogenesis of early onset Alzheimer's disease pathology in DS patients. APP is processed through at least two different pathways. The α-secretory pathway involves α- and β-secretases generating sAPPα and p3, whereas the β-secretory pathway produces sAPPβ and Aβ by the action of β- and γ-secretases (Mills and Reiner, 1999). It appears that in the normal healthy cell, both amyloidogenic and non-amyloidogenic APP processing pathways operate and that a precise balance between these pathways is maintained (Wasco and Tanzi, 1995). If such a balance exists, any increase in expression of APP caused by trisomy 21 overburdens the secretory pathway and shifts the balance towards the amyloidogenic pathway thereby producing more sAPPβ. Analysis of the immunoreactivity of both sAPPα and sAPPβ did not show any appreciable difference between controls and DS indicating that the balance in the two APP processing pathways is still maintained in DS fetuses. Although Aβ was not detected in our setting, Teller and his colleagues detected Aβ42 in half of the fetal DS brain they investigated and none of the controls. In an attempt to know why they failed to detect soluble Aβ in some fetal DS brains, they analyzed APP mRNA expression and found no difference between DS fetuses with and without Aβ (Teller et al., 1996). Had APP overexpression been to account for the detected Aβ, APP mRNA would have at least increased in those fetal DS brains with soluble Aβ. Together these findings point to the fact that overexpression of APP is unlikely to occur in DS fetal brain and the phenotype might be caused by a defect in mechanisms that clear soluble Aβ (Teller et al., 1996) or dysfunction of APP. Indeed, an increasing body of evidence indicates that APP is a functional receptor linked to Go and APP related disease may be a consequence of disruption of its function (Neve et al., 2000). ETS-2 is the first cellular proto-oncogene found on chromosome 21 and encoded by a gene residing in the DSCR (Antonarakis, 1998; Baffico et al., 1989). It is expressed almost in all fetal tissues and is involved in developmental processes (Baffico et al., 1989). The ubiquitous distribution of ETS-2, including the brain (Baffico et al., 1989; Bhat, 1987), and its role in leukemia and organogenesis of the brain, heart and skeletal system, together with the fact that it is encoded on chromosome 21, would make ETS-2 a logical protein to provide evidence for the gene dosage hypothesis and its consequences. In this regard, the finding of slightly increased ETS-2 mRNA in fetal DS (Baffico et al., 1989) and development of skeletal abnormalities resembling to that of DS in transgenic mice overexpressing ETS-2 (Sumarsano et al., 1996) strengthened this view. Contrary to this notion, the present finding shows unaltered expression of ETS-2 in fetal DS brain. This finding complements the previous evidence presented by our group against overexpression of ETS-2 in hearts of children (Greber-Platzer et al., 1998) and brains of adults (Greber-Platzer et al., 1999) with DS, reinforcing the notion that overexpression per se is unlikely to account for the DS phenotype.

Collagens are a large number of heterotrimeric or homotrimeric triple-helical proteins that are present in the extracellular matrix. Collagen type VI appears to be synthesized and secreted as a precursor which may undergo a stepwise conversion to the mature protein (Trüeb and Bornstein, 1984). Few studies were carried out on expression of collagen type VI in DS. Increased expression of collagen α1 (VI) chain in the skin of DS fetuses was demonstrated semiquantitatively by Northern blotting (von Kaisenberg et al., 1998). In contrast to this peripheral finding, we found a significant decrease in fetal DS brain compared to controls. The discrepancy could arise from using either different tissue (skin vs brain) or protein mRNA instability. In addition, von Kaisenberg et al. reported a normalized value against glyceraldehyde-3-phosphate dehydrogenase that assumed to be a marker of constant expression, but was found to be significantly increased in DS brain (Lubec et al., 1999). If such changes occur in the skin, it would undoubtedly affect the reported values. Collagen type VI forms a microfibrillar extracellular network that interacts with basement membranes, fibrillar collagens, and other components of the extracellular matrix as well as with receptors at the cell surface (Brown and Timpl, 1995). In vitro studies suggest that in its native state collagen type VI participate in cell adhesion and migration through interaction with members of the integrin receptor family or the NG2 proteoglycans (Levine and Nishiyama, 1996). Although normally collagen type VI is a heterotrimer composed of the three polypeptide chains $\alpha1$, $\alpha2$ and $\alpha3$ (Chu et al., 1988), other forms that have different combination and stoichiometry have been described (Chu et al., 1998; Furthmayr et al., 1983). Given this, in a situation where there is scarcity in one of the gene products of collagen type VI, it is likely that homotrimeric collagen type VI molecules may be formed. Sequence data predict that these composites would have a lower structural stability than heterotrimeric molecules (Chu et al., 1988) which could, in turn affect the properties of the extracellular matrix. An extracellular matrix altered in such a way thus may interfere in cell adhesion and migration, and contributes to the brain pathology of DS.

To sum up, neither of the investigated gene products of chromosome 21 showed overexpression as predicted from the gene dosage hypothesis. Thus, the current data indicate that the DS phenotype can not be fully explained by the gene dosage effect.

Acknowledgements

We are highly indebted to Red Bull Company, Salzburg, Austria for the generous financial support. E. E. is a scholarship holder of the Austrian Academic Exchange Program, ÖAD.

References

Antonarakis SE (1998) 10 years of genomics, chromosome 21, and Down syndrome. Genomics 51: 1–16

Arai Y, Suzuki A, Mizuguchi M, Takashima S (1997) Developmental and aging changes in the expression of amyloid precursor protein in Down syndrome brains. Brain Dev 19: 290–294

Baffico M, Perroni L, Rasore-Quartino A, Scartezzini P (1989) Expression of the human ETS-2 oncogene in normal fetal tissues and in the brain of a fetus with trisomy 21. Hum Genet 83: 295–296

Berndt P, Hobohm U, Langen H (1999) Reliable automatic protein identification from matrix-assisted laser desorption/ionization mass spectrometric peptide fingerprints. Electrophoresis 20: 3521–3526

Bhat NK, Fisher RJ, Fujiwara S, Ascione R, Papas TS (1987) Temporal and tissue specific expression of mouse ETS genes. Proc Natl Acad Sci USA 84: 3161–3165

Bradford M (1976) A rapid and sensitive method for the quantitation of microgram quantities of protein utilizing the principle of protein-dye binding. Anal Biochem 72: 248–254

Brown JC, Timpl R (1995) The collagen superfamily. Int Arch Allerg Appl Immunol 107: 484–490

Chu M-L, Conway D, Pan T-C, Baldwin C, Mann K, Deutzmann R, Timpl R (1988) Amino acid sequence of the triple helical domain of the human collagen type VI. J Biol Chem 263: 18601–18606

de Haan JB, Wolvetang EJ, Cristiano F, Iannello R, Bladier C, Kelner MJ, Kola I (1997) Reactive oxygen species and their contribution to pathology in Down syndrome. Adv Pharmacol 38: 379–402
Epstein CJ (1995) Down Syndrome. In: Scriver CR, Beaudet AL, Sly WS, Valle D (eds) The metabolic and molecular bases of inherited disease, 7th edn, vol. I. McGraw Hill, New York, pp 749–794
Fountoulakis M, Langen H (1997) Identification of proteins by matrix assisted laser desorption ionization mass spectrometry following in-gel digestion in low salt, non-volatile buffer and simplified peptide recovery. Anal Biochem 250: 153–156
Furthmayr H, Wiedemann H, Timpl R, Odermatt E, Engel J (1983) Electron microscopical approach to a structural model of intima collagen. Biochem J 211: 303–311
Golaz J, Charnay Y, Vallet P, Bouras C (1991) Alzheimer's disease and Down syndrome. Some recent etiopathogenic data. Encephale 17: 29–31
Greber-Platzer S, Schatzmann-Turhani D, Wollenek G, Lubec G (1998) Evidence against the current hypothesis of "gene dosage effects" of trisomy 21: ETS-2, encoded on chromosome 21 is not overexpressed in hearts of patients with Down syndrome. Biochem Biophys Res Commun 254: 395–399
Greber-Platzer S, Schatzmann-Turhani D, Cairns D, Balcz B, Lubec G (1999) Expression of the transcription factor ETS-2 in brain of patients with Down syndrome-evidence against the overexpression gene hypothesis. J Neural Transm [Suppl] 57: 269–281
Griffin WST, Sheng JG, McKenzie LE, Royston MC, Gentleman SM, Brumback RA, Cork LC, Del Bigio MR, Roberts GW, Mrak RE (1998) Life-long expression of S-100β in Down's syndrome: implication for Alzheimer pathogenesis. Neurobiol Aging 19: 401–405
Hatori M, Fujiyama A, Taylor TD, Watanabe H, Yada T, Park H-S et al (2000) The DNA sequence of human chromosome 21. Nature 405: 311–319
Howlett DR, James S, Allsop D, Roberts GW (1995) The biology and molecular pathology of β.amyloid protein. In: Dawbarn D, Allen SJ (eds) Neurobiology of Alzheimer's disease. BIOS Scientific Publishers, Oxford, pp 9–49
Jander R, Raterberg J, Glanville RW (1983) Further characterization of the three polypeptide chains of bovine and human short-chain collagen (intima collagen). Eur J Biochem 133: 39–46
Kang J, Lemaire H-G, Unterbeck A, Salbaum JM, Masters CL, Grezeschik K-H, Multhaup G, Beyreuther K, Müller-Hill B (1987) The precursor of Alzheimer's amyloid A4 protein resembles a cell surface receptor. Nature 325: 733–736
Klewer SE, Krob SL, Kolker SJ, Kitten GT (1998) Expression of type VI collagen in the developing mouse heart. Dev Dyn 211: 248–255
Knupp C, Squire JM (2001) A new twist in the collagen story – the type VI segmented supercoil. EMBO J 20: 372–376
Lamande SR, Shields KA, Kornberg AJ, Shield LK, Bateman JF (1999) Bethlem myopathy and engineered collagen VI triple helical deletion prevent intracellular multimer assembly and protein secretion. J Biol Chem 274: 21817–21822
Langen H, Röder D, Juranville JF, Fountoulakis M (1997) Effect of the protein application mode and the acrylamide concentration on the resolution of protein spots separated by two-dimensional gel electrophoresis. Electrophoresis 18: 2085–2090
Levine JM, Nishiyama A (1996) The NG-2 chondrotin sulfate proteoglycan: a multi functional proteoglycan associated with immature cells. Perspect Dev Biol 3: 545–559
Lubec G, Labudova O, Cairns N, Fountoulakis M (1999) Increased glyceraldehyde-3-phosphate dehydrogenase levels in brain of patients with Down syndrome. Neurosci Lett 260: 141–145
Maraoulakou IG, Papas TS, Green JE (1994) Differential expression of ETS-1 and ETS-2 genes during murine embriogenesis. Oncogene 9: 1551–1565
Mills J, Reiner PB (1999) Regulation of amyloid precursor protein cleavage. J Neurochem 72: 443–460
Neve RL, Finch EA, Dawes LR (1988) Expression of the Alzheimer's amyloid precursor gene transcript in the human brain. Neuron 1: 669–677

Neve RL, McPhie DL, Chen Y (2000) Alzheimer's disease: a dysfunction of the amyloid precursor protein. Brain Res 886: 54–66

Oyama F, Cairns NJ, Shaimada H, Oyama R, Titani K, Inara Y (1994) Down's syndrome: upregulation of β-amyloid protein precursor and tau mRNAs and their defective coordination. J Neurochem 62: 1062–1066

Pellegrini L, Passer BJ, Tabaton M, Ganjei JK, D'Adamio L (1999) Alternative non-secretase processing of Alzheimer's β-amyloid precursor protein during apoptosis by caspase-6 and -8. J Biol Chem 274: 21011–21016

Pritchard MA, Kola I (1999) The "gene dosage effect" versus the "amplified developmental instability" hypothesis in Down syndrome. J Neural Transm [Suppl] 57: 293–303

Shapiro BL (1983) Down syndrome — a disruption of homeostasis. Am J Med Genet 14: 241–269

Shapiro BL (1999) The Down syndrome critical region. J Neural Transm [Suppl] 57: 41–60

Sumarsono SH, Wilson TJ, Tymms MJ, Venter D, Corrick CM, Kola R, Lahoud M, Papas TS, Seth A, Kola I (1996) Ets transgenic mice develop skeletal abnormalities analogous to those found in Down syndrome. Nature 379: 534–537

Teller JK, Russo C, DeBusk LM, Angelini G, Zaccheo D, Dagna-Bricarelli F, Scartezzini P, Bertolini S, Mann DMA, Tabaton M, Gambetti P (1996) Presence of soluble amyloid β-peptide precedes amyloid plaque formation in Down's syndrome. Nature Med 2: 93–95

Trüeb B, Bornstein P (1984) Characterization of the precursor form of type VI collagen. J Biol Chem 259: 8597–8604

von Kaisenberg CS, Brand-Saberi B, Christ B, Vallian S, Farzaneh F, Nicolaides KH (1998) Collagen type VI expression in the skin of trisomy 21 fetuses. Obstet Gynecol 91: 319–323

Wasco W, Tanzi RE (1995) Molecular genetics of amyloid and apolipoprotein E in Alzheimer's disease. In: Dawbarn D, Allen SJ (eds) Neurobiology of Alzheimer's disease. BIOS Scientific Publishers, Oxford, pp 51–76

Wasylyk B, Hahn SL, Giovane A (1993) The ETS family transcription factors. Eur J Biochem 211: 7–18

Weil D, Mattei MG, Passage E, N'Guyen VC, Pribula-Conway D, Mann K, Deutzmann R, Timpl R, Chu ML (1988) Cloning and chromosomal localization of human genes encoding the three chains of type VI collagen. Am J Hum Genet 42: 435–445

Wisniewski KE, Wisniewski HM, Wen GY (1985) Occurrence of neuropathological changes and dementia of Alzheimer's disease in Down syndrome. Ann Neurol 17: 278–282

Yoshikai S-I, Sasaki H, Doh-ura K, Furuya H, Sakaki Y (1990) Genomic organization of the human amyloid beta protein precursor gene. Gene 87: 257–263

Authors' address: Prof. Dr. G. Lubec, Department of Pediatrics, University of Vienna, Währinger Gürtel 18–20, A-1090 Vienna, Austria, e-mail: Gert.Lubec@akh-wien.ac.at

Reduction of nucleoside diphosphate kinase B, Rab GDP-dissociation inhibitor beta and histidine triad nucleotide-binding protein in fetal Down Syndrome brain

R. Weitzdoerfer[1], **D. Stolzlechner**[1], **M. Dierssen**[2], **J. Ferreres**[2], **M. Fountoulakis**[3], and **G. Lubec**[1]

[1] Department of Pediatrics, University of Vienna, Vienna, Austria
[2] Medical and Molecular Genetics Center-IRO, Hospital Duran i Reynals, Barcelona, Spain
[3] F. Hoffmann-La Roche, Gene Technologies, Basel, Switzerland

Summary. Information on the various factors leading to impairments in the developing brain of fetal Down Syndrome patients is limited to few histological reports. We therefore attempted to describe expression levels of proteins in brain using the proteomic technique of two-dimensional electrophoresis with subsequent mass spectroscopical identification of protein spots and quantification with specific software. Cortical tissue was obtained from autopsy of human fetal abortus. Protein levels of GTP-binding nuclear protein ran, guanine nucleotide-binding protein g(o), alpha subunit 2, guanine nucleotide-binding protein g(i)/g(s)/g(t) beta subunit 1, -beta subunit 2, guanine nucleotide-binding protein beta subunit 5, nucleoside diphosphate kinase A, nucleoside diphosphate kinase B, Rab GDP-dissociation inhibitor beta, Rho GDP-dissociation inhibitor 1, biphosphate 3′-nucleotidase, small glutamine-rich tetra-tricopeptide repeat-containing protein and histidine triad nucleotide-binding protein were studied.

Quantification revealed statistically significant reduced levels of nucleoside diphosphate kinase B, Rab GDP-dissociation inhibitor beta and histidine triad nucleotide-binding protein in fetal DS brain as compared to controls.

We conclude that in early prenatal life proteins involved in neural differentiation, migration and synaptic transmission are impaired in DS cortex. These results may help to understand the abundant mechanisms leading to abnormalities in the wiring, structure and function of DS brain.

Introduction

Cells transduce signals by using proteins that serve as a molecular switch. These signalling proteins are localized in the cytosol, in the nucleus, or on a specific membrane. G proteins include heterotrimeric G proteins (Gilman et al., 1987), G proteins involved in protein synthesis (Kaziro, 1978) and small G proteins (Takai et al., 2001). The family of membrane proteins known as guanine nucleotide binding proteins (G Proteins) contains members that are involved in signal transduction. Common feature of all G proteins is the ability, to change from a functionally inactive GDP-binding configuration to a functionally active GTP-binding form. By binding extra-cellular agents to the receptor, the G protein plays a connecting role between the agents, the receptor and the effector proteins. Heterotrimeric G-proteins appear to be involved in almost every kind of signalling in the nervous system. Each heterotrimer G protein is composed of different α-, β- and γ-subunits, which in turn are classified into different subtypes, coupled with effectors including, ion channels, adenylyl cyclase, phospholipase C, phospholipase A2 and phosphodiesterase, depending on the type of the subunit and cell involved. The subunits are implicated in a variety of cellular activities. G_o-alpha subunits, for example, play a major role in $Ca2^+$ channel regulation, motor control, motor behaviour and pain perception. Beta and gamma subunits are also required for the GTPase activity and G protein-effector interaction. Rab, Rho and Ran belong to the major class of monomeric G proteins termed the "small G proteins" superfamily (molecular masses of 20–35 kDa; classified into at least five families: the Ras-, Rab-, Rho-, Sar1/Arf and Ran families (Takai et al., 2001) The Ras family regulates gene expression, while the Rab family regulates intracellular vesicle trafficking. Each Rab protein is localized in organelles and is involved in the vesicle transport of the organelle (Nuoffer et al., 1994; Schimmoller et al., 1998). Rho proteins were shown to regulate cytoskeletal reorganization and gene expression, and Rho dissociation inhibitor (GDI α and β) acts as a molecular switch to transduce an upstream signal to a downstream effector. The Ran family regulates nucleo-cytoplasmic transport and microtubule organization (Fleig et al., 2000). Evidence has also accumulated that they may play additional roles in gene expression. The Rab dissociation inhibitors (GDIs) play an important role in specific delivery of Rab proteins to their target membranes (Takai et al., 2001). In addition, the Rab- GDP/Rho-GDP dissociation inhibitors regulate the GDP/GTP exchange reactions of most Rab/Rho proteins by inhibiting the dissociation of GDP and the subsequent binding of GTP.

The nucleoside diphosphate kinases are enzymes involved in cell proliferation, differentiation, development, tumor progression, and metastasis (Roymans et al., 2000). The enzymes are located in the cytosol and nucleus, and play a major role in catalysing the synthesis of nucleoside triphosphates. As one of the nucleoside triphosphates (GTP) is responsible for activation of G-proteins, the proteins may regulate signal transduction by maintaining the GTP pool. Biphosphate 3′nucleotidase, a protein that hydrolyses both nucleotides and inositol polyphosphates, is an enzyme that belongs to the lithium-sensitive phosphomonoesterase family, which are implicated in manic depression, a polygenic disorder involving multiple signalling pathways (Spiegelberg et al., 1999). Histidine triad nucleotide-binding protein is a cytoplasmic protein that cleaves ADP into AMP (Lima et al., 1997) and inorganic phosphate and may play important role in transcription (Razin et al., 1999).

Defects of transmembrane signalling proteins have been reported in brains with neuro-degenerative diseases, such as Alzheimer's disease. For example, neuropathy target esterase (NTE), an integral membrane protein in vertebrate neurons, causes delayed neuropathy with degeneration of long axons in peripheral nerves and spinal cord (Glynn, 2000). Down syndrome (DS) is associated with developmental abnormalities of the central nervous system that result in mental retardation and age-dependent Alzheimer-type neuro-degeneration. Although published reports indicate aberrant expression of proteins involved in signal transduction processes in DS (Fuentes et al., 2000; Fountoulakis et al., 1999), little is known whether this also happens early in life or not. The present study is therefore undertaken to address this issue. To this effect, the expression pattern of G-proteins, nucleoside diphosphate kinases and biphosphate 3′-nucleotidase investigated in brains of control and DS fetuses using two-dimensional electrophoresis (2-DE) coupled to matrix assisted laser desorption ionization mass spectroscopy (MALDI-MS).

Materials and methods

Fetal brain tissue and brain preparation

Abortus fetal brain sample (cerebral cortex) of controls (n = 7; 1 female and 6 males with mean age of 18.8 ± 2.2 weeks) and DS (n = 8; 2 females and 6 males with mean age of 19.8 ± 2.0 weeks) were obtained from Drs. Mara Dierssen and J. C. Ferreres the Medical and Molecular Genetics Center-IRO, Hospital Duran i Reynals, Barcelona, Spain. All samples were stored at −70°C and the freezing chain was never interrupted. Brain tissue was suspended in 0.5 ml of sample buffer consisting of 40 mM Tris, 5 M urea (Merck, Darmstadt, Germany), 2 M thiourea (Sigma, St. Louis, MO, USA), 4% CHAPS (Sigma), 10 mM 1,4-dithioerythriol (Merck), 1 mM EDTA (Merck), 1 mM PMSF (Sigma) and 1 μl/mg of each pepstatin A, chymostatin, leupeptin and antipain. The suspension was sonicated for approximately 30 seconds and centrifuged further at 150,000 × g for 45 minutes to sediment undissolved material. The protein concentration of the supernatant was determined by the Coomassie blue method (Bradford, 1976).

Two-dimensional gel electrophoresis

Two-dimensional gel electrophoresis was performed essentially as reported (Langen et al., 1997). Samples of 2 mg protein were applied on immobilized pH 3–10 non-linear gradient strips in sample cups at their basic and acidic ends. Focusing started at 200 V and the voltage was gradually increased to 5,000 V at 3 V/min and kept constant for a further 24 h (approximately 180,000 kVh totally). The second-dimensional separation was performed on 9–16% SDS gradient polyacrylamide gels. The gels (180 × 200 × 1.5 mm) were run at 40 mA per gel. After protein fixation for 12 h in 40% methanol, containing 5% phosphoric acid, the gels were stained with colloidal Coomassie blue (Novex, San Diego, CA, USA) for 48 h. Molecular masses were determined by running standard protein markers (Gibco, Basel, Switzerland), covering the range 10–200 kDa. pI values were used as given by the supplier of the IPG strips. Excess of dye was washed out from the gels with H_2O and the gels were scanned in an Agfa DUOSCAN densitometer (resolution 200). Electronic images of the gels were recorded using Photoshop (Adobe) and PowerPoint (Microsoft) software.

Matrix-assisted laser desorption ionization mass spectroscopy (MALDI-MS)

MALDI-MS analysis was performed as described elsewhere (Fountoulakis and Langen, 1997) with some modifications. The spots were excised with a spot picker and placed into 96-well microtiter plates. Each spot was destained with 100 μl of 30% acetonitrile in 50 mM ammonium bicarbonate and dried in a speedvac evaporator. Each dried gel piece was rehydrated with 4 μl of 3 mM Tris-HCl, pH 9.0, containing 50 ng trypsin (Promega, Madison, WI, USA). After 16 h at room temperature, 7 μl of H_2O were added to each gel piece and the samples were shaken in for 10 min. Four μl of 50% acetonitrile, containing 0.3% trifluoroacetic acid and the standard peptides des-Arg-bradykinin (Sigma, 904.4681 Da) and adrenocorticotropic hormone fragment 18–39 (Sigma, 2465.1989 Da), in water were added to each gel piece and shaken for 10 min. The application of the samples was performed with a SymBiot I sample processor (PE Biosystems, Framingham, MA, USA). 1.5 μl of the peptide mixture was simultaneously applied with 1 μl of matrix, consisting of a saturated solution of α-cyano-4-hydroxycinnamic acid (Sigma) in 50% acetonitrile, containing 0.1% trifluoroacetic acid. Samples were analyzed in a time-of-flight mass spectrometer (Reflex 3, Bruker Analytics, Bremen, Germany). An accelerating voltage of 20 kV was used. Peptide matching and protein searches were performed automatically. The peptide masses were compared to the theoretical peptide masses of all available proteins from all species. Monoisotopic masses were used and a mass tolerance of 0.0025% was allowed. The algorithm used for determining the probability of a false positive match with a given MS-spectrum is described elsewhere (Berndt et al., 1999).

Quantification of proteins

Protein spots were outlined (first automatically and then manually) and quantified using the ImageMaster 2D Elite software (Amersham Pharmacia Biotechnology). The percentage of the volume of the spots representing a certain protein was determined in comparison with the total proteins present in the 2-D gel.

Statistical analysis

Software calculated the percentage of spot volume in each particular gel after subtraction of background values. Non-parametric Mann-Whitney U-test calculated between-group differences. Level of significance was set at $P < 0.05$. If the spot of a protein was not detectable a value of 0.01 was inserted to make statistical analysis possible.

Results

Identification and quantification of proteins

Protein spots were analysed by MALDI-MS, following in-gel digestion. Peptide masses were matched with the theoretical peptide masses of all proteins of the SWISS-PROT database. We used internal standards to correct the measured peptide mass thus reducing the windows of mass tolerance and increasing the confidence of identification. In this study 12 proteins were quantified using the Image Master 2D Elite software (see Fig. 1). Expression levels of proteins were determined as a percentage of spot volume in the area

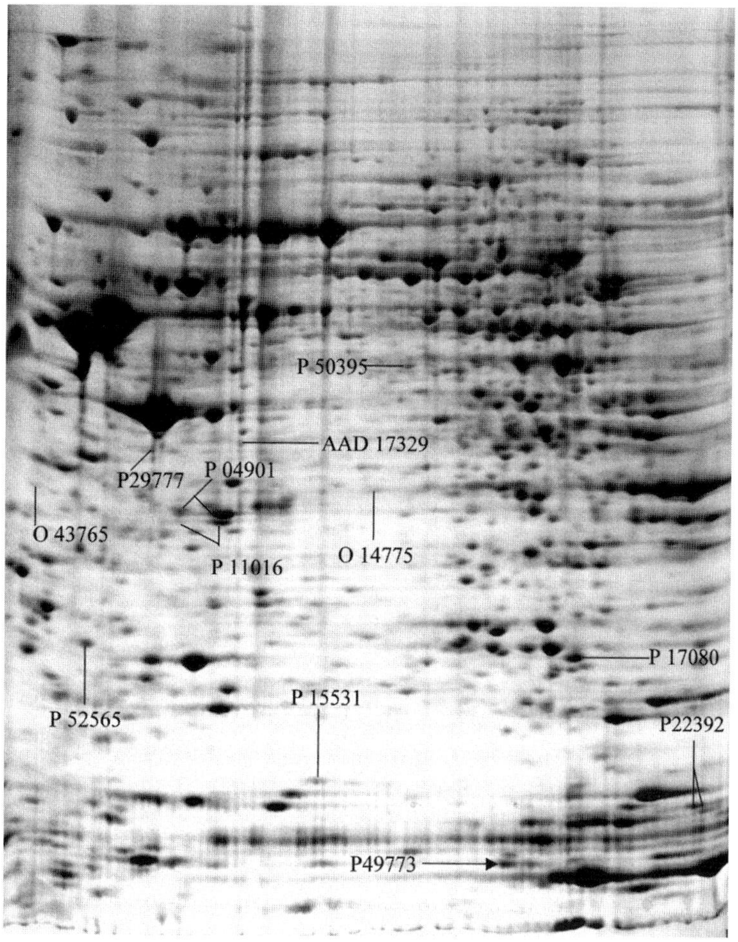

Fig. 1. Identification of proteins measured: *P17080*: GTP-binding nuclear protein ran; *P29777*: Guanine nucleotide-binding protein g(o), alpha subunit 2; *P04901*: Guanine nucleotide-binding protein g(i)/g(s)/g(t) beta subunit 1; *P11016*: Guanine nucleotide-binding protein g(i)/g(s)/g(t) beta subunit 2; *O14775*: Guanine nucleotide-binding protein beta subunit 5; *P15531*: Nucleoside diphosphate kinase A; *P22392*: Nucleoside diphosphate kinase B; *P50395*: Rab gdp-dissociation inhibitor beta; *P52565*: Rho gdp-dissociation inhibitor 1; *AAD17329*: Biphosphate 3′-nucleotidase; *O43765*: Small glutamine-rich tetratricopeptide repeat-containing protein *P49773*: Histidine triad nucleotide-binding protein

of interest. SWISS-PROT accession numbers and names of the quantified proteins with MALDI information are given in Table 1.

The levels of GTP-binding nuclear protein ran, guanine nucleotide-binding protein g(o) alpha subunit 2, guanine nucleotide-binding protein beta subunit 5, nucleoside diphosphate kinase A, Rho GDP-dissociation inhibitor 1, biphosphate 3′-nucleotidase and small glutamine-rich tetra-tricopeptide repeat-containing protein each of them represented by a single spot were comparable between DS and control samples.

Guanine nucleotide-binding protein g(i)/g(s)/g(t) beta subunit 1 was represented by two spots but only one could be quantified as the second could not

Table 1. The protein spots were excised from the gels, digested with trypsin and the peptides generated were analysed by MALDI-MS as stated under Materials and Methods. The identified protein spots are characterized by their SWISS-PROT accession numbers. The approximate observed MW and pI values, as well as the matching are given

Number	Protein name	MW	pI	Peptide matches
P17080	GTP-binding nuclear protein ran	24,509	7.11	6
P29777	Guanine nucleotide-binding protein g(o), alpha subunit 2	40,429	5.75	6
P04901	Guanine nucleotide-binding protein g(i)/g(s)/g(t) beta subunit 1	38,151	5.87	8
P11016	Guanine nucleotide-binding protein g(i)/g(s)/g(t) beta subunit 2	38,048	5.87	7
O14775	Guanine nucleotide-binding protein beta subunit 5	39,532	5.98	6
P15531	Nucleoside diphosphate kinase A	17,308	6.13	6
P22392	Nucleoside diphosphate kinase B	17,400	8.64	4
P50395	Rab gdp-dissociation inhibitor beta	51,088	6.15	5
P52565	Rho gdp-dissociation inhibitor 1	23,249	4.85	6
AAD17329	Biphosphate 3′-nucleotidase	33,713	5.49	6
O43765	Small glutamine-rich tetratricopeptide repeat-containing protein	34,269	4.62	7
P49773	Histidine triad nucleotide-binding protein	13,776	6.96	5

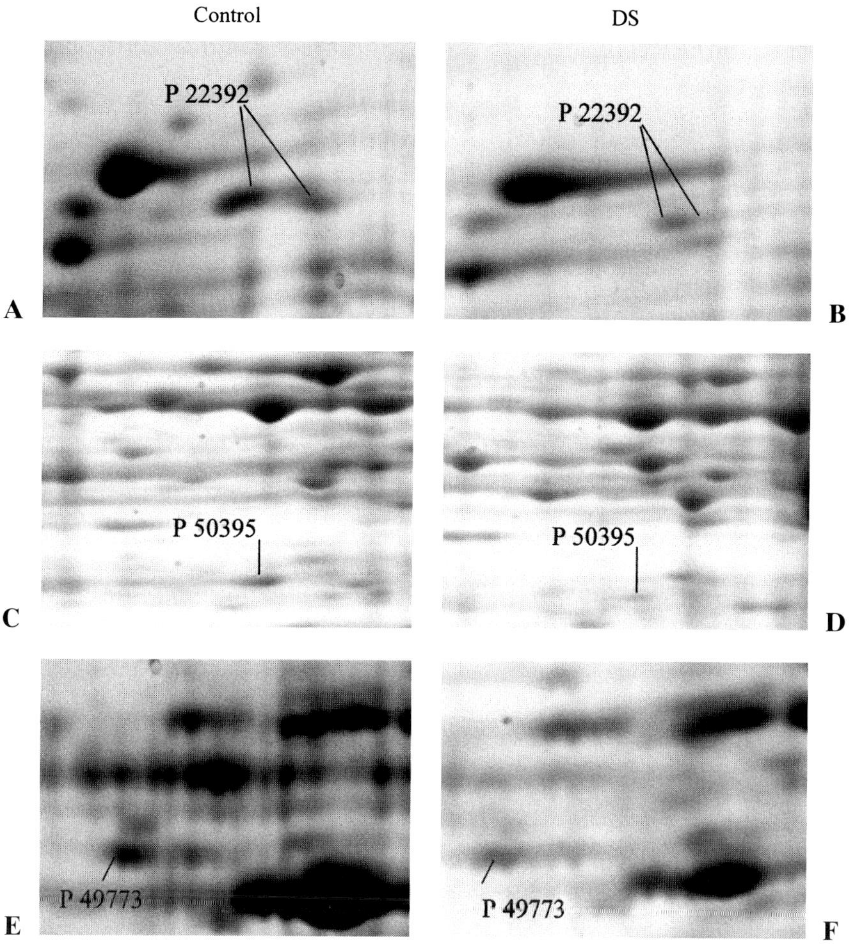

Fig. 2. A NDPK-B in fetal control brain (arrow) **B** NDPK-B in DS samples showed reduced levels. **C** GDI-2 in control cortex (arrow) **D** GDI-2 levels in DS brains were decreased. **E** Levels of HINT in control brain (arrow) **F** HINT levels in DS individuals were reduced

be regularly and clearly separated. Expression levels in DS brains were normal as compared to control samples.

Levels of one of the two spots assigned to guanine nucleotide-binding protein g(i)/g(s)/g(t) beta subunit 2 did not show significant difference between DS and control samples. The other spot could not be statistical evaluated as it was detected in too few samples.

The two spots assigned to nucleoside diphosphate kinase B showed a significant decrease of the protein levels in DS samples compared to controls ($P = 0.009$; $P = 0.020$) (see Fig. 2A,B).

Rab GDP-dissociation inhibitor beta represented by one spot showed reduced levels in DS individuals compared to controls (see Fig. 2C,D).

Histidine triad nucleotide-binding protein showed decreased levels in DS brain as compared to control samples ($P = 0.035$) (see Fig. 2E,F).

Means and SD are given in Table 2.

Table 2. Protein levels in fetal DS and control brains. The investigated protein levels were determined as the percentage volume of the protein compared to total proteins present in the gel part using the ImageMaster 2D Elite software. In comparison with control, $^{*}P < 0.05$, $^{**}P < 0.01$ were considered significant

Frontal Cortex Proteins	DS means ± SD	Control means ± SD
GTP-binding nuclear protein ran	3.09 ± 1.52 (n = 8)	8.29 ± 5.27 (n = 7)
Guanine nucleotide-binding protein g(o), alpha subunit 2	0.36 ± 0.16 (n = 8)	0.73 ± 0.51 (n = 7)
Guanine nucleotide-binding protein g(i)/g(s)/g(t) beta subunit 1	0.59 ± 0.57 (n = 7)	0.96 ± 0.54 (n = 7)
Guanine nucleotide-binding protein g(i)/g(s)/g(t) beta subunit 2	1.18 ± 0.58 (n = 7)	1.83 ± 0.76 (n = 7)
Guanine nucleotide-binding protein beta subunit 5	2.30 ± 1.48 (n = 7)	1.81 ± 1.16 (n = 7)
Nucleoside diphosphate kinase A	3.44 ± 1.29 (n = 7)	4.88 ± 2.02 (n = 6)
Nucleoside diphosphate kinase B (NDPK-B) (spot 1)	2.50 ± 1.51** (n = 8)	5.84 ± 2.58 (n = 7)
Nucleoside diphosphate kinase B (NDPK-B) (spot 2)	0.90 ± 0.68* (n = 8)	2.83 ± 1.98 (n = 6)
Rab GDP-dissociation inhibitor beta (GDI-2)	1.36 ± 0.95* (n = 8)	2.59 ± 1.16 (n = 7)
Rho GDP-dissociation inhibitor 1	0.99 ± 0.35 (n = 8)	0.95 ± 0.30 (n = 7)
Biphosphate 3′-nucleotidase	0.26 ± 0.14 (n = 8)	0.28 ± 0.17 (n = 7)
Small glutamine-rich tetratricopeptide repeat-containing protein	0.17 ± 0.14 (n = 6)	0.10 ± 0.09 (n = 7)
Histidine triad nucleotide-binding protein (HINT)	0.59 ± 0.18* (n = 8)	0.99 ± 0.29 (n = 7)

Discussion

2-D-electrophoresis with subsequent mass-spectrometrical analysis of protein spots unambiguously identified GTP-binding nuclear protein ran, guanine nucleotide-binding protein g(o), alpha subunit 2, guanine nucleotide-binding protein g(i)/g(s)/g(t) beta subunit 1, -beta subunit 2, guanine nucleotide-binding protein beta subunit 5, nucleoside diphosphate kinase A, nucleoside diphosphate kinase B, Rab GDP-dissociation inhibitor beta, Rho GDP-dissociation inhibitor 1, biphosphate 3′-nucleotidase, small glutamine-rich tetra-tricopeptide repeat-containing protein and histidine triad nucleotide-binding protein.

The major findings of this study were that protein levels of nucleoside diphosphate kinase B (NDPK-B), Rab GDP-dissociation inhibitor 2 (GDI-2) and histidine triad nucleotide-binding protein were significantly decreased in fetal Down syndrome brains as compared to control samples.

Nucleoside diphosphate kinases (NDPK) are enzymes, which are ubiquitously expressed, catalysing phosphoryl-group transfer between nucleoside di- and triphosphates through a conserved histidine as phosphorylated intermediate. The two isoforms NDPK-A and NDPK-B investigated in this study are 88% homologous and closely linked on chromosome 17q21 (Backer et al., 1993). They are encoded by the nm23 (non-metatstatic) gene and also known as nm23-H1 (NDPK-A) and nm23-H2 (NDPK-B) proteins. Nm23 genes have been implicated to play an important role in a variety of basic cell functions such as cellular proliferation (Keim et al., 1992; Ohneda et al., 1994), differentiation and apoptosis (Venturelli et al., 1995) as well as inhibition of tumor metastasis and differentiation inhibition in leukemic cells (Okabe-Kado et al., 1995).

NDP kinases play a key role in maintaining intracellular nucleotide concentrations (Postel et al., 2000) as for instance GTP concentration. GTP binding induces the generation of free α-subunits by causing the dissociation of the α-subunits from its β- and γ-subunits and from the receptor. Free α-subunits, bound to GTP, are functionally active and directly regulate a number of effector proteins. A decrease in NDP kinases may lead to a decrease in GTP and influence the activity of effector proteins, like adenylyl cyclase, phospolipase C, phospholipase A_2 and phosphodiesterase. Our results of impaired NDPK-B levels in fetal DS brain may be connected to observed reduction of accumulated cyclic AMP (cAMP) in response to these beta 1-adrenergic ligands in DS (Reed and Ozand, 1985) and to the impaired basal production of cAMP in the hippocampus of trisomic mice (Dierssen et al., 1997). There is evidence that a down- regulation of cAMP is involved in pathological glia activation which is discussed as a common pathogenic factor for progressive nerve cell damage in Alzheimer dementia (Schubert et al., 2000) and thus might play a role in Alzheimer-like neuro-degeneration occurring in adult DS patients.

Besides this catalytic function NDPK-B has the ability to stimulate gene transcription important in cell growth and differentiation. This ability was confirmed by the finding that NM23-H2/NDPK-B is located in the nucleus associated to chromatin (Kraeft et al., 1996). As Postel et al. reported, NDPK-B is a transcription factor for c-myc oncogene (Postel et al., 1993, 1996). Furthermore, it is shown that NDP kinase B is a DNA-binding protein and may regulate gene expression through its DNA-binding activity (Postel, 1999; Postel et al., 2000; Raveh et al., 2001), with a single strand specificity (Agou et al., 1999).

Decreased NDPK-B expression has been reported in high metastatic potential tumor cells (Stahl et al., 1991) although to lesser extent than nm23-H1 and mutations of the nm23-H2 gene were shown to occur frequently in mouse leukaemia (Joosten et al., 2000), a hematopoietic malignancy for which DS patients in advanced life bear a higher risk than controls.

We propose that our findings of reduced NDPK-B levels in fetal DS brains may represent or lead to abnormalities in neural differentiation and in modelling and wiring of the brain of DS patients described in various studies (Epstein, 1995; de la Monte, 1999; Wisniewski, 1994) as protein kinases are also needed in guidance processes of the growth cone, a sensory structure receiving directional information from the surrounding but also a motor structure leading to axonal elongation (Sanes and Jessell, 2000).

Rab GDP-dissociation inhibitor 2 (GDI-2) located on chromosome 10p15, is a regulatory protein supporting membrane release of Rabs (Shisheva et al., 1999), which are a group of small G-proteins with essential role in vesicle formation, vesicle docking and membrane fusion (Schimmoller et al., 1998). The efficacy of interaction with Rabs is suggested to be modified by tyrosine phosphorylation of GDI-2 (Shisheva et al., 1999).

Importance of the GDIs in cell function was shown by depletion of the GDI gene in yeast, which led to various transport defects in the cell. Furthermore D'Adamo et al. showed that GDI has an important role in neuronal function as mutation in the gene encoding GDI-1, which is the brain specific isoform, is responsible for X-linked mental retardation (D'Adamo et al., 1998).

Taken together these functions suggest that decreased levels of GDI-2 found in the present study in fetal DS brains may represent one of the multiple factors leading to impaired vesicle transport at synapses and mental retardation in DS patients.

Histidine triad nucleotide-binding protein (HINT) is a member of the histidine triad (HIT) superfamily of nucleotide-binding proteins, including HINT homologues and fHIT, a Hit protein encoded on the fragile site of chromosome 3 involved in tumor suppression (Siprashvili et al., 1997). Brzoska et al. mapped HINT to 5q31.2 by fluorescence in situ hybridisation and showed its homology of 96% to bovine and 53% to maize members. Furthermore they report that it is localised to structures of the cytoskeleton (Brzoska et al., 1996). Its enzymatic activity of ADP cleavage into AMP and inorganic phosphate was described by Lima et al. (1997). A possible function of the HIT proteins may be found in the enabling of interaction between transcription factors and activators as negative regulation of the microphthalmia transcription factor (mi) (Razin et al., 1999) and an interaction between Cyclin-dependent kinase 7, which is phosphorylating the C-terminal of the large subunit of RNA polymerase II in the basal transcription factor TFIIH (Korsisaari and Mäkelä, 2000) were described.

Our findings of reduced HINT levels in fetal DS brains therefore may represent or lead to dysfunction in transcription.

We conclude that as early as in the second trimester reduced protein levels of NDPK-B, GDI-2 and HINT may be responsible for and/or represent impairments in neural differentiation, migration and synaptic transmission as well as in transcription in fetal Down Syndrome brain.

Acknowledgment

We are highly indebted to the Red Bull Company (Salzburg, Austria) for generous financial support.

References

Agou F, Raveh S, Mesnildrey S, Veron S (1999) Single strand DNA specificity analysis of human nucleoside diphosphate kinase B. J Biol Chem 274: 19630–19636

Backer J, Mendola C, Kovesdi I, Fairhurst J, O'Hara B, Eddy RL, Shows T, Mathew S, Murty V, Chaganti R (1993) Chromosomal localisation and nucleoside diphosphate kinase activity of human metastasis-suppressor genes NM23-1 and NM23-2. Oncogene 8: 497–502

Berndt P, Hobolm U, Langen H (1999) Reliable automatic protein identification from matrix-assisted laser desorption/ionisation mass spectrometric peptide fingerprints. Electrophoresis 20: 3521–3526

Bradford M (1976) A rapid and sensitive method for the quantitation of microgram quantities of protein utilizing the principle of protein-dye binding. Anal Biochem 72: 248–254

Brzoska P, Chen H, Levin N, Kuo WL, Collins C, Fu KK, Gray J, Christman M (1996) Cloning, mapping and in vivo localisation of a human member of the PKCI-1 protein family (PKCNH1). Genomics 36: 151–156

D'Adamo P, Menegon A, Lo Nigro C, Grasso M, Tamanini F, Bienvenu T, Gedeon A, Oostra B, Wu SK, Tandon A, Valtorta F, Balch W, Chelly J, Toniolo D (1998) Mutations in GDI1 are responsible for X-linked non-specific mental retardation. Nature Genet 19: 134–139

De la Monte S (1999) Molecular abnormalities of the brain in Down syndrome: relevance to Alzheimer's neurodegeneration. J Neural Transm [Suppl] 57: 1–20

Dierssen M, Vallina I, Baamonde C, Garcia-Calatayud S, Lumbreras M, Florez J (1997) Alterations of central noradrenergic transmission in Ts65Dn mouse, a model for Down syndrome. Brain Res 749: 238–244

Epstein C (1995) The metabolic and molecular bases of inherited disease In: Scriver SR, Beaudet AL, Sly WS, Valle D (eds) Down Syndrome (Trisomy 21). McGraw Hill, New York, pp 749–794

Fleig U, Salus S, Karig I, Sazer S (2000) The fission yeast ran GTPase is required for microtubule integrity. J Cell Biol 151: 1101–1112

Fountoulakis M, Langen H (1997) Identification of proteins by matrix-assisted laser desorption ionisation-mass spectrometry following in-gel digestion in low-salt, non-volatile buffer and simplified peptide recovery. Anal Biochem 250: 153–156

Fountoulakis M, Cairns N, Lubec G (1999) Increased levels of 14-3-3 gamma and epsilon proteins in brain of patients with Alzheimer's disease and Down Syndrome. J Neural Transm [Suppl] 57: 323–335

Fuentes J, Genesca L, Kingsbury T, Cunningham K, Perez-Riba M, Estivill X, de la Luna S (2000) DSCR1, over expressed in Down Syndrome, is an inhibitor of calcineurin-mediated signalling pathways. Hum Mol Genet 9: 1681–1690

Gilman A (1987) G proteins: transducers of receptor-generated signals. Annu Rev Biochem 56: 615–649

Glynn P (2000) Neural development and neurodegeneration: two faces of neuropathy target esterase. Prog Neurobiol 61: 61–74

Joosten M, Vankan-Berkhoudt Y, Van den Broek M, Beijen A, Lowenberg B, von Lindern M, Delwel R (2000) The gene NM23-M2 is frequently mutated in Leukemia. Exp Hematol 28: 1491

Kaziro Y (1978) The role of guanosine 5′-triphosphate in polypeptide chain elongation. Biochim Biophys Acta 505: 95–127

Keim D, Hailat N, Melhem R, Zhu XX, Lascu I, Veron M, Strahler J, Hanasch S (1992) Proliferation-related expression of p19/nm23 nucleoside diphosphate kinase. J Clin Invest 89: 919–924

Korsisaari N, Mäkelä T (2000) Interaction of Cdk7 and Kin28 with Hint/PKCI-1 and Hnt1 Histidine triad proteins. J Biol Chem 275: 34837–34840

Kraeft S, Traincart F, Bourdais J, Mesnildrey S, Veron M, Chen LB (1996) Nuclear localisation of nucleoside diphosphate kinase type B (nm23-H2) in cultured cells. Exp Cell Res 246: 355–367

Langen H, Roder D, Juranville J, Fountoulakis M (1997) Effect of protein application mode and acrylamide concentration on the resolution of protein spots separated by two-dimensional gel electrophoresis. Electrophoresis 18: 2085–2090

Lima C, Klein M, Hendrickson W (1997) Structure-based analysis of catalysis and substrate definition in the HIT protein family. Science 278: 286–290

Nuoffer C, Davidson HW, Matteson J, Meinkoth J, Balch W (1994) A GDP-bound of rab1 inhibits protein export from the endoplasmic reticulum and transport between Golgi compartments. J Cell Biol 125: 225–237

Ohneda K, Fukuda M, Shimada N, Ishikawa N, Ichou T, Kaji K, Toyota T, Kimura N (1994) Increased expression of nucleoside diphosphate kinases/nm23 in human diploid fibroblasts transformed by SV40 large T antigen or 60Co irradiation. FEBS Lett 348: 273–277

Okabe-Kado J, Kasukabe T, Hozumi M, Honma Y, Kimura N, Baba H, Urano T, Shiku H (1995) A new function of Nm23/NDP kinase as a differentiation inhibitory factor, which does not require its kinase activity. FEBS Lett 363: 311–315

Postel E (1999) Cleavage of DNA by human NM23-H2/Nucleoside diphosphate kinase involves formation of a covalent protein-DNA complex. J Biol Chem 274: 22821–22829

Postel E, Berberich S, Flint S, Ferrone C (1993) Human c-myc transcription factor PuF identified as nm23-H2 nucleoside diphosphate kinase, a candidate suppressor of tumor metastasis. Science 261: 478–480

Postel E, Weiss V, Beneken J, Kirtane A (1996) Mutational analysis of NM23-H2/NDP kinase identifies the structural domains critical to recognition of a c-myc regulatory element. Proc Natl Acad Sci USA 93: 6892–6897

Postel E, Abramczyk B, Levit M, Kyin S (2000) Catalysis of DNA cleavage and nucleoside triphosphate synthesis by NM"§-H2/NDP kinase share an active site that implies a DNA repair function. Proc Natl Acad Sci USA 97: 14194–14199

Raveh S, Vinh J, Rossier J, Agou F, Veron M (2001) Peptidic determinants and structural model of human NDP kinase B (Nm23-H2) bound to single-stranded DNA. Biochem 40: 5882–5893

Razin E, Zhang ZC, Nechushtan H, Frenkel S, Lee YN, Arudchandran R, Rivera J (1999) Suppression of Microphthalmia transcriptional activity by its association with Protein kinase C-interacting Protein 1 in mastcells. J Biol Chem 274: 34272–34276

Reed W, Ozand P (1985) Adrenergic sensitivity of cultured murine trisomy 16 cells. Dev Pharmacol Ther 8: 200–208

Roymans D, Willems R, Vissenberg K, De Jonghe C, Grobben B, Claes P, Lascu I, Van Bockstaele D, Verbelen J, Van Broeckhoven C, Slegers H (2000) Nucleoside diphosphate kinase beta (Nm23-R1/NDPKbeta) is associated with intermediate filaments and becomes upregulated upon cAMP-induced differentiation of rat C6 glioma. Exp Cell Res 261: 127–138

Sanes JR, Jessell TM (2000) The guidance of axons to their targets. In: Kandel ER, Schwartz JH, Jessell TM (eds) Principles of neural science. McGraw Hill, New York, pp 1063–1086

Schimmoller F, Simon I, Pfeffer S (1998) Rab GTPases, Directors of vesicle docking. J Biol Chem 273: 22161–22164

Schubert P, Morino T, Miyazaki H, Ogata T, Nakamura Y, Marchini C, Ferroni S (2000) Cascading glia reactions: a common pathomechanism and its differentiated control by cyclic nucleotide signalling. Ann NY Acad Sci 903: 24–33

Siprashvili Z, Sozzi G, Barnes L, Mc Cue P, Robinson A, Eryomin V, Sard L, Tagliabue E, Greco A, Fusetti L, Schwartz G, Pierotti M, Croce C, Huebner K (1997) Replacement of FHIT in cancer cells suppresses tumorigenicity. Proc Natl Acad Sci USA 94: 13771–13776

Shisheva A, Chinni S, DeMarco C (1999) general role of GDP Dissociation Inhibitor 2 in membrane release of Rab proteins: modulation of its functional interactions by in vitro and in vivo structural modifications. Biochem 38: 11711–11721

Spiegelberg B, Xiong JP, Smith JJ, Gu R, York J (1999) Cloning and characterization of a mammalian lithium-sensitive bisphosphate 3′-nucleotidase inhibited by inositol 1,4-bisphosphate. J Biol Chem 274: 13619–13628

Stahl J, Leone A, Rosengard A, Porter L, King C, Steeg P (1991) Identification of a second human nm23 gene, nm23-H2. Cancer Res 51: 445–449

Takai Y, Sasaki T, Matozaki T (2001) Small GTP-binding proteins. Physiol Rev 81: 153–208
Venturelli D, Martinez R, Melotti P, Casella I, Peschle C, Cucco C, Spampinato G, Darzynkiewicz Z, Calabretta B (1995) Overexpression of DR-nm23, a protein encoded by a member of the nm23 gene family, inhibits granulocyte differentiation and induces apoptosis in 32Dc13 myeloid cells. Proc Natl Acad Sci USA 92: 7435–7439
Wisniewski K, Kida E (1994) Abnormal neurogenesis and synaptogenesis in Down Syndrome brain. Dev Brain Dysfunct 7: 289–301

Authors' address: Prof. Dr. G. Lubec, Department of Pediatrics, University of Vienna, Währinger Gürtel 18, A-1090 Vienna, Austria, e-mail: gert.lubec@akh-wien.ac.at

Alteration of gene expression in Down's syndrome (DS) brains: its significance in neurodegeneration

A. Sawa

Department of Psychiatry and Behavioral Sciences, Department of Neuroscience, Johns Hopkins University School of Medicine, Baltimore, MD, USA

Summary. Several groups have reported pro-apoptotic alteration of gene expression in Down's syndrome (DS) brains. Aged DS brains manifest a similar neuropathology to Alzheimer's disease (AD), including the presence of senile plaques (SP) and neurofibrillary tangles (NFT). Although it is controversial if neurodegenerative processes play a pathological role in DS brains, evidence such as cortical neurons from fetal DS brains showing vulnerability to cell death when compared with neurons from control subjects supports this point of view.

In this chapter, we review the reports that demonstrate pro-apoptotic alteration of gene expression in DS brains. In addition to the pathogenic genes on chromosome 21, such as amyloid precursor protein (APP) and CuZn-superoxide dismutase (SOD1), other genes which associate with p53, or with processes for protein folding have been frequently found.

Abbreviations

AD Alzheimer's disease, *ALDH* aldehyde dehydrogenase, *ALS* amyotrophic lateral sclerosis, *APP* amyloid precursor protein, *DS* Down's syndrome, *ELISA* enzyme linked immunosorbent assay, *GAPDH* glyceraldehyde-3-phosphate dehydrogenase, *GPX* glutathione peroxidase, *JNK* Jun N-terminal kinase, *mRNA* messenger RNA, *NAIP* neuronal apoptosis inhibitory protein, *NFT* neurofibrillary tangle, *PIGs* p53 induced genes, *ROS* reactive oxygen species, *RT-PCR* reverse trascriptase-coupled polymerase chain reaction, *SAGE* serial analysis of gene expression, *SOD* superoxide dismutase, *SP* senile plaque

1. Introduction

Down's syndrome (DS) is defined by trisomy of chromosome 21, showing distinct characteristics such as mental retardation, and premature development of the pathologies of Alzheimer's disease (AD) (Mann, 1988). In aged

DS brains, abundant senile plaques (SP) and neurofibrillary tangles (NFT) exist. Although there have been debates to segregate dementia from underlying mental retardation, at least some portion of DS patients exhibit deteriorated mental status with aging. The mechanism underlying these abnormalities at the molecular level remains to be elucidated. Several lines of evidence suggest that aberrant neurodegenerative process may exist in aged DS brains. Over-expression of genes located with a segment of chromosome 21, termed the DS locus (de la Monte, 1999), may play roles in the pathological mechanisms. It is reasonable to assume that the DS locus is associated with alteration of gene expression directly and indirectly, which leads to a variety of DS pathologies.

First of all, we will discuss the relevance of neuronal cell death in DS pathology. To address this question, it is important to evaluate alteration of gene expression in DS brains, with a particular interest in genes associated with cell death. We will overview current methodologies for evaluating gene expression both at messenger RNA (mRNA) and protein levels. Then, we will list the genes whose expression are reportedly altered in DS brains, and discuss an interpretation of current results and future directions.

2. Neuronal cell death in DS: does neuronal cell death really play roles in neuropathology of DS?

DS brains are small in several regions (Coyle et al., 1986), however, it is still controversial if neuronal cell death underlies such abnormalities or not. At least two lines of evidence described below may support a possible involvement of neuronal cell death in DS brains (Busciglio and Yankner, 1995; Mann, 1988; Wisniewski et al., 1985).

First, cortical neurons from fetal DS brains showed vulnerability to cell death when compared with neurons from age-matched control subjects (Busciglio and Yankner, 1995). DS neurons exhibit a three to fourfold increase in intracellular reactive oxygen species (ROS) and elevated levels of lipid peroxidation that preceded neuronal death. Increased susceptibility to cell death in cultured cells from patients with other neurodegenerative conditions was also reported (Sawa, 1999a). Biochemical measurements of lipoperoxidation in fetal brains of DS support the finding from cortical neuron culture from fetal DS brains (Brooksbank and Balazs, 1984). However, Hayn and her colleagues (Hayn et al., 1996) show no elevation of lipid peroxidation and SOD1 activity in aged DS patients.

Second, DS shares some pathology with AD (Mann, 1988; Wisniewski et al., 1985). Most DS brains over 40 years old show SP and NFT, hallmarks of AD. The main component of SP is the Aβ peptide, which is cleaved from amyloid precursor protein (APP) (Goldgaber et al., 1987; Kang et al., 1987; Tanzi et al., 1987). In DS brains, Aβ42, a longer form of Aβ crucial to AD pathology, is detected (Iwatsubo et al., 1995; Mann and Iwatsubo, 1996). The toxicity of the Aβ peptide was proposed in vitro, or cultured cells (Yankner et al., 1989; Yoshikawa et al., 1992; Yamatsuji et al., 1996). These lines of evidence suggest that mechanisms of cell death for AD may also occur in DS brains. However, it is still not entirely clear if the AD like pathology in DS brain really contributes to neuronal cell loss. Instead, evaluation of gene expression, with an emphasis on cell death associated genes, in DS brains may provide new clues to this question (Sawa, 1999b).

In the next chapters, we overview the methodology to evaluate gene expression both at mRNA and protein levels, and summarize the genes whose expressions are reportedly altered.

3. Methodologies to evaluate gene expression

(A) mRNA level

Northern blotting, RNA protection assay, and S1 nuclease mapping has been used for monitoring mRNA levels. Quantitative reverse transcriptase-coupled polymerase chain reaction (quantitative RT-PCR), due to its high sensitivity, is a powerful strategy for analyzing molecules at low expression level. These techniques are used when we have particular molecules of interest. Recently these techniques have also been used to confirm the results obtained from systematic detection methods for various gene expression described below.

In the early 1990s, several polymerase chain reaction-based methods had been introduced for systematic detection of genes whose expressions differ among cells or tissues. Liang and Pardee (1992) developed a method, differential display, to separate and clone individual mRNAs by means of PCR using a set of oligonucleotides primers. One primer is anchored to the polyadenylate tail of a subset of mRNAs, and the other is short and arbitrary in sequence so that it anneals at different positions relative to the first primer. The mRNA subpopulations defined by these primer pairs are amplified after reverse transcription and resolved on a DNA sequencing gel. When multiple primer sets are used, reproducible patterns of amplified cDNA fragments are produced. A method called serial analysis of gene expression (SAGE) also allows the quantitative and simultaneous analysis of a large number of transcripts (Velculescu et al., 1995). Polyak and her colleagues (Polyak et al., 1997) analyze p53 induced genes (PIGs) using the SAGE technique.

According to the progress in Human Genome Project, DNA microarrays hold the greatest promise in the studies for comparing gene expression between tissues or cells (Celis et al., 2000). Microarrays are usually made by deposition of cDNA spots on a solid support like a coated glass surface.

(B) Protein level

Western blot analysis and enzyme linked immunosorbent assay (ELISA) are classical ways to evaluate protein level for candidate molecules. Immuno-

Table 1. Methods to evaluate gene expression both at messenger RNA and protein levels

1) Messenger RNA level
Candidate approach
Northern blotting
RNA protection assay
S1 nuclease mapping
Quantitative RT-PCR
Systematic approach
Subtractive hybridization
Differential display
Serial analysis of gene expression (SAGE)
DNA microarray
2) Protein level
Candidate approach
Western blotting, immunoblotting
Enzyme linked immunosorbent assay (ELISA)
Immunohistochemistry
Systematic approach
Proteomic methodology
(two dimensional gel eletrophoresis combined with mass spectrometry)

histochemistry can be used for monitoring protein expression in specific subsets of cells, however, this technique is not very quantitative.

For more systematic evaluation and comparison, several new technologies, termed proteomics, have become available to analyze global patterns of protein expressions (Celis et al., 2000). Proteins are the functional molecules, whose difference in profiles between tissues or cells may directly reflect biological relevance. The proteomic technology includes two-dimensional gel electrophoresis to resolve proteins, and equipment for quantifying and identifying proteins such as phosphorimager, scanners, and mass spectrometry.

The above techniques are summarized in Table 1.

4. Genes whose expressions are changed in DS brains: their implication for neuronal cell death

Several genes were reported to have altered expression, using various techniques described above. We describe genes which are potentially associated with neuronal cell death here. The up-regulated genes include not only the genes on chromosome 21, as a direct effect of trisomy, but also the genes on other chromosomes.

(A) Amyloid precursor protein (APP)

The amyloid precursor protein (APP) gene is located in chromosome 21. The level of APP mRNA in DS brains is significantly higher than controls (Oyama et al., 1994). APP is a precursor of the Aβ peptides. At least in vitro, Aβ has cytotoxicity associated with oxidative stress (Yankner et al., 1989). Overexpression of APP leads cytotoxicity in some, but not all, cell lines and conditions (Yoshikawa et al., 1992; Yamatsuji et al., 1996). Some transgenic mice, which express full length or cytoplasmic region of APP, show accelerated aging and apoptotic neurons (Games et al., 1995; LaFerla et al., 1995). Some familial forms of AD occur by distinct mutations in APP gene (Chartier-Harlin et al., 1991). Taken together, dysregulation of APP may play roles in neuronal cell death in DS brains, although the precise mechanisms for cell dysfunction and death by aberrant metabolism of APP remains to be elucidated.

(B) CuZn-superoxide dismutase (SOD1)

CuZn-superoxide dismutase (SOD1) gene is also located on the DS locus. SOD1 is a key enzyme in the metabolism of oxygen free radicals or reactive oxygen species (ROS) by catalysing the dismutaion of superoxide anions to hydrogen peroxide. Hydrogen peroxide is then neutralized to water and oxygen through the actions of either glutathione peroxidase (GPX) and/or catalase. In DS brains, SOD1 level is elevated, which may lead to overproduction of hydrogen peroxide (Gulesserian et al., 2001). It is postulated that an imbalance of SOD1 to GPX and/or catalase may increase susceptibility to oxidative stress. Indeed, DS neurons exhibit a three to fourfold increase in intracellular reactive oxygen species (ROS) and elevated levels of lipid peroxidation that preceded neuronal death (Busciglio and Yankner, 1995). In humans, mutations in SOD1 gene cause some familial forms of amyotrophic lateral sclerosis (ALS) (Rosen et al., 1993). These lines of evidence suggest that some dysfunction and/or overexpression of SOD1 may lead to neurodegenerative conditions.

(C) Genes associated with apoptotic cell death

Several groups have reported alteration of gene expression, which is associated with apoptotic cell death. Although no direct evidence has been obtained which demonstrates neuronal apoptosis in DS brains, these changes indirectly suggest neurodegenerative process in DS brains.

p53 is a key transcription factor for neurodegenerative conditions (Morrison and Kinoshita, 2000). p53 up-regulates pro-apoptotic gene, bax, and Fas/APO-1, and down-regulates anti-apoptotic gene, bcl-2. Two groups have reported elevated expression of p53 in DS brains (Seidl et al., 1999b; de la Monte et al., 1998). Consistent with these findings, we reported an apoptotic index, the ratio of bax to bcl-2, is increased in DS brains, especially in the temporal lobes, where the pathology is the most severe (Sawa et al., 1997b). Elevation of Fas/APO-1 expression was also reported (Seidl et al., 1999; de la Monte et al., 1998). Accordingly, p53 may play roles in DS brains.

Decreased expression of transcription factor junD was recently reported (Labudova et al., 1998). Although Jun N-terminal kinase (JNK) has possible signaling linkages between junD and p53, it is not clear if junD down-regulation can be interpreted in the p53 paradigm or not.

p53 also up-regulates a set of genes which are associated with oxidative stress, named PIGs for *p*53-*i*nduced *g*enes (Polyak et al., 1997). ROS production follows PIG-3 overexpression. It is intriguing, as involvement of oxidative stress in DS pathology is highly suggested (Iannello et al., 1999). In one report, a gene for a mitochondrial enzyme, the complex I, NADH3, is significantly decreased (Krapfenbauer et al., 1999). Mitochondrial enzyme deficiencies have already been reported in DS: significant reductions of mitochondrial monoamine oxidase, cytochrome oxidase, and isocitrate dehydrogenase were reduced in DS platelets. Such impaired mitochondrial redox systems would be compatible with observation of increased oxidative stress and higher vulnerability to cell death in DS, as described above.

Glyceraldehyde-3-phosphate dehydrogenase (GAPDH) had been believed merely as a glycolytic enzyme, however, recent works demonstrated its role in the nucleus for apoptotic cell death (Sawa et al., 1997a). In DS brains, increased GAPDH expression was reported (Lubec et al., 1999b). It remains to be elucidated if this change is relevant in apoptosis, or if it is just reflecting some change in glycolysis or energy metabolism in DS brains.

Alterations of 14-3-3 (Fountoulakis et al., 1999), DNaseI (Schatzmann-Turhani et al., 1999), neuronal apoptosis inhibitory protein (NAIP) (Seidl et al., 1999a), and c-fos (Greber-Platzer et al., 1999) expression were also reported. The changes were interpreted in the context of apoptotic cell death in the original articles.

(D) Genes associated with misfolded and insoluble proteins

The expression of HSP70 seems to be changed in DS brains in an age dependent manner. It is downregulated in fetal DS brains (Kitzmueller et al., 1999), and upregulated in aged DS brains (Yoo et al., 1999), when compared with age-matched control subjects. HSP70 plays a very important role in neurodegenerative conditions. In a model for polyglutamine diseases, over-expression of HSP70 can suppress the neurodegeneration in vivo (Warrick et al., 1999). HSP70, as a chaperon protein, participates in diminishing misfolded proteins, which are prone to be insoluble.

Formation of SP and NFT may accelerate neuronal cell death. Reduction of aldehyde dehydrogenase (ALDH) in DS brains was reported using proteomic technology (Lubec et al., 1999a). ALDH is a key enzyme in fructose, acetaldehyde, and oxalate metabolism and represents a major detoxification system for reactive carbonyls and aldehydes. The decreased ALDH levels in the DS brain may result in accumulation of aldehydes which can lead to the formation of SP and NFT reflecting abnormally cross-linked, insoluble proteins.

The genes listed here are summarized in Table 2.

Table 2. Genes whose expressions are reportedly altered in DS brains: in the context of neuronal cell death

Gene	Method used	Change	References
APP	S1 nuclease mapping	increase	Oyama et al. (1994)
SOD1	proteomics	increase	Gulesserian et al. (2001)
p53	immunoblotting	increase	de la Monte et al. (1998)
	ELISA	increase	Seidl et al. (1999)
bax, bcl-2	quantitative RT-PCR		
Fas/APO-1	immunoblotting	increase	de la Monte et al. (1998)
	ELISA	increase	Seidl et al. (1999)
JunD	Western blotting	decrease	Labudova et al. (1998)
GAPDH	2D gel analysis	increase	Lubec et al. (1999)
14-3-3γ, ε	proteomics	increase	Fountoulakis et al. (1999)
DnaseI	subtractive hybridization	increase	Schatzmann-Turhani et al. (1999)
NAIP	subtractive hybridization	decrease	Seidl et al. (1999)
c-fos	quantitative RT-PCR	increase	Greber-Platzer et al. (1999)
NADH3	differential display	decrease	Krapfenbauer et al. (1999)
HSP70	subtractive hybridization	decrease	Kitzmueller et al. (1999)
	Western blotting	increase	Yoo et al. (1999)
ALDH	subtractive hybridization and proteomics	decrease	Lubec et al. (1999)

5. Discussions and future directions

As described above, several reports from different groups have suggested the existence of alteration in gene expression associated with neuronal cell death.

In general, we frequently face some difficulty in reproducibility and high variability, when we handle clinical specimens. Postmortem delay of autopsy brains directly influences the quality of mRNAs and proteins (Fountoulakis et al., 2001). The discrepancy among reports may mainly stem from such technical issues. Accordingly, it is very important how to interpret and organize various kinds of results for the alteration. The results from animal models help for the interpretation. Several animal models are now available for DS. The trisomy 16 (Ts16) mouse is a classical model for DS (Coyle et al., 1988). Mouse chromosome 16 contains a cluster of genes and loci on human chromosome 21. Ts65Dn was generated through a reciprocal translocation, which can be used to produce segmental trisomy for mouse distal chromosome 16 (Davisson et al., 1990; Reeves et al., 1995). Brain atrophy, especially in the cerebellum, is found in Ts65Dn mice. Therefore, it is worthwhile trying to obtain gene expression profiles from these DS model mice. The comparison of the profiles from mice and DS patients may bring us insights as to how to interpret the variability of results from human specimens.

The comparison of gene expression profiles between DS and AD would also be important, as they share part of the important neuropathology (Wisniewski et al., 1985; Mann, 1988).

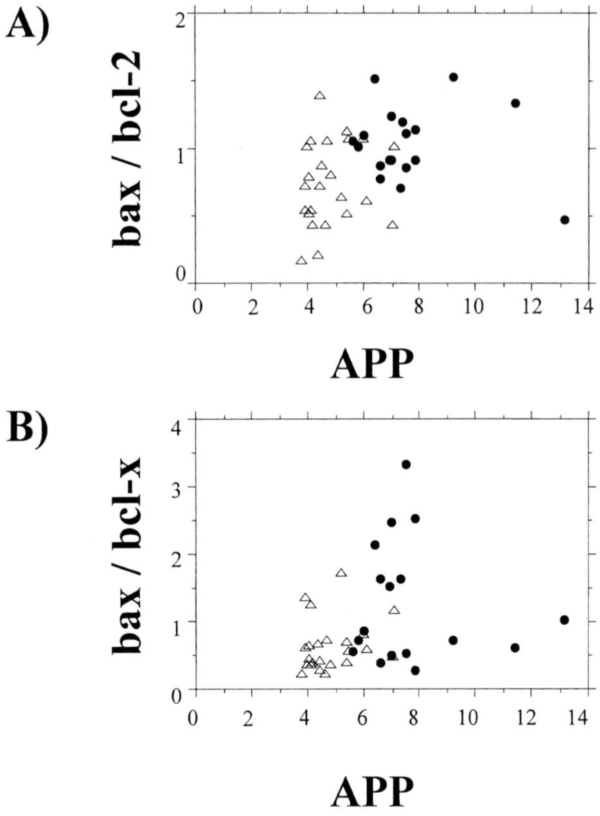

Fig. 1. Significant correlation between the APP expression and the ratio of bax/bcl-2. The expression level of APP in each sample was calculated by nuclease S1 protection analysis in our previous study (Oyama et al., 1994). There was a significant correlation between APP expression and apoptotic indices; The Spearman's correlation coefficient were 0.426 for APP and bax/bcl-2 ($P < 0.01$), and 0.404 for bax/bcl-2 ($P < 0.05$). No correlation was found between the indices and other genes in brain including tau and MAP2 (data not shown) (from Sawa et al., 1997b, with permission)

Our group reported that change in the ratio of bax to bcl-2 showed some correlation to the level of APP in each DS brain, suggesting that APP may be upstream of these changes (Sawa et al., 1997b) (Fig. 1). The upregulation of the ratio (bax to bcl-2) or Fas/APO-1 may be interpreted as a downstream effect of p53 (de la Monte et al., 1998; Seidl et al., 1999b). Such an organization would be important to clarify a chain of events, which may be associated with death signaling in DS brains.

Acknowledgements

I would like to thank L. K. Mawer for preparation of the manuscript. I would like to thank Mr. Z. Kaminsky for careful reading of the manuscript.

References

Brooksbank BW, Balazs R (1984) Superoxide dismutase, glutathione peroxidase and lipoperoxidation in Down's syndrome fetal brain. Brain Res 318: 37–44

Busciglio J, Yankner BA (1995) Apoptosis and increased generation of reactive oxygen species in Down's syndrome neurons in vitro. Nature 378: 776–779

Celis JE, Kruhoffer M, Gromova I, Frederiksen C, Ostergaard M, Thykjaer T, Gromov P, Yu J, Palsdottir H, Magnusson N, Orntoft TF (2000) Gene expression profiling: monitoring transcription and translation products using DNA microarrays and proteomics. FEBS Lett 480: 2–16

Chartier-Harlin MC, Crawford F, Houlden H, Warren A, Hughes D, Fidani L, Goate A, Rossor M, Roques P, Hardy J, et al (1991) Early-onset Alzheimer's disease caused by mutations at codon 717 of the beta-amyloid precursor protein gene. Nature 353: 844–846

Coyle JT, Oster-Granite ML, Gearhart JD (1986) The neurobiologic consequences of Down syndrome. Brain Res Bull 16: 773–787

Coyle JT, Oster-Granite ML, Reeves RH, Gearhart JD (1988) Down syndrome, Alzheimer's disease and the trisomy 16 mouse. Trends Neurosci 11: 390–394

Davisson MT, Schmidt C, Akeson EC (1990) Segmental trisomy of murine chromosome 16: a new model system for studying Down syndrome. Prog Clin Biol Res 360: 263–280

de la Monte SM (1999) Molecular abnormalities of the brain in Down syndrome: relevance to Alzheimer's neurodegeneration. J Neural Transm [Suppl] 57: 1–19

de la Monte SM, Sohn YK, Ganju N, Wands JR (1998) P53- and CD95-associated apoptosis in neurodegenerative diseases. Lab Invest 78: 401–411

Fountoulakis M, Cairns N, Lubec G (1999) Increased levels of 14-3-3 gamma and epsilon proteins in brain of patients with Alzheimer's disease and Down syndrome. J Neural Transm [Suppl] 57: 323–335

Fountoulakis M, Hardmeier R, Höger H, Lubec G (2001) Postmortem changes in the level of brain proteins. Exp Neurol 167: 86–94

Games D, Adams D, Alessandrini R, Barbour R, Berthelette P, Blackwell C, Carr T, Clemens J, Donaldson T, Gillespie F, et al (1995) Alzheimer-type neuropathology in transgenic mice overexpressing V717F beta-amyloid precursor protein. Nature 373: 523–527

Goldgaber D, Lerman MI, McBride OW, Saffiotti U, Gajdusek DC (1987) Characterization and chromosomal localization of a cDNA encoding brain amyloid of Alzheimer's disease. Science 235: 877–880

Greber-Platzer S, Balcz B, Cairns N, Lubec G (1999) c-fos expression in brains of patients with Down syndrome. J Neural Transm [Suppl] 57: 75–85

Gulesserian T, Seidl R, Hardmeier R, Cairns N, Lubec G (2001) Superoxide dismutase SOD1, encoded on chromosome 21, but not SOD2 is overexpressed in brains of patients with Down syndrome. J Invest Med 49: 41–46

Hayn M, Kremser K, Singewald N, Cairns N, Nemethova M, Lubec B, Lubec G (1996) Evidence against the involvement of reactive oxygen species in the pathogenesis of neuronal death in Down's syndrome and Alzheimer's disease. Life Sci 59: 537–544

Iannello RC, Crack PJ, de Haan JB, Kola I (1999) Oxidative stress and neural dysfunction in Down syndrome. J Neural Transm [Suppl] 57: 257–267

Iwatsubo T, Mann DM, Odaka A, Suzuki N, Ihara Y (1995) Amyloid beta protein (A beta) deposition: A beta 42(43) precedes A beta 40 in Down syndrome. Ann Neurol 37: 294–299

Kang J, Lemaire HG, Unterbeck A, Salbaum JM, Masters CL, Grzeschik KH, Multhaup G, Beyreuther K, Muller-Hill B (1987) The precursor of Alzheimer's disease amyloid A4 protein resembles a cell-surface receptor. Nature 325: 733–736

Kitzmueller E, Labudova O, Rink H, Cairns N, Lubec G (1999) Altered gene expression in fetal Down syndrome brain as revealed by the gene hunting technique of subtractive hybridization. J Neural Transm [Suppl] 57: 99–124

Krapfenbauer K, Yoo BC, Cairns N, Lubec G (1999) Differential display reveals deteriorated mRNA levels of NADH3 (complex I) in cerebellum of patients with Down syndrome. J Neural Transm [Suppl] 57: 211–220

Labudova O, Krapfenbauer K, Moenkemann H, Rink H, Kitzmuller E, Cairns N, Lubec G (1998) Decreased transcription factor junD in brains of patients with Down syndrome. Neurosci Lett 252: 159–162

LaFerla FM, Tinkle BT, Bieberich CJ, Haudenschild CC, Jay G (1995) The Alzheimer's A beta peptide induces neurodegeneration and apoptotic cell death in transgenic mice. Nature Genet 9: 21–30

Liang P, Pardee AB (1992) Differential display of eukaryotic messenger RNA by means of the polymerase chain reaction. Science 257: 967–971

Lubec G, Labudova O, Cairns N, Berndt P, Langen H, Fountoulakis M (1999a) Reduced aldehyde dehydrogenase levels in the brain of patients with Down syndrome. J Neural Transm [Suppl] 57: 21–40

Lubec G, Labudova O, Cairns N, Fountoulakis M (1999b) Increased glyceraldehyde 3-phosphate dehydrogenase levels in the brain of patients with Down's syndrome. Neurosci Lett 260: 141–145

Mann DM (1988) The pathological association between Down syndrome and Alzheimer disease. Mech Ageing Dev 43: 99–136

Mann DM, Iwatsubo T (1996) Diffuse plaques in the cerebellum and corpus striatum in Down's syndrome contain amyloid beta protein (A beta) only in the form of A beta 42(43). Neurodegeneration 5: 115–120

Morrison RS, Kinoshita Y (2000) The role of p53 in neuronal cell death. Cell Death Differ 7: 868–879

Oyama F, Cairns NJ, Shimada H, Oyama R, Titani K, Ihara Y (1994) Down's syndrome: up-regulation of beta-amyloid protein precursor and tau mRNAs and their defective coordination. J Neurochem 62: 1062–1066

Polyak K, Xia Y, Zweier JL, Kinzler KW, Vogelstein B (1997) A model for p53-induced apoptosis. Nature 389: 300–305

Reeves RH, Irving NG, Moran TH, Wohn A, Kitt C, Sisodia SS, Schmidt C, Bronson RT, Davisson MT (1995) A mouse model for Down syndrome exhibits learning and behaviour deficits. Nature Genet 11: 177–184

Rosen DR, Siddique T, Patterson D, Figlewicz DA, Sapp P, Hentati A, Donaldson D, Goto J, O'Regan JP, Deng HX, et al (1993) Mutations in Cu/Zn superoxide dismutase gene are associated with familial amyotrophic lateral sclerosis. Nature 362: 59–62

Sawa A (1999b) Neuronal cell death in Down's syndrome. J Neural Transm [Suppl] 57: 87–97

Sawa A, Khan AA, Hester LD, Snyder SH (1997a) Glyceraldehyde-3-phosphate dehydrogenase: nuclear translocation participates in neuronal and nonneuronal cell death. Proc Natl Acad Sci USA 94: 11669–11674

Sawa A, Oyama F, Cairns NJ, Amano N, Matsushita M (1997b) Aberrant expression of bcl-2 gene family in Down's syndrome brains. Brain Res Mol Brain Res 48: 53–59

Sawa A, Wiegand GW, Cooper J, Margolis RL, Sharp AH, Lawler Jr JF, Greenamyer JT, Snyder SH, Ross CA (1999a) Increased apoptosis of Huntington's disease lymphoblasts associated with repeat lengh-dependent mitochondrial depolarization. Nature Med 5: 1194–1198

Schatzmann-Turhani D, Labudova O, Yeghiazaryan K, Rink H, Hauser E, Cairns N, Lubec G (1999) Overexpression of DNAse I in brain of patients with Down syndrome. J Neural Transm [Suppl] 57: 353–362

Seidl R, Bajo M, Bohm K, LaCasse EC, MacKenzie AE, Cairns N, Lubec G (1999a) Neuronal apoptosis inhibitory protein (NAIP)-like immunoreactivity in brains of adult patients with Down syndrome. J Neural Transm [Suppl] 57: 283–291

Seidl R, Fang-Kircher S, Bidmon B, Cairns N, Lubec G (1999b) Apoptosis-associated proteins p53 and APO-1/Fas (CD95) in brains of adult patients with Down syndrome. Neurosci Lett 260: 9–12

Tanzi RE, Gusella JF, Watkins PC, Bruns GA, St George-Hyslop P, Van Keuren ML, Patterson D, Pagan S, Kurnit DM, Neve RL (1987) Amyloid beta protein gene: cDNA, mRNA distribution, and genetic linkage near the Alzheimer locus. Science 235: 880–884

Velculescu VE, Zhang L, Vogelstein B, Kinzler KW (1995) Serial analysis of gene expression. Science 270: 484–487

Warrick JM, Chan HY, Gray-Board GL, Chai Y, Paulson HL, Bonini NM (1999) Suppression of polyglutamine-mediated neurodegeneration in Drosophila by the molecular chaperone HSP70. Nature Genet 23: 425–428

Wisniewski KE, Wisniewski HM, Wen GY (1985) Occurrence of neuropathological changes and dementia of Alzheimer's disease in Down's syndrome. Ann Neurol 17: 278–282

Yamatsuji T, Matsui T, Okamoto T, Komatsuzaki K, Takeda S, Fukumoto H, Iwatsubo T, Suzuki N, Asami-Odaka A, Ireland S, Kinane TB, Giambarella U, Nishimoto I (1996) G protein-mediated neuronal DNA fragmentation induced by familial Alzheimer's disease-associated mutants of APP. Science 272: 1349–1352

Yankner BA, Dawes LR, Fisher S, Villa-Komaroff L, Oster-Granite ML, Neve RL (1989) Neurotoxicity of a fragment of the amyloid precursor associated with Alzheimer's disease. Science 245: 417–420

Yoo BC, Seidl R, Cairns N, Lubec G (1999) Heat-shock protein 70 levels in brain of patients with Down syndrome and Alzheimer's disease. J Neural Transm [Suppl] 57: 315–322

Yoshikawa K, Aizawa T, Hayashi Y (1992) Degeneration in vitro of post-mitotic neurons overexpressing the Alzheimer amyloid protein precursor. Nature 359: 64–67

Authors' address: A. Sawa, MD, PhD, Department of Psychiatry and Behavioral Sciences, Department of Neuroscience, Johns Hopkins University School of Medicine, 725 N. Wolfe Street, Baltimore, MD 21205, U.S.A., e-mail: akira@jhmi.edu

Springer-Verlag
and the Environment

WE AT SPRINGER-VERLAG FIRMLY BELIEVE THAT AN international science publisher has a special obligation to the environment, and our corporate policies consistently reflect this conviction.

WE ALSO EXPECT OUR BUSINESS PARTNERS – PRINTERS, paper mills, packaging manufacturers, etc. – to commit themselves to using environmentally friendly materials and production processes.

THE PAPER IN THIS BOOK IS MADE FROM NO-CHLORINE pulp and is acid free, in conformance with international standards for paper permanency.